How to Keep Your VCR Alive

VCR Repairs Anyone Can Do

Second Edition

by

Steve Thomas

Worthington Publishing Company
Tampa, Florida

Warranty Warning

Warning: Opening or otherwise modifying your VCR may void any manufacturer's warranty still in effect on the machine.

Warning: Dangerous voltages and currents may exist inside VCR's, even when the power switch is turned off. Make certain that the VCR is unplugged from the power outlet when opening, trouble-shooting, or repairing it. Only trained technicians should trouble-shoot the VCR when it is plugged in. Always unplug it before working on it or touching the insides.

Library of Congress Cataloging-In-Publication Data

Thomas, Steve, 1942-
  How to keep your VCR alive.

  1. Video tape recorders and recording — Maintenance and repair — Amateurs' manuals. I. Title.
  TK6655.V5T47   1990   621.388'337   89-40618
  ISBN   0-9618359-5-8

Worthington Publishing Company
P.O. Box 16691
6907-202 Halifax River Drive
Tampa, Florida   33687-6691

© 1990   Stephen N. Thomas

20   19   18   17   16   15   14   13   12   11

International Standard Book Number: 0-9618359-5-8
Library of Congress Catalog Card Number: 89-40618

Technical Consultant: Brook Negusie
Editor: Robert Bell
Cartoonist: Troy DeVolld
Printed in the United States of America

Trademark Acknowledgements:

Listed below are all terms used or mentioned in this book that are known, or believed to be, trademarks or service marks.

Aiwa, Akai, Beta, Brocksonic, Canon, Circuit City, Colt, Curtis Mathes, Daewoo, Daytron, Denon, Donald Duck, Dual, Dynatech, Electrodynamics, Elmo, Emerson, Fisher, Funai, General Electric, Goldstar, Harmon Kardon, Hitachi, Ikko, Instant Replay, J.C. Penny, J V C, Jensen, K L H, K M C, Kenwood, Kodak, Lloyds, Logik, MCM, Magnavox, Marantz, Matsushita, Memorex, Midland, Minolta, Mitsubishi, Montgomery Ward, Multitech, N E C, Olympus, PBR, Panasonic, Pentax, Philco, Pioneer, Portland, Quasar, R C A, Radio Shack, Randix, Realistic, Sampo, Samsung, Sansui, Sanyo, Scott, Sears, Sharp, Shintom, Singer, Sound Design, Sony, Supra, Sylvania, Symphonic, T M K, Tatung, Teac, Technics, Teknika, Toshiba, Totevision, Unitech, Vector Research, Video Concepts, Yamaha, Zenith.

Such terms have been appropriately capitalized throughout the book. The author and publisher cannot authenticate the accuracy of this list or notation. The appearance or use of such terms in this book should not be regarded as affecting the validity or ownership of any trademark or service mark.

Note: To use this book effectively, begin by reading the *Introduction*!!!

Table of Contents

Appendices

Supplement

Introduction

How to Use this Book

Video cassette recorders are marvelous family home entertainment and educational sources. They permit the whole family to watch taped movies and instructional programs in the comfort of the home, at less expense and with greater safety and convenience than going to a theater.

Families with a VCR are not at the mercy of a theater to set the time to start the movie, and there is no need to miss part of the movie while stepping out to make popcorn in the kitchen. The program can be stopped and restarted at will. And if part of the show was missed while playing it on the VCR, or if part of a crucial scene went by too fast to be understood, the video tape can be quickly rewound, and the part that was missed be replayed.

Finally, a VCR can be used to record one program while you watch another program at the same time, so that you never need to make a difficult choice between two good programs or movies on at the same time: with a VCR, you can watch both by taping one and then watching it later, while watching the other live at the time it is broadcast. Chapter 1 explains how to do these things with your VCR.

Owners of VCRs do not have to wait for the movie they want to watch to be shown in a local theater. They can drop by their neighborhood video tape rental store and choose what they want to see when they want to see it. But people are not going to rent tapes if their VCR is malfunctioning. Tape rental outlets do not have customers if their customers' VCRs are not working.

Service, Maintenance, and Repair

The problem is that VCRs tend to require lots of maintenance and repair, and servicing a VCR can be expensive. *Consumer Reports* magazine found on the basis of their studies, that the typical cost to fix a VCR is $80. The truth is, however, that 50 to 70 percent of all VCR repairs involve only simple steps of cleaning or replacing simple mechanical parts,

servicing that owners can do easily, once they know exactly how to do it. This means that lots of time and money can be saved by doing your own servicing.

This book shows you how to do all these simple jobs yourself, leaving only the difficult jobs requiring advanced knowledge and equipment to the professional service technicians. Besides saving lots of money, you often can save time by repairing your own VCR quickly the same day that the trouble appears rather than being forced to wait days or weeks for a shop to do it.

And this book shows you how to hook up your VCR correctly, how to clean it, and how to diagnose problems that arise inside the VCR so as to determine whether you can fix it yourself, or whether it is one of the less common problems that will require sending the VCR out for professional service. Clear, step-by-step procedures are given for making the repairs. This book focuses especially on the most common problems in the most popular models.

The parts that you will need for the jobs that you can do yourself are given, including part numbers so that you can obtain the parts easily, and a list of parts sources for most VCRs appears at the end. The book also shows you where to obtain the few inexpensive tools that you may need by mail order if they are not available where you are locally. Even if you use only the chapter on how to clean your VCR, or the instructions how to remove a stuck cassette safely without damaging anything, this book will save you money.

Where to Start

First, finish reading this Introduction, and determine the category to which your brand belongs from the table later in this Introduction. Then proceed as follows:

To find out how to hook up a VCR, begin by reading Chapter 1.

To remove a stuck cassette in circumstances where the first priority is just to get the cassette out of the machine without damaging anything, begin by reading Chapter 2, "How a Video Recorder Works," which explains the simple terminology used in the chapters that follow. Then turn directly to Chapter 7, if it is a top-loader, or Chapter 8, if it is a front-loader.

To service or repair a VCR, or to check out a used machine that you are considering buying, begin by reading Chapter 2, Then read Chapter 4, "Basic Universal VCR Check-Out Procedure," which will enable you to analyze the problem, and determine whether it requires professional service, or whether it is one of the majority of problems that you can fix yourself.

Do NOT try to fix your problem by looking it up in the Table of Contents, because VCR repair is more complicated than this. The actual hidden cause of your problem may be treated in an earlier chapter, so use the Chapter 4 check-out procedure to find out what chapter to use.

This check-out procedure will tell you which subsequent chapter contains the step-by-step procedure to fix your problem. You do not need to read all the chapters in this book to fix your problem. You only need to read a few of them. You only need to read Chapter 2, and then Chapter 4, which will tell you which one of the following chapters you need to read to fix your problem. This book is large because it covers so many different problems in various makes and models, each of which takes a few pages.

Appendices at the end of the book tell you where to get parts and additional service literature for all makes and models of VCR, how to open a VCR cabinet, and provide other needed information.

If you have any difficulty following the step-by-step procedures in the chapters that follow, or if you discover any important omissions or errors, please write me, the author, c/o Worthington Publishing at the address on the inside front page.

Never Start by Oiling

Here, at the beginning, may be a good place to remark that problems in VCRs are almost NEVER corrected by putting oil on parts. In years of working on VCRs, I have only seen two cases where oil was needed, but I have seen *many* VCRs whose problems were made worse by owners putting oil in the wrong places.

A great many mechanical problems in VCRs are caused, in fact, by *too much* natural oil: oily deposits come from the environment and need to be removed to make the VCR work correctly. Oil from the

environment getting on critical parts is a major cause of VCR malfunctions. So, DO NOT BEGIN BY PUTTING OIL ON PARTS PRIOR TO READING THE PROCEDURES IN THIS BOOK. And *NEVER, NEVER SPRAY LUBRICANTS OR OTHER OILY MISTS INTO A VCR!* A VCR is not a bicycle. Instead, follow the procedures in this book.

Unplug the VCR

If you are about to repair a VCR, I suggest that you now unplug the machine, and leave it unplugged while you read at least the next two chapters (or for at least ten minutes). When you plug it back in again, you may be delighted to discover that the trouble has entirely disappeared and that the VCR works perfectly again — this happens sometimes.

The reason for this miracle cure is that it sometimes happens that a momentary surge either on the power line or TV cable to which a VCR is connected causes gibberish (bad data) to be stored inside the memory of the little computer inside the VCR, causing the machine to malfunction. Unplugging for ten minutes or more may permit the bad information to disappear out of its memory, so that the unit again functions correctly.

You do not even need to understand this explanation. Just unplug the machine for ten minutes, and then plug it back in again to see if it now works. Merely turning off the power switch on the VCR is not enough, however, because the little computer is actually still receiving power even when the power switch is off. You need to completely unplug the VCR from the power outlet.

Where to Look for Information on Your VCR

The difficult task of writing a universal VCR repair manual that do-it-yourselfers can use to repair most problems in most VCRs is simplified somewhat by the fact that just as many parts may be identical in certain models of different cars, like Chevrolet, Oldsmobile, and Buick, so too many parts may be interchangeable in different brand-names of VCRs. In automobiles, some parts, like batteries, tires, and light bulbs may even be interchangeable between a Chevrolet and a Ford, and likewise in VCRs, some parts may be interchangeable even between machines manufactured by different companies.

The information and procedures in this book are specially organized around these similarities. To find specific information about your VCR, find its brand name in the left-hand column of the table that follows, and look under the name(s) alongside it in the right-hand column. When more than one name appears on the right, it means that for some machines with the brand on the left, you look under the one name, while for other models of the same brand, you look under the other name. The abbreviation for this name used with parts numbers is shown after the name in parentheses. ("Matsushita" is the Japanese name of the company that makes Panasonic; it is sometimes abbreviated "(Mat)," so (Mat) = (Pan).) For RCA table models whose model number has a *middle letter* later than "K" in the alphabet like (like "VMT250"), and all RCA portables, look under "Hitachi" or "Samsung"; for RCA table models whose middle letter is "K" or earlier (like "VDT650"), information

is filed under "Panasonic." For other brands for which more than one category is listed, you'll need to look under each category listed until you can put your VCR into one of the listed categories. Appendix II contains addresses and phone numbers for all parts sources.

One brand asked not to be listed in this table, because the company did not want you to have this information about the machines they are selling. In the future, it is recommended that you purchase only the brands listed here.

FOR BRAND NAME:	LOOK UNDER:
Akai	Mitsubishi (Mit)
Broksonic	Shintom (Snt)
Canon	Panasonic (Pan)
Circuit City	Shintom (Snt)
Colt	Shintom (Snt)
Curtis Mathes	Panasonic (Pan)
Daytron	Daewoo (Dae)
Dynatech	Funai (Fun)
Emerson	Emerson (Ems), Mitsubishi (Mit), Goldstar (Gds)
Fisher	Fisher (SFC)
Funai	Funai (Fun)
GE	Hitachi, Panasonic (Pan)
Goldstar	Goldstar (Gds)
Harmon Kardon	decks: Mitsubishi (Mit), N E C (NEC)
Hitachi	Hitachi (Hit)
Instant Replay	decks: Panasonic (Pan), Hitachi (Hit), & Funai (Fun)
J C Penney	Hitachi (Hit), Panasonic (Pan), Goldstar (Gds)
J V C	J V C (JVC)
K L H	Shintom (Snt)
K M C	Sharp (Shp)
Kenwood	J V C (JVC), N E C (NEC)
Kodak	Panasonic (Pan)
Lloyds	N E C (NEC), Funai (Fun)
Logik	Shintom (Snt)
Magnavox	Panasonic (Pan)
Memorex	Goldstar (Gds)
Midland	Samsung (Sam)
Minolta	Hitachi (Hit)
Mitsubishi	Mitsubishi (Mit)
Montgomery Ward	Sharp (Shp), Panasonic (Pan)
Multitech	N E C (NEC), Funai (Fun)

FOR BRAND NAME:	LOOK UNDER:
N E C	N E C (NEC)
Panasonic	Panasonic (Pan)
Penneys	See J C Penney
Pentax	Hitachi (Hit)
Philco	Panasonic (Pan)
Portland	Daewoo (Dae)
Quasar	Panasonic (Pan)
R C A	Panasonic (Pan), Hitachi (Hit), Samsung (Sam)
Sampo	Sampo (Smp)
Samsung	Samsung (Sam)
Sansui	Sansui (San)
Sanyo	Sanyo (SFC)
Scott	Emerson (Ems)
Sears	Hitachi (Hit), Fisher (SFC)
Sharp	Sharp (Shp)
Singer	Shintom (Snt)
Sony	Sony (Son)
Sound Design	Funai (Fun)
Supra	Akai (Aka), Samsung (Sam) Funai (Fun), Supra (Sup)
Sylvania	Panasonic (Pan)
Symphonic (Sym)	Funai (Fun), N E C (NEC)
T M K	T M K (TMK)
Teac	J V C (JVC), Funai (Fun)
Teknika	Panasonic (Pan), Teknika (Tek) and other (?)
Technics	Panasonic (Pan)
Toshiba	Toshiba (Tos), J V C (JVC), Shintom (Snt)
Totevision	Goldstar (Gds), Samsung (Sam), Daewoo (Dae)
Unitech	Samsung (Sam)
Vector Research	N E C (NEC), Funai (Fun)
Video Concepts	Mitsubishi (Mit)
Zenith (Znt) VHS	J V C (JVC)
Zenith (Znt) Beta	Sony (Son)

Other Abbreviations and Terminology Used in this Book

Beta	= the home recording system developed by Sony and used in many other brands too
C C W	= counterclockwise — that is, rotating in the direction opposite to that in which the hands of an old-fashioned clock rotate, as viewed from the front
C W	= clockwise — that is, rotating in the same directions as the hands on an old-fashioned

clock rotate, as viewed from the front

left side = left side of the VCR as viewed from the front when it is right-side up
right side = right side of the VCR as viewed from the front when it is right-side up
V C R = video cassette recorder
V H F = conventional television channels 2 through 13
V H S = Video Home System, the recording system most widely used in home recorders
U H F = conventional television channels 14 through 80

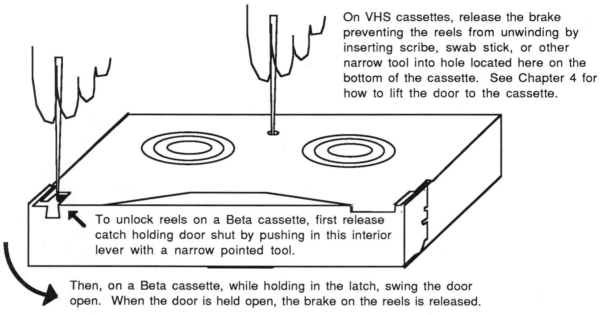

On VHS cassettes, release the brake preventing the reels from unwinding by inserting scribe, swab stick, or other narrow tool into hole located here on the bottom of the cassette. See Chapter 4 for how to lift the door to the cassette.

To unlock reels on a Beta cassette, first release catch holding door shut by pushing in this interior lever with a narrow pointed tool.

Then, on a Beta cassette, while holding in the latch, swing the door open. When the door is held open, the brake on the reels is released.

How manually to release the internal brakes that prevent tape spools from unwinding inside cassettes

"Have you got the VCR hooked up yet, Dad?"

Chapter 1

How to Connect a VCR to a TV Set and Antenna or Cable

To keep it clear and simple, let us proceed in a series of steps. First come the simplest connections just to play and view a prerecorded tape, such as rented movie. Then more connections will be added so that programs arriving on a TV antenna or cable can be recorded and played back. Then we will set up to copy material from a tape in one VCR onto a second tape in another VCR. Finally, how to record one program while watching another different program at the same time will be explained.

An initial, short, step-by-step outline of each procedure will precede a more detailed explanation of all steps and terminology.

1. How to Connect a VCR to a TV Set to Play a PreRecorded Movie orTape

SUMMARY

1. Unplug everything. If an antenna is connected to the TV set, disconnect it. Connect the "RF out" terminal of VCR to the TV antenna terminal with a piece of cable. You may need to use a matching transformer or "balun."

2. With VCR and TV set unplugged, select either channel 3 or 4 on VCR as output channel.

3. Set TV set channel selector to same channel number as VCR output.

4. Plug in and turn on power to both VCR and TV set. Set "TV/VCR" switch on VCR to "VCR" position.

5. Insert cassette into VCR and push "Play."

6. On a few GE and other VCRs, it is necessary to set the VCR speed control to the same speed (SP, LP, or SLP or EP) as the tape is recorded in, but on most VCRs this is done automatically.

7. Adjust fine tuning control on TV set as necessary to bring in picture.

8. Adjust tracking control on VCR for best picture.

DETAILED EXPLANATION

Look at the back of your VCR. You will see one or two threaded metal female connectors about 7mm (5/16 inch) in diameter that look like this

F-type Connector

These are called "F-type connectors," or "F-type terminals," a term that will be used frequently in this chapter. Note the little hole in the center.

Phono Plug Connectors

F-type connectors should not be confused with "RCA phono plugs," often called for short, just "phono plugs," which look like this:

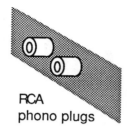

RCA
phono plugs

Phono plugs are the same as the connectors for hooking together stereo components. Phono plugs have a smooth outer surface, while F-type connectors have threads on the outside for screwing on the cable connectors.

Phono plug connectors are used to connect two VCRs together to copy tapes, to connect a stereo VCR to a stereo set when you do not have a stereo TV set, and to connect a VCR to a special type of TV set called a "video monitor."

Cables with F-Type Connectors

The cables that connect to F-type terminals are called "coaxial cables" or "F-to-F connectors" in

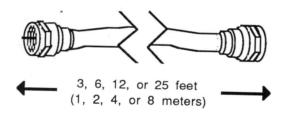

3, 6, 12, or 25 feet
(1, 2, 4, or 8 meters)

F-to-F type Coaxial Cable

video stores. Either end of such a cable will screw onto the F-type connector at the back of your VCR.

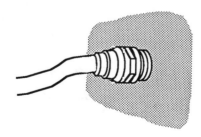

Attached F-type Connector

You can also get adapters that screw onto the ends of the cable so that you can just push on the connectors instead of screwing them on, but these are not advisable unless you expect to be connecting and disconnecting the cables several times a day. The screw-on type give better connections over the long haul.

There is a small stiff solid wire in the center of the terminals at the ends of the cable, and there is a single small hole in the center of the F-type connector attached to the back of your VCR. To attach an F-type connector correctly, the little stiff center wire in the cable must go INTO the little hole in the center of the connector as you push the two together before you begin screwing the threaded parts together. If this center wire is bent, you will need to straighten it before connecting the cable to the terminal. If this center wire is broken off or missing, you will either need to get a new cable or get someone to cut back the cable and attach a new threaded terminal.

"In" and "Out" Terminals

On most VCRs there are two F-type terminals:

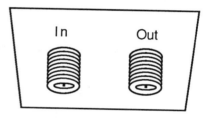

The terms "In" and "Out" refer to the direction of flow of programs relative to the VCR. Programs come into the VCR (from cable or antenna) through the terminal marked "In" (or "RF in" or "Ant.In"). Programs come out through the terminal marked "Out" (or "RF out" or "Out to TV"). Since we are presently hooking up to play a tape, we are now mainly interested in the "Out" terminal, through which the material on the tape will come out to the TV set.

VCRs with Only One Terminal

On some portable VCRs, there is only one "Out" F-type terminal located on the side instead of the back of the unit. Also, on "Play Only" units, there is only one "Out to TV" F-type connector.

Connect from VCR "Out" to TV set

Wherever it is, the most common way to connect a VCR to a TV set is by a piece of special cable one end of which connects on one end to the F-type connector on the VCR marked "Out" (or "VHF Out," or "Out to TV," or "RF out"). In most cases, you will need at least one cable with an F-type connector at the end to connect a VCR to a TV set. Usually, a cable with

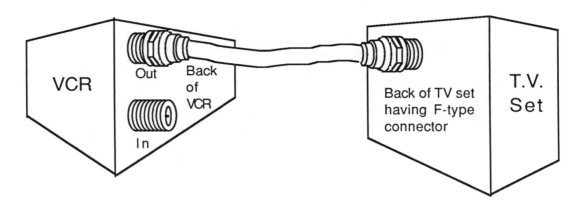

type connectors at both ends is used, either by itself or together with another accessory, called a "matching transformer," or "balun," described later in this section.

When a "Balun" is Unnecessary

The kind of connecting cable and accessories needed depends on the design of your particular TV set. If you have a new TV set, it may have one or more F-type connectors on the back like on the back of your VCR. In that case, only a cable with F-type connectors at both ends is needed. Simply screw the second end of the cable onto the threaded connector on the TV set, as shown at the top of the page, and you are hooked up. When we know that the connections are made like this and it is unnecessary to show all the details of the connectors in illustrations, this same hook-up can also be pictured as shown on the next page as a simplification.

When a Balun is Needed

If your set has only screws on the back marked "VHF" and "UHF,"

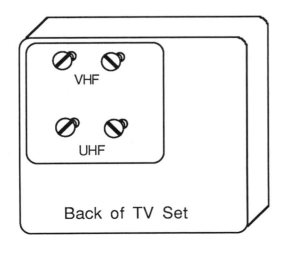

then you will need an additional accessory called a "75 ohm to 300 ohm matching transformer," or

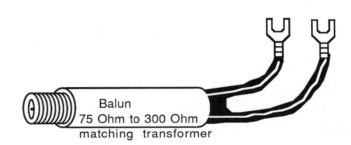

"balun," to connect to your TV set. A "balun," or "matching transformer," has an F-type connector at one end, and two wires coming out of the other end. Make sure that you get one that looks like this with two wires coming out of one end, and a single F-type terminal with threads on the outside on the other end.

After connecting one end of the F-type cable to the VCR, connect the other end of it to the F-type connector on the matching transformer, and then attach the two wires coming out of the other end of the matching transformer to the two screws marked "VHF" on the back of the TV set, as shown in the illustration. It does not matter which wire goes under which of the two screws, but do not attach the two wires to the screws marked "UHF," because that would not work. The other end of the cable should be connected to the "Out" F-type connector on the VCR.

Simplification of Drawings

To simplify the illustrations in the rest of this chapter, I will not try to include a picture of the balun attaching to the TV set whenever one is used, but will just show a cable attached to directly the TV, even though a balun may be used. So, a simplified picture of this hook-up will look the same as the drawing for a direct F-to-F hook-up, even though a balun goes at the place circled.

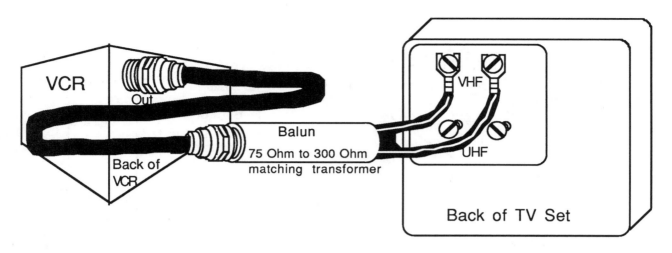

Hooking a VCR to an older type TV set by means of a Balun (Matching Transformer)

Stereo Sound Connections

If you have a MTS-type stereo VCR and a stereo TV set, then the connections described above will automatically give you stereo sound at the TV set. But if you have a stereo VCR and a nonstereo TV set, then then you can still get stereo sound if you connect the audio output phono jacks from the VCR to a pair of input phono jacks on a nearby stereo record player or hi-fi. Here is how to do this.

Begin by finding an unused pair of input jacks on your stereo set. They might be labeled "aux. in" or "tape in" or "AM/FM" — whatever you're not currently using on your stereo. The only connectors on the stereo set that will not work are any labeled "output" or ". . . phono in."

Connect one stereo cable from the "Left channel

out" on the stereo VCR to one of these jacks, and connect a second cable from "Right channel out" to the other jack. Turn down the volume control on the nonstereo TV. Turn on the stereo, and set the channel or source selector switch to the position corresponding the label on the pair of jacks to which you connected the VCR. Adjust the position of the speakers to produce the effect you like best.

Setting the Controls

Now we come to setting the controls on the VCR and TV set. In order to play a prerecorded tape, you must set a number of controls. For the moment, let me just list them, so that you can understand the necessity for the explanation in the paragraphs that follow:

Basic connections to play a tape
with a TV set that has an F-type
input terminal

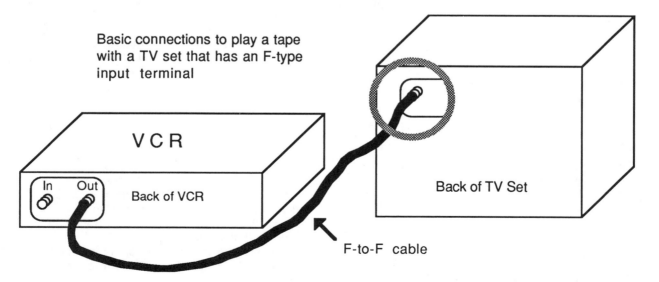

F-to-F cable

Simplified drawing of VCR-to-TV connection of any type

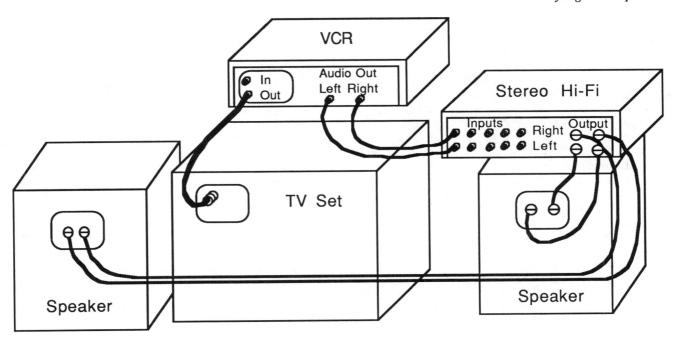

Using a Stereo Hi-fi to Get Stereo Sound from a Stereo VCR and nonstereo TV set

To set things up and make everything work to play a tape, you must:

1. Set the output channel of the VCR to channel 3 or 4. (This is not the same as the output terminal.) You do NOT set this on the TV set — you set it on the VCR. It's a little switch somewhere on the back, or side, or bottom of the VCR.

2. Set the channel selector of the TV set to the output channel of the VCR and adjust the tuning on the TV set to tune in the little station broadcasting out of the VCR.

3. Use the TV/VCR switch correctly.

You need to do all this correctly in order to play a tape on the VCR and see it on the TV set. And you will need to change the settings of some of these controls if you want to use your TV set instead to watch a program being broadcast from a local TV station or sent down the cable to you.

So, you must operate some controls in order to play tapes. The controls are not as complicated as flying an airplane, but they are more complicated than operating any other machine in the house. Playing a prerecorded tape is the simplest thing you can do with a VCR — it gets more complicated when you go on to recording tapes and more advanced topics. Part of what is explained in this section is designed also to help when you advance to the next level of VCR operation.

I will try to explain everything very clearly and simply, but it must be understood that there is a certain level of complication inherent in this

equipment that no one can remove. In order to operate the controls, and avoid confusion, it helps to understand what the controls do, obviously.

Setting the Channel for the VCR to Broadcast On

Somewhere hidden on the back, or bottom, or side of the VCR is a tiny switch that has just two positions, one marked "3" and the other marked "4." I am NOT talking about the channel selector switch on the VCR. I am not talking about the controls that you use to select the channel that you want to view on either the VCR or the TV set. I am talking about another, different switch that you need to set maybe only once, when you first are getting everything connected and into operation. Look on the back, or bottom, or maybe the top or side, of the VCR until you find this switch. It is called the "output channel selector switch," but probably this name will not be written beside it. Usually it will just say "3" and "4" beside it, perhaps together with the word "channel" or the abbreviation, "Ch."

Do not confuse this little switch that controls the OUTPUT channel on which your VCR will broadcast its programs with the *channel selector* switches on the front of the VCR used for choosing the INPUT channel of program material you want to record or view off the air. These will be discussed later, when we come to making a recording and other topics. The *channel nput selector(s)* on the front of your VCR are recognizable as offering at least channels 2 through 13, plus probably additional UHF channels. The output selector switch that we are setting right now, in contrast, has only two choices or positions, "3" or "4," and is used to tell the VCR which channel to output the programs on.

When you put the VCR's "Output Selector Switch" in the position marked "3," the show on the tape is sent out of the VCR on channel 3. When you put the "Output Selector Switch" in the position marked "4," the program on the tape is sent out of the VCR on channel 4. If there is a television station in your locality broadcasting on channel 3, the clearest picture is usually produced by setting this switch to the position marked "4." If there is a television station in your locality broadcasting on channel 4, it usually produces the clearest picture to set this switch to the position marked "3."

Remember the number of the channel to which you set this switch, because that is the channel number to which you will need to tune your TV set to see tapes played on the VCR. Maybe you should write down the number that you set the output switch to, so that you do not forget it.

Set "TV/VCR" Switch to "VCR"

You have connected the VCR to the TV set, selected either channel 3 or channel 4 for the VCR's little transmitter to send the program out on, and set your TV set's channel selector to the same numbered channel, so far — right? Now you can plug in both the VCR and TV set and turn on their power switches.

Before you can start watching the tape, the "TV/VCR" switch must be set to the position marked "VCR." This switch will be on the front, or top, of your VCR somewhere. It may say "Tape/VCR," or "TV/VCR," or "TV/VTR," or "TV/Video," or just "VCR." The words used in the label for this switch vary from model to model, but all video recorders have this switch, however they may label it. (One exception: If your VCR is only a player, and not also a recorder, then possibly there may be no such switch, and you can ignore this step.)

Put this switch into the position marked "VCR" (or "VTR" or "Video" or "Tape") -- that is, put it in

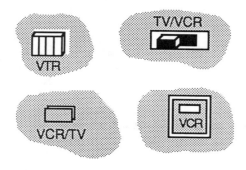

Different TV/VCR Switches

the position that is NOT marked "TV." If the switch is marked just "VCR," push it so that the light saying "VCR" comes on. On some models, tiny letters that say "VCR" appear up with the clock display in a window when you put this switch in the VCR position.

This is one of the most important operator controls on the VCR. People who do not know how to use it will often have troubles and frustrations trying to get their VCRs to do what they want. Later, when we discuss connecting the VCR to an antenna or cable, the purpose of the other position of this switch will be explained. For now, just remember to put this switch in the position NOT marked "TV" in order to watch a tape playing on the VCR.

This should be remembered, because this switch will need to be moved later to perform other functions, and also because, on *some* VCRs having an electronic, rather than mechanical, TV/VCR switch, the switch automatically resets itself to the "TV" position whenever power to the VCR is turned off or interrupted, so that this switch *always* needs to be set to the "VCR" (or "Video" or "VTR") position *every time* the user wants to watch a tape.

And sometimes children (or gremlins in the VCR) change the position of this switch when no one is looking. So, if you are getting no picture, always check the position of this switch.

Tuning the Television Set to the Little Station in the VCR

If you told your VCR to broadcast its little signal on channel 4, for example, then obviously you must set the TV channel selector on your TV set to receive channel 4. On the other hand, if you pushed the little switch on your VCR to tell it to output on channel 3, then you will need to tune your TV set to channel 3 also, in order to watch whatever you are playing on the VCR.

If your TV set were tuned to any other channel, it would not let through the channel on which your VCR will now output the show when playing a tape. Thus, the channel selector on your TV set must be tuned to the same number channel as you tell your VCR to output its tiny signal on — otherwise you will not see and hear the program recorded on the tape that is playing.

A common source of confusion is the fact that when you watch, record, or playback a copy of, programs reaching the VCR from an outside cable or antenna source, the little transmitter in the VCR still always sends out the program on only one channel, 3 or 4, as you determine by setting the position of the little output switch, even though the VCR originally received this program coming in from a channel with a different number.

Insert Cassette and Push "Play"

Having completed these steps, insert a recorded cassette into the VCR, push "Play," and in a moment, a picture should appear on your TV receiver.

Adjust TV Set Fine Tuning

The VCR has its own little miniature TV station inside, broadcasting over the cable to your TV set. As far as your TV set is concerned, what is coming into it is just like what it receives over the airwaves from big TV broadcast stations. So, you may need to adjust the fine tuning on your TV set just as you would if you were tuning in a broadcast station, as necessary for best picture.

Adjust VCR Tracking Control

All VCRs have something called a "Tracking Control." It looks like some kind of round knob or wheel sticking out through a hole in the lower front or top of the cabinet, except on the newest models, where it is found only on the separate remote control. The position of the tracking control may need to be readjusted every time a new tape is played (unless the tapes were made on precisely similar machines).

Do not be afraid to adjust the tracking control. It is easy to put it into the position that gives the best picture, so you do not need to worry about misadjusting it. Rut it all the way to one end, and then all the way until it stops at the other end, and then put it into the position that gives the best picture. When you have the problem of lines appearing across the picture, the first thing to try is adjusting the tracking.

On the new models where the tracking control is on the separate remote control, usually there are two buttons. On these models, ress either one of the buttons, and hold it down and watch the picture. If the picture becomes good, stop. If the picture does not get good, try pressing the other button. Go back and forth between pressing the two buttons, one and then the other, until the picture is the way you want it to be.

Solving Problems When Playing

If no picture appears, check the following:

1. It may be that the tape has a region with nothing on it at the point where it is now in the VCR. Prerecorded tapes, especially movies, often have a region at the beginning with no picture recorded. To check this, push Stop, then Fast-forward and let it wind some of the tape from the reel on the left to the

reel on the right, getting somewhere into the middle of the movie. Then try pushing "Play" again. If the picture comes up, you can rewind by trial and error back to the place where the program begins.

2. Make certain that you have the piece of cable connected to the F-type connector at the back of the VCR marked "Out," and NOT "In."

3. The fine tuning on your TV set may need further adjustment in order to tune exactly to whichever channel you set your VCR to output on. Try adjusting the fine tuning on your TV set in one direction, and then in the other direction, to see if you can tune in the movie. Naturally this is done while the VCR is running the tape in the Play mode.

4. Try slowly rotating the knob or wheel marked "tracking" all the way in one direction, and then all the way in the other direction, to see if this brings up the picture.

5. The cable that you used to connect the VCR to the TV set may be bad. Try another piece of coaxial cable, or find some other way to test for sure that the piece of cable you are using is OK.

6. There may be a trouble in the VCR. Try setting it up to tune in a broadcast station as described in the next section. If you can tune in a station, come back and test the Play function again.

If you have followed all the instructions, and if the TV set works when connected straight to an antenna or cable, and you get NO picture when you try to tune in a station, then probably there is a malfunction in the VCR.

Rewinding the Tape

One of the most confusing things to beginners is the way that VCR designers save their companies money by having the same control button perform several different functions. When you want to rewind the tape quickly, so that you can play it again or return it, hit "Stop" first, then press the button labeled something like "Rewind" or "Rewind/-Review." If you do not press "Stop" first, then pressing the Rewind button makes the machine go into "reverse search" where it plays the tape, but in the reverse direction, looking on the screen like a movie run backwards.

2. How to View and Record Programs
from an
Antenna or Cable

© 1990 Stephen N. Thomas

BRIEF PRELIMINARY OUTLINE OF PROCEDURE

1. In addition to connecting the VCR to the TV set as described in Section 1, connect the antenna or line from the cable company to the VHF input terminal on VCR.

2. Turn on the power to the VCR and the TV set.

3. If you desire only to *view* a program that is being broadcast at that time, without recording anything, you can simply set the "TV/VCR" switch to the "TV" position, and tune in the station's channel number with the channel selector on the TV set.

But if you wish to start *recording* a program, then set the "TV/VCR" switch (on the VCR) to the "VCR" position, and continue as follows.

(If you want to *watch one program while recording another program at the same time*, then you need to go to the more advanced Section 4 of this chapter. But you may wish to finish covering the basics in the section you are now reading first.)

4. Tune the TV set to whichever channel (3 or 4) the VCR is set to broadcast on.

5. Set the VCR's channel selector (one of the VCR's controls) to the number of the channel carrying the program material you desire to record.

6. Adjust the fine tuning on the VCR for the best picture and sound quality. Also adjust the fine tuning on your TV set, if necessary. The picture and sound should come up on the TV set.

7. Select desired tape recording speed (SP, LP, or EP/SLP).

8. Insert a blank tape with intact "record safety tab," rewind if necessary, and put VCR into "Record" mode.

DETAILED EXPLANATION

After the channel selector and fine tuning on the TV set have been adjusted so that tapes played on the VCR show correctly on the TV screen, you can make further connections to permit record and view programs as they arrive on the antenna or cable. It is unnecessary to unhook the connections that you made in order to play a tape, but we will make additional connections.

First, it will be explained how to hook everything up when an ANTENNA is your source of programs, and then next, how to connect to a CABLE SYSTEM.

Connecting a VCR to an Antenna

If your source of broadcast program material is an antenna, the antenna will need to be disconnected from your TV set (as was already done in order to connect the VCR to the TV set), and now connected instead to the VCR. A VHF antenna needs to be connected to the F-type connector marked "In from Ant," or "VHF in," or "RF In," at the back of the VCR.

It is not connected to the plug that looks like a phonograph plug, nor to the UHF screw-down connectors.

Round-type Line from Antenna

If the wire coming from your antenna is already a round cable with a single stiff, solid strand of wire running down the middle and a threaded connector at the end, then you can screw it directly to the F-type

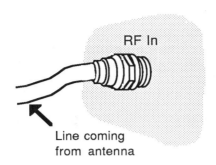

Line coming
from antenna

RF In

connector at the back of the VCR, remembering to make sure that the solid strand of wire in the center of the connector goes into the tiny hole in the center of the connector on the back of the VCR.

Flat Twin-lead from Antenna

However, you will need to use a "balun," or "75 to 300 Ohm matching transformer," if you have a "rabbit ears" antenna, or if you have an external antenna, attached to what technicians call "300 ohm cable," which looks like this

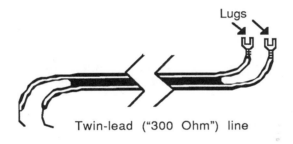

Lugs

Twin-lead ("300 Ohm") line

To connect this to the F-type connector at the back of your VCR, you will need a special adapter which you can purchase at a video store, or at an electronics parts store. Just tell the salesperson that you want "a matching transformer to connect flat-type TV antenna cable to the threaded connector at the back of a VCR."

These connectors are not expensive, and come in many shapes, such as the one shown in the next illustration already connected to the end of a piece of

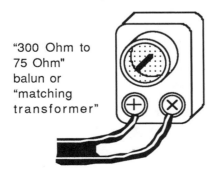

"300 Ohm to
75 Ohm"
balun or
"matching
transformer"

flat-type TV antenna wire. This is another type of balun or matching transformer, but with opposite kinds of terminals than the transformer described earlier in Section 1.

If you are using the flat-type antenna cable and it has not already been split back, cut through some plastic at the end to separate the two strands for about two inches (50mm), and then strip the insulation off about one half inch (12mm) at the very end of each multi-wire strand, twist it slightly so that the little wires form two little twisted ropes of wire, unscrew the little screws on the transformer-connector part way, wrap one twisted little rope of wires around the shaft of each screw, and screw down each tightly.

Lugs

If you want a neater job, you can obtain small metal connectors (called "terminal connectors," or "lugs," available at electronics parts stores) to attach to each twisted strand of wire.

Push the Connector onto the F-terminal

Attach the other end of the transformer-connector to the F-type terminal marked "In from Ant" or "VHF in" or "RF In" at the back of the VCR. The antenna is now connected to the VCR.

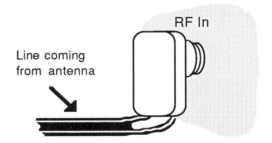

RF In

Line coming
from antenna

UHF Connections

If you also have a UHF antenna with its own piece of twin-lead, you can simply connect it directly to the two screw terminals marked "UHF In" on the back of the VCR. Then run a second piece of twin-lead directly from the two screw-terminals marked "UHF Out" on the VCR to the two screw-terminals marked "UHF In" on the TV set.

The Complete Simple Hook-Up

The piece of coaxial cable running from the VCR to the TV set that was installed earlier should remain in place, of course.

When simplified by omitting details of the connectors, the total hookup can be pictured as follows:

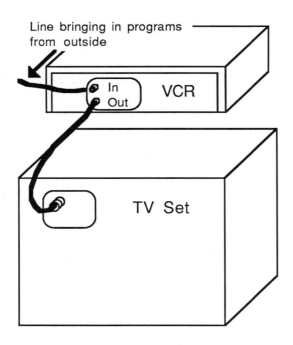

Line bringing in programs from outside

In
Out

VCR

TV Set

Complete Simple Hook-Up

The Tuner/Channel-Selector's Job of Traffic Control

The channel selector plus the fine-tuning adjustments constitutes a "tuner." It has a part that you use to select the number of the channel that you want to receive by pushing a button or turning a knob, and it may have an additional control that you use to bring in the picture and sound clearly (called the "fine-tuning adjustment").

The tuner or channel selector in your TV set acts like a super highway cop, letting through only the channel with the number you choose. The tuner does not look at the program contents.

So, if you want to watch today's baseball game, you tune to the channel with the number that is carrying the game. If you want to watch a certain movie, you tune to the channel with the number that is carrying it. And so on.

In metropolitan areas, think of the traffic cop as stopping twenty or thirty lanes on the freeway, and letting only the desired channel carrier pass.

The VCR also has its own tuner-traffic-cop which must be set to the channel that you desire to receive.

The VCR's Tuner

The important step of *tuning the VCR* to receive broadcast stations will be explained after taking care of the people on a cable system in the next few paragraphs.

Connecting to a Cable System

If you are on a cable system, the way to connect the cable to your VCR will depend on whether or not you have a "cable-ready" VCR. A "cable-ready VCR" can tune directly to the special channels (frequencies) used by cable companies. A VCR that is not "cable-ready" cannot handle these special channels.

Is the VCR Cable-Ready?

You may already know whether your VCR is "cable-ready," because if so, this fact is probably mentioned in the operator's manual or other material that came with the VCR. Many cable-ready VCRs will say "Cable TV," or "CATV" somewhere by the tuning controls, but some SHARP, MONTGOMERY WARD, and other brands do not tell you anywhere on the machine whether or not they are cable-ready.

If you cannot find any information on your machine and if you do not have the owner's manual, you can try following the procedures in the section on hooking up a cable-ready VCR. If you cannot directly tune in the cable channels, then either your VCR is not cable-ready, or there is something wrong with it.

Next try the procedure explained later in this section for non-cable-ready VCRs. If that works, then you have a non-cable-ready VCR.

Connecting the Cable to a Cable-Ready VCR

If your VCR is cable-ready, simply connect the threaded connector on the end of the cable wire coming out of your wall directly to the F-type connector on the back of your VCR marked "VHF Input," "RF In," or "In from Antenna." (Do not worry if it says "antenna" on the VCR — it's the same thing.) The hookup looks exactly the same as the picture shown earlier of "the complete simple hookup" of an antenna to a VCR. If there is no threaded connector at the end of the wire that the cable company left when they installed your cable, you will need to call the installer back to attach one for you.

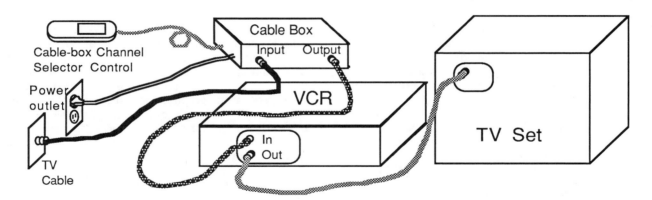

Connecting a VCR that is Not Cable-ready to a Cable System

If the cable company left you a cable box, leave it out of the hook-up, but hold onto it. If your TV set is not also "cable-ready," you will need to use the cable box later to watch one channel while recording another channel. You can use your cable-ready VCR to select the cable channel you want, but you may still need the cable box to control additional channels beyond the maximum number your VCR can handle, as explained later.

Connecting the Cable to a Non-Cable-Ready VCR

If you subscribe to a cable service, but your VCR is not "cable-ready," then you need to put a special "cable box" (what technicians call a "downconverter") between the line coming in from the cable company and your VCR.

Connect the line coming in from the cable company to the input of the cable box, and connect another piece of F-to-F type cable from the output of the cable box to the "RF In," or "In from Ant." terminal on the back of the VCR. If you had the cable before you got a VCR, you probably had a piece of this wire connected from the cable box to your TV set, which you disconnected when you hooked up your VCR to play a tape in Section 1. If you have not already used it, you can use it for this purpose, or buy another piece at the video store.

Your "cable box" also probably has a power cord that needs to be plugged into the power outlet in the wall (and it may have a fourth insulated wire that comes out to some sort of dial or control that can be used to select the channel you wish to view, or if it does not have a wireless remote control).

Tuning the VCR to the Channel to be Recorded

The following paragraphs are mainly about the channel selector on the VCR, and not about the one on the TV set. We are trying to get ready to use the VCR to record a movie or other broadcast program.

The TV set has a "tuner" in it to act as a traffic cop to let through only the desired channel carriers, and stop all the others from coming through. For the same reason, in order to record programs, a VCR also has its own separate "tuner-traffic-cop." You now have two (or more) channel selectors to set to the right channels and fine tune.

This can be confusing at first, because the various channel selectors usually need to be tuned to entirely different channels, with different numbers, in order to do what you want. The channel selector in the VCR needs to be tuned to the incoming channel carrying the program you desire to record.

But in order to view the material on the channel being recorded and to make certain that the VCR is getting it right, the TV set must be tuned to whichever channel the little station inside the VCR has been set to output on. (This will be either channel 3 or 4.)

Using the "TV/VCR" Switch

The function of this important switch is obscured by the fact that different manufacturers give different names to it. Most manufacturers call it the "TV/VCR" switch, but some call it the "TV/Video" switch, while others call it the "TV/VTR," or the "Tape/TV" switch, or the "Video" switch.

In every case, whatever its name, in its "TV" position, it has the effect of connecting the "Output" F-type terminal of the VCR directly to the terminal on the VCR marked "In," so that the VCR simply passes along unchanged whatever material comes into it on *all* the channels carriers together, without changing anything to different channel carriers. In other words, with this switch in the "TV" position, it makes it like the TV set were connected *directly* to the wire from the antenna or cable with no VCR in between.

When this switch is in the "TV" position, channel carriers with all the different numbers go straight from the antenna or cable to the TV set . The VCR is still sitting there, of course, with the wires connected to it, but when this switch is in the "TV" position, the VCR is just acting as a highway expressway for all the different channel carriers to travel across together.

In the other position, regardless of whether it is called "VCR," "VTR," "Video," or "Tape," when the VCR is playing a tape, this switch causes only the program material on the tape that is currently playing to be sent out of the VCR to the TV set on channel 3 or 4, so that the tape can be watched as it plays. Or, if no tape is playing at the time, but the unit's power switch is turned on, putting the "TV/VCR" switch in the "VCR" position causes the machine to take whatever program material is arriving on the channel to which it is tuned, and "convert" it to the output channel, sending it out to your TV set on whichever of the channels, 3 or 4, you selected earlier.

Terminology

To save words in the following discussion, we will simply refer to the other position of this switch (that is, the position NOT marked "TV") as the "VCR" position, even though it has a different name on some machines. Owners of VCRs on which the other position of this switch is marked "Video" or "VTR" or "Tape" should understand "VCR" in this discussion to refer to the switch position marked with this other name on their machines.

A "Cable Box" also Converts Channels

A cable box "converts" program material from one channel to another. Program material being received on some high cable channel, say, for example, cable channel 30, can be converted to a different lower-numbered channel carrier for convenient viewing — say, maybe, VHF channel 2. That is, it converts the program material from the chosen cable channel to one single low-numbered VHF channel for the subscriber's convenience.

This way, subscribers can leave their television sets tuned to this one low-numbered channel only, and choose the cable material they want to see by setting the cable-box selector to the number of that cable channel. For example, suppose a subscriber wants to see the program on cable channel 35. The subscriber sets the cable-box selector switch to 35, and the cable box takes the material on this cable channel and sends it out to the television set (or VCR) on the channel that the box converts everything to.

A second reason for the cable box is that cable companies put the programs onto special odd high-numbered carriers that the traffic cops in older VCRs and TV sets cannot handle. Just as an old-fashioned traffic cop would be unprepared to handle air-traffic at an airport in such a way that a selected air carrier would land correctly, so too the tuner in an older VCR or TV set is not prepared to handle special cable "high band" traffic in such a way that a selected carrier can get through correctly. Newer VCRs and TV's have special tuner-controllers in them that can handle all types of traffic.

Most cable boxes output only on one channel, usually either 2, 3, or 4. Suppose that a cable box outputs on channel 2. Then it will send the program selected by the subscriber out on channel 2. Naturally, this means that the subscriber must have his receiving equipment tuned to channel 2 to record it.

Or, if the cable box outputs on channel 4, then the receiving equipment must be tuned to channel 4, to record it. So, when a cable box is connected to the input terminal on VCR, the VCR must be tuned to the channel (2, 3, or 4) on which the cable-box outputs before the VCR can record the program.

So far, I have described *what* your equipment must do to record a tape (namely, use the VCR's tuner-traffic-cop to select and tune in the program you desire to record). Before I go over *how* to do this, in case you are uncertain, or have trouble, to reinforce your understanding so far, let me give you a tricky question to answer.

Fun Quiz

Suppose that someone wants to record the program from channel 9, and sets the TV/VCR switch to the "TV" position, tunes the TV set to channel 9, but leaves the VCR tuner-cop set to channel 7. *Question:* What channel will he see on his TV set while he runs the recorder (thinking he is recording channel 9) and what channel will he see later when he plays back the tape? Try to figure out the answer before reading the next paragraph.

Answer

Answer: He will see channel 9 on his TV set while recording, because it is among all the channels being sent on the bypass directly to the TV set, and his TV set's tuner is set to select channel 9. This may lead him to think mistakenly that he is successfully recording channel 9.

Meanwhile, unknown to him at the time, the VCR is actually recording the program that the VCR tuner-cop is letting through, which is channel 7, not channel

9, in this example. So when he plays back the tape, he will be baffled and disappointed to discover that he recorded channel 7 rather than channel 9. Or, if there is nothing on channel 7 at the time, he will see only snow and hear noise when he plays back the tape.

This is why it is necessary to understand the functions of the various controls — to prevent such disappointments.

Adjusting the VCR Tuning Controls

Owning both a TV set and a VCR, you have two (or if you use a cable box, three) pieces of equipment on which you must set the channels and tuning correctly to get the desired result. This is confusing and sometimes makes it hard to set all the controls right.

Your TV set needs to be tuned to whichever channel (3 or 4) the tiny transmitter in your VCR broadcasts on, as determined by the little switch discussed earlier in the section on playing a tape.

The VCR itself must be tuned to the channel you wish to receive and record. If you are operating from a cable box, this will be simply the single channel on which your cable box outputs everything (which might be channel 2, or 3, or 4). If you have a cable-ready VCR, or use an antenna as your outside program source, then the VCR must be tuned directly to the desired channel.

And the selector for the cable box, if you use one, will need to be set to give you the desired cable channel.

With a Cable Box

If you subscribe to a cable service and have a cable box, the situation may be that your TV set needs to be tuned to channel 4 (because this is the channel on which you instructed your VCR to output or broadcast its programs), while your VCR needs to be tuned to channel 3 (because this is the channel on which your cable box outputs all programs). This is just an example to illustrate a typical situation.

Or, for similar reasons, your TV set might need to be tuned to channel 3, while your VCR might need to be tuned to channel 2. Or, it could happen that the TV and VCR both needed to be tuned to the same channel number, say both to 4.

Choice of VCR Output Channel

With a VCR, you now own a little TV station — the little one inside your VCR. And as the owner of the station, you have the privilege of selecting whether

it will broadcast on channel 3 or channel 4.

You probably have no control over the channel on which programs come out of your cable box — because this is determined by the cable company — but you can choose whether your VCR outputs on channel 3 or channel 4.

Avoiding Interference Lines

As explained earlier, the best results usually are obtained when the VCR is set to output on a channel that is not used by your cable box or local stations as their output channel, but you can experiment with both settings, comparing the results using channel 3 with the results using channel 4, and remembering to change the channel selection and fine tuning on your TV set when you change the output channel choice switch on the VCR.

If an antenna is your source of programs, then you will need to set the channel selectors and fine tuning adjustments on your VCR for the best reception of the VHF and UHF station channels available in your locality. You might need to set your VCR so that it will tune in channels 3, 8, 10, and 13, and UHF channels 16, 28, and 44, for example, while your TV set might need to be tuned to receive say, channel 4, if that is the channel on which you have instructed your VCR to broadcast its program material, as explained previously (plus the other active channels in the list, if you want to be able to watch one channel while recording another, as explained later).

Second Quiz Question: If the VCR output selector is set to send programs out on channel 4, what are the two different ways in which you could view local channel 10?

Answer: You could view local channel 10 in two ways. You could tune the VCR to channel 10, put the TV/VCR switch in the "VCR" position, and tune your TV set to channel 4, using your VCR as the tuner-selector, with channel 10 being converted to channel 4 inside the VCR and the program coming out of the VCR on channel 4 when the "TV/VCR" switch is in the "VCR" position. Or you could put this switch in the "TV" position, and view the same program material directly via the bypass by tuning your TV set directly to channel l0.

Fine-tuning Your VCR to a Desired Incoming Channel

Now we come to explaining how to set the tuning on the VCR so that its traffic cop will let through the channels you desire. It is assumed that you have finished connecting everything together as described earlier. Plug in and turn on the power to everything.

Tuner/Camera/Aux Switch in "Tuner" Position

Most VCRs also have another switch somewhere marked "tuner /camera" or it may be marked "tuner/aux." This switch has two or three positions. It must be set to the position marked "tuner" in order to record a program from the antenna or cable, (In case you are interested, the other position of this switch, marked "camera" or "auxiliary," is used only when you want to record from a video camera, or want to copy a tape playing on another VCR. To record from your own video camera, or to copy a tape from a second VCR, as explained later section in this chapter, you would need to set this switch to one of its other positions.) For now, make sure that this switch is set to the "tuner" position, the position used when you want the VCR to receive something arriving on the antenna or cable.

On some VCRs, this switch sets itself automatically, but do not expect this. The "tuner/-camera/auxiliary" control is like the control on a telephone with several lines coming into it. If you want to use Line 1, you have to push the button for line 1, but if you want to be connected to another line, then you have to push the button for that other line. Even if there is nothing arriving on Line 2, you still need to push the button for line 1 to talk on Line 1, because the telephone cannot figure this out for itself.

Similarly, you have two (or three) possible ways for program material to come into your VCR, and most VCRs need to be told explicitly that you want to watch what is coming from the tuner, and not to from a camera.

Note: On a few VCRs, there is no such switch; if this is your situation, skip this step.

More Gremlins

If, one fine day, your VCR suddenly seems not to tune in any station, when it had been working perfectly the day before, always remember to check the position of the "tuner/camera" or "tuner/auxiliary" switch — A child or gremlin may have switched it to the other position. Don't laugh — this happens all the time, and people bring their machine into the shop and pay an expensive minimum bench fee, when it needed only to have this switch put back into the "tuner" position.

Deceptive Channel Number Indicators

With your VCR now connected to the antenna or cable as a program source, it still must be *set and*

fine-tuned to the channel you want to receive. You still must tune your VCR. The next section begins to explain how to do the tuning on different types of VCRs.

Do not make the mistake of assuming that simply because the channel number "4," or "7," or "13," is illuminated on the front display of the VCR, for example, it is correctly tuned to channel 4, or 7, or 13. Unless you are extremely lucky, or have a VCR that came from the factory already correctly tuned, or have one of the latest self-tuning models, you may need to adjust the rough and fine tuning on your VCR just as you do occasionally on your television set.

Different Types of VCR Tuning Controls

Different VCRs vary in how their fine tuning is adjusted. Use the following paragraphs to identify the type of tuning system in your VCR, and then turn to the section that follows explaining how to adjust it.

Do not confuse the "tuner" with the "tuner/-camera/aux" switch discussed earlier.

The earliest VCRs had a round rotating dial in combination with a rotary switch that selected the channels. This was called a "turret" tuner, because the control knob was shaped like a turret. This system was used on machines with metal push-down buttons that looked like the machine in this picture:

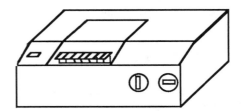

If your machine looks like this, turn to the section entitled "How to Adjust Turret-type tuners" to read about how to adjust its tuner.

The second generation of VCRs had a large number of tiny "thumb wheels," one for each channel

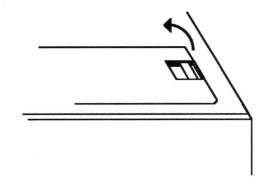

button, each of which separately adjusts only one channel. These are usually located either on the front panel of the cabinet, or on the top, under a little trap door.

When you open the door or remove the lid, they might look like this, for example:

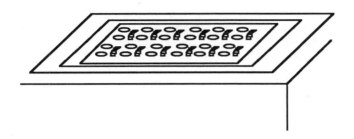

Thumb Wheel Type Fine Tuning Adjustments

There is one thumb wheel for each channel. On some models, the thumb wheels are on the front, behind a little door, rather than on the top.

Electronic Tuning

On the next generation of VCRs with electronic tuning, the system is like the thumb wheels, except that to fine tune each channel you just push in one of two little buttons. You could think of one little button as operating an electric motor that rotates a hidden internal thumb wheel in one direction for you, while pushing the other button makes the motor turn the wheel the opposite way. The internal electronic circuitry actually works in a different way, and uses no motor, but it sometimes helps initially to visualize what you are doing in this way.

Tuners that Automatically Set Themselves

On some VCRs, there is only one button to push to set all channels. You just push this button, and the circuits inside the VCR register all the different channel carriers coming into the VCR, and go through and set the tuning for each one. If your VCR is like this, it is best to consult the owner's manual for the proper procedure if you have trouble, because there are so many differences from model to model.

Frequency Synthesized or "Phase-Locked Loop (PLL)" Tuners

VCRs with "quartz" or "phase-locked-loop" tuners require no fine tuning. They come permanently preset to tune to all standard channel carriers. They only require that you tell them by throwing a switch whether you want to tune to broadcast channels (VHF

and UHF), or to the special cable channels (CATV). And they may require that you go through and tell them which channel numbers you want included in the menu, and adding or deleting channels. If your VCR is like this, it is also best to consult the owner's manual for the proper procedure, because there are so many differences from model to model.

Setting the "AFT" Switch

If the VCR has an "Automatic Fine Tuning" (or "AFT") button, it is usually best to switch it to the "Off" position while you do the tuning. Then switch it back to the "On" position, after tuning in the channel(s) that you want as well as you can. In some cases, when you fine tune by hand, you may only be able to get a black-and-white picture, but the color will come up when you push on the AFT switch. Occasionally, I have encountered machines where the reception was so weak that they could only be tuned at all with the AFT switch in the "On" position — in such a case, you have no choice and must tune with the automatic fine tuning switched on.

How to Adjust Turret-type Tuners

To adjust turret-type tuners, first rotate the channel selector rotary switch so that the number of the desired channel is displayed. This approximately tunes in the channel carrier having this number. To finish the job, push *in* the second knob that is mounted together with the rotary switch,

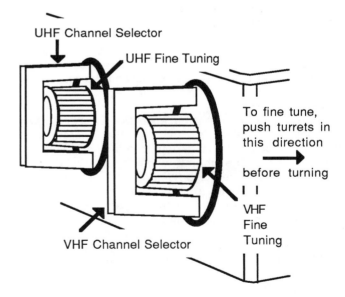

and turn it slowly in one direction until a picture comes up clear.

If nothing comes up, try rotating the fine tuning in the opposite direction. If you still get nothing, try turning the channel selector through other channel numbers until something comes up on one of the numbers.

How to Adjust Thumb-wheel and Electronic Tuners

Earlier we learned how to *identify* thumb-wheel and electronic tuners. Now we need to learn how to *set and adjust* them

Start by picking out a channel that you wish to start trying to receive. If you have a cable-box, this will be the channel on which your cable box outputs (2, 3, or 4, probably). If you are using an antenna, start with the channel of the local station that seems to have the best picture where you are located. This might be VHF channel 5, 6, 7, 8, 9, 10, 11, 12, or 13, for example. If you have the cable attached to a cable-ready VCR, you might start by picking a number that is the same as a cable channel with which you are familiar (like HBO) — or just pick any of the channel numbers on the VCR at random).

Push the button that turns on the light on the front panel behind the number .

If you have separate thumb-wheels, find the thumb wheel correlated with this channel light. On some machines, there will be a tiny alphabetic letter ('a', 'b', 'c', etc.) printed beside this light, and the same alphabetic letter is printed beside the thumb-wheel that goes with this channel light. On other VCRs, there are no alphabetic letters, but the thumb-wheels are all in a straight line like the channel lights. The first thumb-wheel from the left goes with the first light, the second thumb-wheel goes with the second light, and so on. Each thumb wheel is used to tune one channel only.

If you have electronic tuning, locate the two buttons that you press to tune in the two directions. Often one has a plus (+) beside it, and the other has a minus sign (-).

Do not assume that because the number showing in front of the light is, say, '4', that therefore the VCR is automatically tuned to channel 4. This is a mistake. On the thumb-wheel type system, this channel can be tuned to *anything!* It could be tuned to channel 7, or 13, or to UHF channel 44. With the thumb-wheel system, you can use each thumb-wheel to set each channel position on the front panel to anything within the range of the machine. The illuminated numbers are merely suggestions of the channel number to which many people will want to set each thumb-wheel, but you can set them to anything and even buy kits with different plastic numbers that you can set in front of the little lights to show whatever numbers you tuned those channel buttons to.

Set Range Selector Switch

Before you start rotating the thumb-wheel or pressing the button to tune to the desired channel, however, you first must use a second tiny selector switch to tell the VCR the *range* in which the channel you desire falls. Usually, there are three choices: "low" (for channels 2 to 6), "high" (channels 7 to 13), and "ultra high" (UHF, above 13).

To tune directly to the special cable channels on some "cable-ready" VCRs, you set this switch to the low or to the intermediate position. On cable-ready VCRs, on the little door over the thumb-wheels, or beside the channel range selector for this position, there may be printed something like "Cable TV A-1, 7-13 UHF, J-W — V_H" or "V_{MHS}."

On some models, the range selector switch is a straight-line slide switch.

On other models it is a tiny rotary switch that you can turn to point to one of the choices.

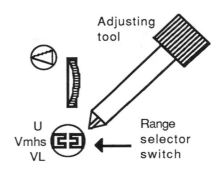

In this illustration, the circular-type range-selector switch is pointing at the middle setting, "Vmhs." If it were rotated slightly clockwise, it would point at "U" (representing UHF). Turned counterclockwise, it would set the range to the lower VHF channels ("V_L").

Often a small plastic screwdriver comes with the VCR, held by a plastic clip somewhere in the same recessed area as the little switches, each of which has a tiny slot to turn it. But if you have no such tool, you can use a small screwdriver, a nail file, or just a a fingernail to turn the switch.

Set the selector in the position pointing to the range that includes the channel you desire to receive. For example, if you want to tune to channel 4, set it to the "low" position; if you want to tune to channel 10, set it to the "high" position; if you want to tune to a UHF channel, set it to the "UHF" position.

Also, if the VCR has a switch marked "Automatic Fine Tuning," or "AFT," it is generally best to put it in the "off" position while tuning, as mentioned earlier.

Now you are finally ready to tune to the desired channel. Double-check to make certain that you are turning the thumb wheel that corresponds to the position lighted by the button. Rotate this thumb wheel a few turns in one direction, and then, if no picture appears, turn it back twice as many turns in the opposite direction.

Rotating Adjusters

On some VCRs, the thumb wheels move circular adjusters that go around like a hand on a clock. With this type, as you continue to turn the thumb wheel in one direction, the pointer goes all the way around and returns to the position from which it started. In order to tune a channel, you do not need to reverse direction unless you want to.

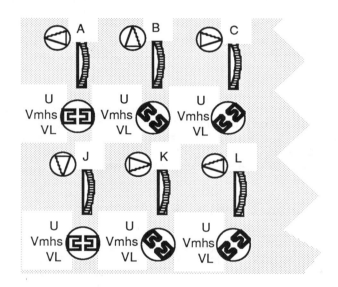

Clocklike Thumb Wheel Adjusters With no End-points

Straight-line Adjusters

On other models, there is a limit to how far the thumb wheel can be moved in either direction. When this limit is reached, the thumb wheel must then be rotated in the opposite direction in order to run the tuning over its complete range in the other direction to the opposite extreme end-point.

Usually a tiny moving part can be seen travelling up and down as you turn the thumb-wheels one direction or the other. If you go all the way to the top or bottom, they stop moving until you start turning the thumb-wheel in the opposite direction.

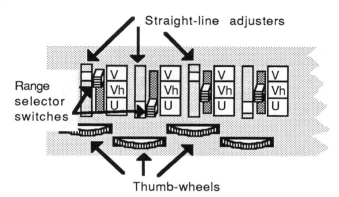

Tuning Adjusters with Stops at the Ends of their Ranges

Individually Tuned Channels

Each different channel position must be individually tuned to one desired station. When you have done that, you push the button to illuminate the next channel number window, find its correlated thumb-wheel, and start tuning it.

What Will be Seen

At some points, you should at least see differences in the patterns and noise on the TV set as you turn the adjustment. If you are connected to a signal source like an antenna carrying many channels, you may see one picture come up at one position, then disappear as you continue turning the thumb wheel, and then another picture come up from another channel and then disappear, and so on.

You are actually tuning through all channels in the range selected, so different channels in that range can come up and disappear at different points in the tuning adjustment. The different pictures and sound that you observe are the programs on the different channels that are active on your area in the selected range. These may include channels other than the one you desire. For example, you may be attempting to get channel 9, but as you turn the thumb wheel, you may see channel 7 come and go, then 9, followed by 11 and 13 (assuming stations broadcast on these channels in your area).

Avoiding Confusion

This characteristic of the "separate thumb-wheel tuning system" is confusing at first. On the old-fashioned "turret-type" rotary channel selectors, when you turned it so that it said "9," for example, then it really was set to channel 9, and you only needed to make small adjustments with the fine-tuning dial to sharpen up the picture, bring in the color, etc.

But on the thumb-wheel system, the button could be pushed that lights the number "9" and yet the VCR actually be tuned to channel 2, or 7, or 13, or even to a UHF channel with a higher number. The number lighted on the display means practically nothing initially. The channel to which the VCR is actually tuned depends on the setting of the correlated range selector switch and thumb-wheel. By turning these to various positions, you can make the VCR be tuned to any VHF or UHF channel you wish, even though the lighted number still (falsely) says, for instance "9."

It Must be Done by You

If you want the VCR *actually* to be tuned to VHF channel 9 when the lighted display says "9," then you must produce this effect yourself by adjusting the range selector and thumb wheel until the machine is tuned to channel 9 when the lighted display says "9." The same goes for all the other channel positions too, of course.

Figuring Out Which Channel is Up

If the VCR is connected to the output of a cable box, then you need only to tune to the single channel (for example, 2, 3, or 4) on which the cable box outputs the cable program material you select with the cable box selector. There is just one channel to tune in.

But if you are connected to an antenna, or if you have the cable from a subscription cable service hooked directly to the input terminal of a cable-ready VCR with no cable box in between, then you will need to be clever to figure out which of the different programs that you see come and go as you turn the thumb-wheels is on the channel you are trying to tune in so that you know that you should stop there.

A person with a lot of time to kill could stop and watch each channel until a station-break identification comes up, and find out in this way which channel the VCR currently is tuned to.

Or one might be able to tell to which local station the VCR is tuned at the moment from the content of the program.

Fast Method

Or one can note the *content* of the program, then put the "TV/VCR" switch in the "TV" position, go to your *TV set* and turn *its* channel selector through all its numbered positions until you see the same program come up, and note the number of the channel selector switch on your TV when this happens. This would be the actual number of the channel to which you had the VCR tuned when you saw the same material (assuming your TV set's channel selector is accurate).

If you use this fast method, when you move back and forth, remember BOTH to return the "TV/VCR"

switch to the "VCR" position, AND to put the *Television set's channel selector switch* back to the channel 3 or 4 position on which your VCR is outputting, or you might not receive anything. (NOTE: This "fast method" will *not* work if you have a TV set that is *not* "cable-ready" connected to a cable-ready VCR which is directly hooked to the cable service, because a non-cable-ready TV set cannot tune directly to the cable channels when you put the "TV/VCR" switch in the "TV" position.)

If the channel you are receiving with a given position of the thumb wheel on the VCR is not the channel that you desire to receive when you push the corresponding button, then continue turning the thumb wheel and checking by trial and error, until you finally get the position set correctly for the desired channel.

Once is Enough

Fortunately, you should only need to run this procedure once for each channel that you wish to receive, because once done, each setting should stay tuned so that next time you need only push the button to receive that channel. However, the fine tuning thumb wheel may need a little readjustment every few months, so it is important to know how to do it yourself.

As mentioned, if you have a cable box and only desire to see what is coming out of it, then you need only tune one position of the VCR channel selector buttons to the single channel on which your cable box outputs everything. However, you may need to readjust its fine-tuning too every few months.

Problems with Reception

If you cannot get, or tune to, any channel at all, check to make certain that a signal is actually coming into the VCR input terminal by putting the "TV/VCR" switch to the "TV" position and testing to see whether you can pick up anything on the TV set by running its channel selector through all its various channels.

If a picture can be picked up this way, then a signal is reaching your VCR, but it is either too weak for the VCR to pick up, or there is still something incorrect in the setting of the switches on the VCR, or there is something wrong in the VCR.

Recording a Program

Once you have adjusted the tuning on the VCR and TV set so that you get good picture and sound quality with the "TV/VCR" switch set in the VCR position, you can record a program. To record a program, you will need a video cassette that still has the erasure safety tab in place, such as a new blank tape.

This safety tab must be in place, because if it is broken off, the VCR has a sensor that detects this fact and prevents the VCR from going into the record mode, a safety feature designed to permit owners of valuable tapes to safeguard the program material on these tapes from accidentally being erased and recorded over

On VHS cassettes, the record safety tab is located along the back edge of the cassette, and looks like this

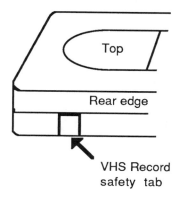

Top

Rear edge

VHS Record safety tab

Once the plastic safety tab is broken off, the VCR is supposed to be prevented from ever recording on this tape.

If the record safety tab has been broken off, but you are certain that you do want to erase this tape and use it to make a new recording, you can fold a piece of thin cardboard (like a matchbook cover) into a little rectangle that will fill the recessed rectangular area, and put a piece of tape over it to hold it in place. This trick usually will fool the mechanism inside the VCR that feels whether this tab is in place or broken off.

On Beta cassettes, the record safety tab is located on the bottom of the cassette box at one end, near the hinge for the door

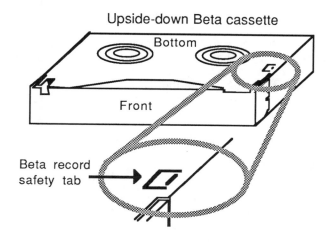

Upside-down Beta cassette

Bottom

Front

Beta record safety tab

Beta Cassette with Erasure Safety Tab Still Intact

With the power turned on, check the quality of picture and sound with VCR/TV switch set in the "VCR" position. Adjust as necessary. Insert the cassette, set the desired recording speed, and start the VCR in the record mode. As a safeguard against accidental destruction of program material, on many VCRs it is necessary to push BOTH the "Record" and "Play" buttons at the same time in order to make a recording.

After recording, the tape can be rewound and played like any pre-recorded tape.

Correct Use of the TV/VCR Switch to Check Recording Quality

Remember that in order to check whether a VCR is correctly tuned to the channel you desire to record, you must put the "TV/VCR" switch in the "VCR" position and use the TV set to verify that the picture and sound are good.

If you try to check the picture and sound quality while recording by watching the TV set while this switch is in the "TV" position, you might see a good picture on the screen while recording, but have a terrible recording when you try to play it back.

This is because in the "TV" position, you are only seeing channels that are bypassed around the VCR. On the other hand, in the "VCR" position, the TV screen shows the picture that the VCR has selected, seen as the VCR is receiving it. In the "VCR" position, therefore, any need for readjustment of the VCR tuning will show up in the picture on the TV screen if the VCR fine-tuning needs adjustment. So, never try to monitor the quality of a recording in process with this switch in the "TV" position. Always put it in the "VCR" position to check the recording as you are making it.

VCR/TV Switch in "TV" Position

When the "TV/VCR" switch is in the "TV" position, the TV set is connected directly to the wire bringing the program material into the VCR. Here it is as if the VCR were not even doing anything. When the switch is in the "TV" position, in effect, the VCR is taken out of the path between the TV set and the signal source (except for the little part inside that connects the two F-terminals together).

Monitoring Reception when Recording

In the "TV" position, the same TV channels are still coming into the VCR, but the VCR could actually be mistuned to the desired channel, or it could even be tuned to a completely different channel instead by mistake, and these problems would not appear on the TV screen (except later, when you tried to watch the tape you had made), because when this switch is in the "TV" position, the TV set is only looking at the signal that is arriving at the VCR, before it is processed by the VCR.

In this position of the switch, the TV set is NOT monitoring the *results* of the VCR's reception and tuning of the incoming signal. So, before making a recording, to check to make certain that the VCR is properly tuned to the signal coming into it, put the "TV/VCR" switch in the "VCR" position and look at the quality of the picture on the TV set.

Timer   Recording

The owner's manual for your VCR should explain how to use the timer and clock to make unattended recordings of programs that are broadcast at inconvenient viewing times, enabling you to view them later at more convenient times. There are so many different timers on various models, all with different programming procedures, that trying to present them in this book would completely fill it with information that is already available in owner's manuals. If you have no owner's manual for a VCR, try contacting the supplier at the address listed in Appendix II.

3. How to Copy or "Dub" the Contents of One Tape onto a Second Tape Using Two VCRs

You should always make a back-up copy of irreplaceable family tapes or other priceless tapes in your collection to protect against the possibility of a VCR malfunction, or an active child, accidentally damaging the original tape some day.

Protect   Priceless   Tapes

Right now would be a good time for you to break off the plastic safety tabs on all video cassettes containing priceless, irreplaceable tapes. This will make it more unlikely (but not totally impossible) that they will ever get erased accidentally.

Two   Machines

To make a duplicate copy of a tape, two VCRs are needed, one to play the tape being copied, and a second to record the copy. If you do not already have two machines, you will need to rent or borrow a second one. The machine on which the tape being copied is played can simply be a player; it need not have recording capability.

Check   Out   the   Other   Machine   First

If the second machine is only a player with no recording capability, then you have no choice: you must use it to play your priceless tape, and do the recording on your old reliable VCR.

Let's check out the second machine first, however. Before you insert your priceless tape into a

strange VCR, it is wise to test it by inserting a new blank cassette (say, the one on which you plan to record the copy) — and putting it into the "Play" mode to run several inches of known good tape through it in order to check to make certain that the other machine does not have some problem that will damage the priceless tape that you are trying to protect. Most types of tape damage can be seen visually on a close inspection of the surface of the damaged tape.

Now rewind only PART WAY, NOT all the way back. Eject the tape. Press the little square button on the side of the cassette. Lift the protective plastic door over the exposed part of the tape in the cassette, as explained in Section 12 of Chapter 4. Now visually inspect this tape very closely to make certain that the second machine is not putting any wrinkles, creases, edge curls, or dents in the tape, or damaging it in any other observable way, as also explained in Chapter 4.

If the machine is damaging the tape, then obviously you should not put your priceless tape into this machine. Instead, either repair the defect in the second machine using the procedures given later in this book, or return the machine and exchange it for another one.

How to Connect Two Machines Together for Copying

It is possible to copy tapes by simply connecting the F-type connector labeled "VHS out" or "Out to TV" on the machine that will play the tape to be copied to the "VHS in" (or "In from Ant.") F-type connector

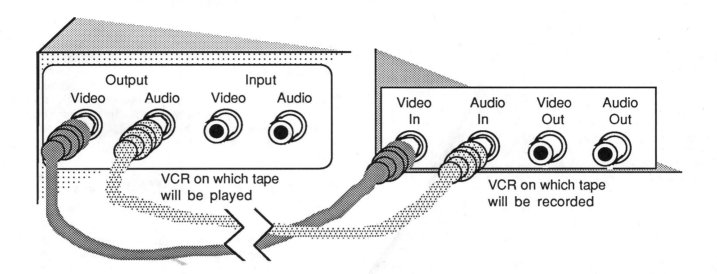

on the other recorder containing the tape onto which the material is to be copied.

Use Stereo Cables for Best Results

However, you can get greatly improved quality if you connect the "video out" and "video in" RCA phono jacks of the VCRs together with stereo cables, and likewise connect with the "audio out" to the "audio in" jack, because if you do this, the picture and sound go directly from one tape to the other tape, without getting bumped around while being sent out in a carrier.

Pictures of RCA "Phono-plug" connectors appeared at the beginning of Section 1 of this chapter. These phono plug connectors are located in different places on different models. On some machines, they are on the back, while on others, they are on the front, or side. To connect them together you will need a set of "stereo" cables with male phono plugs on all four ends.

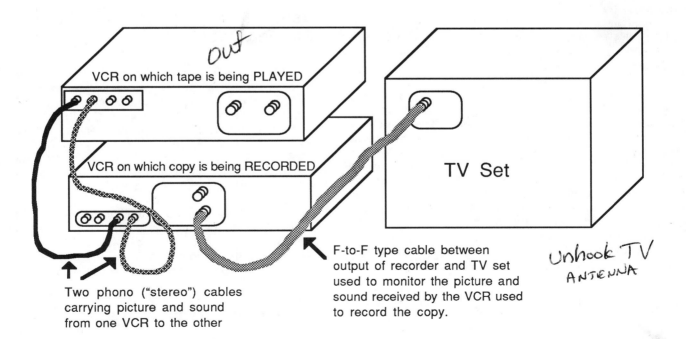

Two VCRs Connected to Each Other, and to a TV Set to Monitor the Copying of the Contents of One Tape onto Another Tape

Monitoring the Recording Process

Whichever method you use, check the quality of the sound and picture as received by the second VCR by connecting your TV set to the output of the VCR that will be doing the recording,

Before making the recording, make some tests to check everything out. For best recording and playback quality, you should do the recording in the same VCR that you will be playing it back on — that is, in your VCR, and not the borrowed machine. This usually will give you the best quality, when you play back the copy later on your VCR.

However, there is a small element of risk in doing it this way, because the condition of the borrowed machine, in which you will be playing your priceless tape, may be unknown. You would not want to put your priceless tape into a defective machine and have the tape get damaged.

So, begin by making several preliminary tests: as explained earlier, check to make certain that the borrowed machine will not damage a new blank tape put into it, check the quality of the picture and sound that gets sent from one VCR to the other, and check to make certain that the recording process will run OK.

Check Second VCR's Reception of Picture and Sound from First VCR

If you have not already done so, next connect the two VCR's together for copying from one to another, and connect a TV set to the output of the machine that will do the recording, so that the quality of the sound and picture as received from the first VCR can be checked.

As mentioned, in order that the end-product will look best when played on your own VCR, your VCR should be used to do the recording. If your VCR is already connected to your TV set, fine — that saves you from having to run an F-to-F coaxial cable from the "VHF out" terminal of your VCR to the antenna input of your TV set. Otherwise make this connection, and check to make certain that your TV set is tuned properly to the output from the VCR by putting any good recorded tape into it, pushing "Play," and adjusting the fine tuning on your TV set for best picture and sound (with the VCR's "TV/VCR" switch in the "VCR" position, of course).

Make the Connections

Now connect the second VCR — the one you borrowed or had to rent, maybe — to your VCR. To record on your VCR and play your priceless tape on the second machine, you will need to connect the output from the second machine to the input of your machine.

As mentioned, the preferred method of connection, for best picture and sound quality, is to use RCA phono connectors (the same type of cables used to connect stereo components) to connect the "video out" phono jack of the second VCR to the "video in" phono jack of your machine, and another cable to connect between "audio out" on the second VCR and "audio in" on your machine.

Is There a "Tuner/Auxiliary" Switch?

If you use the stereo-cable method, and if your machine is like many VCR's, you will need to push a switch somewhere to tell it to take its input from the video input jack rather than from the F-type "VHF in" terminal.

So, if you have the two VCR's connected together by RCA phono cables instead of a coaxial cable, find the "camera/tuner" switch on the VCR that is hooked to the TV, and set it to the "camera" position.   If this switch has three positions, "camera/auxiliary/-tuner," set it to the "auxiliary" position. If there is no such switch, then probably you have one of the common models in which this switch is automatically thrown to the "auxiliary" position by the action of pushing in the stereo cables.

Using F-to-F Cable Instead of Stereo Cables

Alternatively, you can hook the two VCR's together with a single piece of coaxial cable, connecting one end of the coaxial cable to the F-type connector marked "VHF in" (or "In from ant.") on your VCR, and connecting the other end to the terminal marked "VHF out" (or "Out to TV") on the second VCR. However, the results of using this method generally are not as good as the results when stereo cables are used, as described earlier.

If you have the two machines connected through their F-type connectors by a coaxial cable instead of stereo cables, set the channel selector on the recording VCR to the channel with the number (3 or 4) on which the playing machine is set to broadcast.

Adjust the Two VCR's

Turn on the power for both VCR's and the TV set. Put the "TV/VCR" switch on both machines to the "VCR" position, so that the TV set can be used to verify that the second VCR is receiving good picture and sound from the first machine. Tune the TV set's channel selector and fine tuning to the channel number (3 or 4) on which your VCR broadcasts.

To check the quality of the transmission of the program material from the borrowed VCR to your VCR, put a prerecorded tape into the borrowed VCR, and push "Play." You should see and hear the program material on the TV set connected to your VCR. If you are using the single-coaxial-cable method, you probably will need to adjust the fine tuning on the recording VCR for best picture coming from the playing VCR.

Adjust the "tracking" control on the playing VCR for best picture.

Make a Short Test Recording

When you have gotten everything adjusted so that picture and sound are good, put the blank tape into your VCR, set the counter to zero, set the desired speed, and start it making a test recording. Make a short recording, stop both VCR's, rewind the tape in the recording VCR, and play it back in the same machine to check the picture and sound quality.

If it is acceptable, rewind both tapes to the desired starting point.

If the quality of the recording is bad, see the section "Solving Problems in Dubbing" below.

Go for a Copy

If the results were good, you are now ready to make the copy. Wind both tapes back to the beginning, push "Play" on the second VCR and start your VCR recording. You may wish to monitor the picture and sound quality on the TV set as the tape is being copied. When the copying process is completed, rewind the copy, hit "Play," and check its picture and sound quality.

Duplicating Priceless Originals

If you often play your priceless tape, you might wish to consider storing it and playing the new copy instead. The average home video tape begins to deteriorate irreparably after about fifty runs through the VCR. If instead of the priceless tape, you play a copy, then you can use the preserved master copy to make more duplicates later when the first duplicate that you just made wears out. The same goes for any tape that the kids play again and again. Duplicates of

the original master copy are usually better than duplicates of a duplicate.

Solving Problems in Dubbing

If the test tape was damaged by the second VCR, then you will need either to repair the VCR using the procedures in the chapters after Chapter 4, or else exchange the defective VCR for another machine that does not damage tapes.

If the test tape was not damaged, but the quality of the picture or sound was poor when you tried to play your priceless tape on the second VCR even after you tried adjusting the tracking control, then it is possible that the two VCR's are not interchangeable. Sometimes tapes made on one VCR that play fine on it, will not play back properly on some other VCR models, even after you adjust the tracking.

Try making the recording on the other VCR and see whether it will play back OK on your VCR. If so, then use your VCR to play the priceless tape, and use the second VCR as the recorder. Connect the two together as described before, except with their positions reversed, each recorder in the place of the other. If you still do not get a good recording, substitute another VCR for one or another of the machines until you find a compatible pair. Just because you cannot make one pair of VCR's work for copying does not mean that no other pair would work.

Due to differences in design, some combinations of different models (even different models made by the same manufacturer) do not work very well together, just as some combinations of VCR's and TV sets do not work well together. If this is the source of the problem, there is no solution except to obtain a different VCR to use as the second machine.

First, though, let's take a moment to check out some other possibilities.

Rainbow Problems

If the problem is that the newly recorded tape seems to have a kind of rainbow effect appearing vertically over the picture when you play it back, this usually is due to failure of the erasing system to erase the previous material fully before recording the new material.

If there is only a little bit of rainbow at the beginning, the problem can be avoided during recording by letting the recorder run a few seconds before starting to input the material you want to record. You will need to go back and rerecord the material to do this, obviously.

If the rainbow runs through the entire recording, try erasing the entire cassette with a "bulk eraser" (which you can purchase inexpensively) before recording on it. If you want to avoid this incovenience in the future though, the recorder needs to be taken to a shop for repair.

Copyguard Jitters

If you are trying to duplicate copyrighted material (such as a prerecorded movie) that plays well on your TV set, but when you try to play back the copy you have made, the picture periodically rolls and jitters as if the "vertical hold" adjustment on your TV set needed adjustment, then you may have run

afoul of the "copyguard" system used by owners of copyrighted material to prevent people from making unauthorized duplicate copies for which they have not paid additional royalties.

To explain the exact details of how this electronic copyguard system works would go beyond the scope of this book, but the net effect is that although the video tape containing the guarded material will play properly on most (but not all!) VCR-TV set combinations, unauthorized copies of the guarded tape made on most video recorders will not play properly, and hence are useless.

A few VCR models are rumored to be immune to this protection system. The actual extent to which copying copyrighted material for personal use truly is barred by a correct interpretation of the copyright laws is a controversial legal topic outside our expertise.

4. How to Watch One Program While Simultaneously Recording a Different Program

Have you ever been in the situation where two different programs that you wish to watch are being broadcast on different local channels at the same time?

If you receive your programs from an ANTENNA, the "TV/VCR" switch on your VCR can be used easily to view one program while, at the same time, recording the other program for later viewing.

If you are on a cable system, read this section, then read the section on CABLE.

Proceed as follows

First, decide which programs you want to record while watching the other channel. Suppose you want to record channel 8. Turn on the VCR and TV set, put the "TV/VCR" switch in the "VCR" position. Set the channel selector switch on the TV set to the channel

number on which your VCR broadcasts its output (either channel 3 or 4). Tune the VCR to the channel (for example, channel 8) carrying the program you wish to record. Check picture and sound quality on the TV set, and start the VCR recording this program (or set the timer to start recording the tape at the time it will be broadcast).

Now change the "TV/VCR" switch to the "TV" position. This causes all the stations being picked up by your antenna to be sent directly into both your VCR and your TV set. Use the channel selector on your TV set to tune the second channel that you wish to watch at the same time (say, channel 13).

While you are watching this program on channel 13, your VCR will be recording the program on the other channel (channel 8, in this example). Later, you can rewind the tape and view other program, recorded from channel 8.

5. How to Find Out Whether the Format of Your Subscription Cable System Will Permit You to Watch One Program While Simultaneously Recording a Different Program on Another Channel

The Question

If your source of program material is a subscription cable system that provides its subscribers with some sort of "cable box," then whether or not you can watch one program while recording another depends on the type of cable system you have. Some types of cable system work by sending into your home only one channel at a time. This is called an "addressing cable system."

"Addressing" Cable Systems

With an "addressing" type of cable system, when you choose a channel on the cable box, a signal goes out to the company's equipment outside your home which responds by sending only the chosen channel down the cable into your home.

On this type of system, you cannot watch one program while recording from another channel at the same time, because only one channel is coming to you at any time. You are not getting two channels at the same time. No second channel is coming in for you to watch while recording the first channel. If you have this type of cable system, there is no way to watch one program while recording another, short of urging your local government to change the terms of the franchise it granted to the cable company.

Nonaddressing Cable Systems

You can watch one program while recording another only if you have the type of cable system that sends many cable channels down the cable into your home at the same time, and uses your cable box to enable you to select the one that you want to watch. Since programs on all channels are present on the cable coming out of your wall, it is possible for you to watch one channel while recording another.

With a nonaddressing-type cable system, you can watch any channel while recording any other — every combination is possible (assuming the channels are not *scrambled).* But to accomplish this without "cable-

ready" equipment, you will need a special hook-up, as described in the next section.

Scrambled Channels

If some of the channels come to you "scrambled" and your cable box serves to unscramble those you have paid for, you can do everything except watch one scrambled channel while recording another scrambled channel (unless you have two cable boxes, in which case you can even do this).

What Type of Cable System Is Yours?

The first step, then, is to find out which type of cable system you are on. If you have trouble finding someone who understands your question, call your cable company and ask to speak to the "Technical Department."

Avoiding a Brush-Off

Remember that technicians hate to answer the type of question you are going to ask, because they are afraid that you will expect them to explain in detail over the telephone how to make all the connections, wiring, tuning, etc., to do what you want to do. Unaware that you have this book to show you how to make the connections and knowing from experience how difficult this can be to explain for the first time to a nontechnician, especially without pictures, they are going to try to get rid of you instead of answering the question.

One Way to Find Out

One way to get around this psychological obstacle and obtain the technical information you need is by asking them a slightly different question which they will answer more readily, and from which you can deduce the answer you need to your real question.

Don't ask them, "With your cable system, can I watch one channel on my TV while recording another channel on my VCR," Instead try something like saying the following: "I am considering purchasing a cable-ready TV or a cable-ready VCR. If I connect the cable coming out of my wall from your company directly to a cable-ready machine, can I use it to select channels on your system?"

Even though you are not really considering such a purchase, they probably *will* answer this question because they want no trouble from customers who purchase "cable-ready" units only to discover that they will not work with their cable system.

If a "cable-ready" TV or VCR *would* work on your cable system, then you can record one channel while watching another, even though you do not have cable-ready equipment, using the accessories and procedures explained later in this chapter.

If they say that "A cable ready unit will work on everything but the scrambled channels," then you know that you can do everything but watch one scrambled channel while recording another scrambled channel at the same time.

But if they say "A cable-ready unit cannot be used at all to select channels with our cable system," then you know that you cannot watch one channel while recording another until you and others lean on your local government to force the cable company to change its equipment and system.

6. How to Watch One Program While Simultaneously Recording a Different Program on Another Channel from a Nonaddressing Subscription Cable System

If your source of program material is a subscription cable system that provides each subscriber with a nonaddressing cable box as determined in the previous section, you can record one cable program while watching a different program on another cable channel at the same time.

Doing so is slightly more complicated than it would be if your signal source is an antenna, but of course, you probably get a much bigger selection of choices with the cable than you would if you had only an antenna, so there is some compensation for your trouble.

Is Your Equipment "Cable-Ready"?

Obtaining the very best results requires a special "cable-ready" VCR and a cable-ready TV set (or two cable-ready VCR's and a non-cable-ready TV).

Even if your VCR does not say "cable-ready" on the front, it might still be cable-ready. If it says "105 channel" or "107 channel" on the front, probably it is cable-ready. If it says "Cable," or "CATV" or beside thumb-wheel tuning adjusters, or if there is a switch on the front or top that mentions "CATV" in one of its positions, then the VCR is cable-ready. The same generally applies to TV sets.

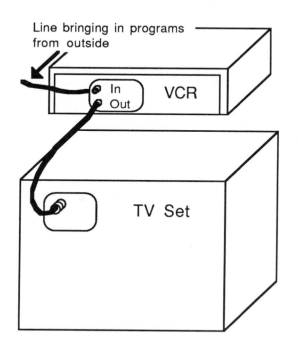

Line bringing in programs from outside

In
Out   VCR

TV Set

Hook-Up to Watch One Channel While Recording Another Using Cable-Ready Equipment

Use of Cable-Ready Equipment

With both a cable-ready VCR and a cable-ready TV set, to watch one program while recording another, you need only hook the cable to the VCR input, and run another cable from the VCR's output to the TV set's input as described earlier in this chapter. While the VCR is recording whatever is on the cable channel to which its selector is tuned, push the "TV/VCR" switch to the "TV" position to bypass everything directly to the TV set, and then use the TV set's channel selector to tune to the other program you desire to watch at the same time.

Solving the Problem with Equipment that is Not Cable-Ready

Acceptable results can be obtained with a non-cable-ready VCR and TV together with less expensive accessories like a "block converter," a "splitter," and an "A/B Switch."

Splitters

A "splitter" is a device that takes the signal coming in on one line and "splits" it, to send it down two (or more) different lines at the same time.

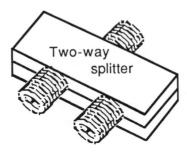

Splitting reduces the power of the signal that is sent out on the different lines, unless you use an "active" (rather than "passive") splitter, which plugs into a power outlet and amplifies the signal as well as splitting it.

A/B Switch

If you wish to be able to connect a unit first to one source of programming, and then disconnect it quickly and reconnect it to another source, you can use an "A/B Switch." If you buy an A/B switch, it is wise to get a good one, because poorly made A/B switches are often a source of interference lines (herringbone and moire) in the picture that you cannot get tuned out.

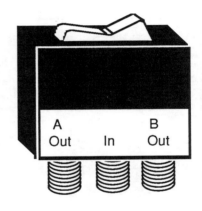

A/B Switch

The Problem to be Solved

This additional equipment is needed because channels with program material come into your home from the cable system on the cable wire on special carriers which the tuners in non-cable-ready VCR's and TV's cannot handle. Just as most car radios cannot tune to short-wave radio stations because they are outside the range to which they can tune, non-cable-ready VCR's and TV's cannot tune directly to the frequencies used by cable companies to transmit cable channels.

What a Cable Box Does

This is why the wire from the cable company is run into a "cable box" supplied by the cable company with the subscription. This box contains complex circuitry that takes the program material from high-frequency cable channels, according as the subscriber chooses by means of the selector connected to the box, and "down-converts" it to another lower channel to which everyone's TV's and VCR's can tune — usually either channel 2, 3, or 4. The cable box converts the special channels used by the cable company to VHF channels to which conventional, non-cable-ready VCR's and TV sets can tune.

The Chosen Channel

A VCR or a TV set connected to the output of the cable box only receives the program put out on the one channel carrier always sent out by the cable box. This is why, in this situation, the VCR (or TV) is always left tuned to the same one channel (say, channel 2, 3, or 4), while the viewer changes channels with the channel selector switch attached to the cable box. The channel selector attached to the cable box determines which of the cable carriers will have its picture and sound transferred and sent out of the cable box on the single VHF carrier.

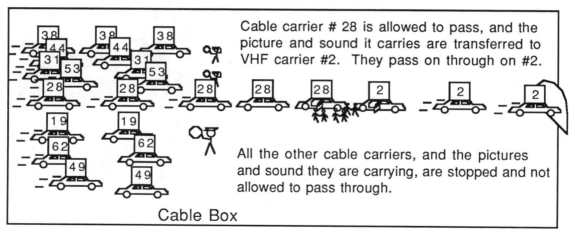

Cable carrier # 28 is allowed to pass, and the picture and sound it carries are transferred to VHF carrier #2. They pass on through on #2.

All the other cable carriers, and the pictures and sound they are carrying, are stopped and not allowed to pass through.

Cable Box

What Happens Inside a Cable-company Cable-Box

The channel selected with this switch always is converted to the same one channel carrier by the cable box and sent out of it on that same carrier. For example, if an old Fred Astaire movie is playing on cable channel 28 and the customer turns the cable-box selector switch to 28, the Fred Astaire movie is converted for example, to the normal VHF channel 2 carrier and sent out of the cable box on channel 2. If the customer's VCR is tuned to VHF channel 2, it will receive the Fred Astaire movie on channel 2 (as it comes in), as if the movie had originally been broadcast on channel 2.

A basketball game simultaneously being broadcast on cable channel 31 will not also be present on the channel 2 output from the box, because the box only converts and outputs one channel at a time.

But if the subscriber's cable box selector switch is changed from number 28 to number 31, the Fred Astaire movie will disappear from the channel 2 output from the box, and the basketball game will appear instead coming out of the box on channel 2. Channel 2 is used here only as an example, of course; the cable box might output on channel 3 or 4 instead, depending on the equipment used by the particular cable company in your locality.

Now You See It

Thus without changing the tuning on the VCR or the TV set, the Fred Astaire movie will disappear from the cable box output, and the basketball game will come up instead when the selector switch attached to the cable box is changed to a different number.

Focus on the Problem

You can now appreciate the problem that must be solved in order to *watch* one channel while *recording*

another channel when a cable service is the source of programs and you do not have "cable-ready" equipment.

The possibility of watching one program while simultaneously recording another program for subsequent viewing requires that both of the two channel carriers be coming into the VCR at the same time, so that the VCR can record one while passing the other(s) along to the TV set.

No Problem with an Antenna

This requirement is automatically fulfilled when the VCR is connected to an antenna in a locality where two or more stations are active.

In that situation, all the local stations are present on the wire coming into the VCR's input terminals at the same time, so the VCR's tuner can be set to pick out one station and copy its material onto a tape, while all the various stations' different channel signals simultaneously are passed along to the TV set, which can then be tuned to a different station than the VCR is tuned to. (It is assumed here that the VCR's "TV/VCR" switch is now set to the "TV" position as explained earlier.) In this situation, the viewer of the TV set easily can watch a program different from the one that the VCR is recording.

No Problem with Cable-Ready Equipment

With cable-ready equipment, you can leave the cable-box completely out of the hook-up, and feed all the cable channels directly into the VCR. Since all the many cable channels are coming into the VCR at the same time, you can choose to record one, while using the "TV" position of the TV/VCR switch to bypass everything directly to the TV set. If the TV set is also cable-ready, then you can use its tuner-traffic-cop to select another channel that you want to watch at the same time the VCR is recording a different program.

The Problem with Non-Cable-Ready Equipment and a Subscriber's Cable Box

But when you do not have cable-ready equipment and you are using a cable box, there may be only one program coming into your VCR at any given time. (I say "may be" because some cable companies permit local stations also to come out of the box on their normal frequencies along with the one cable channel chosen by the subscriber's selector switch.)

If the Fred Astaire movie is on the selected channel, then probably the basketball game is not coming out of the cable box at the same time at all. If the basketball game is selected, then the Fred Astaire movie is not reaching the VCR from the cable on any channel for the same reason. Since only one of the two programs is being sent out of the cable box at a given time in this situation, it is impossible to watch one while simultaneously recording another.

With a cable box, you do not have two channels coming to the VCR at the same time, so you cannot record one while watching another without doing something special. If you merely hook the cable box to your VCR and try to watch a second channel while recording another, you will see nothing. How to get around this problem?

A Good Question

At this point, someone may ask, "Why not simply eliminate the cable box, run the cable into a 'splitter,' connect one output of the splitter to the 'VHF input' on the VCR, run the line from the other splitter's output through a matching transformer to the VCR's UHF input, use the VCR's UHF tuner to select, say cable, channel 28 to record the Fred Astaire movie, connect the VCR's 'UHF Out' terminal to the TV set's 'UHF In' terminal, put the 'TV/VCR' switch in the 'TV' position to pass everything along to the TV set, and use the TV set's UHF tuner to select channel 31 to view the basketball game?"

Same Number, but Different Carriers

This excellent question shows a good understanding of all the relationships and equipment. And the idea would work except for one small but crucial fact: the channels that the cable company may call "2," "3," ..., "15," "16," ..., "28," "29," etc., are not at the same carrier frequency as the familiar VHF and UHF channels that conventional, non-cable-ready TV's and VCR's denote with the same numerals.

A Confusing Situation

Cable companies may take the same numerals and assign them to totally different channels. So, what the cable company calls "channel 28," for example, is totally different from what a conventional TV or VCR recognizes as UHF "channel 28." It is very confusing to have the same numeral used for two completely different channel frequencies, but that is how it is.

"Cable Channels" vs. Plain Old Channels

Video people often distinguish between the two as "Cable channel 28" versus just plain "channel 28, UHF." Prefixing the word "cable," or the letters "CATV," to the front of the phrase "channel 28" indicates that the different cable carrier with the number 28 is being referred to, rather than UHF channel 28. Also, technicians may use alphabetic letters rather than numbers to denote cable channels ("Cable channel A," "Cable channel B," etc.).

The subscriber's cable box converts the special cable channels down to one lower ordinary VHF channel like 2, 3, or 4, as previously explained. But if you hooked the cable directly to a conventional non-cable-ready VCR or TV, you probably would be unable to receive the cable channels at all, even in the UHF mode, because the cable channel carriers are so different from anything that the conventional unit's tuner can select or receive. Only "cable ready" units can tune directly to these special cable channels.

Possible Solutions

However, there are several possible ways to watch one channel while recording another without cable-ready equipment.

Using Two Cable Boxes

If you have a way with words, or influence, you might be able to get the cable company to give you a second cable box. Then you could use one cable box to feed to the VCR what you want to record, while using the second cable box to send another channel to your TV set for simultaneous watching.

You will also need a splitter to send the cable input to the two boxes, and an A/B switch to choose between the two sources for your TV set. When you

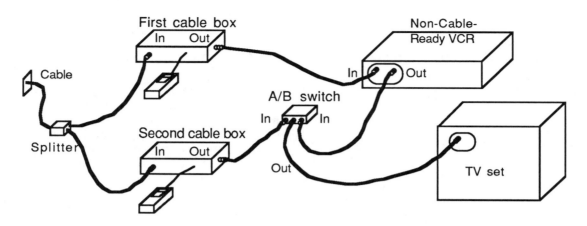

Using Two Cable Boxes to Watch One Channel While Recording Another Without Cable-Ready Equipment

put the A/B switch in one position, you watch what is coming out of the VCR. When you switch it to the other position, you see what is coming out of the second cable box.

Solving the Problem with a Block Converter

If neither your TV set nor your VCR is "cable-ready," and you do not have a way to get two cable boxes, one economical way to add the capability to record one cable channel while watching another at the same time is by purchasing an inexpensive accessory called a "block converter."

Block Converters

A block converter takes the program material arriving on all of the different cable carriers at the same time, and converts each of them simultaneously

to a different conventional UHF channel of the sort to which conventional VCR's and TV's can tune directly.

For example, the Fred Astaire movie on cable channel 28, which the conventional set cannot pick up directly, might be converted to UHF channel 64, to which the non-cable-ready VCR or TV set can tune directly in its UHF mode. Simultaneously, the basketball game on cable channel 31 might sent out of the block converter on UHF channel 67, to which the conventional VCR or television set again can tune.

Similarly, the program material on each different special cable channel is sent out on a different standard UHF channel of the type that can be directly tuned in by a conventional TV or VCR's UHF selector.

You can attach your cable directly to the input of the block converter rather than to the cable box supplied by the cable company. These various programs are sent out of the block converter's "UHF Output" on many UHF channels at the same time, on

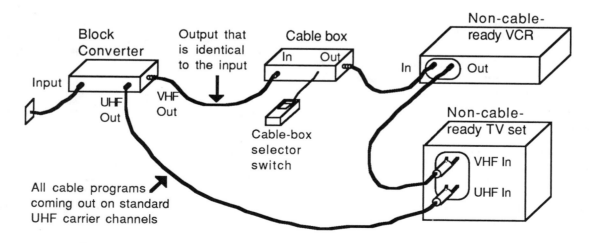

Using a Block Converter to Record One Channel While Watching another Channel with Non-cable-ready Conventional VCR and TV Set

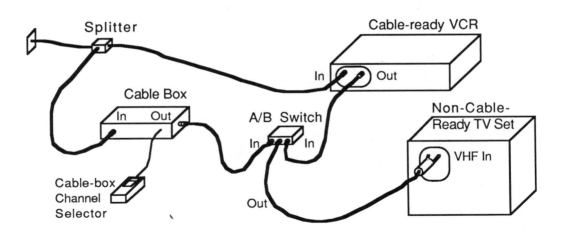

Connections to Watch One Channel While Recording Another Using a Cable-Ready VCR and a Non-Cable-Ready TV Set

the same one wire, that you can attach at its other end to the "UHF In" terminals of the non-cable-ready TV (or VCR) through a balun or matching transformer. In the preceding illustration, this wire is connected directly to the TV. Now by setting the TV set's channel selector to the UHF range, and adjusting the fine tuning, all the cable channels can be viewed by switching from one UHF channel to another.

Meanwhile, all the same programs still on their special cable carrier channels are coming unchanged out of the block converter's "VHF Output" terminal and are being sent to the subscriber's cable box. The subscriber's cable box converts the channel chosen with the cable-box selector switch to either channel 2, 3, or 4 and sends this out to the VCR input.

The VCR now can record the program chosen with the cable box selector switch, while you are viewing a different cable channel coming into the TV set on UHF. When you want to playback the tape (or look at what the VCR is recording while it is doing so), change the TV set's channel selector to the VHF channel that the VCR outputs on (that is, VHF channel 3 or 4, as determined by the output switch on the back or bottom of the VCR).

Block converters tend to drift slightly off the center of the channel as time passes, but if you have one with a tuning control on it, you can adjust it so that it is well-received each time you watch it. Doing this will readjust all the UHF channels coming out of the block converter at the same time.

Solving the Problem Using a Non-Cable-Ready TV Set and a Cable-Ready VCR

If you have a non-cable-ready TV set, but a cable-ready VCR, you can use a splitter plus an A/B switch to watch one channel while recording another, without needing a block converter. Hook it up as shown in the illustration above.

Feed one of the two identical outputs from the splitter into the input of the cable-ready VCR, and connect the other to the cable box. Tune the VCR directly to the special cable carriers, and use it to record the desired program.

Meanwhile, the cable box will convert whatever other cable channel you desire to watch into an output carrier (channel 2, 3, or 4) to which your conventional TV set can tune directly. Use the A/B switch to choose whether to watch the output from the VCR (on channel 3 or 4, as you choose) or the output from the cable box, as chosen with the cable selector switch. You will need to change the channel selector switch on the TV set as well as the A/B switch each time you change back and forth.

If the VCR has fewer channel-selector buttons than the cable has channels, set up the VCR to tune in the channels that you are most likely to want tok record, and watch the others live when desired via the cable box connection directly to the TV.

Chapter 2

How a VCR Works

The names and jobs of important parts that you need to know in order to use the information in the chapters that follow, and important principles of operation that apply to all home video cassette recorders, are presented in this chapter.

1. Video Tape Cassettes

As you know, VCR's are used to record television picture and sound (called "video" and "audio") onto a long strip of magnetic tape wound on spools inside a cassette box. This whole assembly, consisting of a plastic box with enclosed tape, is called a "video cassette," or simply a "cassette," for short, throughout this book where it is understood that we are talking about video cassettes.

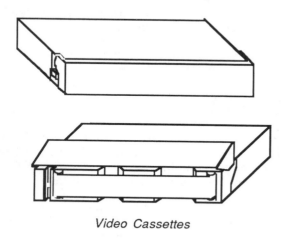

Video Cassettes

Video cassettes come in different shapes and sizes, corresponding to different designs of home video recording machines. The most common types are VHS, Beta and 8mm or VHSC used in the new combination camera-and-recorders ("camcorders").

Each VCR is restricted to using only one type of video cassette. You cannot use a Beta cassette in a VHS machine, for example, or a Beta cassette in a VHS machine. Within these basic categories, it is possible to buy video tapes of different lengths in the same size plastic cassette box. For example, you can buy a VHS cassette of the standard length (called "T120") or extra long ("T160), which is 1 1/2 times the standard length.

Longer cassettes obviously can hold more hours of program material. Extra-long tapes are made to fit into the same size cassette box by making the tape thinner, which causes some machines to have trouble playing extra-length cassettes. It is wise to buy only the standard-length tapes, unless you must have the extra length for some good reason.

2. Video Heads

All the complexities of the process by which the picture and sound are put onto this tape would take a long time to explain, but the basic idea is that the picture is quickly broken down into tiny parts, and these parts of the picture are then represented by microscopic areas magnetized into the tape as it moves past the crucial "video heads," that do this job.

The head mounting design used in the Beta format looks like this:

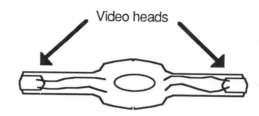

Beta Video Heads

The design used in VHS machines looks like this:

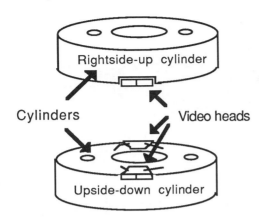

VHS Video Heads

The rightside-up heads are normally positioned as shown in the top half of the illustration. But if you removed them and turned them upside-down, they would look as shown in the bottom half of the drawing.

Separate heads make a magnetic representation of the *sound* on the same tape as it passes by. This is how video and audio information are recorded onto the tape.

When the VCR is put into "Play," electric motors cause the tape and heads to pass by each other exactly as they did previously during the recording of the program onto the tape. Now the same tiny magnetized areas of the tape produce little electrical signals in the video heads, which travel to circuits in the VCR that convert these signals back into patterns that are then sent out to the TV set which uses them to reproduce the picture and sound that was originally recorded.

You can already see why these heads must be clean. Any dirt or deposits that get onto the heads causes bad communication between the tape and the heads. Dirty heads may cause the program to be poorly recorded and play back with horizontal white streaks, or with a white lacey overlay, or even just a picture screen full of snow.

It is not "dirt" or dust from the outside environment that causes "dirty heads" as much as it is metallic material from the tape itself rubbing off onto the heads, leaving deposits that gradually accumulate over weeks and months, until finally the heads need to be cleaned in order to restore a good picture.

Since video heads are extremely delicate and precisely positioned, they must be cleaned by special procedures using only certain special cleaning materials. Improper cleaning techniques or tools, including wet-type head-cleaning tapes, can damage them so that they must be replaced which is

extremely expensive.

The correct way to clean video heads is explained in detail in Chapter 3. Read Chapter 3 before attempting to clean the heads in a VCR, but finish this chapter first.

Accumulated deposits are only one cause of "head problems," as they are called. The metallic particles on the tape also act like very fine sandpaper, gradually wearing away the metal from which the heads are made. Finally, a time comes when the heads have gotten worn down so far that cleaning the heads can no longer restore a good picture. When this happens, either the heads must be replaced (expensive), or else you need to get a new VCR (also expensive). Dry-type head-cleaning tapes also grind away the heads.

Since the main cause of dirty and worn heads is the material from which the tapes are made, obviously the brand of tape you use can make a big difference in length of time you can go between head cleanings, and also, in the life expectancy of the heads.

Off-brand, bargain-basement video cassettes are usually no bargain in reality, because the money saved in the purchase price is lost many times over in the time and expense involved in extra head cleanings, or even head replacement. Even some expensive, well-regarded video tapes that give a good picture also, unfortunately, cause bad problems with the heads.

At the present time, one brand good for heads is the middle-priced Fuji or Panasonic line, length T120 (VHS) or L750 (Beta). I do not recommend the use of chrome tapes at all.

3. Cleaning a VCR

Heads can also be ruined by improper head cleaning techniques, and wet-type head-cleaning cassettes can permanently damage video heads in two ways.

First, users sometimes fail to allow enough time (24-hours) for the heads to dry completely before attempting to run a tape through the machine after using a wet-type head cleaner. As explained later, these video heads are spun against the tape at high speed from a position in a cylinder around which the tape is part way wrapped.

If these spinning parts are anything less than completely dry, the tape may stick to them by surface adhesion and dislodge a video head from its delicately aligned position. Once this happens, the video head assembly is ruined and must be replaced. It is impossible for even the best-equipped shop to realign a video head.

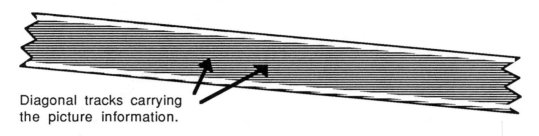

Diagonal tracks carrying
the picture information.

Recording Tracks Across Magnetic Tape

Another way that wet-type head-cleaning tapes can damage a VCR cannot be blamed on user error. On wet-type head-cleaning tapes, the material that runs through the VCR and past the heads is made slightly rough, in order to scrub the heads as it passes by. This roughness can damage the delicate video heads. The edge of a video head can catch in the head-cleaning tape as the head spins rapidly past, bending the head enough to ruin it.   This produces damage that may be more expensive to repair than the value of the VCR. These dangers are greatly reduced by cleaning the heads the only correct way, gently by hand, using the procedures in Chapter 3.

Cleaning Other Parts

Another reason for cleaning a VCR by hand is that, for proper operation on many VCR's, various little rubber wheels, belts, and other drive surfaces not even touched by so-called "head-cleaning" tapes, also need to be cleaned free of traces of oil that naturally settle on them from the air. In a proper cleaning, oily deposits from the environment that are not even reached   by a "head-cleaning" tape are removed by hand.

The heads are only one place where metallic deposits left by the tape cause problems. Deposits also accumulate on the "tape guides" and other parts that control the movement of tape. Deposits so small that they are almost invisible to the naked eye can disturb the smooth motion or exact position of the tape as it passes by enough to adversely affect both the picture and sound. When you remove the top of the cabinet and observe the path of the tape as a VCR is playing, you can see the many surfaces over which the tape passes that need regular cleaning.

It sometimes is difficult to remember all these crucial locations when the tape is not playing, especially because some of these parts change position when the tape is unloaded from the VCR. In order to learn how to clean a VCR, and also learn how to perform other maintenance and repair procedures, you need to know the names and locations of these important parts, which will now be explained.

4. Spinning Video Heads

The VCR is the most complicated piece of equipment ever placed in the home. Its complexities are due to the enormous amount of information that must be stored on a tape in order to record a TV program.

If the designers tried to use same kind of process that is used in an old fashioned audio tape recorder, where the mechanism simply moves the tape past stationary heads, the tape would need to move at extremely high speed (like more than 23 feet per second) in order to be able to "write" on to the tape all the video information required to copy the TV picture as it changes.

It is physically impractical to move tape this rapidly, and even if it were feasable, such a great length of tape (forty miles!) would be required to hold, for example, a two hour movie, that the average family would need to set aside a large room for the sole purpose of storing their video tapes. The reels would be like truck tires. Obviously this would be impractical.

This problem is solved by spinning the video heads diagonally across a slowly moving tape in successive lines, as illustrated above.

This amazing feat is accomplished by mounting the video heads in recessed areas in a metal cylinder, and wrapping the tape at a slight angle part of the way around the cylinder and then spinning the heads rapidly while slowly pulling the tape past them.

At least two heads usually are used to do this, so that a second head can begin scanning a new diagonal line across the tape just as the first head runs off the tape at the end of its diagonal track.

In the Beta system, only the heads and their support move, contacting the tape through a slot cut around the side of the stationary cylinder.

In the VHS sytem, the entire top part of the cylinder spins with the heads attached to it. The

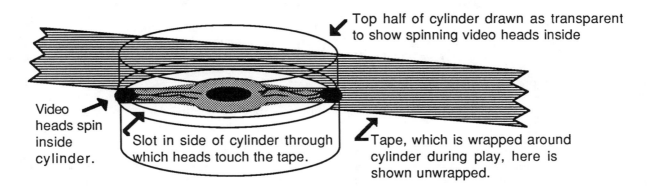

Top half of cylinder drawn as transparent to show spinning video heads inside

Video heads spin inside cylinder.

Slot in side of cylinder through which heads touch the tape.

Tape, which is wrapped around cylinder during play, here is shown unwrapped.

Spinning Beta Heads Contacting the Tape through a Slot in the Side of the Cylinder

heads then run across the tape in a diagonal pattern as explained earlier. In operation, the tape is wrapped more than half-way around the upper and lower cylinder. The drawing shows the tape unwrapped for clarity.

In order to record and play tracks that run diagonally across the tape, the cylinder with the video heads is mounted in the VCR at an angle.

Knowing this, you will not become concerned like the nice lady who came into our shop worried because she had looked inside her VCR and noticed that the cylinder was crooked.

In order to wrap the tape around the cylinder at an angle, some other parts must be mounted at an angle too. People who are aware of this fact won't do like the truck driver who brought a VCR in to our shop after taking a pair of pliers and trying to straighten all the parts that are supposed to be bent (what a mess!).

Since the spinning heads cannot contact the moving tape correctly unless the tape snuggles flat around the cylinder and moves smoothly, an almost invisible tiny fragment of tape deposits stuck to the cylinder from a disintegrating, worn-out, or damaged tape obviously could cause a major problem in

recording and playing back tapes. It can also cause slow or wavering sound.

Many VCR's have more than two video heads in the cylinder. These additional heads may be used to optimize performance when the tape is moving at slower speeds in "extended play" (like EP, LP) modes of operation, or for special effects like "freeze frame", or even to record high fidelity sound in "hi-fi" models.

5. Tape Path Control

When playing or recording, the video tape must always remain wrapped around the cylinder at a precise exact height, angle, and pressure, controlled by various mechanical parts. These parts occasionally malfunction and cause problems which you can fix once you know what to look for.

If the tape runs around the spinning cylinder too high, too low, at the wrong angle, or with a wrinkle in it, serious disturbances in the picture will result. The angular path that the lower border of the tape is supposed to follow around the cylinder is indicated by a small ridge formed in the metal of the lower part of the cylinder. When the tape passes around the cylinder in the correct position, the edge of the tape

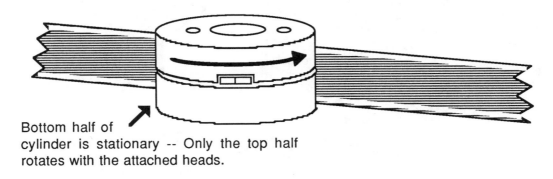

Bottom half of cylinder is stationary -- Only the top half rotates with the attached heads.

Spinning VHS Heads and Upper Cylinder Contact Tape

should butt exactly against this ridge, and not twist, curl, or ride up onto the edge of this ridge.

The position of the edge of the tape as it runs can be raised or lowered, when necessary, by means of fine adjustments of the height of one or more round roller guides the tape passes before and after the cylinder.   These rollers are called "guides" or "pins."

On VHS machines, two of them are called "P-guides" The guide that controls the height of the tape at the point where it begins to pass around the cylinder is called the "entrance guide" or "entrance P-guide," and the guide that determines its position at the point where it leaves the cylinder is called the "exit guide" or "exit P-guide."   There usually are additional guides to control the path of the tape between the cassette and the heads.

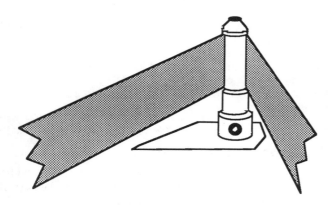

The position of the entire tape, as it runs around the cylinder, is raised or lowered when these guides are moved up or down.

The guides are mounted on threaded shafts so that they can be lowered by being screwed in a clockwise direction, and be raised by being turned in a counter-clockwise direction, but only after another little "set screw" that locks the adjusting mechanism in position has first been loosened.

Roller guides come in different design configurations, but on all, the length of the space for the roller must exactly equal the width of the tape. If it does not, something is wrong.

The simple special tools for making P-guide adjustments are described in Appendix I, and the exact "when and how" of making these adjustments is explained in Chapter 18.   (Do not even even dream of touching these adjustments until you have read the procedures in Chapter 18!)

6.   Erase Head

A VCR has other heads besides the video heads. Before the tape gets to the spinning video heads, it passes a stationary head that erases any previous program material recorded on the tape.   This *erase*

head obviously should operate only in the "Record" mode, and not when the VCR is playing back a recorded tape.

The large illustration shows a perspective view of the position of the erase head, audio, and control

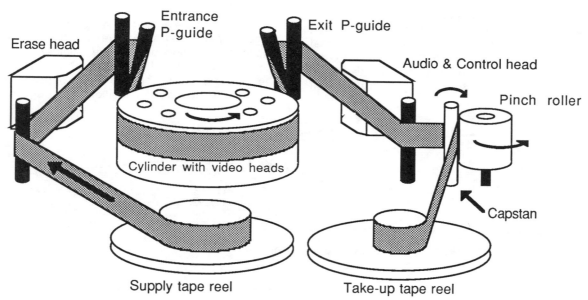

Path Followed by Tape in VHS Video Recorders

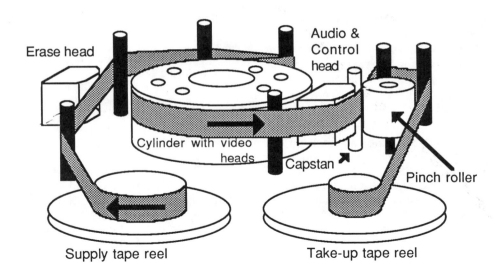

Erase head

Audio &
Control
head

Cylinder with video
heads

Capstan

Pinch roller

Supply tape reel

Take-up tape reel

Path Followed by Tape in a Typical Beta Video Recorder

head, tape guides, cylinder with video heads, capstan, and pinch roller along the tape path in VHS type machines.

There are, of course, many details omitted from this drawing, and differences among different makes and models, but this diagram represents the basic concept common to all VHS machines.

Beta VCR's have more variation in the layout of parts, but the Beta illustration shows a typical configuration

In other Beta models, the audio/control head, capstan, and pinch roller may be positioned toward the left rear rather than toward the right front, but the basic principles and concepts are the same. On a Beta, the tape always is wrapped most of the way around a stationary cylinder that contains the spinning heads.

7. Audio & Control Head

After passing the spinning video heads and the exit guide, the tape next crosses a combination "audio and control head."

This single unit contains a small head for audio recording and playback that is embedded in the top of the polished surface over which the tape passes.

And it also contains another small head embedded at the bottom of the same surface that records a periodic pulse that enables the VCR to precisely control the speed of the movement of the tape, and regulate various other functions, on playback.

This head is sometimes called the "audio/control head," or "audio-control head," or just "A/C head" for short.

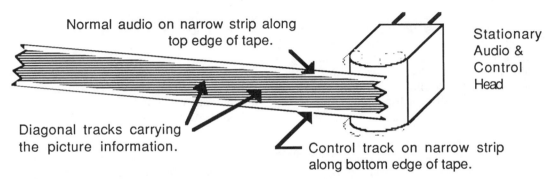

Normal audio on narrow strip along
top edge of tape.

Stationary
Audio &
Control
Head

Diagonal tracks carrying
the picture information.

Control track on narrow strip
along bottom edge of tape.

Tape Passing the Stationary Audio & Control Head

Like the head on a simple audio tape recorder, the angle at which this head meets the tape is crucial for good sound.   Problems with sound are covered in Chapter 20

8. Tracking Control

Since the pulses recorded on the top track on the tape determine, among other things, when the electronic circuitry inside the VCR switches back and forth between one video head and the other, as well as the exact position of the rotating video heads at any given instant, the distance along the tape between the cylinder and the audio/control head is another critical adjustment.

Since small variations in this distance from one VCR to another are inevitable, all VCR's have a front panel "customer tracking control" to enable the user to advance or delay these pulses to compensate for small variations in distance so as to get the best picture when playing back tapes recorded on other machines.

This is why the tracking control sometimes must be adjusted to eliminate horizontal stripes or bands that look like scratches, or scratchy lines in the picture, when you play a rented tape or a tape originally recorded on someone else's machine.

Whenever you get a bad picture, first try adjusting the tracking control.   The tracking control is usually a plastic knob or wheel located on the front (or

Some Typical VCR Tracking Control Knobs

top) of the cabinet, sometimes behind a little door, that turns a short distance back and forth to stopping points with a tiny click or "detent" in the midway position.   On newer machines, it is located on the separate remote control, and consists of two pushbuttons, as shown in the picture on the right.

Even if the tape you are trying to play was recorded on the same machine, when you are getting a bad picture, always try adjusting the tracking control. Unknown to you, a child, visitor, or video gremlin, may have turned this knob to a different position, so that you need to readjust it.

Don't Be Afraid to Move Controls

For that matter, do not hesitate to operate any of the controls visible on the outside of the VCR.   An amazing number of VCR's brought into technicians' shops for repair work actually needed only to have the customer adjust the tracking control or the channel tuning, or flip a switch from the "camera" to the "tuner" position, and so on.

This is like someone bringing a car radio into a radio shop when it needed only to be tuned to the station.   People who neglect to attempt first to correct the problem by adjusting the controls usually end up paying a technician bench charge as well as losing the use of the VCR during its time in the shop.

9. Tape Speed Controlled by Capstan and Pinch Roller

The tape must move past the various heads at a precisely controlled speed. If it moves at the wrong speed, or the speed varies, you will hear a problem in the sound, and you may see the picture go alternately good and bad.   The speed of movement of the tape is primarily controlled by a rotating metal shaft called the "capstan," against which the tape is tightly pressed by a free-rolling rubber cylinder called the "pinch roller."

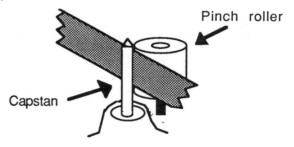

Capstan and Pinch Roller

The capstan shaft is turned from underneath, either directly by an electric motor, in some models, or indirectly via a belt from a motor.   The exact speed of the capstan motor is supposed to be precisely controlled by complex electronic circuitry.

The pinch roller is mounted on a movable assembly.   When the VCR loads the tape to play it, the mechanism moves the pinch roller over and causes it to press the tape against the capstan.

If the tape fails to run smoothly at the correct speed because of some trouble in this circuitry or motor, you will probably need to go to a shop for service, unless it is only dirt or a belt that has come off or broken underneath, or even a motor, which you can replace yourself.

But there are several things you can do yourself first to correct many common causes of problems with tape movement. One common cause of slow or uneven tape movement is deposits building up along the tape path, on the heads and rollers, on the capstan and pinch roller, causing the tape to drag or catch. You can fix this by cleaning all these parts using the procedures in Chapter 3.

The pinch roller must be dull black with no shiny glazed deposits on its surface, and it must press the tape tightly against the capstan in order for the capstan to control the speed of the tape correctly. If the pinch roller fails to press the tape firmly against the capstan, usually a safety sensor detects this failure and shuts the whole system down before the tape gets damaged, so a characteristic symptom of this problem often is "tape will not play" rather than "tape plays too slow or at wrong speed."

The pinch roller should move to press the tape tightly against the capstan as one of the last steps in the process of what is called "tape loading," explained next.

10. Tape Loading System

In a VCR, the tape must follow a complicated path, passing between the capstan and pinch roller and winding around various guides, rollers and heads. How does the tape get threaded around all these parts correctly?

Manufactures of consumer electronics products know that most customers do not like to thread tape by hand. Consumers prefer the convenience of being able simply to insert a cassette box and push a button.

In a simple audio cassette recorder, it is possible to record and play the tape without pulling it away from the plastic box holding it, but this is impossible with home video tape recorders because the tape needs to be physically wrapped part way around the cylinder with spinning video heads, as just explained.

For home video recorders, therefore, fantastic mechanisms have been designed to pull the tape out from the cassette box and wrap it around the cylinder containing the spinning heads. This is called the "loading mechanism."

Troubles with this mechanism, including many simple problems you can easily fix yourself, are among the most common cause of VCR's refusing to play or record tapes, damaging tapes, and failing to eject a cassette already in it. To correct these problems, it is necessary to know how the tape-loading mechanism should work, and how to diagnose various common malfunctions.

Let us start with the metal, or metal-and-plastic, container, or enclosure, into which cassettes go when you first put them into the machine. This is called the "cassette basket."

In a "top-loading" VCR, the cassette basket rises up out of the top of the machine, a cassette is inserted into it, and you push it down. In a top-loader, you can see part of the cassette basket when it springs out the top. In a "front-loading" VCR, the basket is located out of sight inside the VCR behind the front cabinet panel, and the cassette goes in through a little door or flap, but you can see part of the basket if you open the little front door through which the cassette goes in, and peek inside.

When a cassette is placed in the basket of a VCR, part of the basket should push a small plastic button on the edge of the cassette box to release the latch holding closed the plastic cassette door that protects the exposed tape along one side of the cassette.

On VHS cassettes, this button is on the right side of the cassette.

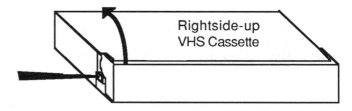

On VHS cassettes, the cassette door release button is on the side of the cassette box.

Location of VHS Cassette Door Latch Release

On Beta cassettes, the latch is a lever hidden in a notch on the bottom of the front edge of the cassette:

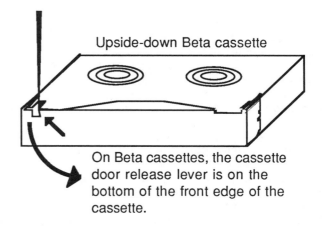

On Beta cassettes, the cassette door release lever is on the bottom of the front edge of the cassette.

Location of Beta Cassette Door Release Lever
(here shown upside-down for clarity)

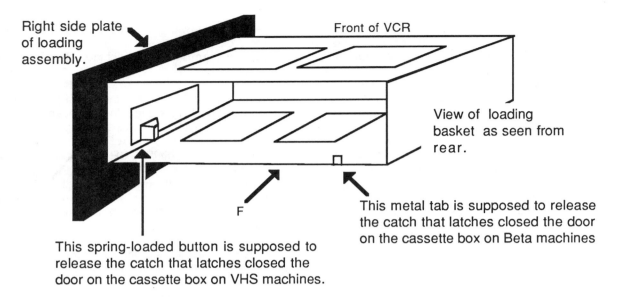

Right side plate of loading assembly.

Front of VCR

View of loading basket as seen from rear.

F

This spring-loaded button is supposed to release the catch that latches closed the door on the cassette box on VHS machines.

This metal tab is supposed to release the catch that latches closed the door on the cassette box on Beta machines

Composite Illustration Showing the Cassette Door Release System on VHS and Beta VCR's

The button or hidden lever is pushed by part of the cassette basket.

The position of both as seen from the rear is shown above in a single illustration to save space, although, of course, a VHS machine would have only the side-release, and a Beta would have only the front release tab.

Another part of the mechanism inside the VCR actually opens the cassette door on its hinges as it goes down into the machine. If this mechanism is inoperative, then obviously the VCR will not play or record tapes, and possibly not even accept them.

11. The Reel Table and Tape-Loading Machinery

The big square piece of metal, the chassis or big assembly, on which the cylinder, heads, rollers, capstan, cassette basket, and other mechanical parts, are mounted is called the "reel table" in this book. Some writers about VCR's call each of the reels a "reel table," but in this book, the term "reel table" will be used to refer to the metal table on which these reels and other parts are mounted.

As a cassette goes into the VCR, it fits over a number of crucial parts mounted on the reel table. Some of these parts extend up into the cassette box beside the tape as the cassette box settles in a precise position on top of four short metal pins permanently fixed to the reel table. These pins reach through holes in the bottom of the basket to support the cassette box when it is in the VCR.

Knowing what to look for makes it easier to diagnose and repair another common VCR problem: namely, trouble with the tape loading mechanism.

The names and functions of these parts will be explained separately for the two systems, Beta and VHS. If you are only interested in Beta machines, skip the next section and go directly to Section 13. If you are interested in the VHS format, continue with Section 12.

12. Tape Loading Process in VHS Machines

In VHS machines, the loaded cassette goes down over:

(A) Two reels that go into the center of the two spools around which tape is wound inside the cassette.

(B) A center post that extends up into the cassette releasing a catch inside the cassette box that otherwise would prevent the spools on which the tape is wound from turning. When this catch is released, the spools can turn.

(C) Two movable tape "P-guides" with rollers used to control the position and height of the tape wrapped around the cylinder, as described previously. When you push "Play" or "Record," a motor is supposed to move these guides and rollers along two tracks, pulling the tape out away from the cassette box and up to the cylinder, and wrapping the tape around it.

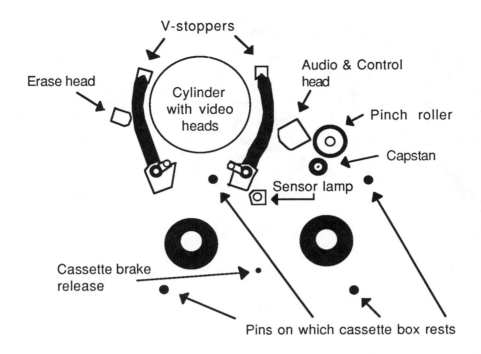

Top View VHS Reel Table

At the end of their tracks, they come up against "V-block stoppers," or "V-stoppers," whose critical position is exactly set using a precision jig at the factory, and which never can be returned to their correct position if they are ever loosened or moved. NEVER, NEVER try to move or adjust the position of the V-stoppers!

The cassette also goes down over (D) the sensor light, and (E) the tape tension control, which will be discussed in the section titled "Safety Systems."

The cassette box goes down over (F) the capstan too, which is separated from the pinch roller by a space into which the tape fits. When the cassette goes down into the VCR, the tape itself slides between the pinch roller and the capstan. The capstan goes between the tape and the cassette.

The P-guides slide on their tracks to pull some tape out of the cassette, moving it to a position past the erase head and the audio/control head, and causing it to be wrapped around the spinning cylinder and video heads. Then, as a final step, the tape-loading mechanism should push the pinch roller against the capstan, with the tape pressed in-between tightly

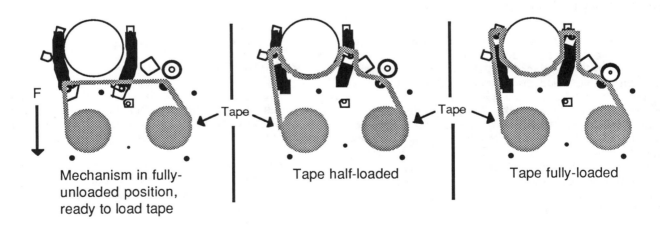

Mechanism in fully-unloaded position, ready to load tape

Tape half-loaded

Tape fully-loaded

P-Guides Sliding on their Tracks Toward the Rear of the Machine to Load a VHS Tape, Wrapping it Around the Cylinder (Outside Cassette Box Omitted from Drawing for Clarity)

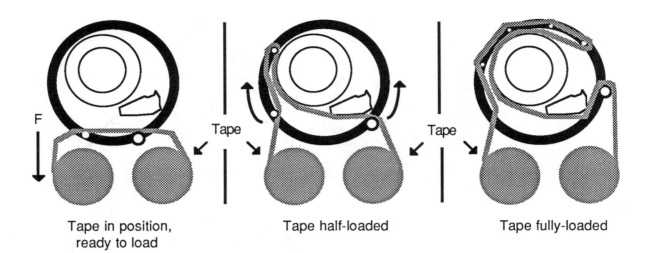

Tape in position,   Tape half-loaded   Tape fully-loaded
ready to load

Loading Rings Running on their Circular Track Around the Cylinder to Load a Beta Tape,
Wrapping it Around the Cylinder (Outside Cassette Box Omitted from Drawing for Clarity)

and the motor drive should start rotating the capstan and the take-up reel.

13. Tape Loading Process in Beta Machines

On Beta machines, rollers mounted on a big circular ring (shown in black) extend up inside the loaded cassette when it is loaded into the VCR. A motor rotates the rings on this circular track that runs around the cylinder with the video heads in such a way that the tape is wrapped almost completely around the cylinder, and also runs past the audio/control head. The illustration shows the later design. A diagram of the slightly different earlier Beta design appears in Chapter 14.

14. Special VCR Terminology

The term "cassette" refers to the entire plastic box containing the tape and other parts. The term "tape" refers to the long strand of flexible material wound on spools inside the cassette. People often use the term "tape" loosely to refer to what I am calling the cassette, but it will help clarify all the explanations in this book if we are careful to use the term "tape" only to refer to the flexible material inside the cassette, and use the term "cassette" to refer to the whole cassette box, including the tape. If I sometimes forget myself in the following pages and say "tape" when I should say "cassette," please understand me to mean "cassette."

The term "load" is used to refer to two different processes in a VCR, depending on whether we are talking about "loading a cassette" or "loading a tape." The phrase "loading a cassette" refers to the initial process in which a cassette is brought into the

VCR and set down on the reel table. The different phrase "loading a tape" is used in this book to refer to the subsequent process in which the P-guides or loading ring pull a loop of tape out from the cassette and wrap it around the cylinder with the spinning video heads. The *cassette* must be loaded into the VCR before the machine can load (thread) the *tape* into the tape path.

The term "unload" also is used to refer to two different processes, depending on whether we are talking about "unloading the cassette" or "unloading the tape." *Unloading a cassette* is ejecting the cassette from the machine. *Unloading the tape* is pulling the loose loop of tape out of the VCR tape path and winding it into the cassette box. The machine must unload the tape before it unloads the cassette, since otherwise the loop of tape that is still threaded in the tape path will get broken off when the cassette is unloaded.

15. Positions of VHS Tape-Loading Mechanism

There are also special terms that are used to describe the different positions of the moving parts in a VCR. Different words are used to refer to the different positions of the tape-loading mechanism.

The phrase "fully-unloaded position of the tape loading mechanism" refers to the position that it should be in at any time that a cassette is inserted or ejected. On a VHS machine, when it is in the "fully-unloaded position," the P-guides are all the way forward, as far as they will go, as shown in the next illustration. When there is a cassette loaded in the machine, but the tape-loading mechanism is in the fully-unloaded position, the area enclosed by the rectangle in the illustration is hidden underneath the

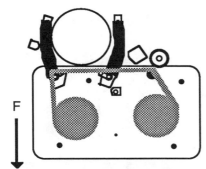

Fully-unloaded position of
tape-loading mechanism

cassette box, so what you would see if you looked into a VHS machine when it is in the fully-unloaded position would look more like this.

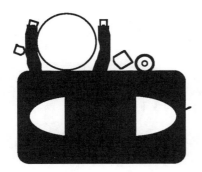

Fully-unloaded position of
tape-loading mechanism

The black object is the top of the cassette box.

Actually, above the top of the cassette box would be some pieces of the cassette loading mechanism that is holding the VHS cassette, so what you would see would look more like this:

Fully-unloaded position of
tape-loading mechanism

Here the gray parts represent parts of the cassette loading mechanism holding the cassette box. If you did not know what is underneath what you see here, you would have difficulty understanding or recognizing when the mechanism is in the fully-unloaded position.

When a VHS tape-loading mechanism is in the "half-loaded" position, it looks as shown below, with a cassette blocking the view of the reels:

What you see here would be partly obscured by the metal of the cassette-loading mechanism. Imagine the gray area on the previous illustration cut out and pasted on top of the black cassette box in this drawing to get a more accurate picture of what you would actually see.

When the tape-loading mechanism is in the "fully-loaded" position, it would look like this:

In this illustration, I have shown the mechanism of the cassette-loading assembly as well as the cassette box covering part of the reel table.

16. Positions of Beta Tape-Loading Mechanism

Similar terminology applies to Beta machines, except, of course, that we have a loading ring instead of P-guides, and the reels are hidden underneath a Beta cassette box. When a new-format Beta is in the fully-unloaded position, it looks like this:

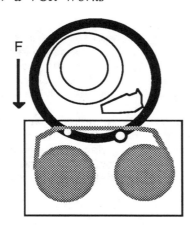

When there is a cassette loaded in the machine, but the tape-loading mechanism is in the fully-unloaded position, the area enclosed by the rectangle in the illustration is hidden underneath the cassette box, so what you would see if you looked into a Beta machine when it is in the fully-unloaded position would look more like this:

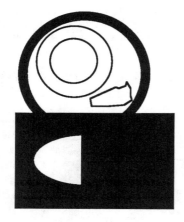

What you see here would be partly obscured by the metal of the cassette-loading mechanism. Here is a picture of what it might look like:

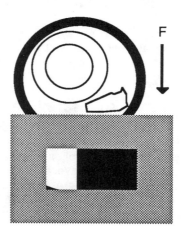

When a Beta tape-loading mechanism is in the "half-loaded" position, it might look like this:

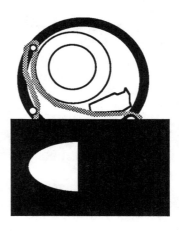

Here the cassette-loading mechanism is omitted from the drawing for clarity, but the black cassette box is shown.

When a Beta tape-loading mechanism is in the "fully-loaded" position, it might look like this:

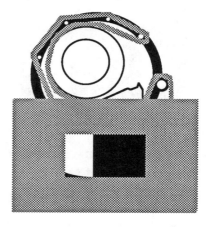

In this illustration, I have shown a typical housing of the cassette-loading assembly as well as the cassette box covering part of the reel table.

Pictures of the loading process in early Beta machines appear in Chapter 14. To see what they look like in various positions with a cassette in the VCR, imagine a black rectangle positioned over the reels in each of the stages of loading shown in Chapter 14.

17. Common Tape-Loading Problems

With some idea what to look for now, you are ready to learn to diagnose and repair another very common type of VCR problem: trouble with the tape loading mechanism. The characteristic symptom of this problem is that when a cassette is put into the machine and the "Play" or "Record" button is pushed, the mechanism begins to load the tape, starting to go through the process just described, but then it stops, reverses the process, unloads the tape, goes back to the "Stop" position, and stops.

Often, the loading process will be 99% completed before it "aborts," coming so close to 100% completion that it is necessary to observe the process several times very closely (with the cabinet open and the topside of the reel table exposed, of course) before you notice perhaps that the pinch roller is not moving quite all the way into firm contact with the capstan, or perhaps that the cylinder or the take-up reel is not starting to turn.

A common cause of this symptom is some failure in the mechanism underneath the reel table, such as a belt (a "loading belt") slipping on a pulley attached to the little "loading motor" that is supposed to drive the mechanism as it runs through the loading process. Chapter 14 explains in detail how to troublshoot and repair such a problem. I mention it here because it is important to understand a VCR's normal operation in order to repair it.

18. Safety Systems

After reading the previous section, someone might ask why the loading mechanism reverses its direction and unloads when a slipping belt or other problem prevents it from fully loading the tape properly. How does the VCR know that a malfunctioning mechanism has prevented full completion of the loading process? How does it know to return itself to the "stop" position, and shut itself off?

The answer is that a little computer, a "micro-computer," or another safety logic and control circuitry inside the machine's "System Control," monitors the tape-loading process.

If a malfunction is detected by a sensor, the little computer sends new instructions out to the motor(s) and/or other subsystems to reverse the loading process, unload the tape, reel it safely back into its cassette (if it can), and shut down operations. This is done because if the VCR were to try to play a tape when some of the mechanism is incorrectly positioned, the tape would not play properly, and might be permanently damaged.

Understanding this feature should help everyone be more sympathetic to a VCR. I have seen owners of a VCR become so frustrated at their machines for starting to play and then shutting down that they whacked it with a karate chop, or kicked it, smashing the cabinet.

This behavior was foolish not only because the

cost of replacing the broken cabinet parts greatly added to the cost of replacing the slipping loading belt, but also because the VCR was punished for doing its owner a favor by shutting down and unloading the tape instead of possibly ruining the cassette by trying to play it when the VCR could not load it properly.

Auto Stop

When a VCR automatically goes back to the "Stop" position because its safety circuitry senses a malfunction somewhere in the mechanism that could damage the tape, it is called "Auto Stop."

For example, if the pinch roller does not press the tape tightly against the capstan, or the capstan fails to start rotating for any reason, the tape will stand still rather than move along the tape path as it should.

If the VCR did not have the auto-stop protective feature, the tape could remain stationary against the spinning video heads long enough for the heads to wear through the delicate layer of metal particles on the tape, ruining it. To prevent such an accident, the VCR is designed to shut itself down automatically when it detects a failure in the mechanism that should load the tape and run it correctly.

Sensors for the Tape Reels

Other safety circuits guard against other dangers. For example, when a tape is moving forward through the VCR, the take-up spool inside the cassette box needs to rotate too, in order to reel in the tape after it has passed by the various heads and capstan. This is accomplished by having both spools fit down over projections on two round plastic "reels" located in the reel table in such a way that when these reels are turned by motors, either directly or indirectly via wheels, gears, or belts, their motion causes the spools inside the cassette box to rotate in the desired direction. The spool on the left side, which holds all the tape after rewinding and before playing it, is called the "supply-side" spool. The one on the right side, onto which tape is wound after it has gone past all the heads, is called the "take-up" reel.

Suppose, for example, that the mechanism that is supposed to rotate the take-up reel while playing tapes fails to operate correctly, so that the take-up reel inside the cassette does not reel in the tape coming off the capstan. In that event, this tape would spill out inside the machine, probably getting dirty, creased, and maybe caught and torn in some other part of the mechansim. Thus, a "sensor" is used to detect whether the take-up reel is turning as it should. If the take-up reel is not rotating properly, System Control detects the failure by means of this sensor, shuts the VCR down, and reels the loose tape back onto the other reel, the "supply reel."

When you see a VCR start to do everything correctly to load a tape including even pressing the pinch roller against the capstan and starting to rotate it to move the tape, except that the take-up reel and and spool sitting over it never starts to turn, as you can observe through the window in the cassette box, and then the machine goes into Auto Stop, you should troubleshoot the mechanism that is supposed to rotate the take-up reel, as also explained in Chapter 14.

You are ready now for a more exotic problem. What if a malfunction occurs NOT in the mechanism that drives the capstan or take up reel, but in the sensor or safety control circuit that is supposed to be monitoring the correct operation of the mechanisms? That is, what if a sensor itself fails? In that event, the system-control circuitry of most VCR's — its little brain — is again designed to put the system into Auto Stop. The machine shuts down because with its safety-monitoring system inoperative, tapes are not protected against possible future malfunctions in the mechanical parts that the broken safety system was suppose to monitor.

Defects in the safety circuits that are suppose to detect proper capstan and take-up reel rotations often require electronic troubleshooting equipment and techniques beyond the scope of this book, but sometimes they involve simple problems like a broken or slipping belt that runs from the take up reel out to the front panel counter which the VCR uses to verify that the take-up reel is indeed moving. This is also covered in Chapter 14.

19. Sensors to Detect When the Tape Comes to an End

The little computer that controls the basic operations inside a VCR also needs some way to tell when the end of the tape has been reached, so that it will stop trying to move the tape, and not stretch or break it, or put undue wear on its own mechanism, by continuing to try to move the tape after it has come to its end and can go no further.

Beta End-Sensors

In Beta machines, this function is accomplished by placing two special magnetic sensors close to where the tape exits and re-enters the cassette box after the tape is loaded around the loading ring and cylinder. In the cassette box, a short length of special metalic tape is spliced to each end of the magnetic tape on which the programs are recorded.

When this metalic strip passes one of the tape-end sensors, the change is signaled to the little brain, which then knows that one end of the tape has been reached and can act appropriately. If the VCR was fast-forwarding or rewinding the tape, this fast motion ceases. If the VCR was playing or recording a tape, the machine stops when it senses that the end of the tape has been reached, and on some models, even automatically starts rewinding the tape.

This Beta system is extremely reliable, and almost never needs repair.

VHS End-Sensors

In VHS machines light-sensitive devices are used as end-sensors. A short length of transparent plastic leader is spliced to each end of the magnetic tape inside the cassette, and a small incandescent or infra-red lamp is centrally mounted on the reel table, like a tiny lighthouse, in a position in which it fits inside a hole provided for it in the tape cassette box.

Small internal and external windows, or passageways for light, are molded into the plastic cassette box so that the light from this lamp can pass through them out to the right and left, toward the two ends of the cassette box, and shine through the transparent plastic leader when the tape is unwound all the way to either end.

Now when the tape ends, the light that could not pass through the opaque part of the recording tape suddenly can shine through the transparent leader at the end. This light strikes an optical sensor that responds to the light by sending a signal to the little computer in the VCR, telling it that the end of the tape has been reached, so that it can make appropriate changes in the movement of the tape.

The supply-side sensor is located near the left end of the cassette from which the tape emerges during play, and the take-up end-sensor is located near the end of the cassette into which the tape returns, so the system can tell which end has been reached by noticing which of the two sensors saw light.

For this function, early VHS models used specially-made incandescent light bulbs that produced visible light, but most later models use devices that produce invisible infra-red light. The incandescent lamps burn out about every two to four years, and when they do, the VCR brain senses the failure of the lamp and prevents the machine from operating at all, putting it in complete Auto Stop condition, again to protect tapes from damage.

When an incandescent sensor lamp burns out, the VCR usually will do nothing on command except eject the tape. (On some Sanyo-Fisher VCR's the dew lamp will come on, but blinking instead of shining continuously). This is another situation in which fustrated VCR owners have been known to bash their VCR's rather than recognizing that the problem is a burned out sensor lamp, which is easily replaced using the part numbers and procedures given in Chapter 13.

Appendix V explains how to "fool" the micro-computer in the VCR into thinking that a tape cassette is in the machine, when actually no tape (and even no cassette basket) is present, and so going through its loading and playing motions. This trick is useful in diagosing some mechanical malfunctions.

20. "Cassette In" Detectors in Front-loading VCR's

When you insert a cassette into a front-loading VCR that is powered up, the force of the inserted tape moves the basket mechanism so that a small switch gets "thrown," which tells the little computer to turn on the power to the motor to pull the cassette in, and set it down correctly on the reel table. When the cassette is lowered to the correct position on the reel table, another little switch gets thrown, which tells the little computer to shut off the cassette-loading motor, because the cassette is already in all the way.

Likewise, when "Eject" is pushed, the micro-computer is supposed to activate the mechanism to unload the tape (if it was playing or recording at the time), rewind any loose tape back into the cassette safely, and then run the motor to the front-loading assembly in the opposite direction, to lift the cassette off the reel table, and carry it back out to the front of the machine. When the cassette reaches the fully-ejected position, the position of the basket mechanism is supposed to trip a switch that reports this to the little computer, which then shuts off the motor.

Many VCR problems involve these switches and the connected mechanical parts. The basket mechanism may get misaligned so that the switches send false reports to the brain. For example, it may tell the little computer that a cassette is inside the VCR, when actually no cassette is in the machine. Or it may fail to activate the mechanism to pull in the cassette.

Or the little computer may get signals from some sensors that contradict what other sensors are telling it. For example, the end-sensors may tell it that the machine is empty, while a switch on the front-loader is telling it that a cassette is in the machine. In such a case, the little computer gets so confused that it just shuts down most operations. Once we understand this, we can straighten the problem out. Treatment of these disorders is covered in Chapters 8, 10, and 11

21. Tape Back-Tension

With the capstan pulling the tape along past various heads, a certain pull in the other direction against this movement is needed as the tape unwinds from the supply reel, in order to put enough tension on the tape to make it closely contact the heads. This "back- tension" as it is called, is provided either by a motor or by a flexible felt-lined brake pad that rubs against the supply reel.

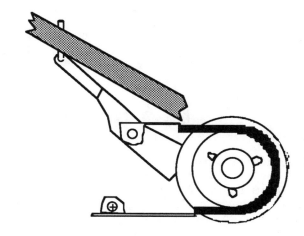

Back-Tension Band

One end of the back-tension band is connected to a mechanical arm with a vertical pin mounted on it that rubs against the side of the tape. The mechanism is arranged so that when the tape goes slack, the arm moves in a way that tightens the band around the reel, increasing the back-tension, and taking the slack out of the tape. If the tape gets too tight, it pushes the arm in the other direction, loosening the band around the reel and reducing the back-tension on the tape. The result of a correctly-operating system is that the tension on the tape remains constant and correct.

Too much back-tension can cause the heads to get dirty sooner than they should, and either too much or too little can also cause problems to appear in the picture.

22. Tape Slack Sensor (Betas Only)

Beta machines have an arm with a little vertical rod that presses against the side of the tape as it stretches from the spools to the rest of the

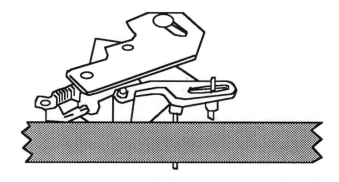

mechanism.If too much slack develops in the stretched tape, this causes the arm to move, which shuts down the machine.

Chapter 3

How to Clean a VCR

1. Tools and Materials Needed

You will need:

NOTE: Chapter 2 should be read before using this chapter

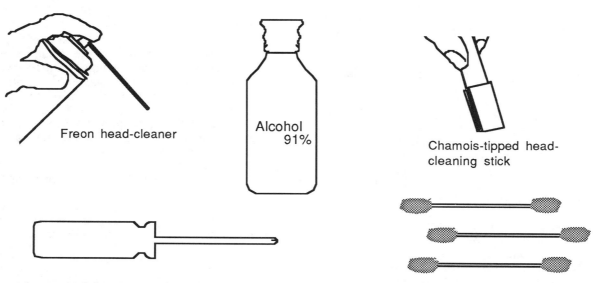

Freon head-cleaner

Alcohol 91%

Chamois-tipped head-cleaning stick

Cross-slot or Phillips screwdriver

Cotton-tipped swab sticks

MATERIALS AND TOOLS NEEDED TO CLEAN A VCR

Freon head-cleaner liquid from an electronics store — Head cleaner in an spray can with a long thin detachable nozzle is more useful for spraying out impossible-to-reach areas and extremely dirty heads, but the plain liquid is usually enough. Do NOT use "tuner cleaner"!

Alcohol — 90% or stronger rubbing alcohol from the drug store. Some sources say that alcohol should not be used to clean video heads, but the service literature of major manufacturers (like, Sony and Hitachi) lists alcohol as acceptable for cleaning video heads, and I have never had trouble using it for this purpose. If I had no other cleaner at hand, and needed to get a machine working immediately, I especially would not hesitate to use alcohol to clean video heads. The special chamois-tipped head-cleaning sticks must be used, however.

Head Cleaning sticks — plastic sticks with the end tipped with chamois or doeskin from the electonics store. (In an emergency, you can use the foam-tipped sticks from Radio Shack, but they are not recommended, and may damage heads.) NEVER use cotton swabs to clean the video heads, as they often will do great damage, destroying the heads.

Sticks with cotton swabs on the end (for example, Q-tips) — to clean parts other than the video heads. Never try to clean video heads with cotton swab sticks or with anything other than special head-cleaning sticks.

Clean rag — lint-free rag for cleaning pinch roller.

Cross-slot or Phillips screwdriver — to open cabinet, and perhaps to unfasten shields and circuit boards over the reel table.

See Appendices I and II for sources for these materials, if you cannot obtain them from an electronics parts store locally.

Warranty Warning

If your machine is still under warranty, you may wish to take it to a factory-authorized service center instead of working on it yourself. Head cleanings are usually not covered under the manufacturers' warranties, but opening or otherwise modifying it may void the manufacturer's warranty on it.

The procedures in this and following chapters are intended primarily for VCR's that are no longer under factory warranty. If your VCR is more than two years old, it probably is no longer covered by the warranty. On most VCR's, the warranty on labor is 90 days, and on parts is two years from date of purchase.

Warning:

Opening or otherwise modifying your VCR may void its manufacturer's warranty.

---

2. Procedure

The first thing to remember is NEVER to use Q-tips or cotton swabs to clean the video heads. You can catch one of the delicate recessed video heads on a cotton fiber and throw it out of alignment, or even break it, causing damage that often costs more than the value of the machine to repair. Sorry to keep repeating myself, but this is very important. I have seen VCR's destroyed by people who tried to clean the heads with cotton swabs, or with wet-type head-cleaning tapes. Most of the manufacturers also recommend against them.

My recommendation on head cleanings is to wait until you see some trouble in the picture before cleaning the heads. I do not clean video heads on a regular schedule, like an audio tape recorder, but only when the need is indicated by OK sound but a bad picture — like lacy horizontal lines throughout the picture, or a picture that looks like watching a scene through a rainy windshield on a car with bad wiper-blades, or just all snow, usually black-and-white, salt-and-pepper. A need to clean the cylinder around the heads, and other parts of the tape path, is indicated by slow or wavering sound.

The reason I recommend this policy is that the spinning heads move so fast, that on many models, they will keep themselves wiped clean, especially if you use tape like the Fuji or Panasonic medium-priced cassettes. (Some other brands, including some expensive tapes, will clog heads in a very short time. You do not always get what you pay for!)

However, the same cannot be said for rubber parts, especially the rubber tires on idlers. These get an oily residue on them from just standing around in the air, and the machine does not tend to self-clean them. The need to clean these parts is indicated when the machine starts to "eat tapes," which is, of course, too late from the standpoint of saving that tape. So the regular cleaning (and replacement) of rubber parts is a good idea.

Cleaning a VCR is not like what someone does in "cleaning a room," for example. In cleaning a room, a main goal is making it look good. In cleaning a VCR, the objective is not to make it look tidy, but to remove deposits of metal oxide, dirt, and oil that hinder its operation. Some of the most crucial points to clean are difficult to see even when looking into the VCR (unless you have been shown where to look, as the following sections explain). When you are done, the insides of the VCR probably will not look much tidier, but hopefully the machine will play better because you have cleaned crucial surfaces.

3. Gaining Access to the Reel Table

Before you can clean the heads and the rest of the machinery, you need to expose it. First, eject any cassette in the VCR, then unplug the VCR and remove the top of the cabinet (see Appendix III for help opening the cabinet).

Caution:

Before you open a VCR, make certain that it is unplugged. Dangerous voltages and currents are present in its power supply. Only trained technicians should work on VCR's while they are plugged in. Unplug for safety before you open up the unit, and before touching inside of machine.

On many machines, there is one or another obstacle between you and the reel table. You will need to remove these obstacles in order to get at the heads and mechanism.

There may be one or more boards with little electronic components soldered to one side (called "circuit boards") and there may be some pieces of sheet metal (called "shields") also. These will usually be held down with some cross-slot screws.

First, make a rough sketch for yourself of what you see prior to beginning disassembly, and then, as you remove each screw, put it into a new paper cup, draw a circle on your sketch in a position roughly corresponding to where the screw was, and write the number of the paper cup inside this circle on the sketch. The purpose of this is to enable you to get the correct screws back into the correct holes later when you go to put the machine back together again. Do not overlook this crucial step of recording the screw positions and the order of disassembly unless you are already such an expert that you do not need to be reading this book in the first place.

Also, be super-careful not to drop any of the screws into the mechanism, because sometimes they can be very difficult or impossible to get back out again, and may jam up gears or short out components on the circuit board below when you plug it in again. If you do lose a screw down into the mechanism underneath, sometimes you can get it back out by turning the machine upside down over a clean bare floor and shaking it for a long period of time, until the lost screw falls out.

The fact that this is quite difficult to do and time-consuming is all the more reason to take every precaution not to drop a screw. Use both hands when unscrewing screws and focus all your attention on the process, like a Zen exercise. Sometimes it helps to stuff a little tissue paper into any openings in the reel table around the screw head while unscrewing the screw, to catch any screw that tries to get away from you. A magnetic screwdriver (see Appendix I) is also helpful, but do not completely trust a magnetic screwdriver to hold onto every screw.

If there is a circuit board over the mechanism that you are trying to reach, it probably has hinges on one edge and is held down by a few brass screws, often with red heads. After these screws are removed, the board may open on hinges. Use a little wooden stick to prop the board up while you clean the mechanism underneath. Be careful to catch this prop stick behind a component on the board that is big and strong enough, and firmly held down, to be capable to holding this force.

If you are cleaning a VHS machine, go to the next section. If you are cleaning a Beta, go to Section 5 now.

4. Cleaning VHS-type Video Heads

Look into the area on the top of the reel table to the rear of where the cassette comes to rest when it is loaded. This is the region into which tape that is pulled out of the cassette by the mechanism goes.

On VHS type VCR's, the "upper cylinder," which contains from two to five "video heads" (depending on the model) looks like this:

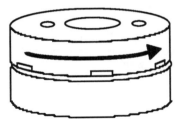

VHS Cylinder with Video Heads

This cylinder, which consists of a stationary lower cylinder and a rotating upper cylinder, is mounted at an angle in the VCR so that it will scan tracks diagonally across the tape.

The video heads are the tiny little parts made from brass and wire of which you can barely see one edge just peeking out from the little recessed areas along the bottom of the upper cylinder. In time, with use, they often get covered with thin deposits of metalic material from tapes. (NOTE: to minimize or prevent this problem in the future, stick to the recommended brands of tape.)

The video heads are cleaned as follows. First, wet a special chamois-tipped head-cleaning stick with

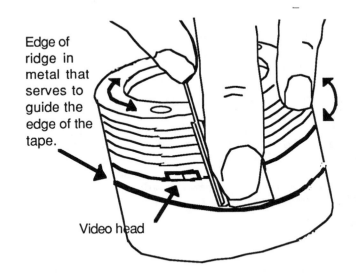

Edge of ridge in metal that serves to guide the edge of the tape.

Video head

Rotating the Cylinder from Side-to-Side Under Special Head-Cleaning Stick Held Stationary to Clean VHS Video Heads

the head-cleaning fluid. Then place this stick flat against the cylinder surface over the slotted or recessed place where a head is. While holding the stick against the surface, rotate the part to which the heads are attached back and forth with your other hand, so that the stick contacts the head under the chamois-covered part of the cleaning stick. NEVER, NEVER move the stick up and down, or side to side. Don't touch the sides of the cylinder with your fingers — Try just to touch the top.

Moving the head under the stick held flat against the cylinder (rather than moving the stick) will help insure that you do not accidentally exert excessive force on the recessed head. Repeat this procedure for each head in the upper cylinder.

NEVER rub a video head in a vertical or diagonal direction, because you may cause it to go out of alignment, and once it is out of alignment, you can never realign them. They are aligned on a special machine at the factory. Video heads differ from conventional audio tape recording heads in being impossible for a repair technician to realign.

If the heads do not come clean (as shown by the picture not coming up clear), you can press a little harder with a finger on the chamois end of the head-cleaning stick while moving the head back and forth as shown in the drawing.

Unless you are interested in Beta's, skip over the next section on Beta head cleaning, and resume at section 6.

5. Cleaning Beta-type Video Heads

On Beta type VCR's, unlike VHS, the heads spin inside the cylinder, without the cylinder spinning. The spinning video heads reach out and touch the tape wrapped around the cylinder through a slot cut all the way around the cylinder. The illustration shows the upper cylinder assembly as transparent so that you can see the hidden video heads inside.

The video heads are the tiny little parts made from brass and wire barely visible peeking through the slot around the side of the cylinder. It may be easier to find them, and see them initially, if you put your finger on the part at the center of the top of the cylinder, and slowly rotate the very center while looking at the slot that runs around the side of the cylinder.

Different models of Beta's will have a different assembly on top of the cylinder, but turning whatever it is will rotate the heads. Do you see something barely visible moving in coordination with the way you are moving the top part that rotates? It is one of the video heads. On most Beta machines, there are two of these video heads, and the other head is just opposite to the one you see. It can be brought into view by rotating the part you are turning 180 degrees (that is, one-half turn).

Beta video heads are cleaned as follows. First, wet a special chamois-tipped head-cleaning stick with the special head-cleaning fluid. Then place this stick flat against the cylinder surface over the recessed place where a head is. While holding the stick against the surface, rotate the part on the top of the cylinder attached to the head(s) back and forth with your other hand, so that the stick contacts the head under the covered part of the cleaning stick. NEVER, NEVER move the stick up and down, or side to side. Moving the head under the stick held flat against the cylinder, rather than moving the stick, will help insure that you do not accidentally exert excessive force on the recessed head. Repeat this procedure for the other head.

Never use anything but a special head cleaning stick to clean video heads. In particular, NEVER use a cotton tipped swab because you might catch one of the heads with a cotton fiber and pull it out of alignment. If that happened, the cost of repairing the damage

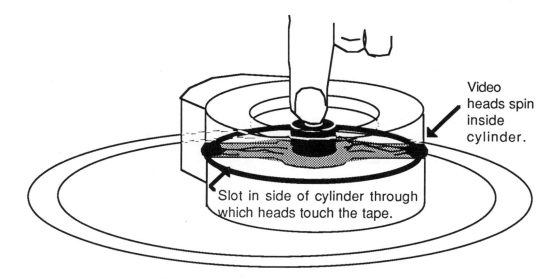

Video heads spin inside cylinder.

Slot in side of cylinder through which heads touch the tape.

Finding Beta Video Heads by Moving Them With a Finger

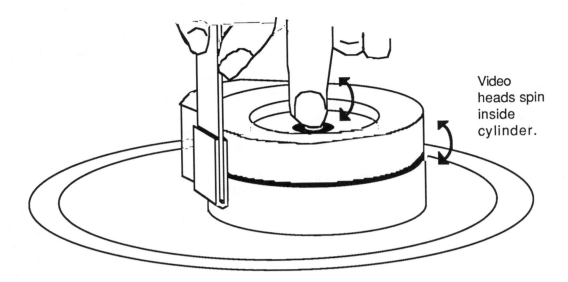

Cleaning the Video Heads on a Beta Machine

might exceed what it would cost to replace the whole VCR. And NEVER rub video heads in a vertical or diagonal direction when cleaning them, because this also might cause them to go out of alignment. You should not press hard. A hundred light strokes is better than three heavy strokes.

6. Spraying Out Clogged Heads

Dirty heads typically cause a snowy picture, or a complete loss of picture. How bad it is depends on how dirty the heads are. If the heads are only slightly dirty, it will look like delicate horizontal streaks scattered through the picture. If the heads are extremely dirty, there may be no picture at all.

Head cleaning is the first thing you should try when confronted with this problem (assuming that you have followed the basic checklist up to the point where you found that the VCR would not play a tape with a good picture.) If one head is not functioning, but the other head is functioning, you may see a lacy smear pattern on the picture, like the smear marks caused by raindrops hitting a dirty car window and making the dirt smear.

If you find that you still have a bad picture with good sound (a condition that indicates dirty heads) after you have finished cleaning and testing, try cleaning the heads a second time in the same way that you did the first time. If that does not solve the problem, then as explained in chapter 14, the next thing to try is unplugging and replugging the little cables connected to the heads, in case there is a bad connection there.

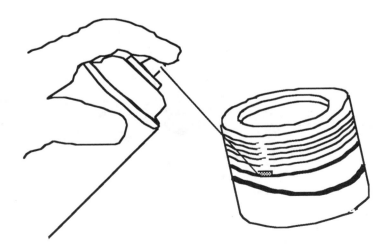

Spraying Out Stubborn Deposits on Video Heads

If this does not solve your problem, it is possible that there are electronic faults that are beyond the scope of this book. But sometimes the problem is just extremely dirty heads that require more ingenious head-cleaning techniques. The next level of attack is to buy head cleaner in an aerosol can with a long thin tube plugged into the nozzle on the can and carefully spray head cleaner for about three seconds onto each video head in turn, and then rub again with a head-cleaning stick.

To do this, you must have head-cleaner fluid in a pressurized spray-can plus a thin plastic tube that fits into the nozzle. This tube comes with most brands of head cleaner when you buy it in a pressurized can.

Put one end of the long tube in the nozzle, and put the other end right up against each video head, and spray head-cleaning fluid into it for about three seconds. Then rub gently with a head-cleaning stick moistened with head cleaner. This will sometimes flush out stuck tape particles that are not removed by other techniques.

It is an Old Wive's Tale that you should never spray out video heads because you might "freeze the heads" with the spray stream and break them, like the belief that you should put a spoon into a teacup before pouring in hot water to absorb some of the heat. I personally know technicians who, as an experiment, have tried deliberately to break video heads by spraying them with half a can of spray cleaner, and have never been able to damage a video head by doing so, even when they tried.

Since VCR technicians have put this warning to the test by deliberately trying to crack old worn-out, discarded video heads by spraying head cleaner on them for much longer than l0 seconds, deliberately trying to see if they can be damaged in this way, and no one I know has ever been able to crack a video head by spraying cleaner onto it, I am pretty skeptical about this tale. And when this technique is used as a last resort, the warnings don't make sense. Maybe those who spread the tale just want you to buy an expensive new set of heads.

Wait about 10 minutes for the heads to dry, then repeat the process. Wait another twenty minutes, using the time to complete the rest of a complete cleaning procedure (belts, idlers, guides, etc.) as presented in the following paragraphs, and then try playing a tape. Be certain that the heads and cylinder are completely dry before trying to play a tape.

You must remember to let the VCR dry thoroughly — say, for 30 minutes — before trying to run a tape through it after spraying out the heads. For if there is still moisture on the cylinder, it might cause the tape to stick to the cylinder, and THAT definitely can break a video head. So, let it dry for thirty minutes before testing again. If you fail to do this, the tape may stick to the heads and do a lot of damage to both the machine and the tape. (Lots of people destroy their VCR's by failing to wait long enough after cleaning the heads with a wet-type head-cleaning tape.)

If the picture is still bad, the next measure to try is alcohol instead of freon on the head cleaning stick. If that works, get rid of your old tapes and switch to new Fuji's or Panasonic's to try to prevent this problem in the future.

If that does not work, and you have tried unplugging and replugging the multi-wire connector from the heads to the rest of the machine, then you probably will need to take it to a shop.

If the problem is definitely in the heads, the shop probably will recommend replacing the heads, although I have seen problems like this fixed by marking the position of the upper cylinder, removing the screws on top, unsoldering and removing the upper cylinder with the heads, soaking the whole thing overnight in a little dish of head-cleaner or alcohol, spraying them out from underneath the next day, and reinstalling them in the same orientation they originally had. (Don't get them on backwards!) But this is an intermediate level technique, beyond the level of this book.

If that does not work, some people remove the upper cylinder assembly as described and put it into an ultrasonic cleaner machine (like that used by jewelers) for about l5 minutes. You can use alcohol or Glass-Plus window cleaner for this. Do not use a glass cleaner with amonia, because it will corrode the heads.

7. Cleaning the Cylinder

Look at the whole region the tape passes over on the upper and lower cylinder carefully. A speck of dirt so tiny that it is almost invisible can cause wavering in the sound, and other problems. If you see any small black tape particles, streaks, or smudges, wipe these off too. If deposits on the cylinder surface do not come off using the freon, try alcohol on a new head cleaner stick. If that fails, you can use cotton swabs, or even a clean rag, so long as you are super-careful not to get near the video heads when you do this.

8. Cleaning any Idlers in VCR

The next important thing to clean is any idler and idler-drive mechanism. The idler is a part used in many machines to turn the reel so as to wind the tape onto the correct reel as the machine runs. If your machine has been "eating tapes" — that is, sometimes ejecting cassettes with tape still hanging out of them — a dirty idler, or idler drive, may be the problem,

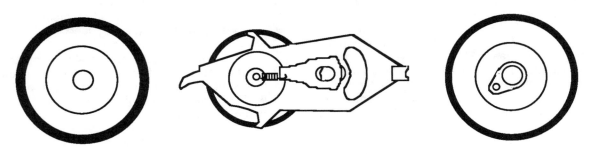

Appearance of Some Typical Idlers

and cleaning it may be the simple solution. Or the idler may need to be replaced, and sometimes doing this saves time in the long run, if you happen to have a replacement idler on hand, but if you do not, let's cleaning it.

Idlers look like little plastic wheels with small rubber tires on them. Although they vary, the basic configuration is a round plastic wheel, from 1/4 inch (6 mm) to 1 1/2 inch (36mm) in diameter, with a small solid rubber tire around it. The illustration shows some typical idlers.

On some machines built by Fisher, the most important idler is hidden deep inside the mechanism where it is difficult to reach for cleaning, but on most other machines, the idler is easily seen by looking down from the top of the machine when the cassette basket is in the ejected position. You may see more than one rubber-tired plastic wheel when you look inside your machine. If your machine has more than one idler, you will want to clean all of them.

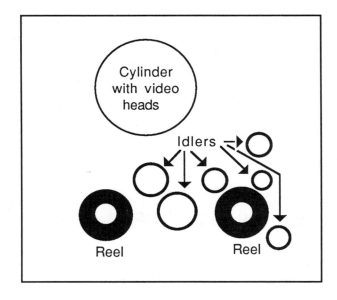

Common Locations of Idlers in VCR's

VCR's may have anywhere from zero to five idlers. Some typical locations of idlers are shown in the illustration. No machine has idlers in all these locations, of course, but many have idlers in one or more of the locations shown. One common place for an idler to be located is on a swinging mechanism located between the two reels — that is, between the "supply reel," on the left, and the "take-up reel," on the right.

Clean the idler(s) by scrubbing the rubber with a cotton-tipped swab dipped in alcohol. Scrub until the black stops coming off onto the cotton. Also clean the nearby round plastic or metal part that contacts the rubber tire and drives it . Many idlers are made to swing from side to side to contact also the edge of one or another of the reels. Clean these reel edges also with the swab and alcohol. Be careful to remove all the oil, and not get any new oil onto these parts. If someone has sprayed oil into the VCR, you will have to spend a lot of time getting it all out.

If you are cleaning a Beta-type machine, skip on down to Section 10 below. If you are cleaning a VHS-type machine, go to Section 9.

9. Cleaning the Tape Path on a VHS Machine

When you look down into most VHS machines, you will see two curved tracks extending around both sides of the cylinder with the video heads. Beside these tracks, or alongside them, the VCR has certain key parts that also should be cleaned. These will be found located in approximately the positions shown in the top view in the next illustration.

Use a cotton swab moistened with alcohol to clean the little ceramic rollers on the tape guides, the angle brackets, tension posts, erase head, impedence roller(s) if any, the audio-control head, the capstan, and the pinch roller. You need only clean the surfaces of these parts that actually contact the tape when it is playing.

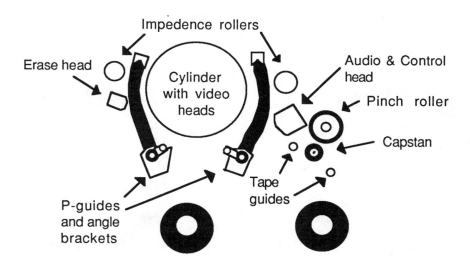

Places to Clean Along Tape Path in a VHS Machine

All of these can be safely cleaned with alcohol, except for the rubber pinch roller, which some authorities recommend cleaning with head cleaner and a head cleaning stick rather than alcohol, while other authorities (with whom I agree) say that alcohol is fine, especially when there is a glazed deposit that needs to be scrubbed off the pinch roller to restore the dull black surface. A clean rag is the quickest tool for cleaning the pinch roller, but cotton swabs are best for the rest.

Do NOT clean off the heavy grease that you may see on the curved tracks on which the P-guides travel in VHS machines, unless you have more of the same grease to replace it. This grease needs to be there to lubricate the tracks for the P-guides to move on.

Cleaning the Capstan

When you clean the capstan shaft, you probably will see little rings of oxide deposit about 3/4 inch (20 millimeters) apart. These develop where the top and bottom edges of the tape run along the capstan. Try to clean off these ridges of deposit completely.

While you are cleaning the capstan would be a good time to put one or two drops of light sewing machine oil on the capstan bearing, especially on Hitachi-built VCR's, where this bearing has a tendency to seize up. If there is a plastic washer that acts as an oil sealer, catch the tip of a scribe or other pointed tool under it, pull it up far enough to get the tip of the oil can under it, and put in one or two drops.

Then slide the little plastic washer back down all the way and make certain that any oil is cleaned off the capstan shaft. Sometimes one of these plastic washers creeps up the shaft and causes curling or

Push oil retainer washer back down as far as it will go.

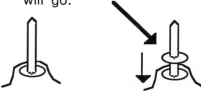

Putting Capstan Oil Retainer Seal Back Where It Should Go

wavy lines to form along the bottom edge of the tape, so make certain that it is pushed down all the way. If there is a metal C-ring instead of a plastic washer, you can just put two drops straight down at the base.

To continue the VHS cleaning, go next to Section 11.

10. Cleaning the Tape Path on a Beta Machine

On a Beta machine, you will want to clean as much of the tape path as you can recognize, concentrating, most importantly, on the audio-control head, erase head, capstan, pinch roller, and guides. It is more difficult to make a picture of the location of these parts, because the mechanism is more complicated and moreover, varies, to some extent, from one Beta machine to the next, and some of the tape guides are inaccessible when the machine is unloaded. But the general illustration shown may be helpful.

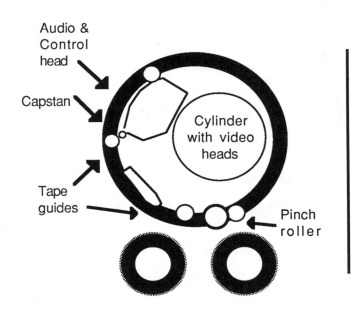

Layout of Reel Table in Early Betas

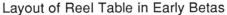

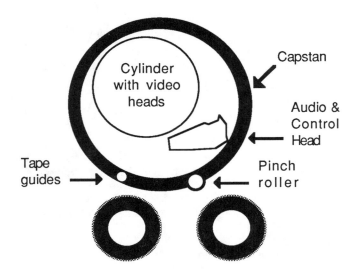

Layout of Reel Table in Later Betas

On some Beta machines, these same basic parts are arranged at other points on the circle around the heads. Imagine rotating the parts around the circle of the mechanism shown in the drawing and you should find them.

The pinch roller is a round cylinder about 1/2 inch (12 mm) to 1 inch (25 mm) in diameter, made out of rubber, possibly with a visible metal center. You will see a shiny area on the sides of it, from pressing against the tape. Scrub it with a head-cleaning stick moistened with head cleaner, or if it cannot be restored to a dull black surface by this means, you can try alcohol on your head cleaning stick, or on a clean rag. Afterwards wipe it with head-cleaning solution to remove the alcohol residue.

The audio and erase heads look like little blocks of metal with a curved shiny surface facing toward the path that the tape follows through the machine. You may want to plug in the machine with the cover off, turn on the power, and load in a Beta cassette to see visually the path that the tape follows when it is pulled out of the cassette by the mechanism, how it goes by the audio and erase heads, and how the pinch roller moves to clamp the tape against the metal shaft of the capstan. Eject the tape, unplug the machine, and clean the capstan, roller, and heads.

You can probably rotate the capstan with your finger so as to clean all sides of its metal shaft. Do not touch it with a metal tool, though, because you might scratch it.

11. Cleaning a Back-tension Band

Many VCR's use a metal band with felt attached as a brake to put the correct amount of back-tension in the tape as it plays. Newer VCR's may accomplish this electronically, but check whether there is a band around the supply reel that looks like this:

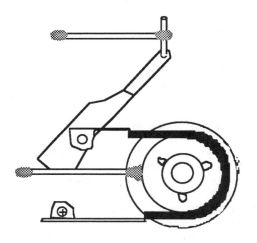

Cleaning a Mechanical Back-tension Band

Clean all the way around the part of the plastic reel that contacts the felt band, as shown with the bottom cotton swab stick in the illustration, and also clean the little post out on the end of the arm that contacts the tape as it goes past. For if the friction from the band

Remove plastic cover

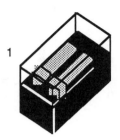

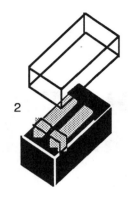

Clean relay contacts with a strip of paper moistened with alcohol or head cleaner

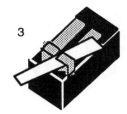

1   2   3

Opening and Cleaning a VCR Relay

against the reel is not smooth and constant, this may produce a wavering in the sound. Check to make certain that the felt band is in the proper position around the reel, and that it has not ridden up out of place too high.

12. Cleaning the Belts Underneath

A few VCR's have some belts on the topside of the reel table, and it would not hurt to run a cotton swab stick moistened with alcohol along the length of these belts to clean them, and also check to make certain that the rubber is alive and tight, and not flabby and worn out.

But even more important are any belts underneath the reel table, because in many designs, these belts run the mechanism that loads the tape, and if they are slipping, the VCR will not even play properly. The difficulty is that in order to reach them, you also need to remove the bottom cover, and possibly remove the front panel piece and a bottom circuit board to get at them. If you need help getting to the belts, see Appendix III. In some cases, it is necessary to remove the front cabinet panel to open up a bottom circuit board located between you and the belts. You may be able to look up your VCR in Appendix VI to tell from the diagram there whether it has belts underneath.

Once you get the bottom of the reel table exposed, you will quickly see any belts there. Clean all the belts with a cotton swab moistened with alcohol, paying special attention to any thin belts (about 1/16 inch (2 mm.) in diameter), because this is the size of most loading belts. A good way to clean oily belts is to remove them completely, push them into the bottle of alcohol, shake it for a while, fish the belt out with a hooked tool, dry it off with a clean rag, and reinstall it.

When cleaning them, check how tight and live they are. If they are loose or flabby, go ahead and clean them, but make a note to order up a replacement for

any tired-out slender belt, because even if the VCR loads tapes OK after the cleaning, it is only a matter of time until cleaning not be enough to make that belt work right. Belts are not expensive, and it is a good idea to have a spare on hand, because you don't know how long it will take to order one. Use Appendices II and VI to get the part numbers and sources to order up replacement loading belts.

Now reassemble everything. Be sure to put each screw and belt back exactly where it was before you took it off.

13. Cleaning Relays

To clean an audio or head-switching relay in one of those rare early Hitachi- or Panasonic-catagory machines that uses a mechanical relay, begin by carefully removing the clear plastic top cover. Catch the tip of a scribe or other sharp-pointed tool between the edge of the plastic cover and the edge of the base, and gently pry the two apart. It usually works best to find the place where you can see a catch molded into the plastic, and start there.

Inside you will see several tiny strips of copper-colored metal. If you look carefully, you may be able to see where tiny metalic contacts are attached to one or both sides of these strips. When the strips bend one direction or the other, these contacts touch, or open up from touching, other stationary contacts. The mating surfaces of these contacts need to be cleaned.

You can try just spraying the contacts with head cleaner from a pressurized can, but the best results are usually obtained by cutting a thin strip of clean paper like typing paper, moistening it with alcohol or head cleaner, and using it to clean the heads. Work the paper strip in between the contacts and then slide the paper back and forth. You may need to push the contacts apart slightly to get the paper in, or if they are already apart, you may need to push them together while you slide the cleaning paper back and forth to clean them off. Replace the cover carefully.

Chapter 4

Basic Universal VCR Check-out and Diagnostic Procedure

NOTE: Chapter 2 should be read before this chapter.

How to Use this Chapter

The following procedure can be used both to diagnose and repair problems in VCR's, and to check out a used VCR before buying it. The hardest problems are the ones listed first below — the rest are more fixable. The procedure will tell you when a problem requires repair techniques beyond the scope of this book.

To find the correct chapter with the information to fix your problem, it is important to follow the check-out steps in the order in which they appear here, without omitting any steps. Otherwise, you may end up going to the wrong chapter. Even though the title of a later chapter seems to describe your problem, the problem may actually be dealt with in an earlier step of the procedure, so make all the tests in the order given here.

Warranty Warning

The cleaning, troubleshooting, and repair procedures in this book are primarily intended for VCR's that are more than two years old, or otherwise out-of-warranty. Opening up the cabinet of your VCR to run the check-out procedure may void the guarantee on new, in-warranty VCR's. But most of the following tests can be made without opening the cabinet if you want to check out a new VCR, or make a preliminary diagnosis of the problem before sending a machine that is still under warranty to a warranty repair center. For the steps in which you need to tell whether the tape is moving, you can do this in most cases by watching the counter.

Safety Warning

Do not touch the insides of the VCR when it is plugged in. Unplug the machine before touching parts inside the machine.

Get ready — Here we go!

1. Position VCR for Easy Access

Unplug machine and place on work bench or other place where you can open it up, but also be able to connect it to a TV set and cable or antenna.

2. Check for Exterior Damage

Visually check power cord, cabinet front and sides for physical damage. Look to see if there is a cassette inside VCR. (You can do this with a front-loading machine by lifting the flap and peeking through the front opening, or reaching in and feeling with your fingers.)

3. Open Up Cabinet

Check to make certain machine is unpluged. Following the procedures in Appendix III, "How to Open VCR Cabinets," open the top of the cabinet and fold up any circuit boards and shields over the reel table. When you have exposed the top of the reel table, return to this point in this basic check-out procedure. (Skip this step if you are running the check-out procedure without opening the cabinet.)

4. Plug in and Observe Behavior

Plug in the machine. Observe whether the clock display glows with a number. If nothing happens, or if something happens and then the VCR shuts off, go to

Chapter 5. (First, though, make certain that the timer switch is not on — that would keep it from working.) On some machines, the numbers will blink, and on others, they will not.

The number displayed must be something that makes sense (for example, "12:00"). If you plug in a VCR and the numbers are whacco—that is, if they are crazy and make no sense — you immediately know that something is wrong with the circuit controlling the display. If this situation exists, then except for the case in the next paragraph, the problem is beyond the scope of this elementary manual, and the VCR must be taken to an electronic repair shop to be fixed.

PANASONIC-category VCR's: If the VCR is in the Panasonic category (see the "Introduction" for a list) and when you plug it in, the VCR mechanism moves continuously back and forth until it shuts itself off, go to Chapter 6 for mode-switch adjustment procedure. If three front panel lights come on together in a combination that makes no sense, see Chapter 5 to find out how to check for a blown fuse .

5. Turn On Power and Observe Response

Now push the power switch. Some display should come on indicating that the power has been turned on. If the VCR has a power light, it should come on. Unless it is set for a camera input, a light indicating the channel to which the VCR is tuned should come on, if it has such a light. On some machines, there will be a noise from the loading motor and loading mechanism, and/or from the cylinder motor. (There may also be noise and movement when you turn the power off) This is normal.

If nothing happens when you push the power switch on, or if a power light comes on and then shortly thereafter shuts off, and the unit will neither play, nor fast-forward, nor rewind the tape, or if a motor runs briefly but the power light will not come on, then go to Chapter 5.

6. Try to Eject Cassette

If there is a cassette in the machine, push "Eject' to raise the cassette basket and eject it. If a cassette is stuck in the machine and it will not eject, go to Chapter 7 if it is a top-loader, or go to Chapter 8 if it is a front-loader.

PANASONIC-category VCR's (see "Introduction" for a list of brands in this category): If the VCR is in the Panasonic category, and when you hit "Eject," it goes into Fast-forward, go to Chapter 6 and adjust the mode switch.

7. Check the "Cassette In" or "Tape In" Light

Except for the earliest machines, most VCR's have a light or indicator that comes on when a cassette has been pulled into the machine fully. Some machines use an illuminated little window with a pattern on it that looks like this:

If the cassette-in light is on when no cassette is loaded in the machine, something is wrong. (On a front-loading VCR, nine times out of ten, the problem is that a switch on the front-loading assembly that is supposed to change position when a cassette is fully loaded is stuck in the wrong position.) If the VCR has a malfunctioning "Tape In" or "Cassette In" indicator that lights when no tape is in the VCR, go to Chapter 11.

8. See Whether the Machine Will Accept a Cassette

With the unit plugged in and power turned on, Insert a cassette into the cassette basket or front loader, and gently try to push it into the machine. Machines that accept cassettes through an opening on the front are called "front-loaders." VCR's that accept cassettes into a basket that rises up out of the top of the machine are called "top-loaders."

If it is a front-loading machine, let the machine pull the cassette into itself, if it will. If the mechanism does not do this, DO NOT FORCE THE CASSETTE IN! There may be something wrong with the mechanism that is supposed to pull it into the machine, and pushing too hard on it could cause expensive damage! If the machine is a front-loader and will not pull the cassette in, go to Chapter 10.

If the VCR is a top loader, put a cassette in, and push the basket down gently. If the basket holding the cassette will not go down, or if it will not stay latched down, go to Chapter 9.

9. Check "Cassette In" or "Tape In" Light Again

If the VCR has a "cassette-in" light or indicator, make sure that it lights up when the cassette goes into the VCR. If not, go to Chapter 11.

If this indicator lights up when no tape is in the VCR, or if the VCR ejects automatically right after loading the cassette, go to Chapter 11 also.

10. Make Certain that Cassette Sits in VCR Correctly

Check visually to insure that the tape cassette is resting evenly on the supports underneath it. It should be horizontal, with none of the corners or edges higher than the others. Unplug and try pressing on the top of the loaded cassette. It should rest solidly down on the reel table, and not be able to move farther down, or in a downward fashion. If not, go to Chapter 12. (Note: Skip this test if you are checking out a machine without opening the cabinet.)

11. Test Fast-forward and Rewind Operation

Plug in again and push the Fast-Forward button, then the Stop button, and then the Rewind button, and check to make certain that the tape moves first in one direction, and then the other. Check this by looking through the window in the cassette, or by setting the display to "counter" and watching the change in the number on the tape counter. (Note: Older machines with mechanical counters do not require setting the display to "counter." On them, the counter should automatically show changes whenever the tape winds or rewinds.) If the tape does not move, or moves only very slowly, or moves with jerks, go to Chapter 13.

If the tape moves for a short time and the machine then shuts down, or the tape does not move at all, go to Chapter 13. See also Chapter 14.

NOTE: This is a very important test. This test is made to insure that the Fast-forward and Rewind controls can be used to pull the tape safely back into the cassette, in order to remove the cassette from the VCR later without damaging the tape, since in

case the unloading mechanism fails to wind the tape fully back onto the supply reel, exposing the tape to damage if you tried to remove it from the machine while tape was pulled out. If fast-forward and rewind functions work, you can always use them to reel any loose loop of tape back into the cassette.

If the tape does not move, or moves only very slowly, or moves with jerks, on a top-loading VCR on which the cassette lid has been removed, try pressing down with your hand on the exposed top surface of the cassette. If it now moves faster, you will need either to reattach the lid to the cassette basket, or else place a (dry) heavy object on top of the cassette to press it down for the test procedure here.   It is probably best simply to reattach the cassette basket lid. If you do so and the tape still does not move as rapidly as it should, then go to Chapter 13.

Remember that if the cassette is fully rewound, the tape will not rewind further.   And if the tape is wound all the way in the other direction, then the tape will not fast-forward.

Also, the tape will not run in a VHS machine correctly if there is a bright light shining on it with the cabinet top removed, because a bright light will activate the photosensor that normally tells the machine when the tape has reached its end by sensing light coming through the clear leader at the ends of the tape. If the machine seems to misbehave, turn down the light over the table or bench, or cover the area over the cassette with a magazine or other object made of nonconducting material to shade it. If it still misbehaves, then you have a genuine problem, and should try going to Chapter 13.

12. Verify that Test Tape is Undamaged Prior to Playing It

Eject the test cassette from the VCR. Visually check a "junk" tape for inconsistencies or damage. By a "junk" tape, I mean a spare tape with nothing of value already recorded on it, not a bad or worn

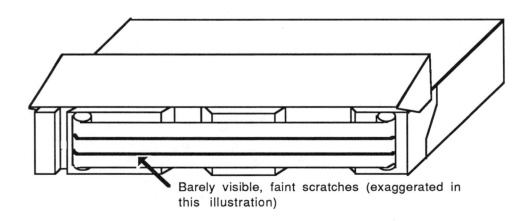

Barely visible, faint scratches (exaggerated in this illustration)

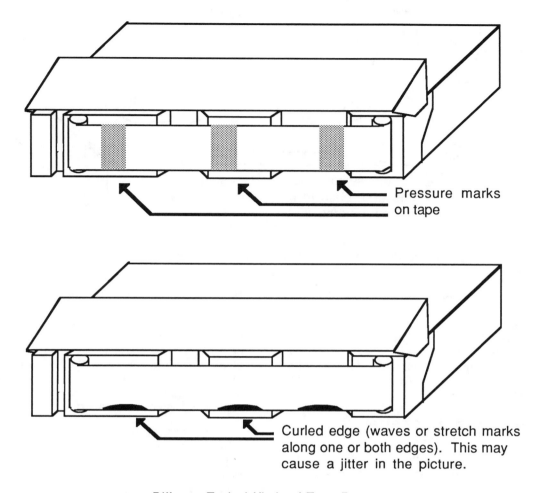

Pressure marks on tape

Curled edge (waves or stretch marks along one or both edges). This may cause a jitter in the picture.

Different Typical Kinds of Tape Damage

cassette. You could use a brand-new cassette for test purposes. For testing purposes, do not use a tape with valuable program material on it, because the tape may get damaged. Inconsistencies or damage includes scratches, creases, folded or wavy top or bottom edge, etc. The illustrations show some common ways in which tape can be damaged.

To check VHS cassettes, push the small square plastic button at the left side of the cassette (seen as viewed looking toward the side with the protective door), and raise the door with your finger so that the surface of the part of the tape visible behind the door can be viewed. Look for waves, or wrinkles, along the top or bottom edge, or horizontal scratches.

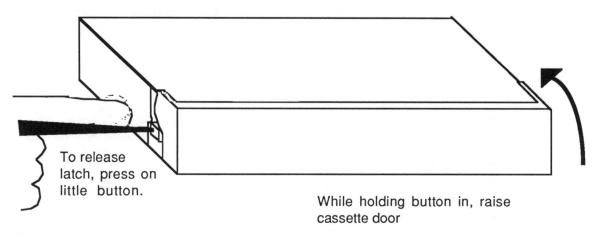

To release latch, press on little button.

While holding button in, raise cassette door

Releasing the Latch to Open the Lid to Check a VHS Cassette for Tape Damage

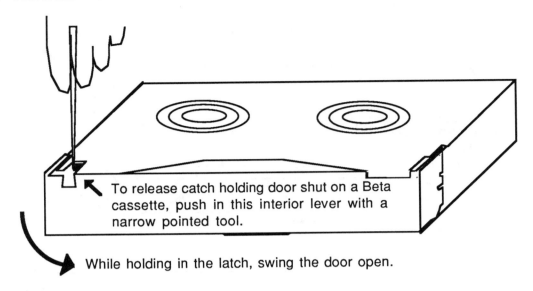

To release catch holding door shut on a Beta cassette, push in this interior lever with a narrow pointed tool.

While holding in the latch, swing the door open.

Releasing the Door Latch to Check a Beta Cassette for Tape Damage

Check Beta tapes by pushing back the little enclosed lever at the bottom right edge of the side with the protective door:
Look for waves, or wrinkles, along the top or bottom edge, or horizontal scratches. As emphasized, use a nonvaluable tape for this test.

13. Again Load Cassette

Insert this cassette into the cassette basket or front loader, and gently try to push it into the machine.

If the VCR will not accept the cassette, go to Chapter 9, "VCR Will Not Accept Cassette — Top Loader," or go to Chapter 10, "VCR Will Not Accept Cassette — Front Loader," whichever is appropriate for your VCR.

If the VCR has a "cassette in" light or indicator, verify that it lights up when the cassette goes down into the VCR. If not, go to Chapter 11.

14. Hit Play Button and Observe Result

Hit Stop, then Play. Does the VCR load the tape into the mechanism, pulling a loop of tape out of the cassette and wrapping it part way around the heads? And if so, does the tape start moving? If not, go to Chapter 14. Remember that it is normal for the tape in a VCR to move extremely slowly. In fact, in SLP or EP, you can barely see the tape move at all, unless you look very closely for a while. (Inside the VCR, the video heads spin rapidly against the tape, creating the same resulting effect as fast tape movement, without the tape having to move rapidly, as explained in Chapter 2.)

Look at the take-up reel for a while, or watch the counter. If the take-up reel is not moving properly, or if it the tape starts to move in Play and then stops, also go to Chapter 14. We are here testing whether the machine is mechanically operating correctly, so that the tape will not get damaged. We are not yet expecting picture and sound.

PANASONIC-CATEGORY: If the VCR is is the Panasonic category (see the "Introduction" for a list) AND when you push "Play" it loads the tape fully, then unloads, then loads again, then unloads, etc., making a clicking sound, then go to Chapter 6 and spray out the mode switch, or if that doesn't fix it, adjust it as explained there.

15. Rewind Partially, Eject, and Check for Fresh Tape Damage

We want to check to see whether the tape is getting bent or damaged as it moves along the tape path during play. We have now run some tape that we know had no marks on it through the tape path. By looking at it, we can see whether the machine is damaging the tape.

To do this, hit Stop and rewind only part way back to the point on the tape where you started.   Do not rewind all the way back, because then you will not be able to tell anything, because the tape that just went through the mechanism will be wound out of sight on the supply reel. Rewind only half way back to the place from which you started it playing.

Hit eject, remove the cassette, open cassette door as explained earlier, and check the tape for any fresh damage. How does the tape that was previously undamaged look now after it has run through the play mechanism once? If the tape has been damaged by the machine, go to Chapters 14 and 15. If tape is left hanging out of the cassette, go to Chapter 14 also. If tape is not damaged, continue as follows.

16. Connect VCR Correctly to Working TV Set

Connect F-to-F coaxial cable between the "out to TV" (or "VHF out") connector on VCR and the "VHF in" (or "Antenna in") terminal on a TV set. NOTE: Some TV's may require a matching transformer (called a "balun") to go between one of the cable ends and the TV set antenna terminal, as explained in Chapter 1. See Chapter 1 for more help hooking up the VCR to the TV set and antenna or cable.

VERY IMPORTANT: It is here assumed that you've already tried connecting your antenna or cable directly to your TV set to verify that your TV set and antenna or cable signal source, are working correctly. There's no sense in wasting your time troubleshooting your VCR, if that's not where the problem is.

17. Connect VCR Correctly to Good Signal Source

If you have not already done so, connect the antenna (or cable, if a subscription cable system is your signal source) to the connector marked "VHF in" (or maybe "in from Ant.") on the back of the VCR, as explained in Chapter 1. If you are using the cable and the VCR is not cable-ready, use a cable-box as explained in Chapter 1.

18. Set VCR Controls to Receive Antenna (or Cable) Input

Some VCR's also have a "tuner/auxiliary" (or "tuner/camera") switch. This switch is for selecting either the CAMERA input jack (sometimes called "aux" or "auxiliary") or the ANTENNA (also called "tuner") as a source of program material. For the following test, set this switch to the postion marked "tuner." NOTE: If a camera is connected to the VCR, disconnect it for the following procedure.

19. Set VCR/TV Switch to "TV" Position and Check Picture

The VCR will have some kind of switch for selecting between "TV" and "VCR." Use of this switch was explained in Chapter 1. Set it first to the "TV" position. Then tune the TV set to an active channel and verify that picture and sound are present.

If a signal comes through OK, go to the next step

© 1990 Stephen N. Thomas

below. But if a picture and sound appear on your TV when the antenna is connected directly to the TV, but not when the antenna is connected to the VCR and the TV/VCR switch is in the "TV" position, then go to Chapter 16. If no picture, or only a very fuzzy picture, appears on your TV set in this setting, you may have trouble in a part of your VCR called the "antenna switcher." Often, this can only be fixed by a shop, but sometimes you can do it yourself as explained in Chapter 16.

NOTE: The function of the "TV/VCR" switch is to control the path of the signal coming out of the VCR and going to the TV, as follows:

"TV" In the "TV" position, the signal goes directly to the television set from the antenna (or cable) without going through most of the VCR's internal circuitry. The program arrives at the TV set on the same channel number that it was on when it came into the VCR (as if your VCR were only a piece of cable connecting the antenna to the TV set).

"VCR" When the switch is in the "VCR" position, the signal is processed through the VCR's internal circuitry, and changed to either channel 3 or 4 before leaving the VCR and going to your TV set.

These different functions might have been clearer if a more appropriate description like "VCR bypass" or "Antenna direct" had been used to label the first position instead of the unclear phrase "TV."

20. Tune TV Set to the Channel that the VCR is Set to Output On

Look for a switch on the back, or bottom, or side of the VCR to see what channel it is set to output on (usually, either channel 3 or 4). Somewhere there will be a small switch, labeled "3" in one position and "4" in the other position. This switch may be located on the back, bottom, side, or top (under a hinged door) of the VCR, depending on the make and model of the machine. NOTE: On portable recorders that come in two parts, this switch is located on the half of the two units that holds and plays the tape.

Set the channel selector switch on the TV set to the same numbered channel as this switch.

21. Set the VCR/TV Switch to the "VCR" Position

Set the "VCR/TV" switch to the VCR position, and set the VCR's channel selector to the number of an active incoming channel. If a picture and sound do not appear now, then go to Chapter 17.

We run this test now, prior to trying to watch a tape play, in order to detect problems in what are called "antenna switching" and "RF modulator circuits" as possible causes of trouble. If the VCR passes this test, then the VCR's little TV transmitter (called its "RF modulator") is good, and the possibility that a defect in these parts is a cause of trouble is eliminated.

22. Set VCR Speed Selector to the Speed of Inserted Test Tape

Insert a known good prerecorded tape (like a taped movie). Make certain that this tape is in a speed (for example, SP) and mode that the recorder should be capable of playing. SP is the fastest speed; EP (also called "SLP") is the slowest speed; LP is a speed intermediate between these extremes. Most prerecorded movies are in SP.

Most machines automatically set themselves to play at the speed at which a cassette put into them was recorded, but to eliminate all doubt with unknown machines, manually set the speed control switch on the front panel to the same speed as the prerecorded tape. This is absolutely necessary on some older machines, including some GE models, and it does not hurt on the others.

23. Hit Play and Check Picture Quality

While watching a local TV channel, hit Play. After a moment, the picture on the screen should switch to the material on the tape. Check the quality of the picture and sound.

If the picture is snowy, or no picture appears, or horizontal bars or streaks appear across the picture, try adjusting the front-panel tracking control.

If the picture remains bad, try pressing down on the top of the cassette with the eraser end of a pencil or a wooden stick. If the cassette moves slightly and the picture goes clear, the problem may be that the cassette is not resting correctly in the machine. If so, go to Chapter 12.

If the picture remains bad, go to Chapter 18.

If the picture plays only in black-and-white, with no color, go to Chapter 19.

If there is no sound or if sound is distorted, go to Chapter 20. If the tape runs through the VCR at a high rate of speed, with Donald-Duck sound, the problem is beyond the scope of this book; you will have to take it to a shop.

PANASONIC-category VCR's: If the VCR ejects the cassette when you hit "Play" on a Panasonic-category VCR, go to Chapter 6 and try adjusting the position of the mode switch.

24. Check Special Effects (Optional)

If the picture and sound are good, then if the VCR offers special effects (like Forward and Reverse Search, Slow motion, Pause, Still Field, Freeze Frame, etc.), check these out. If any of these special effects is bad, you can try cleaning the VCR, following the procedures in Chapter 3. If that does not solve the problem, it is beyond the scope of this elementary book and you will need to take the unit to a shop.

25. Make and Play Back a Short Test Recording from Antenna or Cable

Hit Stop, hit Eject, insert a good blank tape, set the VCR to the slowest speed (EP or SLP, on most), reset the counter to zero, and make a short recording of a local channel on which you are seeing a good picture in the "VCR" mode. Stop, rewind to zero, and playback to confirm the quality of the recording you have just made. Repeat this test for each speed available. If the VCR won't record, go to Chapter 21.

26. Verify Smooth Operation of Mechanism

While the VCR is recording, if the tape is visible, observe to make certain that the take-up reel is running smoothly.

27. Check Clock Operation

To check the clock, hit Stop and try to set the clock, following the procedure in the operator's manual for your model. (If you do not have an operator's manual, try ordering one from the address listed for the brand in Appendix II.) If the clock cannot be set, the problem is beyond the scope of this manual.

28. Check Timer Operation

Set the timer for a timed recording. If the timer cannot be set, the unit will need to go to a shop for repair. On some VCR's, you only set the hour when you want it to come on, and "AM" or "PM"; on other VCR's, you also set the "day" of the week on which you want it to come on. Make a short timed recording, verifying that the timer switches the VCR on and off at the correct time, and that the VCR recorded correctly on the channel for which it was set. NOTE: On some older VCR's, in order to make the VCR come on at the set time, the manual play and record buttons both must be physically pressed after timer switch is set to "on."

29. Check UHF Reception and Tuning

Check to see that the machine performs all functions correctly for UHF channels using the same procedures as above, except setting the VCR channel selector switch to an active local UHF channel. Note: On a VCR that has separate antenna inputs for VHF and UHF, a separate antenna or other signal source must be attached directly to the UHF antenna terminals in order to receive UHF stations.

"No, the problem couldn't be the tape. I've used that tape thousands of times for years and years, and never had any trouble with it."

Important: Use the check-out procedure
in Chapter 4 before using this chapter.

Chapter 5

VCR Does Not Power Up Properly

When you first plug in the power cord of a VCR, even when the VCR's power switch is in the "off" position, you should see or hear something happen. This "something" will vary from model to model, but on VCR's with a clock display, at least some illuminated numerals should light up.

If absolutely nothing happens, the most likely cause is a failure in the power source, or power cord, or a blown fuse or other trouble in the "power supply" inside the VCR, and you should go to Section 1, "VCR is Totally Dead." (The "power supply" is the part of the circuitry that converts the power from the wall outlet into various different lower voltages of a kind that the other parts of the VCR can use.)

If some indication comes on when you plug in, but the VCR will not power up when you push the power switch, or if it does something (and perhaps moves the mechanism briefly), but then the VCR shuts itself off a few moments later, start at Section 2, "VCR Will Not Power Up, or VCR Powers Up, and Then Shuts Itself Down."

Before doing anything else, though, try switching the "Timer" switch to the position opposite to the one it is in now, and try pushing the power switch a couple of times. Then put the timer switch back to the same position in which you found it, and try it a few times more. (It could be that a child or gremlin turned on the timer switch, and that's the whole problem, or that someone touched it accidentally and switched to the timer mode. Perhaps the timer switch is on, but the indicator for the timer switch is burned out, or not showing for some other reason.)

Also, if there is any possiblity that the front of the cabinet could have gotten bumped or hit, look it over carefully for any sign of physical damage, like a crack in the plastic, or a button caught in a depressed position. The trouble may be in a front cabinet button not moving back properly to press the electrical switch on the circuit board behind the front cabinet piece, or a crack or other damage in the circuit board behind the cabinet front.

Safety Warning

All tests here are made with the VCR UNPLUGGED! Do NOT put your hands inside the VCR when it is plugged in, unless you are a trained service technician.

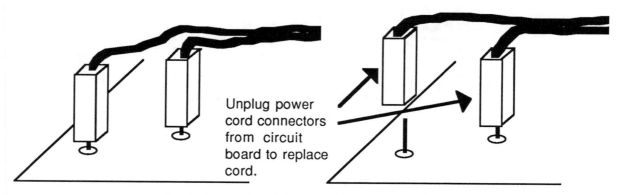

Unplug power cord connectors from circuit board to replace cord.

Replacing a Plug-On Matsushita-type Power Cord

1. VCR is Totally Dead

If the VCR is dead because of trouble in the power supply, or a short circuit elsewhere in the VCR, keeps blowing a fuse or shutting down the power supply, then it is a problem of the kind that is beyond the scope of this basic book. But before turning the VCR over to a repair shop, let us check a few simpler possibilities.

First, of course, double check to make certain that power is present at the outlet into which you've plugged the VCR. You can do this by unplugging a lamp that is working from another wall power outlet and checking to see whether the VCR works when plugged into that same other wall outlet.

Check for Damaged Power Cord

Next, unplug the VCR and check the power cord for any obvious damage. Is there any place where maybe the kids have broken it, or the dog has chewed through it, or it has gotten pinched or broken in any other way? Look and feel carefully with your fingers along the unplugged cord for any place where all the strands of wire inside the rubber insulation might be broken even though the outside insulation is still intact.

Cords most often go bad near the ends where they attach to the plug or termination point, so look especially carefully there. Remove the cabinet top (see Appendix III for help), find the other end of the cord on the inside of the VCR, and with the VCR still unplugged, follow the cord to the place where it attaches to a circuit board, checking for obvious damage.

If the power cord looks and feels good, skip down to the next section below.

If the power cord is bad, it will need to be replaced. Look inside the VCR and carefully follow the same power cord to the point inside the VCR where it attaches to a circuit board. It will either terminate in push-on plastic connectors, or else it will be soldered to the circuit board.

If it terminates in push-on connectors, you need only obtain the part number, order an identical replacement cord, and install it yourself. If it is soldered onto the circuit board, then the repair is beyond the scope of this book, unless you have a soldering iron, and can solder; for more on soldering, see Section 20 of Appendix I.

There are two wires in the power cord, and it is extremely important that you install a new cord with the same wire going to the same place as did each of the two corresponding wires on the old cord — otherwise, there is a danger of people getting a bad shock when they use the VCR. With the cord unplugged, look at the blades on the plug that goes into the wall outlet. Do you see how one blade is made wider than the other blade?

It goes to the wire on the same side in the power cord that runs up to the VCR. Very carefully trace this side into the VCR and make a mark with a felt-tip pen beside where this wire connects to the circuit board. When you install the new cord, make certain that the wire on the side attached to the wider blade on the plug gets attached to this same point inside the VCR, so that the new cord is installed just like the old one was. Double-check this before connecting the other side of the power cord to the other point.

The power cord can be replaced without soldering if it is a Panasonic-category plug-on cord that looks approximately as shown in the illustration. There are several versions, some with round, and others with square push-on connectors. Call the phone number listed for your brand in Appendix II and get the part number for the cord for your make and model. For safety, do not try to substitute any other part.

Wait until you have obtained the exact same identical part before removing the old part, so that the route followed by the old power cord inside the VCR will be fresh in your mind when you go to install the new cord. Notice that the rectangular plastic female connectors at the ends of the power cord are different colors. Put the new color-coded plastic terminals in the same places as the terminal endings with the same color on the old cord.

Soldering for the Adventurous

If you found a defective power cord that is soldered to terminals unlike the plug-on Panasonic-type power cord, then the repair procedure of unsoldering the old cord, and soldering in a replacement cord is more advanced than most of the repairs covered in this book.

However, for more adventurous readers, basic unsoldering and soldering techniques are explained in Appendix I. Although miniature, soldering irons with low-power (25 watt), high temperature (650 degrees

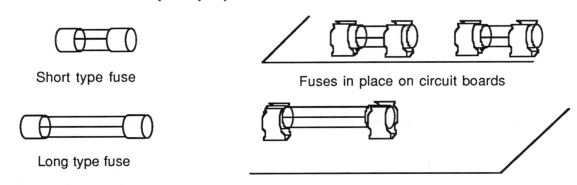

Typical VCR Fuses

F) tips are used by most VCR shops for repairs involving soldering, heavier irons or soldering guns can be used for a few heavier jobs such as replacing the power cord, provided that you are extremely careful not to overheat the terminals to which the power cord attaches, and not to touch any of other part accidentally with the hot parts of the iron.

Attaching New Power Cord

When you attach the new cord, remember to make certain that the wire from the wider blade of the plug at the other end of the cord goes to the same terminal you marked before taking off the old cord. And remember to run the power cord through the hole into the VCR before attaching it.

It is a good policy to disassemble the VCR far enough to be able to get at the terminals easily rather than endangering some other part (or your fingers) by trying to work with a soldering iron in a location that is too cramped.

While the soldering iron is still cool, before plugging it in and heating it up, experiment to find the best route to follow in inserting the soldering iron to reach the terminals without touching or melting anything else.

You will certainly need forceps or needle-nosed pliers to wrestle the wires loose the moment the solder turns liquid. Be careful not to splatter any hot liquid solder on the circuit board or other parts where it might cause a short circuit or other trouble. Double check to make certain that the VCR is unplugged before touching any of the internal parts.

Check for Blown Fuse

By far, the most common cause of a "nothing happens when plugged in" symptom is a blown fuse. A fuse can be caused to blow by another internal malfunction inside the VCR that causes the unit to draw too much current. It also can be caused to blow by a momentary "voltage spike" from the power line, caused by lightning hitting a distant wire, or trouble at the power company, and goodness knows what else.

If the fuse was caused to blow by an internal malfunction, then the replacement fuse should blow too, and this will tell you that you have done all you can, and need to take the VCR to a repair shop. But if the fuse was caused to blow by an external cause, then you may be able to fix your entire problem and have your VCR working again by simply replacing the blown fuse with a new fuse.

Double check to make certain that the VCR is unplugged.

The fuses that you can locate, remove, check, and replace if blown without soldering or need of expensive service literature, will all look like a small, thin clear glass tubes, about 3/16 inch (5 mm) in diameter, with metal caps on both ends held to a circuit board by two metal clamps, one gripping each end of the fuse. They look like some of the types of fuses also used in automobiles, if you are familiar with these.

Fuse Locations

There is no one single location for the fuse or fuses in VCR's. To try to list all the positions of fuses in all the different models of VCR's would already fill a book, but once you know what they look like, you usually can easily find them by attentively looking around inside the VCR at the top along the back and sides of the unit once the top cabinet has been removed. They often are located near the circuit board to which the power cord runs.

Double-check to make certain that the VCR is unplugged before opening the cabinet and looking around inside.

Fuses are usually accessible without removing the bottom cover of the cabinet. Removing the top cover is usually enough. (An exception to this rule

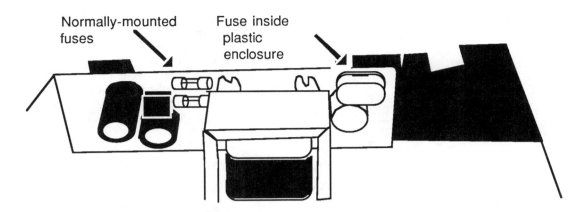

Normally-mounted fuses

Fuse inside plastic enclosure

Plastic Fuse Enclosure

are the Hitachi VT 6800 and the RCA VGP 170 portables, which have most of their fuses on top, but one main line fuse is on the bottom, accessible through a little plastic door held on by two screws.)

On some HITACHI-category VCR's, some fuses are enclosed under curved plastic enclosures that look as shown in the illustration. Pry off the top piece of plastic with a small screwdriver, and you will find a fuse inside that looks like the ones we've already seen.

If there are several fuses, remove no more than one at a time, as an aid to getting different fuses with dissimilar current values back into their proper places. Do not put them back in the wrong clips — they usually cannot be interchanged. Remove only one fuse at a time. Fuses can be removed by prying them loose using the bent end of a scribe or other similar tool. Never pry against the fragile glass part, but only against the metal ends.

Fuse Ratings

It is important to understand that different fuses have different "current ratings." There is a thin, sometimes almost invisible, wire that runs down the center of the glass tube carrying current from one end of the fuse to the other. When the current through the fuse exceeds a specific predetermined amount, it heats this wire enough to melt it, breaking the path for the current to flow, which is what happens when a fuse "blows." A fuse is like a safety valve that prevents the current from exceeding a safe amount.

The fuse is a protective device intended quickly to shut the unit down if some internal malfunction, such as part of it "shorting out," causes the VCR to draw more current than it should. A fuse may also blow to protect the VCR if the unit is plugged into an outlet with too high a voltage, or if the voltage at the

outlet suddenly surges, or rises too high, due to a distant lightning strike on the power lines, or a mistake at the power company, etc. (Sometimes, unfortunately, the fuse does not blow rapidly enough to protect the VCR in this situation, however.)

An approximate value for the maximum voltage that the fuse can handle, and the maximum current it can carry continuously before blowing, are stamped into the side of the metal cap on the end of the fuse. Whenever you replace a fuse, it is crucial that you obtain and install a new fuse with the correct ratings, since otherwise its protective value is lost.

Look inside the glass tube of each fuse. If the little internal wire is broken or absent, the fuse is blown. If you cannot see through the glass because is bluish or blackish on the inside, the fuse is blown. If the fuse is not blown, reinstall it, pushing it back into the holding clips so that both ends are gripped by the metal. (It does not matter which direction a fuse is pointing or oriented, so long as you get each fuse back in the right pair of metal clips.)

It is possible for a fuse to look OK, but have the little internal wire burned through down close to the end in the part hidden under the metal cap, but this is unusual. It is difficult to detect a fuse blown in this unusual way if you have no test equipment. The only way you can do so is by substituting a known good fuse of the same current rating and having the VCR spring back to life when you plug it in again.

NEVER, NEVER try to make a VCR work by shorting across a blown fuse with wire, tinfoil, or any other conductive material! Doing so, for even a fraction of a second, could not only cause very expensive damage to the VCR, even completely destroy it, but it also could start an electrical fire. Blown fuses must be replaced only with fuses having the same maximum current ratings. The current ratings on the side of the fuse might say something like "1 Amp," or "2.5 Amp," or "1.6 Amp," or 1/8 amp, etc.

Fuse Current Rating Printed on Board

The schematic symbol for a fuse, which looks like this,

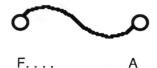

F....   _ _ _ A

Schematic Symbol for a Fuse

is often printed onto one side of the circuit board between the two metal clips that hold the ends of the fuse. The meaning of the numbers that appear after "F" and before the "A" will now be explained.

If the fuse is blown, look to see if there is a number followed by the letter "A" written ON THE CIRCUIT BOARD NEAR the clip that was holding the fuse. For example, if it says "1.5A," this means that a fuse to carry 1 1/2 amps of current should go in that location. This number normally appears in the place where blanks are shown ("_ _ _") in the illustration, and not in the place where dots ("...") are drawn. (The number appearing after "F" where the figure shows dots is only the arbitrary number assigned to the fuse to help service technicians correlate it with their service diagrams; it has no other meaning.) If you see "2A" where the dashes are, this means that a two-ampere fuse should go here.

Current Rating Printed on Fuse

The same number should be faintly printed in the metal at one end of the fuse. If the rating on the blown fuse and the number on the circuit board do not agree, replace the fuse with a new fuse having the value you see printed on the CIRCUIT BOARD, because someone else previously may have installed a fuse with the wrong current value. Do not put in a fuse with a value different from what is supposed to go in that location.

Fuse Lengths

Fuses of the glass-tube type come in different physical lengths, but almost all VCR's use only the same short-length fuse (about 3/16" diameter (4 mm) by 3/4 inch (18mm) in length). The length of this type of fuse is indicated by different special part numbers by different manufacturers. You may need to

know this special number or designation to check over the telephone whether a supplier has the replacement fuse(s) you need in stock. But most parts suppliers will understand if you call this an "F-length" fuse, or a "GMA" fuse, and specify the current. For example, if you need a 1.5 Amp replacement fuse of the short-length type, then a "1.5 Amp F-type fuse" would work. Lots of times they just call them "short-length fuses," or just "short fuses." If you cannot get fuses locally where you are, then try contacting the parts sources listed in Appendix II.

FISHER-category VCR's note: Some VCR's in the FISHER category (see Introduction) use longer length fuses.

Replacement Fuses

With few exceptions, all the fuses used in VCR's are standard values (like .5, 1, 1.5, 2, 3, or 4 amp) for which general-purpose, generic replacement fuses can be used. Some Panasonic and other Matsushita-category VCR's use a 1.6 Amp fuse (Panasonic or Matsushita part number XBA1C16NU100 or RCA part number 149386), for which you should be able to substitute a generic 1.5 Amp fuse while waiting for your supplier to back order and deliver a 1.6A fuse to you.

A "500 mA" fuse is the same as a "0.5 Amp" or "1/2 Amp," fuse. Its Matsushita part number is XBA1C05NU100 or RCA number 149387. Some JVC and Zenith machines used a 0.9 Amp fuse, for which you should be able to substitute a 1 Amp fuse until your supplier can get you an exact replacement 0.9A fuse. Do not be discouraged if the supplier insists on selling you at minimum a box of five. They are not expensive, and it is useful to have spares on hand. If you had taken the machine to a shop when it only needed to have a fuse replaced, the shop might have charged you anywhere from $30 to $90 to replace the fuse, so you are still way ahead.

Installing a New Fuse

To insert a new fuse, first double-check to make certain that the VCR is unplugged. Then simply push the new fuse into the metal clip-ends that grip it, making certain that the fuse is centered properly and gripped on both ends. If it is on a HITACHI-category VCR and had a plastic cover over the fuse, put the cover back over the fuse.

After making certain that nothing is touching the internal works or circuitry of the VCR, get someone to plug in the power cord while you at the same time keep your hands away, but watch the fuse(s) that you just installed. When a fuse blows, usually it flashes for a fraction of a second.

If the fuse does not blow, but the VCR still will not operate properly, you'll need to take it to a shop. Whatever caused the fuse to blow evidently involved other electrical damage too in the VCR.

What If the New Fuse Blows Too?

If the new fuse blows, you can try one more fuse of the same size to double-check if you want, but you are almost certainly just wasting a fuse now, because the indications are that some sort of short ciruit exists inside the VCR, a problem is beyond the scope of this book. If you install a second fuse and it also blows, stop at this point. DO NOT TRY PUTTING IN ANY MORE FUSES! AND NEVER, NEVER PUT IN ANYTHING TO BYPASS THE FUSE!!!

The fuse is not the part at fault, and you could damage an expensive part, or cause other additional damage, or even a fire, if you continue to put in more fuses and blow them. Each time this happens, the circuit is being strained. If this is the situation, you need to take the unit to a good VCR repair technician to get it fixed.

If the new fuse holds and does not blow, then return to the basic check-out and diagnostic procedure in Chapter 4. If the VCR passes all the other tests, then indications are that the VCR is OK and that either the fuse failed randomly, or it laid down its life protecting the rest of your VCR against a momentary surge of voltage from the power line.

2. VCR Powers Up, and Then Shuts Itself Down

Many technicians will agree that one of the most difficult problems to troubleshoot is the case where some lights or other indication of power come on when the VCR is plugged in, but the power light will not come on when the power button is pushed, or the power light comes on by itself and will not go off even when the power button is pushed off, or the power light comes on by itself, then switches off by itself and will not come back on again.

Sometimes such problems are a symptom of serious problems deep in "System Control" (the VCR's little computer, or brain). But often they are entirely the result of simple mechanical failures easily fixed once you know what to do. This section is organized by categories. Look up the category of your VCR in the list at the end of the Introduction, and then look under that name in the list below. If you do not find anything on your category in the list below, then I'm sorry, but I cannot help you. (As a last resort, in such a case, before taking it to a shop, you might try cleaning all the rubber belts and idler tires in the machine following the procedures in Chapter 3 — sometimes this will fix the trouble.)

FISHER:

On the Fisher-category models that have the tape loading motor mounted underneath, near the side edge of the reel table as shown in the illustration, this

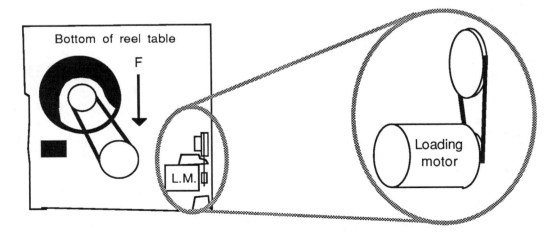

Location of Loading Motor in Many Fisher-Category VCR's

loading motor should move briefly the instant the power cord is plugged into the wall, even before any button is pushed.

If this motor does NOT run, one problem may be that the "chip" that is supposed to drive it (usually an "integrated circuit' with the number LB1649) is burned out, usually due to a worn out, slipping loading belt. Sometimes it destroys the little computer down with it. Replacing these "chips" is beyond the scope of this book.

But if this motor DOES run, but the VCR shuts down, or will not power up, or the red light by the power button goes out and will not come on again when you push the power putton, then you may have caught the problem in time, before it burned out the chip. First, double-check that the timer switch is not on. Then unplug the VCR and try replacing the little belt on this motor. The Fisher part number for this belt is #143-2-7504-00300 or #143-2-7504-00700 (same part), but you can substitute a Panasonic VDV0122 belt, which is usually easier to get and actually works better. It is not necessary to remove the motor to change the belt. Just pull the old belt off, and work the new belt onto the motor and pulley without removing the motor. A scribe or other long, thin tool is helpful here.

If a Fisher-category machine makes a squealing sound when you first plug it in, this usually indicates that this belt needs replacement.

These Fisher loading motors seldom go bad, but sometimes the slipping belt destroys the little plastic pulley, which can be replaced with one from Electrodynamics, part #08-4295 (Edc). The pulley is not available from Fisher, who makes you buy a whole motor (Fisher #4-5254-00250 or equivalently, #4-5254-00251) to get the pulley.

If a Fisher-category VCR will not power up, but the "Cassette In" light comes on with no cassette inside the VCR, unplug and check the position of the top switch (the "cassette-in" switch) above the big plastic gear on the right-hand side of the front-loader assembly. It may have slipped out of position so that the plastic part of the gear that is supposed to press against it when it is in the "fully ejected" position is not doing so correctly. For more on this, see Chapter 11.

If the power light switches on when you push the power button, but the "Stop - Cassette" light blinks with no tape in the VCR and the VCR will not accept a cassette, unplug, open up the cabinet top, and locate the dew sensor. (The "dew sensor" acts to prevent operation of the machine if the inside humidity is too high, since that might make the tape stick to the spinning heads.)

Trace the two little wires from the dew sensor back to the little plug that connects them to a circuit board. Unplug the little plug, and then plug it back in again. Now try again to power up. If this does not fix the problem, as a test, attach a clip-lead to short-circuit between the two little solder blobs on the dew sensor, or use the tip of a screwdriver with an

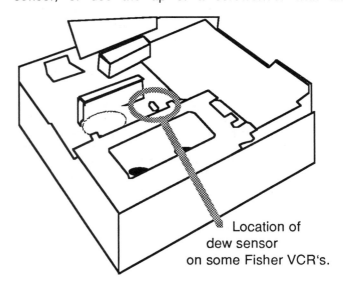

Location of
dew sensor
on some Fisher VCR's.

insulated handle, as shown in the illustration. If the red light stops blinking during the time you are short-circuiting across the contacts, the dew sensor is probably bad.

HITACHI:

If you have the symptom of "no functions," not even preliminary movement wehn powering up, check the switch that is supposed to be switched when a cassette has been fully loaded by the front-loading assembly. Remove the front-loader, disassemble it, realign if necessary, and reassemble it. See Chapter 11 and Appendix VII.

If a motor runs for a while when the VCR is plugged in, but the power light will not come on, try replacing the belt(s) by which the loading motor underneath the reel table drives the tape loading mechanism. (See Appendix VI for location and part numbers and Section 3 of Chapter 14 for how.)

If you have the symptom of "no functions," not even preliminary movement when powering up, check the switch that is supposed to be switched when a cassette has been fully loaded by the front-loading assembly. If the light comes on when you push "Power," stays on briefly, and then switches itself off, the problem may be that the front-loader mechanism is out of alignment so that the wrong sensor switches are being switched in the front-loader, causing the little computer to shut everything down. (This is especially true if it has the one particular model of front-loading assembly shown under Hitachi in Appendix VII.) Remove the front-

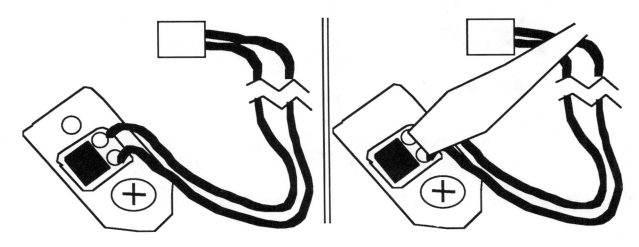

Testing for a Dew Sensor Problem on a Fisher VCR by Shorting Across the Two Contacts

loader, disassemble it, realign if necessary, and reassemble it. If there is a cassette stuck in the VCR, go to Chapter 7 (top-loader) or Chapter 8 (if it is a front-loader) and follow the removal procedures. See Chapter 11 and Appendix VII for additional information..

PANASONIC CATEGORY

Go to Chapter 6.

SHARP, EMERSON, SAMSUNG:

If a Sharp-category VCR with a NIDL-0005GEZZ idler, or a NIDL-0006GEZZ idler (see illustration above), or a model of Emerson or Samsung that also used this idler, will not eject, and/or will not power up, try removing the front-loading assembly (see chapters 8 and 10 for how) and clean or replace the idler (see Chapter 13).

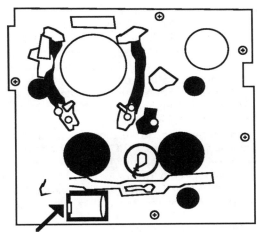

Top surface of the cylindrical housing of small motor sticking up slightly through a rectangular opening in the reel table.

Sharp-category VCR with Tape Loading Motor Underneath the Front of the Reel Table

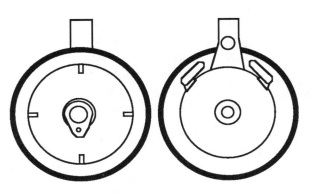

NIDL-0005GEZZ (Shp.)

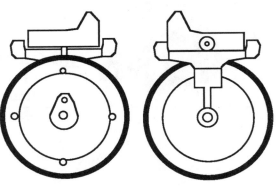

NIDL-0006GEZZ (Shp)

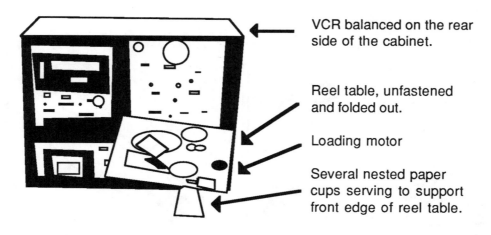

VCR balanced on the rear side of the cabinet.

Reel table, unfastened and folded out.

Loading motor

Several nested paper cups serving to support front edge of reel table.

One Type of Sharp-category Reel Table Folded Out to Get Access to the Loading Motor Underneath

Loading Motor Overshoot Problems

If the problem in a Sharp-category machine is that the power light comes on, but the VCR will not eject the cassette or perform any other functions, then check to see whether the top of the reel table looks as shown in the illustration (when the cassette-loading assembly is manually put into the ejected position or removed). If it does, then it is a common problem on these models for the motor that operates the tape loading and unloading mechanism to overshoot the point where it should have stopped after it unloaded the tape, and go all the way to the end of its excursion and jam there. (If the reel table on your Sharp-category VCR does not have a loading motor in this location, then skip down to the discussion of the Sharp reel table that has its loading motor located at the left rear corner of the top of the reel table.)

The solution is to gain access to the bottom side of the reel table, and manually turn the gears to get the mechanism back into an unjammed midway position, from which you should be able to start it up again, and live happily ever after. Sometimes the easiest way to gain access to the underside of the reel table is to open the bottom of the VCR; general instructions for doing this are in Appendix III. But since the front circuit boards often need to be removed on this model to open up the bottom, it is often easier simply to unplug the VCR and, once the cabinet front and front-loading assembly are removed, remove the few screws holding the metal chassis down to the plastic cabinet, and tip the reel table out carefully, as shown in the illustration with the paper cups. These screws have large heads, and their location on the most common model is indicated by the circled + symbols on the illustration.

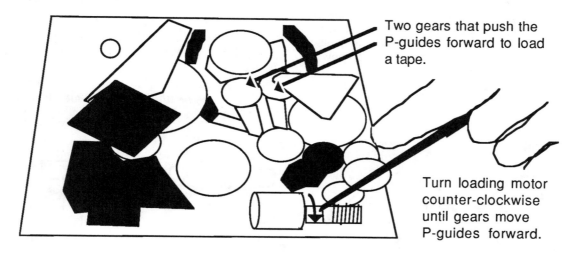

Two gears that push the P-guides forward to load a tape.

Turn loading motor counter-clockwise until gears move P-guides forward.

Manually Running Back a Sharp-category Tape Loading that Overran its Stopping Point and Got Stuck at the End of its Route

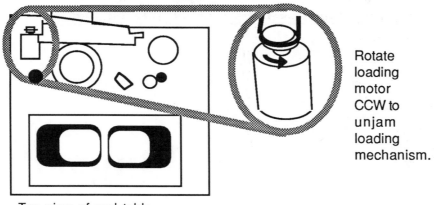

Rotate loading motor CCW to unjam loading mechanism.

Top view of reel table

Unjamming Sharp Loading Mechanism on Models with Loading Motor on Topside of Reel Table

Once you have gotten access to the bottom side of the reel table, find the tape loading motor, and use your scribe or fingers to rotate the motor, or the worm gear attached to it, in a counterclockwise direction. (Note that we are here moving the loading mechanism in the direction to *load* (since it overshot when unloading), and not to unload, as shown in Appendix VI.)

When you see the P-guides start moving along their tracks part way, you can stop. If you have a spray can of head cleaner available, it might be a good idea first to shoot some into the "mode switch" that failed to make contact, as it was supposed to do, when the mechanism reached the stop position.

This may help prevent a recurrence of this problem in the future. Put the thin tube that was supposed to come with the can into the hole for it in the nozzle, put the other end against the small access hole in the mode switch, and spray a little head cleaner into the switch. Put the reel table back into the cabinet, reassemble, plug in, and see whether it works now. If not, and you have also tried replacing (or cleaning) the idler, then you will need to take the machine to a shop.

If you have a Sharp-category VCR that has the loading motor on the top of the reel table at the left rear corner, as shown in the illustration, and the problem is that the clock comes on when you plug in, but it will not power up properly when you hit "Power," then you can check to see whether the problem can be corrected by turning the loading motor counter-clockwise (in the loading direction).

Unplug and turn the loading motor enough turns to cause the P-guides to start to move in their tracks in the loading direction. Then plug in the VCR and see what happens.

If the mechanism moves back to the stop position and operates normally after that, then your problem is solved. But if again it overshoots and thereafter will not power up properly, you will need to take it to a shop. The indications are that there is a problem in system control, perhaps a bad microprocessor.

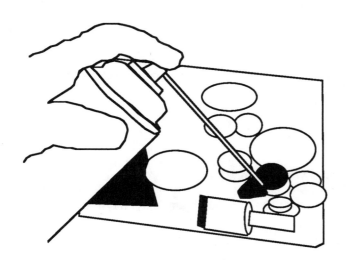

Spray Out a Sharp-category Mode Switch to Clean It

Chapter 6

When VCR is Plugged in, the Mechanism Moves Continuously Back and Forth Until It Shuts Itself Off

VCR's will sometimes exhibit strange behavior. Sometimes the power light will not come on when the power button is pressed, even though it is plugged in, and no motor will run when you push any of the buttons. But if it is unplugged, and then plugged in again, a motor can be heard moving the mechanism for a period of time, and then the VCR shuts off and the power light will not come on when the power button is pressed, even though it is still plugged in. When it is again unplugged, and then plugged in again, it may repeat the same behavior. Strange, isn't it? A motor runs for a little while when first plugged in, and then the machine will not do anything. But if it is left unplugged for some time, then plugged back in again, it may behave properly for a time, and then go back to the same confused behavior.

GOLDSTAR VCR'S:

Some VCR's made by Goldstar will cycle back and forth when a cassette is loaded if the cabinet top is off the machine and light is falling on it. To stop this, replace the top or shield the machine from the light that is affecting the end-sensors.

HITACHI VCR'S:

Front-loading VCR's in the Hitachi category (see Introduction for a list) sometimes get into a state in which the cassette loading motor starts to pull an *empty* cassette basket into the VCR, then stops, then runs the basket back out to the fully ejected position, then stops for a moment, and then repeats the same behavior.

This problem, which may be only intermittent, can be caused by a marginal, or bad, end-sensor, as explained in Chapter 11. If you are experiencing this problem with a Hitachi-built VCR, turn at this time to Chapter 11.

If the loading belts are slipping, this can also cause the machine to run briefly and then shut down. See Chapter 14 to find out how to replace the loading belts.

PANASONIC VCR'S:

If a motor seems to run spontaneously for a while when you plug in a Panasonic VCR (see Introduction for list), and then stop, and then thereafter the VCR will not power up, the problem may have the following story.

The little computer inside the VCR is programmed to check the mechanism of the VCR when power is first applied by running the tape loading motor to move the mechanism first in one direction, and then in the other direction, to make sure that it is free, and not jammed.

If when you push "Play," the VCR loads the tape, then unloads, then loads again (without your pushing the button), unloads, etc., again and again, perhaps making a clicking sound each time, the problem may also involve the mode switch, as follows.

Mode Switch

A slide switch is attached to this mechanism so that it moves along with the mechanism. This switch is called the "mode switch," and it tells the little computer what state the mechanism is in at any given moment. Different positions inside the switch precisely correspond to different positions of the mechanism to which it is attached.

As the mechanism goes from one physical position to another, different contacts are connected inside the mode switch, sending reports of the state of the mechanism to the little computer through wires. This is what is supposed to happen, when everything is working correctly.

Sometimes, however, this "mode switch" gets dirty, or corroded, or slightly out of position, and when this happens, you may see the bizarre behavior described earlier. The little computer sends out a signal to activate the tape loading motor to check to make certain that the mechanism can move properly, the motor moves the the mechanism, but even though the mechanism moves properly, the mispositioned

mode switch fails to send the correct confirming signals back to the little computer. Or in a variation on the same problem, when you push "Play," the little computer sends signals to activate the tape loading motor, which moves the mechanism to load the tape, but it never receives a report from the mode switch indicating the tape is fully loaded, so it reverses the motor to unload.

The little computer is programmed not to give up easily, so it moves the mechanism through its the cycle again, but again it does not get a "mechanism moved properly" confirming signal back from the mode switch. After trying several times, the little computer concludes that it does not know what is wrong with the mechanism, and shuts the whole system down so that if your priceless cassette with pictures of the kids is in the VCR, the little computer will not let anything happen that might damage it. At that point, you are supposed to take it to the shop to be repaired. But if this is the problem, you may be able to repair it yourself without the trouble and expense of taking it to the shop.

Remove the bottom cover, and fold out any circuit board that stands between you and a view of the bottom of the reel table. If it is one of the Panasonic-category machines for which the following procedure is relevant, what you see as you look at the bottom of the reel table will look like one of the following illustrations. If your VCR is like any of the designs pictured, the location of the mode switch is

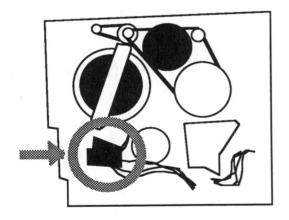

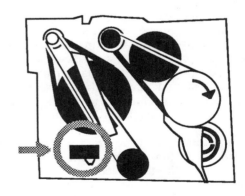

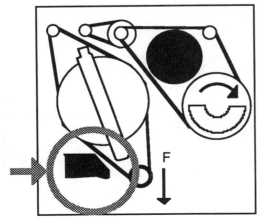

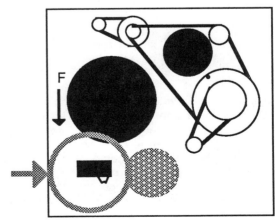

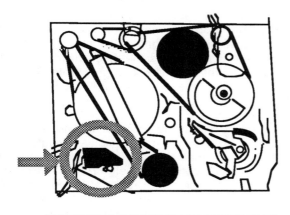

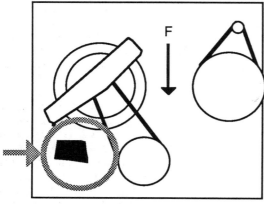

Mode Switch Locations on Common Panasonic-category Reel Tables

indicated by a circle-and-arrow.

First try spraying out the switch with head cleaner through a thin tube (See Appendix I). Make certain that the VCR is unplugged. Direct the spray into any cracks or openings you can see around the edges of the switch without moving the switch. On some models, you can use the pointed tip of a scribe or other thin tool to pry up the black plastic by the sliding white plastic part that is attached to the metal arm coming from the mechanism to make an opening about 1/32" (1 mm) to direct the spray into.

Turn the machine rightside-up to get the liquid to run all around inside the switch. Turn it upside-down again and with the unit still unplugged, use your finger to turn the big loading pulley several turns in the loading direction, and then back to the stop position. Do not force it, however! Do this several times, let the machine dry for twenty minutes, plug in, and see if the problem remains. If it still does, try adjusting the position of this switch, as explained in the next section.

If spraying out the switch fixes the problem, this means that dirty or corroded contacts inside the switch were the cause of the trouble. Technicians have noticed that this problem is most commonly seen in VCR's that are in houses where someone smokes. It may be that deposits settling out of the air from the cigarette smoke form on the metal contacts inside the switch causing the problem. Some technicians remove

the switch, get a new one, pry it open, smear a light grease all over the insides generously, pop it back together again, and reinstall — all to prevent more smoke from getting in and making trouble again.

Adjusting a Panasonic-category Mode Switch

Unplug the machine. Before loosening the screw and moving the switch, it is very important to mark the exact position of the switch prior to moving it, so that you can always get it back to the exact same location, if you accidentally move it too far, or in the wrong direction, and need to go back and restart the procedure. The best way to mark the position of the switch is with a scribe (or other sharp tool like a needle stuck in a cork, or a sharp penknife) to scratch a line in the metal of the reel table as close as possible alongside one (or both, if you can) ends of the switch.

It is generally impossible to scratch a line exactly at the edge of the switch. Usually one must settle for a line that is a slight distance — say, a distance equal to the thickness of the paper cover of a matchbook or playing card — away from the edge of the mode switch. After you have scratched the line as close as possible, look at it carefully and try to memorize exactly its distance from the edge of the mode switch. Fix it firmly in memory.

it still has the same behavior problem as before.

Which direction should you try moving the mode switch first? Usually this problem is solved by moving the switch to the left. (When I say "left" here, I mean to your left side as you view the upside-down VCR.) The distance that the switch needs to be moved is extremely tiny. In my experience, it has never needed to be moved more than the thickness of a playing card or matchbook cover. Be careful not to move it too much.

You may notice a little timing mark, like a small triangle, molded into the plastic case of the mode switch. When the VCR mechanism is in the "Stop" position, this mark is supposed to align with the position of the metal arm held in the little plastic slider along the side of the switch. Typically, it will appear lined up before moving the switch, and it will also appear lined up after moving the switch — so you can see what a small movement we are talking about — roughly, the thickness of a whisker.

The position of this switch is very critical, and if you move it too much, the VCR still will not work. Move it a tiny bit, plug in, and see whether it works. If not, move it a tiny bit more, and retest. If you cannot find any position moving it to the left that makes it work, bring the mode switch back to its original position as marked by the scratch you made earlier, and try moving it in to the right various small distances.

When you find a position in which the VCR starts up correctly, tighten the screw you loosened (not too tight, not too loose), and put a drop of fingernail polish on it like they did at the factory. If you cannot find any position in which it works, you will have to turn the job over to a shop.

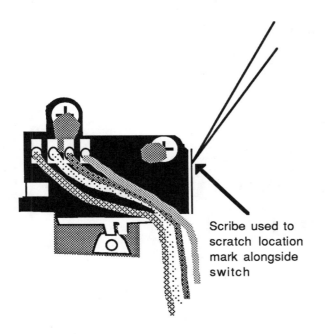

Scribe used to scratch location mark alongside switch

Making a Previous-Location Mark Alongside a Typical Panasonic Mode Switch Before Moving It

Next, loosen the hold-down screw. It goes through an elongated hole or slot in the plastic housing of the mode switch, and usually has a dab of colored paint on it (to keep it from vibrating loose). Loosen it a little, just enough to permit pushing the mode switch slightly to the right, or to the left, as you choose. We will adjust this switch patiently by the same "trial and error" procedure that VCR technicians often use. We will move it a tiny bit, a distance approximately equal to the thickness of a piece of typing paper, tighten the screw, balance the VCR on its end or put it back rightside up, plug in, and see whether it only moves the mechanism once like it should, or whether

Chapter 7

VCR Will not Eject Cassette
(Top-loader)

1. Preliminary Explanation of Terminology

If the VCR on which you are working is a front-loader (that is, if cassettes are inserted and removed through a hinged flap or door located on the front of the VCR), then this is the wrong chapter to be reading. Go to Chapter 8 for front-loaders.

This chapter, Chapter 7, contains special procedures for servicing failure of the eject mechanism in top-loaders only. ("Top-loaders" are VCR's in which the cassette basket lid is located on the top of the machine, and the cassette is supposed to be put into the basket and then manually pushed down into the VCR from the top.) Some of the same information will be repeated in Chapter 8, but many procedures are different for the two types of design.

If you have not already done so, before you do anything else, begin by unplugging the VCR for longer than ten minutes, then plugging it back in and trying again to eject the stuck tape. This sometimes solves the whole problem (for reasons explained in the "Introduction").

The problem of a cassette stuck in a top-loading VCR, or a failure of a top-loading VCR's eject mechanism, will be approached in a series of steps or stages. When a cassette is stuck inside a VCR, it first needs to be removed without damaging either the VCR or the tape itself.

In some cases, we must first finish rewinding back into the cassette a loop of loose tape that the VCR may have failed to reel back into the cassette box. Next, the cassette box with the tape inside must be ejected from the VCR without damaging either the tape or the machine. After this, the next major step will be to try to diagnose the cause of the failure of the eject mechanism and correct it.

Let us begin by quickly reviewing, first for VHS machines, and then for Beta machines, relevant important features of VCR operation previously mentioned in Chapter 2. If you already are familiar with the basic terminology and ideas in the following paragraphs, you can skip over them directly to Section 2 of this chapter.

PANASONIC-CATEGORY PORTABLES:

If the VCR is any of the Panasonic-category portables including Panasonic PV 5800D, PV 8000, Magnavox, Sylvania VC4512SL01, GE 1CVD5025X, and some Curtis Mathes, J C Penny, Quasar, Philco, and Canon portables that come in two parts and that have the model data and manufacturer's warning molded into the bottom of the plastic case, and if the problem is that the deck will not eject, turn directly to the procedure for this problem in Section 8.

Before Ejecting in VHS Machines

For a VHS machine to go from the Play or Record mode to the Stop or Eject mode, it first must go through a process called "unloading the tape." This process is a reversal of the process called "loading the tape" described in Chapter 2.

The two P-guides on sliding tracks that have held the tape wrapped part way around the spinning cylinder and video heads need to move on their tracks back to a position close to the cassette box. At the same time, the pinch roller should move back away from pressing against the capstan, and one reel (usually the supply-side reel, the one on the left), or both reels, should rotate a few turns to pull any loose tape out of the VCR mechanism and wind it back onto a spool inside the cassette box. (Some models follow this step with a final step of rotating the take-up side reel too, just to make certain that all loose tape has been wound back into the cassette box.)

Before Ejecting in Beta Machines

When a Beta VCR goes from the Play or Record mode to Eject, it also must go through its own process of unloading the tape. In a Beta machine, the entire "loading ring" should rotate in a direction opposite to the direction that it rotated when loading the tape, thereby unwrapping the tape from the cylinder containing the spinning video heads.

As this unloading process occurs, the VCR mechanism should cause the supply-side spool inside the cassette box to be rotated counter-clockwise a few turns to reel in the loop of tape pulled out of the cassette during the loading process.

Is the Tape Reeled Back?

Since this book is for everyone, I have no way of knowing whether or not the tape is still pulled out of the cassette box into your machine. If it is not pulled out of the cassette box into your machine, you can skip some of the following steps devoted to getting the tape safely pulled back into the cassette box before ejecting it, since in that case, you do not have a loose loop of tape hanging out into your VCR from the cassete box to worry about.

If the cassette basket that is stuck in your VCR is empty of any cassette, or if there is a cassette stuck in the VCR but it already has all loose tape pulled back inside it and no loose loop of tape hanging out, and if the loading mechanism in your VCR is already sitting in the fully unloaded position, then you can skip straight to Step Three of manually releasing the latch to eject the cassette basket.

But before explaining how to release the latch holding the cassette basket down inside the VCR, we must first take care of the folks who have the problem that a loop of tape is hanging out of the trapped cassette box into their VCR.

Clarification: Every normal video tape cassette has a hinged door along one edge, with a short length of tape exposed and pulled more-or-less tightly along the edge of the cassette box under the door. If you push the small plastic button on the cassette and lift the door, you will see this short stretch of exposed tape. This is not "a loose loop of tape" of the sort that concerns us here. We are NOT talking about tape that is pulled tightly against the side of the cassette box under its door. Instead, we are worried about situations in which ADDITIONAL tape has been pulled out of the cassette into the VCR, making a larger, somewhat loose, loop.

In a VHS machine, this problem situation might appear as shown in the next illustration:

In a Beta machine, the problem situation with a tape pulled part way out of the cassette might look like this:

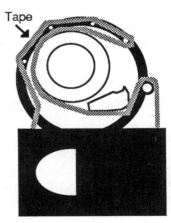

It is assumed that you have already opened the top cabinet of your VCR and observed the situation, so that you know whether or not you are in this situation. If you have not opened the cabinet, do so now. Appendix III provides help in getting off the top of the cabinet.

Problems Removing the Cabinet

If you find yourself in the predicament that you cannot remove the cabinet top without taking off the top lid above the cassette basket in the top loader, and this cannot be removed without ejecting the cassette basket, which your VCR refuses to do, then try and see whether you can get the bottom cover open and unload and reel in any loose loop of tape, and eject the cassette, working entirely from the bottom.

Exception: If the VCR is one of the Panasonic-category portables described in the lengthy section at the end of this chapter, then you may be able to get it to eject by powering up, catching a fingernail under the edge of the cassette basket lid, and lifting up at the same time that you push "Eject."

Avoiding Damage to the Tape

The problem: If a loop of tape is still hanging out of a cassette in your VCR, one problem is that with the cassette in the basket latched down inside the VCR, the hinged plastic door on the front of the cassette is being held open by a special "door opener" located on the reel table inside the machine. If you raise the cassette and basket without first winding the loose loop of tape back into the cassette, the cassette will move up from this door opener, the cassette door will shut too quickly for you to catch it, and when it shuts, it will crease or crimp the tape that is still hanging out of the cassette, damaging the tape irreversibly.

Thus, if the latch holding down the cassette basket were released to eject the cassette before completion of the final steps of the tape-unloading process, the cassette would be ejected with a loose loop of tape hanging out, a loop that would get pinched under the now-open, but then-closed, hinged door of the plastic cassette box, and possibly also caught in the mechanism of the VCR. This would damage the tape.

Avoiding Damage to the VCR

If the cassette basket were simply pried or forced out of the VCR without properly first releasing the catch mechanism latching it down in its stuck position, the cost of repairing the damage to the VCR that would result from such an improper removal would probably exceed the cost of replacing the entire VCR with a new one. After that abuse, the VCR would no longer be worth fixing.

I have seen VCR's come into the shop completely ruined by angry owners who, in their frustration, used improper techniques to force the VCR to eject a stuck tape that could have been removed easily with no damage to the machine if they had used proper removal techniques.

2. Preview of First Steps

Let me first preview the two things that must be done before ejecting the cassette, and then later you can do it as I explain how

Step (1): If the loading ring (in a Beta), or the P-guides (in a VHS machine), are not already back to their fully-unloaded position, the drive mechanism that is supposed to move them back must somehow be turned in order to put the entire mechanism back into the fully unloaded position.

Step (2): If all loose tape is not automatically wound back onto a spool in the cassette when you perform the first step of returning the loading mechanism to its fully-unloaded position, and tape is left hanging out of the cassette, then either (a) one of the spools inside the cassette must be caused to rotate to wind this loose tape back into the cassette, or else (b), if this is impossible for some reason, then the door to the cassette box must be jammed open somehow so that it does not close and damage the tape

when the latch is released ejecting the basket with cassette inside it.

In step (2), it is assumed that you want to avoid damaging the tape when removing the stuck cassette. If the tape is a worthless "junk" tape of no value, then step (2) can be omitted, but step (1) is still necessary to protect the mechanism of the VCR.

This gives you a general idea of what must be done, but I have not yet told you how exactly to do it. With this general idea of what is being done in mind, let us proceed to the specific details of how to do it. The following sections explain how to perform steps (1) and (2) just outlined. (If you can tell from looking inside the VCR, as explained previously, that these two first steps are unnecessary because the tape is already fully unloaded — which is the most common situation — then you can skip over Sections 3 and 4 and go directly to Section 5.

Section 3: Step One: Unloading the Tape-loading Mechanism

As explained in Chapter 2, when a cassette is inserted and a VCR starts to play it, the mechanism inside the VCR pulls a loop of tape out of the cassette and threads it through the tape path, past the various heads, etc. This process is called "loading the tape." The position during play of the little arms or pins that pull a loop of tape out of the cassette is called the "tape-loaded" or just the "loaded" position, for short. Before inserting a cassette, the VCR mechanism should normally be in the "fully-unloaded" position (unless a malfunction has occurred).

When the tape-loading mechanism is in the fully-unloaded position, it should normally be possible to activate the "cassette eject" mechanism, or to move the tape-loading mechanism to a point where the catch holding down the cassette basket will be unlatched. But when the tape-loading mechanism is not in the fully-unloaded position, but instead is in a partially, or fully, "loaded" position, then in most VCR's, the machinery is designed to prevent ejection of the tape, even if you press "Eject."

This design feature is present because, as explained previously, if you ejected a tape before it was fully unloaded from the mechanism, the cassette would be pushed out of the machine with a loose loop of tape still caught inside the VCR, ruining the tape.

So, normally the mechanism inside a VCR must be returned to its fully-unloaded position before the mechanical parts inside will permit the latch on the cassette basket to be released. If you were unaware of this, and forced the latch to release when the mechanism was still in a tape-loaded position, you could do extremely serious damage to the VCR as well as to the tape.

For this reason, before you can operate the eject mechanism, you must make certain that the mechanism is in its fully-unloaded position. If it is not in an unloaded position, you must move it to this position before trying to eject the tape.

In fact, this may be the whole solution to the basic cause of the problem. The tape-loading mechanism may be stuck in the "loaded" position, for some reason, and this may be why it will not eject. For example, a belt on the loading motor may have broken, or may be slipping. Or the little computer

inside the VCR may have gotten confused and lost track of the position of the mechanism that it is supposed to be controlling.

This can happen even when there is no tape inside the VCR! Thus, returning the mechanism to its fully unloaded position (if it is not there already) may be enough to cure the problem, or in doing this, you may discover the true original cause of the problem, like a broken or slipping belt, or a broken piece of plastic or a foreign object jammed in the loading mechanism, etc.

So proceed as follows. Unplug, and using the procedures in Appendix III, open up the top of the cabinet, if you have not already done so, and lift any circuit boards above the reel table. Using the general diagrams in Chapter 2, look at the tape path, tape, and position of the P-guides (VHS) or loading ring and posts (Beta) to determine whether the tape loading mechanism is already in the fully-unloaded position.

If it is already fully unloaded, then go to Step Two on the following pages to reel in loose tape. But if it is not fully unloaded, then you must return the mechanism by hand to its fully unloaded position, and also reel in any loose loop of tape. NOTE: This is NEVER done by pushing directly on the P-guides or loading ring! Instead, you must return the mechanism to its fully unloaded position in the same way that a little motor in the VCR should have done so, namely, by rotating a small pulley, or gear, connected to the tape-loading mechanism.

The problem now is to find the pulley or gear that does this job, and to figure out which direction to turn it. This mechanism is located in different places on different models. On some models, you can reach it from the top of the reel table, while on other models, reaching it will require also opening up the bottom of the VCR to reach the underside of the reel table.

You may be able to figure out where the "loading motor" pulley or gear is on your VCR simply by looking the whole mechanism over carefully, but further specific help is available in Appendix VI, which shows sketches of most common reel tables used in different models by major manufacturers, and in most cases, even shows the position and direction to rotate the loading motor gear or pulley to unload the tape-loading mechanism. Be sure to read the further instructions at the beginning of Appendix VI on how to use the diagrams given there.

If you find the loading-motor pulley or gear but can find no information on which direction to turn it to unload the tape-loading mechanism, then answer this question by trial and error. There are only two possible directions for it to go, clockwise and counterclockwise. Pick one of these directions at random, and very gently start turning the pulley or gear in that direction, while watching what the P-guides or loading ring does. Do this now. If you encounter resistance, or if they seem to move in the direction to load the tape, then stop, and try turning it in the other direction. If they start to move in the direction to return the tape to the cassette, continue gentle rotation until they have moved about 1 inch (25 mm) back from the V-stoppers, and then stop. We need to reel in the loose loop of tape that has resulted from what we have already done before proceeding further.

Section 4.
Step Two: Reel Loose Tape Out of the Tape Path and Back into the Cassette

On a few models, any loose loop of tape will be reeled back into the cassette automatically when you unload the tape-loading mechanism, as done in Step One, but on most VCR's, reeling in the loose tape requires a separate process.

On some VCR's, there is another pulley that you can turn to cause a spool to turn inside the cassette to pull tape back into the cassette as you unload the mechanism. On some of these, it is necessary to move a mechanical arm as well as turning the pulley to reel tape back into the cassette.

What to turn to reel in the loose tape may be shown with an arrow in the diagrams in Appendix VI, and the direction to turn it to reel in tape is the direction the arrow points. On some models, you can wind tape in by turning the pulley in either direction: one direction winds the tape onto the supply reel, and the other direction winds it onto the take-up reel. It does not matter, so long as the tape gets pulled back into the cassette somehow.

On other models, you need to attach a 9-volt battery to another motor, a "reel motor," to wind loose tape back into the cassette.

For many models, this information also is given in Appendix VI. Find out how to reel in the tape on your VCR, and do this now: reel it in until the slack is taken out

VCR's With a Separate Reel Motor

On some models, the tape is wound back onto the reels by a small electric "reel motor." In most cases, you can use a little 9-volt battery with two clip-lead wires (as explained in Section 19 of Appendix I) to operate this motor to reel the tape back into the cassette without plugging in the VCR.

On some models, you can attach the pair of wires from the battery to two exposed terminals on the motor either way, and it will wind the tape back in. On other models, a battery will reel tape in only when the wire from the positive (+) battery terminal is attached to a certain terminal on the motor, and the wire from the other battery terminal (-) is attached to the other terminal. Consult the appropriate diagram in Appendix VI to find out how to handle your model. If no information is available, try connecting the battery one way, and if that does not work, try reversing the connections. It will not harm the motor.

On some VCR's with a reel motor, the terminals to the motor are not accessible, because the wires reaching the motor are covered with insulation, and they disappear into the motor enclosure without ever presenting a bare metal end-point. In such a case, you can attach the clip leads from the battery to sharp pointed tools, like a scribe or a pick, and press these through the insulation of each wire, penetrating the insulation to reach the metal of the wire inside, and power up the motor this way. Or you can stick straight pins through the two wires, and attach clip leads to the pins to power up the motor.

If this fails, you can use instead the trick of jamming the cassette box door open with a stick to protect the loose tape hanging out, and then once the cassette basket is up in the ejected position with the cassette still inside it, reach underneath to turn the reels with your finger to wind the tape back in, as explained later.

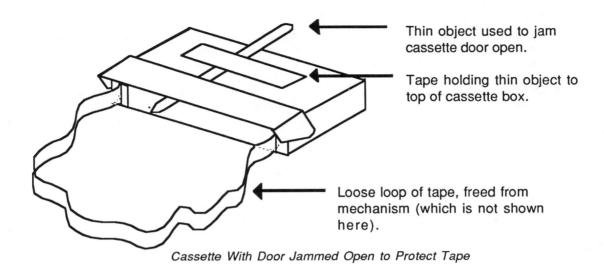

Thin object used to jam cassette door open.

Tape holding thin object to top of cassette box.

Loose loop of tape, freed from mechanism (which is not shown here).

Cassette With Door Jammed Open to Protect Tape

A "Last Resort" Trick

If the "idler" (a rubber wheel that communicates the rotation to the reels) is oily or worn down, or you cannot reel the tape back in for any other reason, you will need to resort to the trick of jamming the cassette door open with a stick, and then, once the cassette basket is ejected and raised with the cassette still inside it, reaching underneath to turn the reels with your finger to wind the tape back in. On a few models, because of their design, it is always necessary to resort to this trick.

Jamming Open the Cassette Door

Use this trick if you could not get the tape to reel back into the cassette, whether due to a malfunction in the mechanism or for any other reason, but you wish to remove the cassette without damaging the tape, and tape is still hanging out of the cassette into the VCR. Jam the cassette door open with a long thin object like a wooden popsicle stick, tongue depressor, fingernail file, a short thin ruler, etc., as shown in the illustration, except that the cassette will, of course, be inside the cassette basket in the top-loading assembly, which is omitted from the illustration for clarity.

Protecting the Tape

This way, the door will not close down on the tape and damage it in the next step when the cassette is ejected together with the cassette basket. If tape is wound through part of the internal mechanism, you will need to use your fingers to free it so that you can eject and reel it in without damaging it.

Go Back and Forth Between Step One and Step Two

Go back and forth between unloading the tape-loading mechanism a little, as described in Step One, then winding back in some loose tape, as just explained here, then unloading a little more as explained in the previous section, and so on.

Don't Go Too Far

Watch carefully and stop when the loading mechanism initially gets back too its fully-unloaded position, because it is possible on some models to go too far and overshoot, in which case the mechanism will still not eject, or on some Sharp-category VCR's, it may go into a position in which, when you try to power it up later, the little computer inside will not be able to figure out where the mechanism is, or what to do next, and so, will shut everything down.* On some Panasonic-category top-loaders, turning the loading pulley all the way to the "Stop" position, and then past this position, suddenly releases the catch holding down the cassette basket, so if you want to avoid damaging the tape, wind the tape back into the reels (Step 2) as you proceed. (On Panasonic-category top-loaders, a weak loading belt is a common cause of failure to eject.)

The first time you manually unload a VCR, probably the easiest and safest thing to do, once the loading posts get back near to the fully-unloaded position, is to try gently pushing the catch latching

---

* When the VCR by itself overshoots when unloading, as was covered in Chapter 5, this is usually due to a dirty or corroded leaf or mode switch.

down the cassette basket, to see whether it releases and lets the basket come up. If the latch does not release, turn the gear or pulley in the unloading direction a little farther, and try again to release the catch. Continue to repeat until the cassette basket ejects, or you encounter resistance, or it becomes clear that it is not going to eject.

Pictures of typical latch mechanisms, some with arrows showing what direction to push them to unlatch, are shown in Step Three later in this chapter. Look ahead, and then return to this point, if you like, but read and follow Step Two first, because another important step — reeling any loose tape back into the cassette — has not yet been covered.

If the Attempt Fails

If at any point it becomes clear that the following procedure is not going to work on your VCR to get it to eject, carefully reassemble everything. Put it back together in a series of steps that is the reverse of the sequence of steps in which you took it apart. Use your disassembly notes and numbered containers, as explained in Appendix III, to get each screw back into the same identical hole it came from.

But before you turn the problem over to a professional shop, there is one more little trick that sometimes works. After the VCR is all reassembled, plug it back into the power outlet, push the power button, press "Rewind," then "Fast-forward," to see whether the machine will wind or rewind the tape as you look through the window on top, or watch the tape counter. If the machine does turn the reels, push "Stop," and then, after a moment, hit "Eject." It is

possible that your efforts got the mechanism unstuck, and that now that the little computer inside is able to do so, it can take over and move the loading mechanism into the correct position to eject the tape.

If this happens, then it may be that the malfunction was a "one-time only" affair, or it may be that it was the effect of some problem that still needs to be fixed in the tape-loading-and-unloading mechanism.

If you observed any problem in the tape-loading-unloading mechanism while you were turning it — like a broken, loose, stretched, or oily belt — then you definitely should correct this problem before you put another cassette into it.

If you observed no such problem but helping the VCR complete its tape-unloading process made it possible for it to eject the tape, than you probably should at least remove and clean any rubber belt connected to the mechanism that you had to turn to unload the loading mechanism. Chapter 3 explains how to clean belts.

If you have the right type of grease for a VCR (see Appendix I), you can also use alcohol or freon spray to clean off the surfaces you saw move against each other when you unloaded the mechanism (like the tracks that the P-guides slide in), and relubricate them. For more on correcting problems in the tape-loading mechanism, see Chapter 14. It would also be a good idea to put a "junk" tape into the VCR and put it through its functions several times before putting a good tape into it. (If it is a VHS machine you can also use an "empty dummy cassette," as explained in Appendix I.)

Section 5. Step Three: Release the Latch Holding Down the Cassette Basket

Finally, the latch holding the cassette basket down in the VCR needs to be located and released, to unlock the cassette basket and permit it to rise up.

To do this on most top-loaders, you will need to remove the bottom cover, and the front of the cabinet (Appendix III explains how), and swing a bottom circuit board out on its hinges to gain access to the latch release.

The latches and procedures for releasing them on different typical VCR's are shown in illustrations.

"Direct Action" vs "Soft Touch" Releases

Cassette basket latches on top loaders can be divided into two categories.

If the latch is supposed to be released by a completely mechanical process when you push the "Eject" button — that is, if pushing the eject button moves a lever that is supposed to move another mechanical part, and that another, and so on, until the catch mechanism is moved and released — then it called a "hard touch" or "direct action" release system. In this case, no electrical circuitry is involved, and it is a straightforward mechanical problem to figure out how to release the latch. Basically, you need to slide or move it in the same way that pressing the "Eject" button is supposed to do.

If pushing the Eject button does not release the catch in a direct, purely mechanical way, but instead moves a switch connected to an electrical circuit, which is supposed to activate a motor or plunger ("solenoid") to move the latch-release mechanism, then it is called a "soft-touch" system. If the catch

is supposed to be released by the turning of a motor or the movement of a plunger when the "Eject" button is pushed, then to get the stuck tape out, you will need to rotate this motor, or push this plunger, in the same direction that the electric power is supposed to do when the machine is operating normally.

Whichever system is employed, we need to release the catch holding the cassette basket latched in the machine and let the basket (and cassette, if it contains one) rise up. If there is a problem with the eject spring(s) or other parts of the mechanism, it may be necessary to help the cassette basket rise up by pulling on it gently.

FISHER

Fisher-category Cassette Basket Latch Release by Pushing in Direction Shown (View shows the bottom of the reel table.)

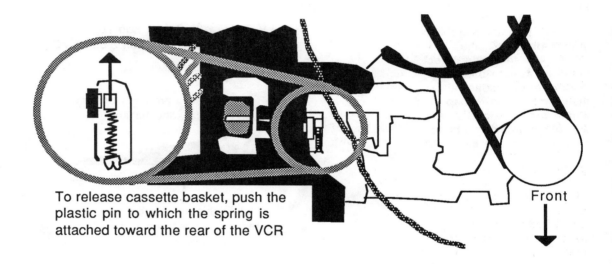

To release cassette basket, push the plastic pin to which the spring is attached toward the rear of the VCR

Front

HITACHI-CATEGORY TABLE MODEL

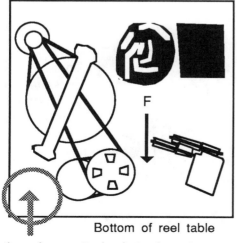

Bottom of reel table

Location of cassette basket release

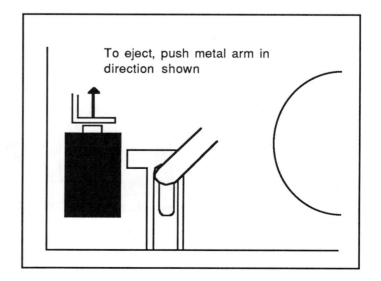

To eject, push metal arm in direction shown

J V C

On early J V C's, as shown in the illustration, the spring that is supposed to press the button that unlatches the cassette door on VHS cassettes sometimes gets tired, or pressed back, so that the cassette door remains closed as the cassette is pushed down into the machine, causing it to jam against the metal finger that is supposed to catch the edge of the descending door and raise it as the cassette goes down. The result can be a cassette jammed in the VCR. If this is the problem in an early J V C, the solution is to press "Eject," pull the cassette basket up, remove the cassette, and then remove and reshape the spring metal strip that is supposed to push the button on the side of the cassette.

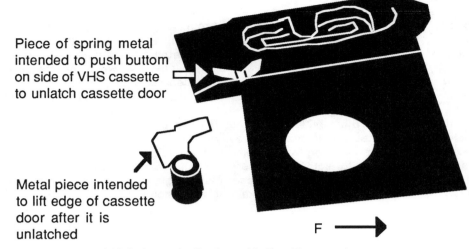

Piece of spring metal intended to push button on side of VHS cassette to unlatch cassette door

Metal piece intended to lift edge of cassette door after it is unlatched

F ➡

J V C Cassette Basket with Top Removed as Viewed from Left Side of VCR

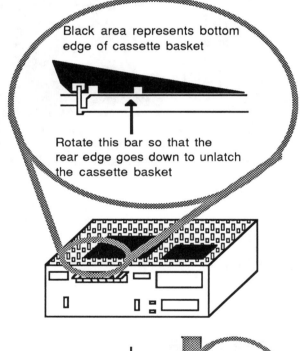

Black area represents bottom edge of cassette basket

Rotate this bar so that the rear edge goes down to unlatch the cassette basket

J V C

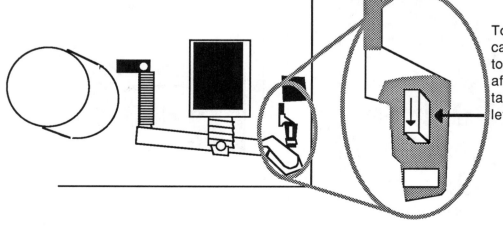

To eject cassette from top-loader after unloading tape, push this lever forward.

PANASONIC-CATEGORY (MATSUSHITA)

On the earliest Panasonic-category machines with the "tank" construction, the cassette basket is simply held down by fingers with rubber rollers that catch on "shelves" molded for this purpose in the plastic cassette lid. The top of the cabinet can be removed while the cassette basket is still down in the loaded position. Once the top is off the cabinet, look in the places indicated by circles on the picture of the exterior cabinet.

Find the rubber rollers and gently pull them back far enough to release the cassette basket.

Later Panasonic-category machines used two different types of "soft touch" or remote control cassette-basket release systems.

One reel table design used a release controlled by an electrical device called a "solenoid," which pulls a plunger when power is applied to it. To make this model release, push the plunger in the horizontal direction. It was mounted at the top front of the reel table in a position where it could be reached with a long thin tool, like a scribe, by reaching down in front of the cassette basket after removing the cabinet top. See Appendix VI for the location of this plunger.

Other Panasonic-category VCR's used the loading motor to release the cassette basket. As the loading motor unloads the tape, an arm slides a switch, called "the mode switch," in the direction shown by the arrow on the diagrams of Panasonic-category reel tables in Appendix VI. As the loading motor continues to turn, it takes this switch past the "Stop" position (where the triangular marks on the stationary and moving parts of the switch align), continuing on to your right (as you view the upside-down machine), until it goes all the way to the right. At this point, it moves a

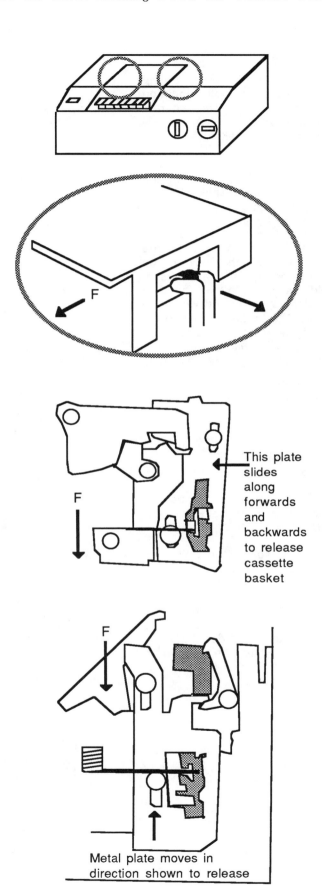

This plate slides along forwards and backwards to release cassette basket

Metal plate moves in direction shown to release

mechanism in the front corner of the underside of the reel table (location indicated by a gray circle on the reel table diagrams in Appendix VI) and this releases the cassette basket.

The actual latch release mechanism appears as shown in the pictures here. If the machine is not moving this mechanism, you can try doing so yourself by taking your finger and rotating the large loading pulley in the direction to unload. When the mode switch lever goes all the way in the direction indicated, the cassette basket latch should release. You will need to stand the machine on its end while you do this, so that the basket can come up. But do not force it. If you encounter resistance as you try to turn the pulley, stop, reassemble, and take it to a shop.

PANASONIC-CATEGORY PORTABLES

If the VCR is any of the Panasonic-category portables including Panasonic PV 5800D, PV 8000, Magnavox, Sylvania VC4512SL01, GE 1CVD5025X, and some Curtis Mathes, JC Penny, Quasar, Philco, and Canon portables that come in two parts and that have the model data and manufacturer's warning molded into the bottom of the plastic case, and if the problem is that the deck will not eject, turn directly to the procedure for this problem in Section 8.

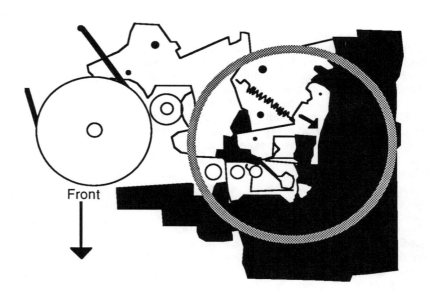

Panasonic-category Cassette Basket Latches Release by Pushing in Direction Shown

Section 6. Step Four: Manually Winding Any Remaining Loose Tape Back into a Jammed Cassette

If no loose loop of tape was left hanging out of the cassette, then pull the cassette out of the basket and proceed straight to Step Five.

But if the door of the cassette box needed to be jammed open with a stick in order to protect a loop of tape pulled out into the VCR because nothing would reel it in while the cassette basket was down (the "last resort" trick discussed in Step Two above), then proceed as follows. Before removing the cassette from the basket, reach under the metal tray on which the cassette box is resting to insert a finger into the opening under either of the spools and turn it. Rotate it slowly until the rest of the loose loop of tape is safely wound back onto one of the spools inside the cassette box. This way, the plastic door of the cassette box can be allowed to close and the cassette pulled out of the basket without damaging the tape.

7. Step Five: Finding the Cause and Curing It to Prevent Recurrence of the Problem

The next step is determining the cause of the original problem. Why did the catch holding the basket latched fail to release? Look carefully first for a foreign object (like a coin, screw, toy, or other small object) that may have fallen into the VCR and jammed the mechanism. If no foreign object is found jammed in the mechanism, then now that you know how to release the latch holding down the cassette basket, you can try to determine the cause of the malfunction by watching the mechanism latch again as you push down an empty cassette basket, and then observing what happens (or fails to happen) when you press "Eject."

On most VCR's, some of the latch-release mechanism is visible only from underneath the reel table, so it will be necessary to stand the VCR on its heavier end, power it up to operate without touching anything inside, and observe the in-and-out (or "up-and-down") action of the cassette basket and the operation of its latch and release mechanism.

There are countless different possible causes of a failure of the latch to release, or failure of the basket to spring up even when the latch is released, so no simple list is possible. But by repeatedly watching your VCR's mechanism continue to fail to eject, and figuring out how it is supposed to work and why it fails to do so, you should be able to find the probable cause of the problem and cure it.

If the latch is supposed to be released by a completely mechanical process when you push the Eject button — that is, if pushing the eject button moves a lever that is supposed to move another mechanical part, and that another, and so on, until the catch mechanism is moved and released — then it is a straightforward mechanical problem to figure out why the mechanism is not moving correctly, and how to cure the problem.

If pushing the Eject button does not release the catch in a direct, purely mechanical way, but instead moves a switch connected to an electrical circuit, which is supposed to activate a motor or plunger ("solenoid") to move the latch-release mechanism, then you need to watch and listen carefully while pushing the Eject button with the power "On" and the empty cassette basket in its down position to determine whether the motor or solenoid is moving or being activated at all. If the catch is supposed to be released electrically by the turning of a motor or the movement of a plunger when the "Eject" button is pushed, does this motor or plunger move, or at least try to move, when the Eject button is pushed? If so, what is preventing its action from unlatching the catch holding down the cassette basket?

If no movement or response at all is observed when the button is pushed, then probably the trouble is an electro-mechanical problem beyond the scope of this elementary manual, and you will need to bring the unit to a VCR shop for repair. However, if pushing the Eject button causes some response in the mechanism, but this response fails to cause the release of the basket latch due to some mechanical fault such as a slipping or broken belt, bent sheet metal on the basket or release mechanism, foreign object in the works, disconnected spring, or broken plastic part, etc., then you probably can fix it by correcting this mechanical problem.

Because the possible causes of a failure of the latch mechanism to release or of the basket to rise

are limitless, no simple single specific procedure can be given for fixing all of them. But with patience, it should be possible for you to perform the five steps to remove the tape cassette safely and diagnose and cure most mechanical causes of failure to eject. Additional information pertaining to typical problems or unique features of various specific models are detailed below, listed alphabetically by manufacturer.

8. Special Procedures

PANASONIC CATEGORY

Procedure for Replacing Cassette Basket in Panasonic-category Portables including: Panasonic PV 5800D, PV 8000, Magnavox, Sylvania VC4512SL01, GE 1CVD5025X, and some Curtis Mathes, J C Penny, Quasar, Philco, and Canon portables.

Symptom: The cassette will not eject from the deck part of a portable unit bearing one of the names listed above, or other Panasonic-category models that have some of the standard manufacturer's caution notice actually molded, or impressed, into the plastic on the bottom of the deck (as contrasted with being printed onto a sticker or plate glued to the unit).

The problem usually is that a part has broken off the cassette basket assembly, and the cassette basket will need to be repaired or entirely replaced. This can be done using the following procedure.   Note, however, that this repair procedure is of intermediate difficulty, somewhat harder than easier procedures presented elsewhere in this book.

Try Pushing Eject and Manually Raising the Lid

It will save some steps later if you can get the machine to raise the cassette basket to the ejected position now, before opening up the machine. Even though it will not eject, if the problem is a broken spring holder, as described below, it may be possible to get it to eject by the trick. Plug in the machine, power up, press "Eject," wait about a second, and then catch a fingernail under the lid of the cassette basket, and try gently to pull it up. If the motor that unlatches the catch holding down the basket is working, it may unlatch the catch so that you can manually pull the basket up, even though the spring is broken. If you do not succeed the first time, try again a few times. If you cannot get a grip on the cassette basket lid, remove the lid as described below, and grip one of the parts of the basket beneath the lid, and try

to lift it after pushing the eject button.

If you still cannot get the basket to come up, stop trying and use one of the tricks described later in this section once the machine is disassembled.

If you do manage to eject the basket, take care in the following steps that it not get latched down again, because after this point, you will not have power to activate the motor that releases the catch, and it is more complicated to release it without power

Disassembly

Unplug machine.

Remove the battery if there is one in the unit.

Remove the two cassette lid screws. They are probably located under two small rectangular rubber covers that you will need first to pry up with a small screwdriver.   Put screws and rubber pieces into a numbered container, and record the container number (#1) and screw location in your disassembly notes. You should be able to remove the cassette lid even if you cannot eject the basket. After removing the two screws, slide the lid forward and up slightly to remove it.

Remove False Cabinet Back

Turn the VCR upside down and remove the two rear-most screws at the back of the bottom, which are actually holding on the false back that encloses the battery. Put in numbered container (#2). Rotate the false back up and remove from VCR.

Remove True Cabinet Back and Top

Turn VCR right-side up again. Look at the back, and find the protruding plastic and metal part to which the battery was attached. It is held on by two screws

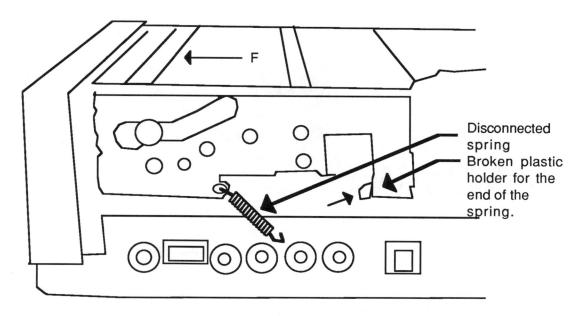

*Broken Plastic End-holder for Cassette Basket-up
Spring on a Panasonic-category Portable Deck*

deeply recessed into two of the four holes that you can see molded into the plastic. Remove these two screws and put into another numbered container (#3). Pull this plastic part back and to the side a slight distance (as far as the wires will permit), so that it will be out of your way.

Remove the two screws holding on the true back and top of the cabinet, which are, of course, molded in one piece. These are the two screws located along the lower edge of the back. Do NOT remove the single screw located by the battery eject button. Put in numbered container (#4)

If there is one screw in each of the sides up near the front, actually going through the side of the front cabinet plastic piece, remove these two screws and put in numbered container (#5).

Remove cabinet top by rotating it up from the back and pulling it away.

Check for Broken Spring Holder

Look at the right side of the cassette basket assembly (right side, as viewed from the front of the VCR). A black spring about 1 1/4 inch (30mm) long is supposed to hook into the little white plastic piece. If the plastic piece is broken, you have found the problem. (If it is not broken, then you apparently have a different problem beyond the scope of this book; you will need to reassemble the VCR and take it to a shop.)

Repair Options

You will need either to repair the broken plastic part, or to replace the entire cassette basket assembly, which involves removing it from the VCR, of course.

This part is very difficult to repair, so most technicians will just replace the cassette basket assembly, but sometimes it is possible to repair it, if you can find a way to drill or melt a small hole through the part of the broken plastic arm that is still attached to the basket assembly — a hole that is big enough that you can hook the free end of the dangling spring through it.

This trick cannot be tried if the plastic arm broke off too close to the metal rivet to which its rearmost end is attached, because the offset molded in the plastic part is necessary to keep the spring from hitting against the side of the basket assembly. But if the plastic part broke off down near the end with the hole to which the spring attaches, you have nothing to loose, and everything to gain, by trying first to solve the problem by drilling or melting a new hole through the end of the broken plastic piece. A soldering iron with a tiny pointed tip, or a very small drill, are suitable tools for making the hole.

If you are going to try this approach, after you drill or melt a hole through the little plastic part, it would be a good idea to reinforce it, because there will now be even more force on it than before, because the spring will be stretched even more. You

Epoxy blob applied to delicate plastic arm to reinforce it against breaking again in the future.

Reinforcing the Plastic Piece that Hooks to the Spring

can adapt the procedure for reinforcement given later in this section. After reinforcing it, reassemble and pray.

Whether you try to repair the basket, or replace it, you should look around and see whether you can find and remove the little piece of plastic that broke off. We would not want it to get caught in some other part of the mechanism and jam it. Sometimes, however, these pieces just never turn up. I don't know where they disappear to.

If you plan to replace the cassette basket assembly, you should order up a new one (Panasonic or Matsushita part #VXAS0594) and have it with you BEFORE you begin further disassembly of the unit. The same Matsushita (Panasonic) part will fit the identical portables sold under the other brand names listed earlier.

It used to be that when you ordered up this part, they sent the entire assembly, including the sensors attached to the sides, but more recently, they have been sending only the subpart of the whole assembly that has the spring and plastic holder. This is probably what you will receive, unless you are shipped something out of old stock. The procedure you follow is slightly different if you have the complete assembly than it is if you have only the partial subassembly to replace. This will be explained in more detail later.

Reinforcing the New Part

Whichever you receive, when you have a new cassette basket assembly in hand, it would be wise to "customize" or modify it in a way that will reduce the likelihood that the same part will break in the same way on the new one — for the truth is that the

Japanese engineers who designed this unit apparently did not realize that many American users like to slam the cassette down, and the plastic part they used to attach this big spring with all its tension is too small to last very long.

Let's beef it up by adding some extra substance to the delicate plastic part in the form of a reinforcement made out of epoxy attached to it.

Proceed as follows: On the new basket assembly, once you have got it, find and clean with degreaser the new plastic part corresponding to the part that broke on the old one, then spray the cleaned part free of residues with some freon from your spray can of head cleaner, if you've got one. Next, mix up some Duro epoxy of the type described in Section 15 of Appendix I, and apply a reinforcing layer around the end that broke, running up to the other end of the plastic piece, as shown in the illustration.

While the epoxy is drying, set the assembly aside, and begin removing the old one from the VCR.

Removing the Bottom Cover

Remove the four or five screws on the bottom cover. Notice where the one short screw goes. Put in numbered container (#6). Work loose the bottom cover and the foil sheet along the back, which is attached to it. Avoid touching the bottom circuit board. Handle the machine by the plastic and metal edges of the chassis.

Remove the Front Cabinet

Remove the plastic cabinet front by gripping the top edge with fingernails and pulling up gently.

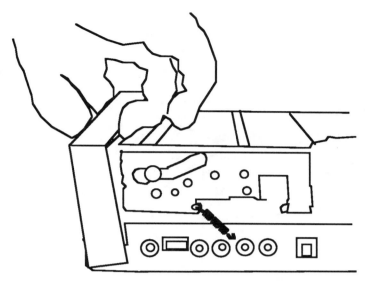

Removing Front Cabinet on Panasonic-category Portable Deck

On some models, there will be one or more multi-wired connectors that need to be disconnected.

Disconnecting the Wires

The ribbon-connector can be disconnected by first drawing a line horizontally across it with a felt-tip pen at the point where it disappears into the black plastic connector attached to the front circuit board, and then gently prying up the plastic piece at the top of the larger plastic piece that is attached to the front circuit board, and into which it disappears. You will be glad later that you first drew this line to show you how deep to insert the ribbon connector when reassembling the unit.

The top part of the plastic piece on the front circuit board into which the ribbon connector goes does not come all the way loose; it only raises up about 1/16 inch (2 mm) and stops, but when it reaches this point, the ribbon is freed and can be pulled out.

The connector(s) with separate wires going into a plastic piece can, in contrast, be freed by pulling on all the wires together, while wiggling it from side to side, until it comes free.

Getting the Cassette Basket into the Up Position

If you were able to get the cassette basket up earlier before you opened up the cabinet, then you can skip over the following procedures and go directly to the step of "Removing the Cassette Basket Assembly." If you were unable to get the cassette basket up before opening the cabinet, then proceed as follows.

Short Cut Method

You may possibly be able to get the cassette basket up by pressing on a little metal piece that is directly connected to the catch release. This piece is a little brass-colored part visible through a tiny hole in the bottom circuit board.

Set the machine on one side and reach into this hole with a scribe or other sharp pointed tool, move the little brass piece, and pull up gently on the cassette basket.

If the basket comes up, skip on down directly to the section under the heading "Removing the Cassette Basket Assembly." If the basket still does not release, proceed as follows:

Ejecting by Turning the Loading Gear

If the preceding procedures failed, you may need to cause the VCR to eject by manually turning a plastic gear that is buried below, inside layers of mechanism and circuit boards.

On many models, an access hole is provided in the bottom circuit board for this purpose. Let's begin by checking to see whether your VCR is one of these models. Gripping the VCR by the mechanical parts on its sides, turn it over, upside-down. Do you see a bottom circuit board with several holes, one of which looks like this:

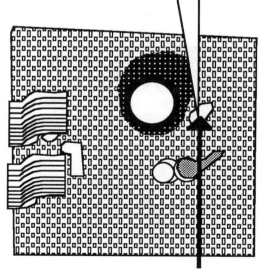

Insert pointed tool here to rotate white plastic loading gear.

Bottom Circuit Board Access Hole

If so, you are in luck. Under this hole, you can see part of a white plastic gear. Catching the point of

Disconnecting the Multi-wired Connector from the Right Side of the VCR

your scribe against the side of this gear, rotate it in a counter-clockwise direction (CCW when viewed from the bottom side that you are now looking at it from). Set the VCR on its side while you do this. Rotate the white plastic gear in a CCW direction until the latch holding down the cassette basket releases. The basket will not "pop" up, because the piece holding the spring that raises it is broken, but it should come free so that you can gently pull the basket up. This gear will turn very stiffly and there are not many raised spokes or ridges on it to push against, so you will need to stick the sharp point of the scribe so that it catches in the flat part of the white plastic gear and push firmly with control.

If your VCR has this access hole, and you get the basket up by turning the white gear, skip the next paragraphs, and go directly to the subsection entitled "Removing the Cassette Loading Assembly."

If your VCR does not have this access hole, then you will need to unfasten and lift the bottom circuit board just enough to be able to reach this gear from

the side with a scribe and rotate it. The next subsection explains how to do this. Skip the next subsection if you have already gotten the cassette basket up.

Releasing the Catch Holding the Basket through a Side Entrance

Unplug 9-wire multiple wire connector at the middle of the top of the right side and disconnect the two-wire connector beside it.

If the bottom circuit board is held in place by locking tabs on the back of the circuit board,free the two at the back by gently pinching them toward each other with one hand, while lifting up on the corner of the board with your other finger, and free two more locking tabs at the front of the bottom circuit board.

Unplug the small two-wire connector at the bottom of the front circuit board, if there is one.

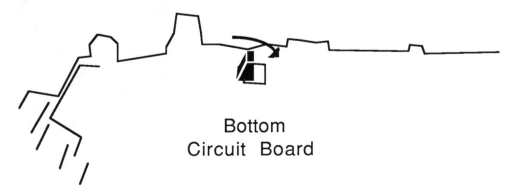

**Bottom
Circuit Board**

Unfastening a Plastic Tab Holding a Circuit Board

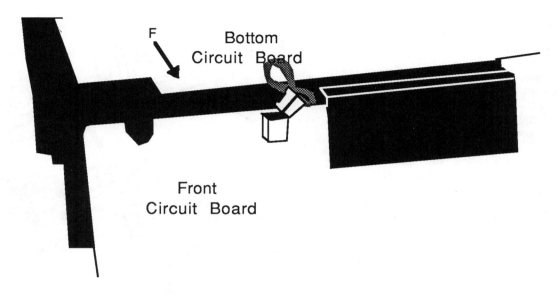

Unplug Connector at Bottom of Front Circuit Board

If the bottom circuit board is held down by four screws, remove them.

Free up the two gray wires nestled into the right side of the mechanism.

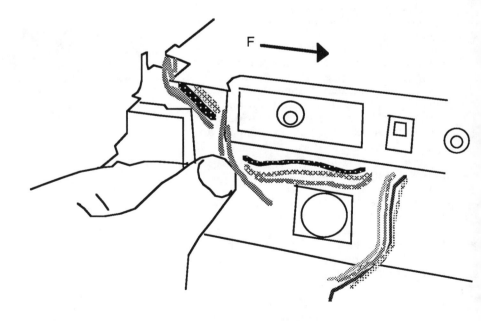

Freeing Up Wires So That Bottom Circuit Board Can Be Raised Slightly

Lift up the bottom circuit board only about two inches (50 mm). Prop it open with a paper container. Be careful not to yank the wide ribbon connector still attached at the front.

Balance the VCR on its left edge or on its end. Look inside the space you have just opened up. You should see a white plastic gear about two inches (50 mm) in diameter, with grooves molded into it. You must rotate this counter-clockwise until the VCR finishes unloading the tape (if it has not already done so) and releases the latch holding the cassette basket down. Push gently but firmly, rotating the white plastic gear CCW until the metal pin comes to the end of the track in the gear. The cassette basket will probably need a little gentle pull to open when the end of the track is reached.

Close up the opening you just made by now reversing the steps above in this subsection.

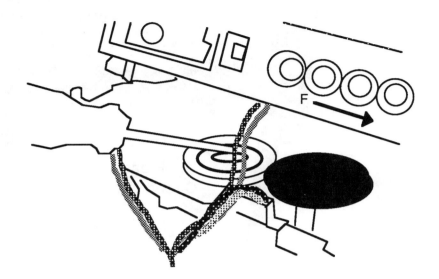

Reaching Inside the Portable to Turn theInternal Cam Gear to Unload & Eject Stuck Cassette

Removing the Cassette Basket Assembly

Set the machine gently back down in the rightside-up position, resting on the bottom circuit board.

Plastic parts with locking tabs hold the front circuit board in place, one in the center, and one at each edge. Pull these three tabs back slightly to free the circuit board and gently swing it down on its two plastic hinges at the bottom. The hinges have a catch in them that will lock when it is folded down about 30 degrees. Rotate the board on past these catches, and these locks will audibly unsnap and let the board come down to almost a 90 degree angle, which is far enough to permit the next step.

Remove two screws at the side of the front plastic piece — that is, the plastic piece behind where the cabinet front was before you removed it. In doing so, be careful not to damage the front circuit board.

Do not remove the screws from the circuit board. Remove only the screws that you can see thread into the aluminum (when you look from the other side, where the cassette basket was before being raised). Put the screws in a numbered paper container (#7).

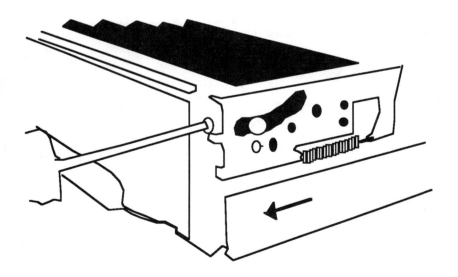

Removing Two Front Screws Threaded Into Cassette Basket Assembly

Remove the single screw, #8 in the diagram, holding the horizontal brace bracket that runs between the back of the cassette basket and the video heads. Be extremely careful not to drop it into the VCR mechanism — a magnetic screwdriver is helpful here. Put screw in separate numbered container.

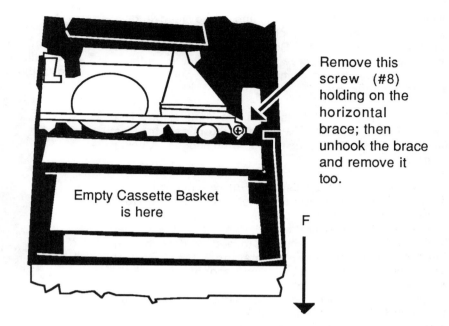

Remove this screw (#8) holding on the horizontal brace; then unhook the brace and remove it too.

Empty Cassette Basket is here

F

Removing Screw Holding on Horizontal Brace

Carefully study how this bracket catches in a little slot on the left end, and how the sheet metal attached to it fits over the parts to the rear of it to the right of the middle. Use a felt-tip pen to write a letter "F" on it with an arrow pointing toward the front of the VCR. When you are sure that you remember how to put it back again, remove the bracket.

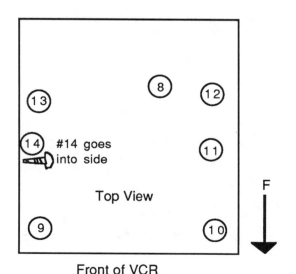

Top View

Front of VCR

Schematic Diagram and Numbering of Remaining Screws to be Removed

Push the cassette basket down slightly — but not enough for it to latch again! — and reach through the holes in the black sheet metal part of the cassette basket assembly with a number 1 Phillips screwdriver to remove the two screws holding the bottom of the assembly down to the reel table. Push DOWN with all your might on each screw BEFORE turning it and WHILE starting to turn it, so that you will not strip the + slot in the head of the screw. These screws are usually screwed in extremely tightly, (the robot who does it must eat Wheaties for breakfast) and it is easy to strip the slot, making a gigantic problem. Lean on it hard, as you start to turn it, with maybe 35lbs (15 kg.) of push. Once the screw has started, you do not need to continue using such strong downward force. These are screws #9 and #10 in the illustration. If you have forceps or tweezers, or a magnetic screwdriver, use them to remove the screws. Be extremely careful not to drop these screws down into the mechanism below. Put them in numbered containers as you remove them.

Do the same with the third screw, #11 in the illustration, located back farther along the right-hand side. Put in a numbered container.

Remove the screw from the plastic piece just to the rear and a little higher up than the screw #11 that you just removed. This is screw #12 in the illustration. Put in separate numbered container.

To prevent loss of the next screw, before loosening it, put tissue paper down in the curved track for the left P-guide and in the hole at the left side of the cylinder. Now remove the screw holding down the bottom left part of the cassette basket assembly. It is down low, at the level of the reel table, going directly into metal. Be extremely careful not to let it drop down into the mechanism below. It is #13 in the illustration, located just behind the erase head and in front of a medium sized molded white plastic part back at the left rear corner. Bear down hard on it when first starting to unscrew it. Remove it with a magnetic screwdriver or tweezers, and put it into a separate container.

There is an end-sensor on the right side. The wires from it go to a plug-in terminal that is accessible enough to enable it to be unplugged. Unplug it now.

What you do next depends on whether your new part is the partial subassembly or the complete assembly, as described earlier. If it does not have any parts with wires attached, then it is the partial subassembly, and you should proceed as described in the next section, "Replacing the Partial Subassembly." If it has end-sensors with little wires attached to the right and left ends, then it is the complete assembly, and you should proceed as described in the longer following section entitled "Replacing the Complete Assembly."

Replacing the Partial Subassembly

The cassette top loading assembly should now be completely detached from the chassis except for wires from parts on the left side. If you are careful, you can detach the subassembly with the broken part, and replace it with the new subassembly, without detaching these wires. This means working in cramped quarters, but it is quicker than detaching the wires and reattaching them. If you decide that you want to get the complete assembly totally separated from the machine to work on it, then go to the section below entitled "Replacing the Complete Assembly," follow its procedure for detaching the rest of the wires, and then return to this point to work on replacing the subassembly.

Compare the new subassembly with the cassette-up unit that you just detached. Do you see what part of the complete assembly is the same as the new partial subassembly. Now look at how it is held on with several C-rings. To replace the subassembly, mark numbers by each C-ring with a felt-tip pen. Use the procedures in Section 9 of Appendix I to remove each C-ring and place it in a container with the same number. Be sure to cover each with a cloth prior to pulling it free, because they do like to fly away and get lost!

Remove the old subassembly, and put the C-rings back at the same corresponding locations, again using the protective cloth.

Go next to the section entitled, "Reinstalling the Cassette-up Assembly."

Replacing the Complete Assembly

Find the small circuit board and black plastic piece with two wires coming to it attached to the left side of the cassette basket.

Remove this screw

You MUST use a tiny Phillips screwdriver of the size smaller than the #1 Phillips for this screw, a screwdriver such as the Xcelite XST-100. It is the size #0 Phillips. Put tissue paper under the screw before you remove it, to prevent it from falling down into the mechanism. It helps to set the unit on its left side so that you can easily bear down hard on the little screw as you start to unscrew it. Believe it or not, removing this part, difficult as it may be, is easier than trying to unplug the wires at the other end, and safer too. Put this tiny screw into a separate paper container (#14). Let the part hang from its wires.

Free the wires from this part that are threaded through and around the left side of the cassette basket, disappearing to some destination down below. Memorize their position as you do so, because you are going to have to restring them when you install the

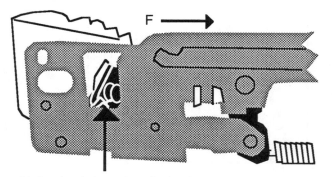

Little circuit board and plastic sensor housing from right side being pushed through opening in side of cassette basket assembly

Pushing Old Sensor Circuit Board Free Through Opening in Side of Cassette Basket

new cassette basket assembly. One set of wires comes from the part that you just detached when you removed the tiny screw, an "end-sensor." The other set of wires comes from the erase head. Push the plastic housing containing the end-sensor through the hole in the white aluminum metal side of the cassette basket assembly.

The trick to getting the black plastic end-sensor enclosure through the hole in the aluminum is to turn the sensor enclosure around so that the round protrusion is facing the corner of the L-shaped hole and the edge with the little plastic guide pin is oriented toward the top of the unit. Use the same trick to put it back through this hole in the new cassette basket assembly.

Push the white plastic block at the corner a little out of the way to get it through. Disconnect and free the wires from where they are caught on, or attached to, the aluminum and the white plastic block. Leave the wires from this left-side end-sensor attached. We are going to install it in the new basket assembly in a minute.

The complete cassette basket assembly should now come free.

A new complete cassette basket assembly comes with end-sensors attached on both sides. Leave the take-up (right) side end-sensor attached but remove the supply-side end-sensor to replace it with the old one whose wires were left still attached for convenience to your VCR. Remove the left end-sensor now from the new basket. Keep it in case you ever need it.

Feed the supply-side end-sensor through the hole in the basket assembly like the hole you just pulled it through in the broken assembly. It helps to work the edge with the little mounting pin through first, orienting the pin toward the corner of the L-shaped hole, then turn the plastic part so that the larger round protrusion faces the corner of the hole and gently force it through.

Reinstalling the Cassette-Up Assembly

This would be a convenient time, by the way, to clean the heads and tape path, before installing the new cassette basket assembly.

Put the new assembly down on the reel table in correct postion. Note that the metal bracket of the assembly goes UNDER the little plastic part from which you removed screw #12 in the illustration, and remember that the wires from the take-up side end-

sensor must extend out the right edge, down and around to the little two-wired plug into which you should plug them. Check the screw-holes on both sides to make certain that you have the assembly properly aligned, before starting to replace the screws.

Attach the little black plastic supply-side end-sensor enclosure back into its proper place, with the mounting pin through one hole, and the protrusion pointing into the VCR from the notch for it in the edge of the aluminum. Put down tissue paper again, and put in the tiny screw you removed as #14. (Save the other little screw from the new basket in case you need it for something.) Rest unit on its left side, and screw down the tiny screw in its proper place.

Put VCR rightside-up again, and plug in the little two-wired connector from the take-up side end-sensor, if you did not do so earlier.

Go back in reverse order through the procedure you just followed for disassembly, replacing each screw in its proper place.

After putting in the two screws on the front, rotate the front circuit board back into place on its hinges, and fasten it with the three locking tabs.

Plug in the multi-wired connector at the side.

If you had to disconnect multi-wire connectors to get in to turn the white cam gear earlier, reconnect these, and fold the two gray wires back into the space from which you removed them. Gently fasten the bottom board down with the four plastic snaps.

The front cabinet piece to which the other end of the ribbon connector is attached, and the cabinet bottom, must be reattached together. First, reattach the cabinet bottom, remembering to put the longer screws through the holes in the plastic, and not into the little metal parts exposed at the rear. Get the attached foil in proper place over the back circuit board before you insert the screws. Remember that the bundle of wires to the battery connector, etc., comes out along the bottom of the back near the corner, and should fit through a narrow gap in the space between the foil and the bottom to which it is attached.

Reattach the front cabinet ribbon connector to the top of the front circuit board by pushing it down into the slot in the top of the plastic part that you raised earlier to disconnect it. Make certain that you are feeding the ribbon into the slot in the top, and not down into the space behind it by mistake. Push it down to the level of the line you drew on it with a felt-tip pen earlier, then while holding it at this depth, push down both ends of the black plastic piece. Test by pulling gently on the ribbon cable, to make certain that it is firmly clamped inside the connector. If not, pull up the top piece of the connector a short distance again, and try again.

Connect the multi-wire connector (or connectors) to the edge of the front circuit board. Make certain that all multi-wire connectors are plugged in all the way, or the unit will not work when you finish assembling it.

Now we need to reattach the front cabinet piece that is attached to the other end of the ribbon cable.

There are little rectangular notches molded in the bottom of the front cabinet piece. Turn the machine upside-down and align these with the little protruding plastic hooks or "hinges" molded to the cabinet bottom, and then, having hooked these, swing or rotate the front cabinet piece so that its top approaches the plastic grippers protruding from the black plastic that is behind the front circuit board. With the plastic catches on the front of the bottom cover put in the places for them at the bottom of the front cabinet piece, swing the front cover into place.

Be very careful, here, because it is easy to break one of the parts at the bottom edge of the front. Take the time to make certain that all the hooks are hooked correctly, and that the other protruding parts and tabs along the bottom edge of the cabinet front piece and front edge of the bottom are correctly aligned, before swinging the cabinet front up to latch at the top. It should swing very freely; if you feel any resistance, back off and realign before trying again. It usually takes several attempts before you get everything properly aligned.

The cabinet front is held up by little protrusions inside the top edge gripping on catches molded in the plastic above the front circuit board along the top edge, so it will still be loose at this point in the

reassembly.

Make certain that all wires are pushed inside out of the way. Work the top cover back into place. Catch the end latches first, then lift up the top edge of the front cabinet with a small flat screwdriver

Using a Small Screwdriver to Get Front Tab on Top Cabinet under Front Cabinet Back Edge

and push the center tab under the edge. Wrestle the right side of the cabinet top together with the plastic molding around the RCA jacks along the right side of the bottom cover. Make certain that the tab on the top piece goes underneath the edge of the bottom piece. Reattach the rest of the cabinet screws, going back throught the earlier disassembly procedure in reverse order, and taking care to get the battery connector reassembled correctly, making certain that the little two wire connector has not come loose at the lower corner of the back.

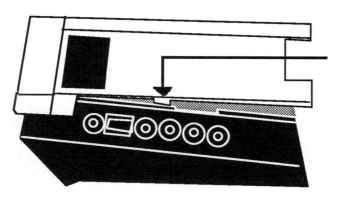

Note tabs on side of top cabinet. They go inside top edge of bottom cabinet

Getting Side Tabs Into the Top Edge of the Cabinet Bottom

Important: Use the check-out procedure
in Chapter 4 before using this chapter.

Chapter 8

VCR Will Not Eject Cassette (Front Loader)

1. Preliminary Notes

Do not try to rotate the motor or move the gears or other parts of the cassette loading mechanism until you have read farther in this procedure. Wait because moving it may interfere with one important diagnostic test that we need to make, and also may damage the tape or mechanism in some circumstances.

NOTE: If the VCR resembles the Panasonic N or O category machines shown in Appendix VI, skip this chapter entirely, and go directly now to the special supplemental section at the very end of this book!

If tapes go into your VCR via a basket that normally pops up from the top of the cabinet, then you have a "top-loading" VCR, and you are reading the wrong chapter. This chapter concerns only VCR's that receive tapes through an opening in the front of the cabinet ("front loaders"). Ejection problems with top loaders are covered in Chapter 7.

2. Front-Loader Terminology

A clarification of terminology will be helpful in the present chapter.

On front-loading VCR's, there are two different "loading" processes, and two different "unloading" processes. One will be called "cassette loading" (and "cassette unloading"), while the other will be called "tape loading" (and "tape unloading").

The term "cassette loading" will be used to refer to the process by which a tape cassette that has been inserted into the front door of the VCR is pulled in and brought to rest on the reel table inside the VCR. "Cassette unloading" is the reverse process in which the tape cassette box is lifted from the reel table and pushed out the same front opening in the cabinet.

In contrast, the term "tape loading" will be used to refer to a process that usually follows "cassette loading." In "tape loading," the mechanism of the VCR pulls a loop of tape out away from the cassette box which is sitting on the reel table, wraps the tape around the cylinder with the video heads, and also threads the tape through the rest of the operating mechanism. "Tape unloading" refers to the opposite process in which this loop of tape is unwrapped from the cylinder, pulled free from the rest of the mechanism, and safely wound back onto the supply spool inside the cassette box. A more detailed description of these processes appears in Chapter 2.

The term "hinged tape cassette door" refers to the hinged door, usuallly made from plastic, attached to one side of the box that holds the tape itself. This "tape cassette door," as it will be called for short, is not to be confused with the hinged door at the opening on the front of the VCR through which tapes are inserted into the VCR, and removed from the VCR. This latter door will be called the "front cabinet door," "front-loader door," or "front flap."

3. The "Complete Cassette-Loading Mechanism"

The term "cassette basket" refers to the moveable metal or plastic enclosure into which the tape cassette directly goes when it is pushed into the front cabinet door of the VCR.

The cassette basket is designed to move somewhat like an elevator, except that it first carries the cassette horizontally into the VCR a short distance before it descends to place the cassette box over the reels on the reel table. In "cassette unloading," this basket moves in the opposite direction.

The stationary housing that moves this cassette basket by means of a motor that drives gears that slide the basket back-and-forth and up-and-down will be called the "complete cassette front-loading assembly," or just the "cassette loading assembly" for short.

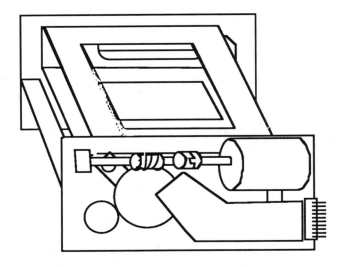

Typical Complete Cassette Loading Assembly As it Looks When Removed from the VCR

4. Complete Cassette-Loading Mechanism Often Are Available as Separate Assemblies

The complete front-loading assembly can be detached from the VCR by unscrewing some screws that hold it exactly positioned on the reel table. In many cases, it can be purchased as one single unit or assembly with a single part number if you cannot, or do not wish to, disassemble, repair, replace, "retime," and reassemble its component parts. (An exception is Panasonic-category machines: separate complete front-loader assemblies are, in general, not available for Panasonic-category VCR's.) Complete replacement is more expensive, of course, but can save time, at least when your supplier happens to have the entire assembly in stock. Call for part number and price.

The complete cassette front-loading assembly is also called the "front-loading assembly," or just the "front loader," for short.

Often, however, a parts supplier will need to back-order a complete assembly, but has the component parts of the loading assembly that fail most frequently (for example, the gears or its little motor) in stock. Also, sometimes the cause of the failure-to-eject problem is not a fault in the cassette-loading mechanism at all, but rather due to a trouble in a circuit on one of the epoxy circuit boards elsewhere inside the VCR. In these cases, replacing even the entire cassette loading assembly will not solve the

problem.

The diagnostic steps later in this chapter will help you determine where the basic trouble lies. Of course, if the cassette loading assembly has been severely deformed with many parts damaged, it may be necessary to replace it with a new one, even if some related component elsewhere in the VCR also needs repair.

The first steps will help you remove the cassette that is will not eject, if there is a cassette stuck in the machine.

5. Overview of the General Strategy to Correct the Problem

The problem of a front-loading VCR not ejecting tapes will be approached in a series of steps or stages. First you will read some background information that you will need, and then we will go to work on the machine.

If a cassette is currently stuck in the machine, the stuck cassette needs to be removed from the VCR without damaging either the tape or the machine.

To do this, we first may need to return the tape-loading mechanism to the fully-unloaded position, and rewind back into the cassette whatever loose loop of tape the VCR may have failed to reel back into the cassette, as it was supposed to do.

Secondly, the cassette will need to be ejected, or otherwise removed, from the VCR without damaging either the cassette or the VCR.

Thirdly, after this has been done, we will try to diagnose the cause of the problem as falling into one of six major categories, and repair the trouble if it falls into one of the five categories of basic problems that can be fixed within the elementary scope of this book, without electronic troubleshooting techniques. In this process, we may also use procedures covered in Chapter 10. Go on now to the next section, 6.

6. Overview of the Possible Causes of the Problem

The basic problem of failure to eject may have any of at least six different causes:

A) The cassette may have been inserted at the wrong angle and gotten jammed in the basket mechanism as a result.

B) The cassette loading mechanism may have itself gotten out of alignment, or gotten warped or mistimed, and consequently jammed with the cassette in it.

C) An electrical switch that needed to be activated by the position of some mechanical part may not have made contact due to corrosion, dirt, "bad contacts," or problems in the mechanism that is supposed to switch it.

D) The cassette loading motor may have gone bad.

E) The motor may have gotten hot and warped some plastic supporting parts, with the result that it is now wrongly aligned to drive the gears to move the tape properly, plus possibly, (D) the motor may now have gone bad too as a result.

F) The cause of the problem may be a failure in other circuitry elsewhere in the VCR that is should supply power to the motor at the right time in the right way to run the mechanism.

If the trouble is due to any of the first five causes, (A) through (E), we probably will be able to repair it with the procedures in this book. If the problem is due to cause (F), it is beyond the scope of this book. Proceed to the next section, 7.

7. Step One: Observe Whether the Tape-Loading Mechanism Currently is in the Fully-Unloaded Position and Whether the Tape is Fully Wound Back into the Cassette

When a VHS machine goes from the Play or Record mode back to the Stop or Eject state, it should go through a process of "unloading the tape" that is basically a reversal of the tape-loading process described in Chapter 2. If you have forgotten the difference between "loading a cassette" and "loading a tape," as these terms are used here, consult Sections 12 through 16 of Chapter 2.

What is meant by "unload"? Answer: Something different is meant by the term "unload" when we speak of "unloading the cassette" than when we speak of "unloading the tape." Unloading the cassette is ejecting the cassette from the machine. Unloading the tape is pulling the loose loop of tape out of the VCR tape path and winding it into the cassette box. We must unload the tape before we unload the cassette, since otherwise the loop of tape that is still lying in the tape path will get broken off when we eject (unload) the cassette.

The phrase "loading a cassette" refers to the

initial process in which a cassette enters the VCR and brought to rest on the pegs and reels. The different phrase "loading a tape" refers to the subsequent process in which the P-guides or loading ring pull a loop of tape out from the cassette and wrap it around the cylinder with the spinning video heads.

The phrase "fully-unloaded position of the tape loading mechanism" refers to the position that the tape-loading mechanism should be in during the time that a cassette is inserted or ejected. When a VHS machine is in the "fully-unloaded position," the P-guides are all the way toward the front of the cabinet, as far as they will go.

Before ejecting a cassette from a VCR, the tape must be unloaded first. This means that the tape must be pulled back from the cylinder and reeled back into the cassette. This must be done before the cassette is ejected ("unloaded") from the machine. We first will try to determine whether your machine had already completed the "tape-unloading" process before it stopped in its present state. If it did not finish unloading the tape, then we will need to complete this process before we can go on to eject the cassette.

Before Ejecting in VHS Machines

For a VHS machine to go from the Play or Record mode to the Stop or Eject mode, it first must go through a process called "unloading the tape." This process is a reversal of the process called "loading the tape" described in Chapter 2.

The two P-guides on sliding tracks that have held the tape wrapped part way around the spinning cylinder and video heads need to move on their tracks back to a position close to the cassette box. At the same time, the pinch roller should move back away from pressing against the capstan, and one reel (usually the supply-side reel, the one on the left), or

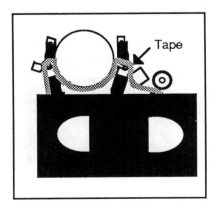

Partly Loaded VHS Machine

both reels, should rotate a few turns to pull any loose tape out of the VCR mechanism and wind it back onto a spool inside the cassette box. (Some models follow this with a final step of rotating the take-up side reel too, just to make certain that all loose tape has been wound back into the cassette box.)

A VHS machine that is not fully unloaded, and still has a loose loop of tape hanging out of the cassette, might look as shown in the preceding illustration.

Before Ejecting in Beta Machines

When a Beta VCR goes from the Play or Record mode to Eject, it also must go through its own process of unloading the tape. In a Beta machine, the entire "loading ring" should rotate in a direction opposite to the direction that it rotated when loading the tape, thereby unwrapping the tape from the cylinder containing the spinning video heads.

As this unloading process occurs, the VCR mechanism should cause the supply-side spool inside the cassette box to be rotated counter-clockwise a few turns to reel in the loop of tape pulled out of the cassette during the loading process.

On Beta VCR's, the big circular loading rings should rotate in a direction opposite to the direction they rotated when loading, the pinch roller should pull back from the capstan, and one or both of the reels should turn, winding the tape that was previously wrapped around the big circular loading rings back onto the spools inside the Beta cassette box.

A Beta machine that is not fully unloaded might look like this:

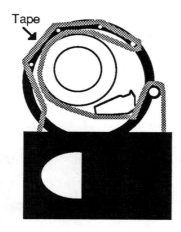

Tape

Partly Loaded Beta Machine

Why You Must Make Certain the Tape is Unloaded Before Ejecting the Cassette

If the cassette-ejection mechanism were activated before the successful completion of this tape-unloading process, the cassette would be ejected with a loose loop of tape hanging out, pinched under the closed door of the plastic cassette box, and maybe caught in the mechanism of the VCR, all of which probably would damage the tape.

And if the cassette basket were simply forced to come out without first putting the VCR mechanism into the correct position for ejecting, it could damage the machine beyond repair. I have seen VCR's come into the shop totally destroyed by frustrated owners who used improper techniques to force their VCR to eject a stuck tape which could have been removed easily with no damage if correct removal techniques had been used instead.

If you have not already done so, remove the top of the cabinet and raise any circuit boards over the reel table that need to be opened to see the place where the cylinder is located. Consult Appendix III, if necessary, for assistance in getting the cabinet open.

If the cassette stuck in the VCR has no loose tape pulled or hanging out of it and the tape-loading mechanism is in the fully-unloaded position, then go straight to the section entitled, "A Quick Cure."

But if, when you look inside the VCR with the cover and circuit boards opened up, you see that the tape loading mechanism is not in the fully unloaded position, and perhaps that some tape is still pulled out away from the cassette, then before removing the cassette, you will need to put the tape-loading mechanism back into the fully-unloaded position as explained in the next section, and either cause one of the spools inside the cassette to rotate to wind the loose tape back into the cassette before ejecting it, or else, if this is impossible, first jam the cassette door open so that it does not close and crease the tape when the cassette is removed, and then gently free the tape from the inner mechanism before trying to remove the cassette.

How to do this will now be explained more fully.

In going through the next steps, while returning the tape-loading mechanism to the fully unloaded position, try to avoid moving the motor in the cassette front-loading assembly if possible, so that we can test it for a "bad spot" later.

Now let's go to work on the machine.

8. Step Two: Return the Tape-loading Mechanism to the Fully-unloaded Position and Rewind Any Loose Tape Back into the Cassette

If you see a loop of tape pulled out of the cassette box and into the VCR mechanism, or if you see that the mechanism is not in its fully-unloaded position (or if you are uncertain whether it is or not), you will need to find the pulley, worm gear, or motor to turn manually to (A) put the VCR into its fully-unloaded position, and (B) wind any loose tape back into the cassette. (The "fully-unloaded position" was explained in detail in Chapter 2.)

Generally, the process will involve (1) making the tape loading mechanism go back to its fully-unloaded position (or "Stop" position), and (2) making the VCR reel the loose loop of tape back into the tape cassette. Basically, you move the mechanical parts in the same way that they would normally be moved by one or more motors turning in the direction that unloads the tape and winds the loose tape back into the cassette box.

You NEVER do this by pushing directly on the P-guides or loading ring! Instead, you must return the mechanism to its fully unloaded position in the same way that a little motor in the VCR should have done so, namely, by rotating a small pulley, or gear, connected to the tape loading mechanism.

The problem is to find the pulley or gear that does this job, and to figure out which direction to turn it. This mechanism is located in different places on different models. On some models, you can reach it from the top of the reel table, while on other models, reaching it will require also opening up the bottom of the VCR to reach the underside of the reel table.

You may be able to figure out where the "loading motor" pulley or gear is on your VCR simply by looking the whole mechanism over carefully, but further specific help is available in Appendix VI, which shows sketches of most common mechanisms used in different models by major manufacturers, and in most cases, even shows the direction to rotate the loading motor gear or pulley to unload the tape and put the mechanism in the "fully-unloaded" position. If you can find your model pictured in Appendix VI, follow the instructions given there to unload the tape. Do this now. Be sure to read the further instructions at the beginning of Appendix VI on how to use the diagrams given there.

It is impossible to give diagrams for every single model, but even if your VCR differs from the many models pictured, looking at how other VCR's are manually unloaded should help you figure out how to do the same on your machine.

If you find the loading motor and observe that a belt that goes from it to some pulleys is off the pulley, broken, or very soft, weak, and worn out, get the part number, clean the pulleys, and replace the belt. Then plug in the machine and see if the loading motor now automatically unloads the tape. If it does not, go on to the next paragraph. If it does unload the tape, then go to the next section below on reeling loose tape back into the cassette ("Reel Loose Tape Out of the Tape Path and Back into the Cassette").

If you can find the loading-motor pulley or gear but can find no information on which direction to turn it to unload the tape-loading mechanism, then you can answer this question by trial and error. There are only two possible directions for it to go, clockwise and counterclockwise. Pick one of these directions at random, and very gently start turning the pulley or gear in that direction, while watching what the P-guides or loading ring does. If they start to move in the direction to return the tape to the cassette, continue gentle rotation. But if you encounter resistance, or if they seem to move in the direction to load the tape, then stop and try turning it in the other direction.

Watch carefully and stop when the loading mechanism first gets back to its fully unloaded position, because it is possible on some models to go too far and overshoot, in which case the mechanism will still not eject, or on some Sharp-category VCR's, it may go into a position in which, when you try to power it up later, the little computer inside will not be able to figure out where the mechanism is, or what to do next, and so, will shut everything down.

On some models, turning one mechanism both unloads the tape and reels the tape back into the cassette. On other models, two separate steps are required to do both.

Reel Loose Tape Out of the Tape Path and Back into the Cassette

As mentioned, on a few models, any loose loop of tape will be reeled back into the cassette automatically when you unload the tape-loading mechanism, but on most VCR's, reeling in the loose tape requires a separate process.

On some VCR's, there is another pulley that you can turn to cause a spool to turn inside the cassette to pull tape back into the cassette as you unload the mechanism. On some of these, it is necessary to move a mechanical arm as well as turning the pulley to reel tape back into the cassette.

On other models, you need to attach a 9-volt battery to another motor, a "reel motor," to reel loose tape back into the cassette, as explained in the next subsection.

For many models, further information on reeling tape back in is given in Appendix VI also.

VCR's With a Separate Reel Motor

On some models, the tape is wound back onto the spools by a small electric "reel motor." In most cases, you can use a little 9-volt battery with two clip-lead wires (as described in Section 19 of Appendix I) to operate this motor to reel the tape back into the cassette without plugging in the VCR.

On some models, you can attach the pair of wires from the battery to two exposed terminals on the motor either way, and it will reel the tape back in. On other models, a battery will reel tape in only when the wire from the positive (+) battery terminal is attached to a certain terminal on the motor, and the wire from the other battery terminal (-) is attached to the other terminal. Consult the appropriate diagram in Appendix VI to find out how to handle your model. If no information is available, try connecting the battery one way, and if that does not work, try reversing the connections. It will not harm the motor.

On some VCR's with a reel motor, the terminals to the motor are not accessible, because the wires reaching the motor are covered with insulation, and they disappear into the motor enclosure without ever presenting a bare metal end-point. In such a case, you can attach the clip leads from the battery to sharp pointed tools, like a scribe or a pick, and press these through the insulation of each wire, penetrating the insulation to reach the metal of the wire inside, and power up the motor this way. Or you can stick straight pins through the two wires, and attach clip leads to the pins to power up the motor.

If this fails, you can use instead the trick of jamming the cassette box door open with a stick to protect the loose tape hanging out, remove the screws attaching the entire front-loading mechanism with the cassette still inside it, remove the whole front-loading assembly, and once it is out of the machine, reach underneath to turn the reels with your finger to wind the tape back in, as explained later. See Appendix III for help getting out Sharp-category front-loaders, or if you have any difficult with other manufacturers. Make certain to remove the right screws.

Go Back and Forth Between Unloading the Tape and Reeling It In

Go back and forth between unloading the tape-loading mechanism a little, then winding back in some loose tape, then unloading a little more, and so on. Do

this now. What to turn to reel in the loose tape may be shown with an arrow in the diagrams in Appendix VI, and the direction to turn it to reel in tape is the direction the arrow points.

On some models, you can wind tape in by turning the pulley in either direction: one direction winds the tape onto the supply reel, and the other direction winds it onto the take-up reel. It does not matter which way you go, so long as the tape gets pulled back into the cassette somehow.

If you cannot get the tape to reel back in, go to Step Six in the section entitled "Removing the Entire Cassette Front-loading Assembly and Manually Rewinding Any Loose Tape Back into the Cassette." But if you can bring the tape-loading mechanism to its fully unloaded position and rewind all loose tape back into the cassette, go to Step Three in the next section, plug in the VCR, power up, and see whether the VCR will unload the cassette.

9. Step Three: A Quick Cure?

I have seen cases in which a VCR suddenly decided to stop misbehaving once it perceived that I had the cover off and was going to deal with it seriously.

I know that it sounds silly, but it will only take a second to make sure that we really have a problem before we go any further. Front-loading mechanisms of VCR's sometimes experience a once-in-a-lifetime seizure that, once its effects are corrected, never recurs. This situation is comparable to a person once having a severe pain which never returns or leads to any other problem. If something like this has happened to your VCR, it may be possible to fix the problem with no further work.

To check this possibility, if there is a cassette stuck inside the cassette loading assembly in the VCR, first look at the cassette carefully without touching it to see if anything looks uneven, lopsided, or out of alignment in the way it is positioned. If the tape cassette has accidentally been inserted upside down, or is crooked in the basket so that some part has caught and jammed against the top, or a side, of the cassette loading mechanism, then the "quick fix" will not work and you will need to unjam the mechanism. (Some brands of tape cassettes have ridges or other external decorative or design features of their molded plastic enclosures that causes them almost always to jam in certain front-loading VCR models — the fault in these cases is in the interaction of the design of the cassette with the design of the VCR — good planning, eh?)

If the cassette appears to be jammed crooked in the front-loading assembly, go to Step Four in the next subsection. If it is not jammed or crooked in the front-loader, go to Step Five in Section 11.

10. Step Four: Straighten Cassette Jammed in Loader

It may be that nothing is wrong with any part of the VCR, and that the entire trouble is only that the cassette got caught and jammed in the mechanism, perhaps due to somehow getting into a crooked position in the basket. Maybe someone did not insert it straight. If so, it is possible that everything will work properly again, and the problem will be corrected, once the jammed cassette is freed, especially if you do it gently. Do not move the mechanism yet unless you can see that it is mechanically jammed.

Make certain that the VCR is unplugged.

If a cassette is stuck in the loading assembly, study carefully its position. If it is crooked, or if a protruding part of the cassette is caught somewhere in the loading mechanism, gently try to work or wiggle it straight and free. Be careful not to bend any part of the loading assembly mechanism with excessive force. If you happened to bend, distort, or break some part of the loading assembly in trying to remove a stuck cassette, you would increase the cost, repair time, and delay required to repair the problem. If the cassette is jammed in such a way, or so tightly, that it is impossible to free it without bending or distorting some part of the loading assembly, then go directly to Step Six in Section 12.

11. Step Five: Plug in the VCR, Power up, and See Whether the VCR will Unload the Cassette

Use a wooden, plastic, or other nonconducting stick to prop open any circuit boards that you needed to open previously to view the reel table, plug in the VCR, press the power switch, push eject (but do not reach inside the VCR), and watch what happens.

If the VCR does not eject the tape right away, immediately pull the power plug from the wall outlet (to protect the motor from overheating). If your top priority is just getting the cassette out, go the section entitled "If Possible, Manually Turn the Cassette Loading Mechanism to the Fully-ejected Position." If the top priority is repairing the VCR, go to the section entitled, "Special Tools and Techniques for Front-Loader Troubleshooting — Using a Battery to Run the Front-Loader."

If the VCR does now eject the cassette under its own power, then probably the trouble is not in the cassette-loading-and-ejecting mechanism at all, but rather is in the tape loading-and-unloading mechanism. (The difference was explained in Chapter 2.) Here is the explanation: to protect tapes, most VCR's are

designed not to carry out an order to eject a cassette unless the tape-loading mechanism is in its fully-unloaded position. So, when the tape-loading mechanism was manually moved into its fully unloaded position, the VCR became able to eject.

Try cleaning all belts and pulleys of the tape loading-unloading mechanism with cotton swabs wet with alcohol, allow to dry thoroughly, plug in, power up, and insert and eject a junk tape a few times to see if it now works. Now try putting it into "Play."

If the unit again fails to load or unload fully, stopping part-way between the fully-loaded and the fully-unloaded position, and again ejecting only after you manually return the mechanism to its fully-unloaded position, then a trouble still exists in the tape-loading mechanism. If there is (are) loading belt(s), or (and) an idler, try replacing it (them), and try again. If the problem remains, go to the section on tape-loading problems in Chapter 14 for more information.

Ejected on First Try

If the VCR ejected the tape on the very first try, even before you moved any of the parts of the mechanism, and you did nothing to move the cassette loading mechanism or rewind the tape, then we have an unexplained phenomenon on our hands.

If it continues to work on successive cassette insertions and ejections, we may have seen another of those miracle cures that sometimes happen to VCR's. Close the cabinet back up again, and use the VCR until the problem recurs.

If the tape loading mechanism was not in the fully unloaded position when you started, and you began by moving the tape loading mechanism to its fully unloaded position manually, then probably the trouble is not in the cassette loading and ejecting mechanism at all, but rather is in the tape loading and unloading mechanism. (The difference is explained in Chapter 2.) To protect tapes, most VCR's are designed not to carry out an order to eject a cassette unless the tape loading mechanism is in its fully unloaded position.

On Sharp-category VCR's, some of these "miracle cures" can be explained as follows. When you turn on the power switch, the first thing that the little computer inside does is to power up the motor that is supposed to move the idler to move the reel, which it can detect by a sensor.

If the reel does not move when it is supposed to, the little computer concludes that the rubber idler is slipping, and will not let you eject any cassette that might be sitting inside the VCR because it is afraid that there may be a loose loop of tape hanging out, and it knows that you would be angry if it ejected under such conditions, possibly damaging the tape.

So, you have to unplug the VCR (clearing the computer's memory) before it will eject after once failing to sense a counter-clockwise (CCW) movement of the take-up reel. If this happened, you probably will need to clean, or preferably replace, the weak idler, as explained in Chapter 13.

If the "quick cure" did not work, go to Section 13.

12. Step Six: Removing the Entire Cassette-Loading Assembly and Manually Rewinding Any Loose Tape Back into the Cassette

If a loop of loose tape is left hanging out of a valuable cassette and the tape could not be wound safely back into the cassette manually by the preceding procedures, and you want to save the tape, or if the cassette is so tightly jammed in the front-loading assembly that you cannot eject it even manually, then it will be necessary to jam the cassette door open, remove the entire cassette front-loading assembly from the VCR with the cassette still inside it, and rewind the tape into the cassette by rotating one of the spools with your fingers until all the loose tape is safely wound back into the cassette.

To do this, jam the cassette door open with a long thin object like a wooden popsicle stick, tongue depressor, fingernail file, a short thin ruler, etc., as shown in the illustration, except that the cassette will, of course, be inside the front-loader assembly, which is not shown in the figure for clarity: This way,

the door will not close down on the tape and damage it when the cassette is removed together with the cassette-loading mechanism. If the loop of tape is wound around part of the internal mechanism, you will need to use your fingers to free it so that you can remove the assembly together with the cassette without damaging the tape.

NOTE: Do not attempt to remove the front-loading assembly from a Fisher model FVH 730, or on the newer Panasonic-category machines labeled with a danger-warning in Appendix VI. On these models, removal of the front-loader is so fraught with danger that even experienced video technicians try to avoid it.

If you can avoid moving the motor and other mechanism of the front-loader as you remove it, this will permit you to make a useful test on it later, once it is removed. But if you accidentally move it, no great harm is done.

Unfastening the Front-loader from the VCR

To remove the cassette-loading assembly, you will need to remove the front cabinet piece, and you will possibly also need to remove the bottom cover in order to get the front cover off. Remember to pull off any tracking knobs or other knobs that run through the front panel before trying to detach the front panel. See Appendix III for further help removing these cabinet parts.

On some Hitachi-category VCR's, there may be a metal brace running across the top of the assembly at

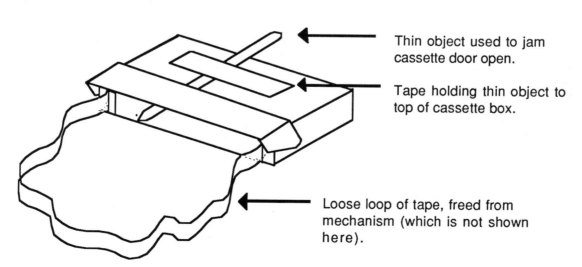

Thin object used to jam cassette door open.

Tape holding thin object to top of cassette box.

Loose loop of tape, freed from mechanism (which is not shown here).

Cassette With Door Jammed Open to Protect Tape As Cassette is Unloaded

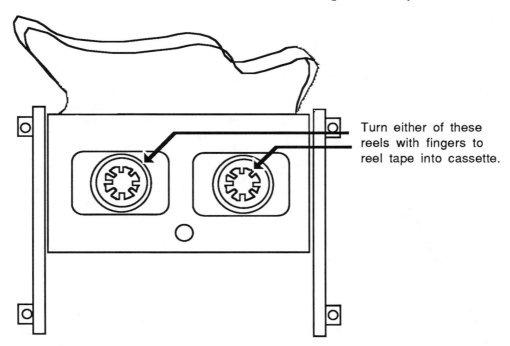

Turn either of these reels with fingers to reel tape into cassette.

Manually Reeling Loose Tape Back into a Cassette Stuck in a Front-Loader Assembly

the front; draw an arrow pointing forward on this part before removing it to help you get it back in correctly on reassembly.

If there is still a loop of tape hanging out, make certain that you have carefully lifted or pulled it free of the mechanism before removing the front-loading assembly.

Once you get the cassette loading assembly out of the VCR with the cassette in it, set it on a table.

If you cannot get the assembly free, you will need to go to Step Nine in Section 16 and try to eject the cassette manually by using your fingers or a scribe to turn the motor that runs the front-loader. Turn it in the direction that makes the basket rise up from the reel table.

Reel in Loose Tape

If tape is hanging out from a cassette stuck in the loading assembly, use your fingers to turn one of the spools to rewind the loose tape back into the cassette. Remove whatever device was used to hold the hinged cassette door open and let the door close gently.

Next we will remove the cassette from the detached front-loader. First we will try to do it with an external battery, then we will try to do it using power from the VCR, and if all that fails, we will manually put the mechanism through its paces to

unload the cassette. Go to the next section.

13. Special Tools and Techniques for Front-Loader Troubleshooting: Using a Battery to Run the Front-Loader

Technicians usually have a piece of equipment called an "external power supply" which can be adjusted to supply power in the desired way to a piece of equipment that they are testing and repairing. We will use a small battery to accomplish the same for our purposes at much less cost. It will be helpful for diagnostic purposes later to be able to connect a small battery directly to the little electric motor that drives the cassette loading assembly on most models to see if this will make it eject the cassette.

There is an excellent reason for trying this, whenever possible, before moving the basket mechanism by hand. For this helps us test to see if the motor has a "bad spot" on it, or is defective in some other way.

A "bad spot" on a motor is one position that it sometimes stops in, but cannot restart from. When a motor with a "bad spot" is rotated slightly from its bad position, it will start up and appear to operate normally, starting and stopping OK — until the day comes when it again happens to stop in that exact same bad position again, at which time, the bad spot prevents the motor from starting up again.

This is the reason for making the battery test, when possible, before moving the motor — For once you move the motor, this test can no longer be relied on to detect a possible bad spot on the motor. But if it is too much trouble to get a battery, or if there is some other reason why you cannot make this test, you can get along without it. But to be able to make this test, proceed as follows.

From a hardware store or other source obtain a little 9-volt battery of the type with a male and female round snap on one end, and a couple of little "clip leads" (wires with little clips on each end). You can use a Radio Shack battery #23-464 and # 278-1157 clip leads, for instance. If you use clip leads with the larger alligator clips at the ends, you can clip one to a scribe and another to a small screwdriver to reach in to apply power to the motor, after you have double-checked to make certain that the VCR is unplugged.

The home-made test equipment is pictured in the illustration.

To apply power from the little battery to the cassette loading motor, you must first find this motor. On most Japanese made VCR's, it is attached to the right side of the cassette loading assembly, but on some Samsung's and other Korean-made VCR's, it is located underneath the reel table, accessible only from below.

And on late-model Panasonic-category VCR's, the cassette loading and unloading is accomplished by the capstan motor, which cannot be run with a battery. These reel tables are pictured in Appendix VI, together with the information that removing a stuck tape from these models pictured with the warning is too difficult to be attempted by anyone but the most advanced technician.

A 9 volt battery will not drive the cassette loading motor on an Emerson VCR model 870 or on early Goldstar machines. Twelve volts are required, and they are difficult to get from a small dry cell. The cassette loading motor on Sharp-category VCR's normally run on 12 volts too, but they will move (slowly) through their paces when powered by a fresh, good 9 volt battery.

We are not quite ready to run this test yet, until called for in the troubleshooting procedure that follows this preliminary explanation, but when the test is called for, proceed as follows. Examine the cassette loading motor. You will see the two wires, or terminals with wires soldered to them, coming from the back end of the cassette loading motor.

When the positive (+) terminal from the battery is connected to one terminal on the motor, and the negative (-) terminal is connected to the other terminal on the motor, the motor should turn, or try to turn, in one direction. If the same wires from the two battery terminals are connected to the same motor in the opposite way, the motor should try to move in the opposite direction.

Rotating in one direction, the motor will tend to pull the cassette basket into the loading assembly and set it down onto the reel table ("cassette loading"). When the motor turns in the other direction, it will tend to move the basket up and out of the VCR ("cassette unloading," also called "cassette ejection").

To repair your problem you do not even need to know that the circuitry inside the VCR activates the cassette loading mechanism in just this way, making it

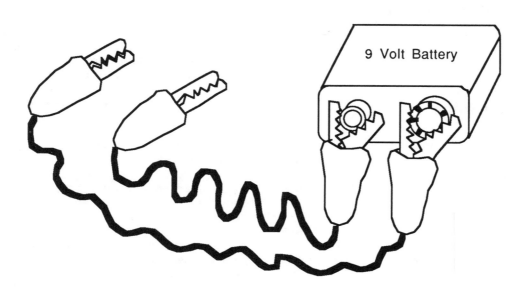

Alligator Clips Attached to Little 9 Volt Battery

"load" or "eject" by reversing the direction or "polarity" in which the voltage is applied across the motor with internal electronic switches, like your reversing by hand the way in which you attach the two battery terminals to the motor.

By trial and error, you can determine which of the two possible ways of connecting the two wires across the motor should cause this motor to turn in the direction that is supposed to cause the mechanism to eject the cassette. If you can reach both terminals of the motor with your "probes" attached to the battery clip leads while the cassette loading assembly is still in the VCR, then you can try first one way, and then the other way, of connecting the probes to the terminals on the motor.

For clarity, the illustration omits the rest of the VCR surrounding the front-loading assembly, and shows attaching a battery to a front-loading assembly that has the motor on the top right rear, the most common position. Your VCR may have the cassette loading motor located elsewhere, but the idea is the same. Also, of course, with the complete cassette front-loading assembly still in the VCR (because we have not yet taken it out) you probably will only be able to see parts of it.*

With one polarity, the motor may or may not do anything. If the motor does not run and move the mechanism, immediately lift the probes away from the motor terminals, and switch their position. With the other polarity, it may start running in the direction to eject the tape. If the motor DOES run and move the mechanism, the motor is probably GOOD, and the problem is not a "bad spot" on the motor.

If the motor does NOT run and move the mechanism when the probes are attached either way, then we do not yet know whether the motor is good or bad. You cannot yet say that it is bad. It might be good, or it might be bad. MORE TESTS will be needed to determine whether the trouble is in the motor or elsewhere in the mechanism.

Sometimes it is too difficult to get access to the cassette loading motor terminals to run this test with the cassette loading assembly still in the VCR. In such a case, you can unscrew and remove the complete cassette front-loading assembly from the VCR (with the tape still inside it), and test it with the battery

---

* The complete cassette front-loading assembly consists of the basket holding the cassette, the sides of the mechanism that move this basket up and down and back-and-forth, and generally a motor that operates this mechanism, as explained at the beginning of this chapter.

with the whole front loader assembly just sitting on the table outside the VCR. If you have not already done so, do not do this yet.

For our procedure, it will not matter which way you attach the wires and probes to the motor first, because we are using a trial-and-error method to determine which way of connecting the wires makes the motor turn so as to eject the cassette.

But before attaching the wires to the motor, let's use a felt-tip pen to make a reference mark on the motor beside one of the terminals, either one. We can call that the "marked terminal" or "terminal A." We will attach one of the probes to that terminal and touch the other probe to the other terminal for a second to see what happens, watching the motor at the same time to see if it starts to rotate in one direction or the other.

If it does, stop and write an entry in your notes to record this observation. Follow the wire from this probe back to the battery and note whether this probe was connected to the male (+) or female (-) terminal on the little 9 volt battery. Then you can record the information on paper by saying something like "Male (+) terminal connected to marked terminal on motor makes it rotate in such-and-such direction (for example, CW or CCW) and eject (or load) the cassette." Really do this; later you may be thankful that you did.

Any time that you are not using this homemade apparatus, disconnect the clip leads from the battery right at the battery terminals. Otherwise the two clips or probes might come into contact with each other on the table when you are not looking, and drain the battery in a few seconds.

This is to be avoided, not only because it would waste the battery and force you to buy a new one, but also because, if you did not see that this had happened, you might later connect the dead battery to the little motor and think that the motor was no good because it did not turn, when actually your new battery was dead. It could lead to an incorrect diagnosis of the problem, and even result in your throwing out a good motor, and buying a replacement unnecessarily.

14. Step Seven: Test to See Whether a 9-Volt Battery Will Drive the Motor to Unload the Cassette

Try applying power to the front-loader assembly from a battery as just described. If nothing happens when you touch the probe, do not hold the probe on. Reverse the connections and repeat the test.

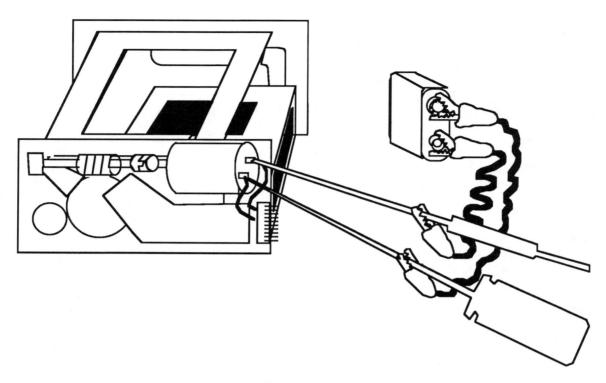

Probes Attached to Battery to Try Powering Up Cassette Loading Motor

There are three possible outcomes: (a) motor will not run at all in either direction; or (b) motor runs and fully ejects cassette; or (c) motor runs a little bit, but cassette basket jams in the mechanism before it gets all the way to the fully ejected position. What to do for each of these is now explained.

(a) If the motor will not run in either direction, go directly to Step Eight, "If No Go, Try Changing the Position of the Motor Slightly and Reattach Battery."

(b) If the mechanism stops or jams in a midway position, remove the probes from the motor terminals and go back and run Step Four, "Straighten and Free Cassette Jammed in Loader."

Notice that when you do get the mechanism back to the fully-ejected position and remove the cassette, you probably will not be able to make the motor and mechanism run in the opposite direction ("cassette loading") with no cassette inside, because most cassette front-loading mechanisms have a mechanical catch that latches the basket in the fully-ejected position until someone inserts a cassette and pushes back gently with enough force to unlatch the catch.

In order to run the mechanism in the inward cassette-loading direction with a battery, it is necessary to unlatch these catches first, either with your fingers, or by inserting a junk test tape. This is

explained more fully in Chapter 10.

(c) If the motor will run in the direction that ejects the cassette, leave the probes touching the motor until it brings the cassette up and pushes it out of the front of the cassette front-loading assembly. If the motor runs the mechanism all the way to the fully-ejected position, disconnect the probes when the mechanism gets all the way to the end of its run and remove the cassette safely. Add to your notes an entry telling which way of running the motor ejects the cassette (for example, "Male battery (+) terminal to marked terminal makes it try to eject cassette").

If you were able to eject the cassette using battery power, your next step depends on the state in which you found the tape-loading mechanism when you began.

If Tape-Loading Mechanism Was Fully-Unloaded When VCR Was First Examined

If you found the tape-loading mechanism in the fully-unloaded position with no loose loop of tape hanging out of the cassette, then go directly to Step Eleven, "Test Whether the VCR Will Now Run the Front-loader under its Own Power."

If Tape-Loading Mechanism Was NOT Fully-Unloaded When VCR Was First Examined

If you did not find the tape-loading mechanism to be in the fully-unloaded position when you opened the VCR, but found it to be fully- or partly-loaded, then possibly the original trouble is in the tape-loading mechanism, and not in the cassette-loading mechanism.

If the VCR is in the Sharp category, or a Samsung that uses a Sharp free-swinging type idler, clean off or replace the old idler before trying to load a cassette.

If the VCR is a VHS machine in the Hitachi, Panasonic, or Sanyo-Fisher category that uses a loading belt, remove the old loading belt, clean the pulleys, and install a new belt before trying to load a cassette. See Chapter 14 and Appendix VI for further information on how to deal with tape-loading and unloading problems.

After correcting possible problems in the tape-loading mechanism, put the cabinet front back on (if you had it off), plug in the VCR, push the power switch, and test to see if the machine will accept a cassette. If it pulls a cassette in OK, push eject to see if it will eject now. If it accepts and ejects a tape properly now, you can return to the universal check out procedure in Chapter 4. Perhaps your whole problem has now been solved.

If it fails either to accept a cassette, or fails to eject a cassette, the trouble is probably either a misaligned mechanism in the cassette front-loading assembly, or bad or mispositioned switches in the assembly, or trouble elsewhere in the VCR. To eliminate the first two possibilities, go to Chapter 10 and follow the front-loader realignment procedure.

15. Step Eight: If No Go, Try Changing the Position of the Motor Slightly and Reattach Battery

If neither way of connecting the probes to the motor results in motor movement, or if you had to skip the battery test, the next step is to find the little shaft leading from the cassette loading motor (usually attached to the cassette-loading assembly) to the pulley or worm gear connected to the cassette loading mechanism. By pushing on it with your finger, scribe, or a similar tool, try to rotate it a half-turn in one direction, and try connecting the probes one way, and then connecting them in the other way. If nothing happens, try rotating the motor in the other direction

one turn and try again both ways of attaching the probes. Note the results.

If the mechanism ejects the cassette, go to Step Ten, in Section 17, "Remove Cassette, Insert Junk Cassette to Test Whether Mechanism will Move a Tape In and Out under Battery Power." If it does not eject, or if it ejects part way and then jams before the cassette is fully ejected, go to Step Nine in Section 16.

16. Step Nine: If Possible, Manually Turn the Cassette Front-Loading Mechanism to the Fully Ejected Position

Look for a shaft or belt running from the little cassette loading motor to some sort of gear mechanism that is supposed to slide the basket back-and-forth and up-and-down on its tracks.

Often there will be a small "universal joint" (that is, a flexible joint between two connecting parts of the shaft) or other little plastic part molded, or otherwise attached, to a metal shaft coming from a motor attached to the cassette loading basket assembly.

On a few models, the mechanism is driven via a belt that comes from an attached motor. On some Samsung and Mitsubishi-category VCR's, the motor is located below the reel table and connected by a belt; in such a case, apply the following instructions to the belt or pulley connecting the motor to the front-loader assembly.

Using an appropriate tool, like your scribe or finger, gently experiment to see if this driving part can be rotated manually, in one direction or the other, to make the mechanism unload the cassette. Try rotating the shaft or pulley in one direction while watching to see whether the mechanism starts to lift the cassette up from the reel table. If it does, continue rotating in the same direction — if not, try rotating it in the opposite direction. The direction of rotation of the motor that unloads the cassette is given for some common front-loading assemblies in Appendix VII, but you can find out the same for yourself by trial and error.

If the horizontal brace across the top at the front of the cassette loading assembly has gotten bent or bowed down so that it is touching or scraping against the top of the cassette as it moves out toward the front door, then you can manually bend it up slightly, straightening it so that the cassette does not hit against it and try again.

If the shaft or pulley resists rotation in both directions, or if the cassette loading mechanism seems itself to jam part way along its route to the fully ejected position, go to Steps Six in Section 12 to remove your front-loader assembly from the VCR (if you have not already removed it), and then go to Chapter 10 to realign the cassette loading assembly.

If the basket carrying the cassette seems to move smoothly along its route toward the ejected position as you turn the mechanism, keep turning it until the stuck cassette is fully ejected.

Remove the valuable cassette, and insert a "junk" cassette (that is, one of no great loss if it got damaged), without plugging in the VCR. Insert the cassette far enough to release any catches locking the basket in the fully ejected position, and hold it there while you start turning the same mechanism in the opposite ("cassette loading") direction.

If the mechanism turns smoothly, run it until the cassette goes all the way in and down to the fully loaded position. Watch especially at the moment that the cassette rounds the corner in its rearward mothing and starts to move downward. If one rear corner of the cassette drops before the other rear corner drops, you probably need to dismantle and retime the cassette front-loader mechanism. Hopefully, yours is one of those covered in Appendix VII. You can read Chapter 10 for background knowledge and general procedures for any front-loader. If the cassette rounds the corner and moves evenly, go to the next paragraph below.

Now reverse directions, and manually run it in the ejection direction about half-way back out. If the mechanism turns freely, next try running it back and forth with a battery as described in the section entitled "Using a Battery to Run the Front-Loader." Disconnect the battery and make a note of whether the battery will run it.

Next, use a stick made from wood or other nonconducting material to prop open any circuit board that you had to raise to get at the cassette loading assembly, plug in the VCR, power up, and see if it will accept a cassette . (If you have removed the front loader, connect it back to the wires you unplugged from it.)

If it will not accept a cassette now, go directly to Chapter 10.   If it will accept a cassette, next test whether it will now eject the cassette correctly.

If it accepts a cassette but again will not eject it, remove the cassette manually as you just did with the other cassette, detach the front-loading assembly from the VCR using the procedure in Step Six, in Section 12, "Removing the Entire Cassette Loading Assembly and Manually Rewinding Any Loose Tape Back into the Cassette." Then go to Chapter10 and following the procedure there, realign it and check it for broken parts.

If it will now both accept and eject the cassette,

return to the check-out procedure in Chapter 4.

17. Step Ten: Remove Cassette, Insert Junk Cassette to Test Whether Mechanism will Move a Cassette In and Out under Battery Power

If the motor now moves and ejects the cassette, then either the mechanism was initially jammed and you freed it, or the motor was stuck on a "bad spot," and should be replaced with a new motor.

To determine which of these is the problem, remove the previously stuck cassette and insert a "junk" cassette. Insert the junk test cassette far enough to unlatch the catches that hold the basket in the fully ejected position.   If you have the front-loader assembly out of the VCR, brace it in an upright position on the table.

Attach battery probes opposite to the way that makes the mechanism eject the cassette, as you noted earlier. If connecting the battery activates the mechanism and makes it pull the tape in, let it run all the way in and down until it stops; then lift the battery probes, reverse their connections and see whether the battery will run the tape back out in the eject direction. If if does, try it several times, and make a note of your observations.

Whatever the outcome, go next to Step Eleven, in Section 18, "Test Whether the VCR Will Now Run the Front-loader under its Own Power."

18. Step Eleven: Test Whether the VCR Will Now Run the Front-loader under its Own Power

Note to readers who have already removed the cassette front-loader from the VCR

If you removed the front-loading assembly from the VCR to reel in tape because the mechanism of the VCR would not unload the tape and reel it back in fully, it would be wise to replace any loading belt(s) and clean, or preferably replace, any idler that is positioned to swing over and contact the supply reel, if you have not already done so, before reinstalling the front-loader. If you need more specifics for your model, see Appendix VI and Chapters 13 through 15.

When reinstalling a front-loading assembly back into the VCR, remembering to reconnect any multi-wired plugs that you unplugged, putting them back in the same way you found them. Refasten at least some of the screws that held the front-loader in the VCR, and reinstall the front cabinet part so as to have buttons to push to try to make it function.

Will the VCR Now Run the Front-Loader?

Plug in the VCR, power up, try to load a cassette, and observe whether the VCR will load and eject a cassette under its own power.

If the VCR loads and ejects cassettes now in repeated tests under its own power, go to Step Twelve, in Section 19.

If the mechanism accepts a cassette but again will not eject the cassette, go back to Step Nine in Section 16 and follow the procedure for manually turning the front-loader's motor, or the mechanism that is supposed to be driven by the motor, to get the stuck cassette out. Then go to Chapter 10 to realign the front-loader assembly and check the switches.

If it refuses to accept a cassette at all, go to Chapter 10 to realign the front-loader assembly and check the switches.

On Sharp-category VCR's that have the NIDL-0005GEZZ or NIDL-0006GEZZ idlers pictured in Chapter 13, the VCR's refusal to accept or eject a cassette can be due to an oily slipping idler (which prevents the reels from being moved during the start-up test that the VCR's little computer runs when it is first turned on, causing it to shut down these functions). To replace this idler, you will need to remove the cassette front-loading assembly (although if there is no cassette in the VCR, you may be able to clean the rubber tire with a very long cotton swab stick moistened with alcohol). To find out how to identify and replace this idler, see Chapter 13.

19. Step Twelve: Results of Testing Operation Under Power

If the VCR now will accept a cassette, pulling it in all the way and setting it down on the reel table, test to see whether it will now eject. Repeat the test several times.

If the VCR is able repeatedly to accept and eject a cassette correctly, return to the check out procedure in Chapter 4. If everything else works too, return the unit to use.

But first, read the next two sections of this chapter, obtain the part number for the cassette loading motor, and be prepared to order up a replacement motor and learn to solder if the same problem recurs.

On the other hand, if the machine again fails to

load and eject a cassette properly, remove the front-loading assembly from the VCR as explained in Section 16, go to the procedure in Chapter 10, realign it, and check all switches.

20. What it May Mean if the Front-Loader Worked Only After Being Moved Slightly

If the front-loader worked only after you slightly rotated the drive shaft of the cassette-loading motor manually, this suggests that the problem was either a jamming of the cassette loading mechanism or a "bad spot" on the motor, discussed in the next section below.

Try inserting and ejecting a cassette again three or four times, allowing 30 seconds of rest between each operation to permit the delicate electrical components to cool so as not to overheat and create a new trouble.

If the VCR passes each test and also will load, play, and unload cassettes correctly, you are faced with the old question that often confronts VCR repair technicians: "Is the machine OK now, or will the same problem recur again in a couple of weeks?"

Since the only way of answering this question for sure is to try the experiment of operating the machine in the normal way over the next few weeks, one reasonable course of action is to check all visible plastic parts of the cassette loading assembly mechanism for warps, put the circuit boards and covers back in place, and try returning the VCR to normal service. If any heat warps in the close vicinity of the motor, or other defects in the plastic parts of the cassette loading mechanism are observed, it would be wise to get the part numbers and order up replacements for these parts.

21. Does the Cassette Loading/Unloading Motor Have a Bad Spot on It?

If the cassette unloading mechanism started only after moving it slightly, you might want to take the precaution of ordering a replacement for the little cassette-loading motor, since the problem likely could be due to what is called a "bad spot" or a "burned spot" on the motor.

This means a position of the motor in which, if the motor happens to stop or come to rest in this position, it cannot start moving again when voltage is applied to it, but if the motor is manually turned to another starting position, it can start up and run OK, even rotating, by momentum, past the bad spot.

But on that dark inevitable day in the future when the motor again happens to stop at this bad spot, it once again will refuse to start up until turned past this spot. The existence of this possible problem is difficult to confirm conclusively, because a person can spend hours starting and stopping, rotating and restarting, a motor, without having it come to rest on its bad spot again. It is like waiting for a roulette wheel to stop at a certain number.

It may be that your motor just happened to land on its bad spot the time it would not start up and operate the eject mechanism, and that it operated after you rotated it by hand (or moved the other mechanism in a way that moved the motor) and pushed "Eject" again, due to the fact that you had turned it past its bad spot.

It also is possible that although a bad spot on the motor was not the original cause of the problem, the motor developed a bad spot when power was applied to it, because the original cause of the problem prevented it from rotating and it heated up. For example, maybe a cassette jammed in the mechanism, and when someone pushed "Eject," power was applied to the motor, which could not turn because the mechanism was jammed, and being locked in that position with power applied to it caused a contact or little "brush" inside the motor to burn.

Confronted with a possible bad spot on the motor, many professional repair technicians simply replace the suspected bad motor on general principles,

reasoning that the customer would rather pay for the part the first time around than need to return the VCR to the shop a week later with the same problem of failing to eject.

However, do-it-yourselfers who are repairing their own VCR's are in a somewhat different situation, and they may feel that it is not too much trouble for them to open up the VCR again if the problem recurs, and they may feel that they would prefer to wait to see if the mechanism again fails to eject before paying any money for a new motor that might actually not be needed.

And even if they must wait weeks for their parts supplier to ship them a replacement motor, in the meanwhile they can always open the cabinet and move the motor a little bit off its "dead spot" each time it fails to start up, in order to use the VCR.

Also, to replace the cassette loading motor, it is usually necessary to remove the whole cassette basket loading assembly, plus the front cabinet cover.

These are the main pro's and con's affecting the decision about what to do when a bad spot is suspected on the motor. Since you have the best knowledge of your own personal wishes, needs, and priorities, you are in the best position to make the final decision whether to replace the motor now, or wait to see if the problem recurs. The wires are probably soldered to the motor too, so to change the motor, you will also need a soldering iron and solder.

Chapter 9

VCR Will Not Accept Cassette (Top-Loader)

A "top loader" is a VCR designed to accept cassettes inserted through an opening on the top of the cabinet. If you are having trouble with a front-loader, then you should be reading Chapter 10, rather than this chapter. A "front-loader" is a machine that accepts cassettes through an opening in the front of the cabinet.

A top-loading VCR has a sub-assembly with a "cassette basket" that holds the cassette with a plastic lid over it that should rise vertically out of the top of the cabinet, receive a cassette inserted from the front side of the raised basket, and then descend and latch when it is gently pushed back down into the VCR. If the problem with your top-loader is that the cassette basket will not rise up when you push "Eject," then you should go to Chapter 7, "Will Not Eject Cassette (Top-Loader)."

In this chapter, the "cassette basket" (which is the holder into which the video cassette goes before it descends into the machine) will sometimes be called just "the basket" for short.

The present chapter assumes that your cassette basket is opened already in the "ejected" position, but that when you try to put a cassette into the basket, it either (i) will not go in properly, or (ii) when you try to push the cassette basket down, it will not descend all the way, or (iii) the basket will not latch or stay in the down position. These three different kinds of problems will be treated separately.

Also it is assumed that you are trying to insert the cassette in the rightside-up position, with the door on the plastic cassette box facing in the correct direction. On all top-loaders, the cassette should be inserted with its hinged door facing toward the rear of the VCR. And it is assumed that you have tried several different cassettes, all of the right type for the machine (VHS or Beta, whichever it is), to eliminate the possibility that the cassette box is itself defective, and that when you tried a variety of cassettes, none was accepted.

1. Problem One: Cassette Will Not Even Go Into Raised Cassette Basket

The most common cause of this problem is some fault in the way the lid is sitting on the sides of the cassette basket underneath.

Look to see whether one or more of the screws holding on the lid might be sticking through on the inside, preventing the cassette box from going in all the way. If you find that the cassette box is hitting the end of a screw in a VCR that has recently been disassembled, chances are that the wrong screw(s) have been put in some places. The shorter screw that was correct for this cassette basket location now may be screwed into some other position with the same hole diameter and threading.

If you cannot find the correct screw for this location mispositioned somewhere else in the VCR, you will need to go to the trouble either of ordering up a replacement, or taking an example of a screw of the correct size (removed from another location on the lid where the screw does not block insertion of the cassette) to a local shop to see if they can be persuaded to let you have a replacement out of the box of miscellaneous lost screws with no home that most shops have.

In the meantime, most VCR's will work OK with one of the cassette lid screws removed, although you may need to set a book or other similar heavy dry object on top of the lid to hold it all the way down when playing a cassette.

By the way, whenever you remove screws from a cassette basket lid, it is wise to cover the whole opening underneath with soft paper or tissue, and to be super-careful not to let a screw drop into the VCR. This little mistake could cause you lots of grief and expense.

If you do drop a screw or other small part into the top of a VCR, you may need to spend a lot of time searching carefully for it, because if you leave it in there, it can damage the circuitry or the mechanism if it gets in the wrong place. Remove the top cabinet and systematically look over every square inch trying to find it. If that fails, hold the VCR upside-down over a large smooth clean floor and shake it until (hopefully) the screw or other loose part falls out.

If a broken cassette lid, or a broken plastic part on the lid, is the cause of the problem, then you may be able to fix it with "superglue," following the instructions that come with the glue. If this does not work, you will need to try to get the part number by contacting the company selling VCR's of your brand, or one of the general parts sources listed in Appendix II, and order a replacement lid for your VCR.

Some cassette lids have two little square holes molded at the front edge, through which two little metal tabs should go. If these locking tabs have come out of the holes, this can cause a cassette-acceptance problem, which can be fixed by carefully removing the screws attaching the lid, repositioning the lid correctly with the tabs through the holes, and putting all screws back in, being careful at all times not to let a screw drop down through the opening and into the VCR.

On the back edge of the cassette basket, little tabs from the sheet metal are bent up to prevent the cassette from going too far toward the back. If one of these tabs has gotten bent so that it stops the cassette in the wrong place, too soon, remove the lid of the basket and use a pair of pliers to bend it gently (and once only!) to the proper position.

2. Problem Two: The Cassette Basket Will Not Go Down All the Way into the VCR

Four possible causes of this problem are:

1) A foreign object has fallen into the VCR mechanism

2) On those few VCR models that have a separate adjustable plastic piece between the front of the cassette basket and the rest of the VCR, this part may have become mispositioned so that it prevents the basket from going down correctly

3) The latch in the VCR that is designed to catch on a part of the basket and hold it down is not open as it should be, but closed, bent, or broken

4) The part of the basket with the pin or arm that is supposed to be caught and held by this latch is bent and needs to be straightened or adjusted a little bit

Let us investigate these possibilities carefully.

3. Foreign Object in VCR

Do not overlook the possibility that a foreign object may have fallen into the VCR and is now preventing the cassette basket from descending and latching properly. This is more frequently a cause of problems than you might think. Pens, coins, bits of paper, and small nuts and bolts are the most common culprits, but over the years, I've removed everything from a child's toy to a medium-sized screwdriver from the inner works of VCR's.

You will need to unplug the VCR, of course, and remove the top cover (see Appendix III for help removing the cabinet) in order to check for the possible presence of a foreign object.

As you will observe, the mechanism of a VCR is so complicated that a small foreign object inside can easily escape your eye if you do not look carefully.

A flashlight is a great help here. In your mind, divide the reel table into little square areas, and search each square carefully. If you find a foreign object, of course, remove it and then check to see whether this solves your problem.

And even if you discover no foreign object, as you check out the other possibilities, continue to bear in mind the possibility that one is there that you've overlooked, and keep your eyes open.

4.   Mispositioned Plastic Cassette In-Pusher on Panasonic-category "Tanks"

Some of the large earliest VCR's with metal frames and push-down mechanical switches carrying Panasonic, RCA, Quasar, Magnavox, Philco, and many other brand names looked like this:

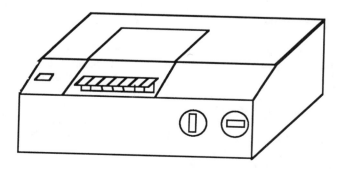

Early Matsushita "Tank" Exterior

If your VCR is not one of these models, you can skip directly to the next section. On these models, a black plastic piece longer than a cassette and about 1/2 inch (12 mm) in thickness was screwed to the metal piece in front of the cassette basket.

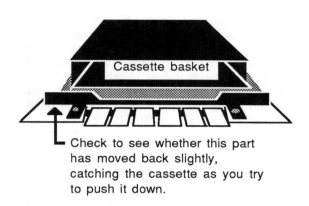

Cassette basket

Check to see whether this part has moved back slightly, catching the cassette as you try to push it down.

Its purpose was to press the cassette all the way back into the basket as it descended into the VCR, but if this plastic part gets displaced slightly toward the rear of the machine from its correct position, the front of the cassette may catch on it and not go down properly. If this may be the cause of your problem, do the following.

Remove the back cover and the top cover (see Appendix III for help). With a scribe or other sharp tool, scratch lines close around the heads of the hold-down screws, and also scratch a close outline of this plastic part in the sheet metal to which it is fastened, so that you can easily put it back in the same position in case we are wrong, and it was in the correct position.

This way, we can easily put it back in the same position, in case it was already positioned correctly.

Now remove the screws fastening this part down to the metal, being super-careful not to let any of them fall into the innards of the VCR.

After lifting the part free and setting it aside, put a cassette into the basket and, holding the cassette pressed in gently with your thumb, try pushing the cassette and basket down into the VCR.

If it now goes in OK, this was your problem, and you will need to reposition the plastic part a tiny bit closer to the front of the VCR when you reattach it.

But if the same trouble still persists, then put the plastic part back in the same identical position when you reattach it, and go on to the next possibility.

5. Trouble With the Latch or Pin

It will be most efficient to consider the remaining two possibilities together. Either the latch that is supposed to hold the basket down by catching a pin or arm that is attached to the bottom of the cassette basket is closed or has something else wrong with it, or else the pin, or its supporting arm on the basket, have gotten slightly bent, so that the mouth of the latch misses it, or it hits on something else instead.

This latch in the VCR, and the part of the basket that it is supposed to catch, have almost as many different locations as there are different models of top-loaders, so it is difficult to tell you exactly where it is located on your machine, but you should be able to figure it out quickly by looking for some sort of pin or arm on the bottom of your cassette basket when it is in the "up" position. A few typical configurations of such catches are shown below.

Look to see whether either part seems bent or broken. Try to double-check by finding a good angle view of what happens as you gently try to push the empty basket down into the VCR.

If the pin appears not to be heading correctly into the mouth of the catch, use a pair of needle-nosed pliers to bend it gently so that it will go into the mouth

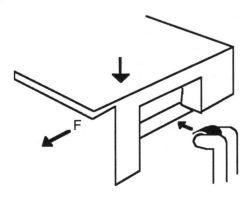
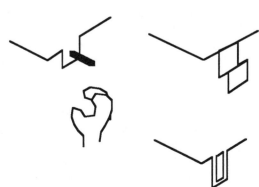

Some Typical Top-Loader Cassette Basket Latches

of the catch. When bending sheet metal, remember that it will break if you bend it back and forth a lot. Try to do it right on the first try.

If, on the other hand, you observe that the mouth of the latch is not open to receive the catch from the basket, then you will need to find a way to get it working properly. A bent catch on the basket that needs straightening may have been the culprit that pushed the latch closed with nothing held in it, but you will still need to figure out a way to open the latch in order to use the VCR.

The first trick to try is the simplest. Plug in the VCR, power up, and push "Eject." If the latch does not open, treat this now as a problem with ejection and go through the procedures in Chapter 7, "VCR Will Not Eject Cassette (Top-Loader)." Return to this point in the present chapter when you have gotten the latch open. If the latch opened when you pushed "Eject," go directly to the next paragraph.

Unplug and try to watch the catch as you gently attempt to push the basket down. This is sometimes difficult to do, because the crucial events occur hidden beneath the cassette basket as it descends. In some cases, you can get a better view by removing the plastic lid of the cassette basket and/or opening up the bottom of the VCR to expose the underside of the reel table.

If you still cannot watch the action of the latch and catch as the basket comes down, you will have to try to deduce what is going wrong, or what happened, from such indirect evidence as scratch marks on the latch or catch, the direction in which the catch was bent, etc. The number of possible mispositions is small enough that, if all else fails, you can simply run through all the possible readjustments until you find a position of the catch that goes correctly into the mouth of the latch and stays held properly until you press "Eject."

Once you've gotten the basket to latch down, check to make certain that the VCR plays and records cassettes properly, because if the latch and catch do not hold the basket quite low enough, cassettes will not sit correctly on the resting pins in the VCR, and as a result, the cassette will not play and record correctly.

Important: Use the check-out procedure in Chapter 4 before using this chapter.

Chapter 10

VCR Will Not Accept Cassette (Front-Loader)

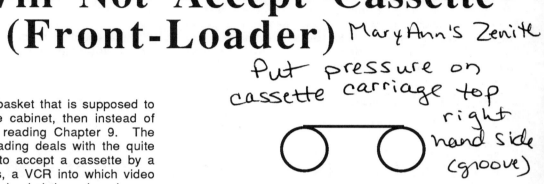

Mary Ann's Zenith
Put pressure on cassette carriage top right hand side (groove)

"Cassette In" Symbol Used in the Front Panel Display on Many VCR Models

If your machine has a basket that is supposed to rise up out of the top of the cabinet, then instead of this chapter, you should be reading Chapter 9. The chapter that you are now reading deals with the quite different topic of the refusal to accept a cassette by a "front-loading" VCR — that is, a VCR into which video cassettes are supposed to be loaded through a door or slot in the front cabinet.

On "front-loaders," when a cassette is pushed gently part way into the front panel opening, a mechanism is supposed to be activated that automatically takes hold of the cassette and pulls it into the VCR and sets it down over the reels on the reel table.

Naturally, you are inserting the cassette pointed the correct way. On most models, the correct way to insert a cassette is with the window side up, the side with the two round spools showing through round holes pointed down, and the side with the flap or door going in first, but on VCR's that accept cassettes inserted end-first, you must insert the cassette in such a way that the slot on the bottom of the cassette (the side with the round holes) is correctly oriented. End-loaders will never accept some off-brand tapes that do not have this slot molded into the plastic — incompatible designs.

If the VCR that is failing to respond as it should when you try to insert a cassette is one of the models that requires you to turn on the power switch first, I assume that you have already double-checked to make certain that the indications are that the power has come on. If your VCR has a "cassette in" indicator, check whether it might be indicating that the machine mistakenly thinks that there is already a cassette inside it. On many VCR's, the "cassette in" indicator is an illuminated figure that looks as shown in the next illustration. If this or any "cassette in" indicator is lighted when no cassette is in the VCR, the machine probably is not accepting a cassette because

it mistakenly "thinks" that it already has a cassette loaded inside it. If this is the situation, read Chapter 11 before continuing with this chapter.

If the power switch is on and the machine has powered up normally, yet it still will not accept a cassette, this indicates that there is some trouble with either the cassette loading-assembly or with the circuitry located elsewhere in the VCR that is supposed to control the front-loader.

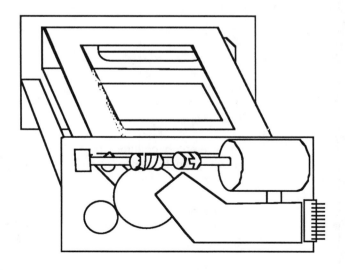

Typical Cassette Front-loading Assembly

The total "cassette loading system," consists of the following:

A) The mechanism of the cassette loading assembly, including some switches, the cassette basket, the cassette loading motor, and the surrounding structure that supports it all. Here is a picture of a typical kind of cassette front-loading assembly:

B) The additional electronic circuitry located elsewhere in the VCR and connected to the cassette loading assembly by wires, that should respond to the insertion of a cassette into the front of the cassette basket by applying power to the cassette loading motor, causing it to turn which pulls the cassette into the VCR, and setting it down in the correct position over the two reels and then shut off.

If the whole problem is in the cassette loading assembly (A), then you probably can fix it yourself (except that changing out the motor or switches requires soldering, an intermediate-level skill), but if the trouble is in the external controlling electronic circuitry, then it is beyond the scope of the methods in this book, and you probably will need to bring it to a shop (unless the problem is just a blown fuse elsewhere in the VCR). Fortunately, most of the time when a VCR will not accept a cassette, the problem is in (A), the front-loader assembly. (It is also possible for a problem in (A) to cause a problem in (B), or vice versa.)

As usual in repairing VCR's, most of the time is spent determining where exactly the cause of the trouble lies.

1. Step One: Is the Cassette Basket in the Fully-Ejected Position?

The loading assembly will not accept your cassette if it is not resting in the fully-ejected position. It must be in the fully-ejected position before it will accept a cassette.

When power is applied to the cassette loading assembly, the little electric motor is supposed to rotate, turning gears that move the cassette basket back-and-forth, and up-and-down, in the cassette loading assembly. When power is connected in one direction, it causes the motor to spin in a direction that "loads" the cassette, pulling it into the VCR and setting it down on the reel table. If the connections are reversed so that power is applied to the motor in the opposite direction, the motor is supposed to turn the other way and "eject" or "unload" the cassette by raising the basket up from the reel table and moving it horizontally out to the front of the VCR.

Let us try to determine whether the trouble is in this cassette-loading mechanism.

Unplug the VCR and remove the top of the cabinet, if you have not already done so. If there is a circuit

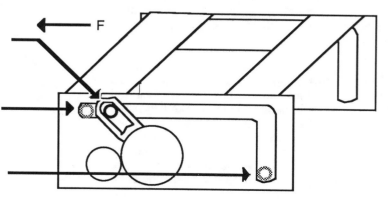

Moving pin from cassette basket in an intermediate position.

Fully ejected position for moving pin from cassette basket.

Fully loaded position for moving pin from cassette basket.

Side view of Cassette Loading Assembly with Motor and some Mechanism Removed to Show L-shaped Slot

board covering the top of the cassette front-loading assembly, unfasten it and prop it up with a stick made from wood or other nonconducting material. See Appendix III for help in removing the cabinet top if necessary.

The cassette-loading assembly is the metal-and-plastic assembly located inside and behind the opening on the front cabinet through which tape cassettes are supposed to be inserted. The sides of this assembly extend down to the reel table to which they are attached. Inside the assembly is a tray or "cassette basket" (called "the basket," for short) that is supposed to move along on L-shaped slots in the housing on both sides of the assembly, as shown in the illustration.

There should be a protruding pin attached to the basket on each side extending through these slots and into a slot in an arm that can move the pin, and with it, move the basket, back-and-forth and up-and-down along the L-shaped path. This part is called the "wiper gear" in Appendix VII. The illustration shows a simplified picture of the front-loader as seen from the right side. The left side is similar.

Do not touch the cassette loading mechanism yet, but observe carefully the position of the basket, and especially this pin. Go on to the next step if the basket is in the "fully ejected" position — that is, up and near the front of the front-loader.

NOTE: On some models, when the cassette basket is in the fully ejected position ready to receive a tape,

the pins on both sides of the basket are all the way forward, right up against the forward end of the L-shaped slots in which they slide, while on other models, when the basket is correctly positioned in the fully-ejected position, there may be a tiny bit of space left empty at the very end of the slot. But in all models, the pin on the right side and the pin on the left side should be the same distance from the end of the slot when fully ejected.

If the cassette basket is in the fully ejected position, all the way, or close to all the way, to the front of the L-shaped track, go directly to Step Two, "Will It Load a Cassette Under Battery Power?"

But if you find the cassette basket to be situated at any position other than the fully ejected position — for example, down on the reel table in the fully-loaded position, or in the middle, half-way between the two extreme end-positions in the L-slot — then plug in the power cord, turn on the power switch, and try pushing "Eject."

If the cassette loading motor shows signs of trying to turn, but the basket in the cassette loading assembly does not move, then unplug immediately and go directly to Section 9, "Clean the Switches Attached to the Front-Loader," and the following sections, and try realigning the front-loading assembly.

If, on the other hand, the VCR responds by moving the cassette basket to its fully-unloaded position, unplug and proceed to Step Two.

Section 2. Step Two: Will It Load a Cassette Under Battery Power?

If you have not already done so, read the section of Chapter 8, entitled, "Special Tools and Techniques for Front-Loader Troubleshooting: Using a Battery to Run the Front-Loader," which explains how to connect a battery to the cassette loading motor. Then return to this point.

Obtain a 9 volt battery and clip leads as described. If no battery is available, go to Step Three.

Find the little electric motor that is supposed to move the cassette basket. Usually it is attached to

the cassette loading assembly, but on a few models, it is located down under the reel table and connected to the front loader via a belt that goes to a little pulley.

Make certain the VCR is unplugged. Push a "junk" test cassette through the front panel opening into the cassette basket gently so as to move the cassette basket enough to free the catches latching it in the fully-ejected position, and at the same time, connect the probes from the battery to the cassette loading motor terminals, first in one polarity, then try reversing the leads if nothing happened when they were connected the first way.

If you don't have enough hands to do all these things at once, get someone else to hold the cassette in while you try attaching the battery first the one way, and then the other way. Observe the result. If the mechanism starts to pull the cassette in, leave the battery connected until the cassette basket and tape go all the way to the other end of the L-shaped slot and stop. If it can do this, next try reversing the connections and observe whether it can run the cassette back out to the fully ejected position.

If the mechanism moved and started to pull the cassette into the VCR under battery power, then you know that the little motor is OK.

If it moved for a ways, but stopped before it got to the other end of the L-shaped slot and would continue no further under battery power, then go to the section entitled, "Aligning and Retiming Front-Loading Assemblies," and the following sections, remove the front-loader, and realign it.

If it went all the way to the end, and back again (when you reversed the connections), then the loading motor is OK and probably the front-loader mechanism is correctly aligned, and the problem is either in the "cassette in" switch(es) or elsewhere in the VCR. Read the section entitled, "Aligning and Retiming Front-Loading Assemblies," and follow the directions in the following sections for checking the operation of the switches in the front-loading assembly.

If the basket did not move, then there is either a problem in the motor or the mechanism is jammed or misaligned. Go next to the test in Step Three.

3. Step Three: Does the Cassette Loading Motor Suffer from a "Bad Spot"?

Look at what connects the loading motor to the gears and other parts that are supposed to move the basket.

In most cases, the connection will be via a shaft that comes from the motor and goes to a flexible mechanical connector (called a "universal joint") attached to a small "worm gear" that meshes with the edge of a larger-diameter gear.

With the VCR unplugged, experiment to find a way manually to rotate this shaft in one direction or another a little bit. Usually this can be done by catching the tip of a scribe or small screwdriver on the edge of the universal joint and pushing gently to rotate it. If the motor is connected to the rest of the mechanism by a belt and pulleys, use your finger to rotate the pulley attached to the motor a half turn.

Now repeat the test described in Step Two.

If the motor still will not start, go to Step Four.

If the motor now will start, the indications are that the motor has a "bad spot" on it.

A "bad spot" is a particular position from which, if the motor comes to rest in that position when it stops rotating, it cannot start up again — but it can start from other resting positions.

If the VCR now accepts a cassette, eject it, and try it again. If the motor again fails to start (at least until you turn the motor to a slightly different position and try again), then the problem is bad enough that the motor is going to need to be replaced.

On the other hand, if the VCR now seems to accept cassettes as it should, the "bad spot" is not yet really bad. Days, weeks, even months may pass before the machine again refuses to accept a cassette, at which time you may be able again to fix the problem temporarily by removing the cover again and rotating the motor slightly.

The question of when the motor has become bad enough that the trouble justifies replacing it is a personal matter that depends on the relative values you assign to cost versus convenience. With some front-loaders, you also may be able to improvise a solution to the problem by removing the motor, turning it 180 degrees around its axis of rotation, and reinstalling the whole motor in this rotated position, effectively changing the position of the burned spot.

If you learn how to solder, you can install a new motor yourself. Section 17 of this chapter provides more information on how to install a new cassette loading motor.

4. Step Four: Can the VCR Now Power Up the Cassette-Loading Motor?

Use a stick made from wood or other nonconducting substance to prop open any circuit boards that you had to lift up to gain access to the front loader. Plug in the VCR, and turn on the power. If the cassette basket starts to move, wait until it stops moving before trying to insert a cassette. Then try to insert a "junk" cassette used for testing purposes, to see whether the VCR now will accept it.

If the motor still does not start up, go to the next step below.

If the motor shows signs of trying to turn, but the mechanism of the cassette basket does not move, immediately unplug and go directly to the section entitled, "Aligning and Retiming Front-Loading Assemblies," and following sections to find out how to disassemble and realign a front-loading assembly.

If the VCR now accepts a cassette, loads it fully, and ejects on command, then the indications are that you have a "bad spot" on your loading motor.

5. Is Insertion of the Cassette Failing to Activate the "Cassette Load" Switch to Turn on the Cassette Loading Motor?

What tells the VCR's little brain to turn on the power to the cassette loading motor to make it pull the basket (plus the cassette inside it) into the VCR? Answer: When the cassette is inserted, it pushes against one (or more) tiny electrical switch(es) mounted in the cassette loading assembly, and this sends a signal to the VCR's little brain that a cassette has been inserted and that it should start the motor to pull it in and set it down on the reel table.

It is possible that your VCR will not accept a tape because something is wrong with the "cassette in" switch(es). The switch(es) may have gotten bent, broken, or out-of-position, or the electrical contacts inside the switch may be corroded or dirty. We will try to take care of this when we remove and realign the front-loading assembly next.

6. Aligning and Retiming Front-Loading Assemblies

Appendix VII gives detailed step-by-step procedures for aligning and retiming some of the most common cassette front-loading assemblies, but limitations of space in this book prevent listing

specific step-by-step procedures for all the various designs used in all the different models. Fortunately, however, this is not a big problem, for once the general resemblances and similarities in all the front-

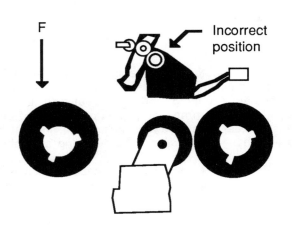

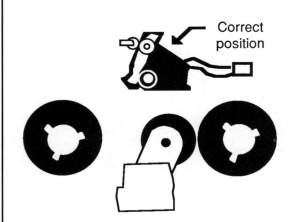

Sharp reel table and sensor lamp
as seen with cassette front-loading
assembly removed from VCR.

Sharp reel table and sensor lamp
as seen with cassette front-loading
assembly removed from VCR.

Putting the Sensor Lamp Back into Correct Position on a Sharp-category VCR

loaders are explained, most of them can be realigned and retimed with no service literature to show you the specific steps.   Understanding these universal similarities will help you fix your front-loader.

If a VHS front-loader starts to load the cassette, but jams part way in, check the little spring-loaded button or protrusion on the right-hand side that is supposed to press the button on the side of the cassette to unlatch the door on the cassette over the tape. Make certain that it is not caught pushed back in. Make certain that it is releasing the door catch, so that the door can come up.   If it is not, find the cause and correct it.

On Sharp-category VCR's, if the front-loader accepts the tape but jams part way in, check the little brass leaf-springs located at the underside of the top of the front door opening to make certain that they are pressing down on the tape with the right amount of force.   If the springs are bent down so that they press too hard on the top side of the cassette, then the cassette basket may start to be pushed back and activate the loading motor before the cassette has gotten quite all the way in, causing the rear edge of the cassette to jam against the edge of the door housing as it starts to load the cassette. If this is the problem, bend them up a tiny bit.

If they have   gotten bent up, in the other

direction,   so that they do not grip the tape firmly enough, the cassette may start to slide back out as the basket starts to move, and again jam against the edge of the loading assembly as the basket tries to move in and down.   If this is the problem, bend them down a tiny bit.

Also on Sharp-category VCR's, check to see whether the plastic base of the sensor lamp may have come loose or twisted out of its correct position causing the lamp to hit against the bottom of the cassette, rather than go into the hole molded for it in the cassette box, as shown in the illustration. If this has happened, reach in and move the sensor back to its correct position, and try again.

Also on Sharp-category VCR's that have not the idler (rubber wheel) shown in the picture above, but rather the NIDL-0005GEZZ or NIDL-0006GEZZ idlers pictured in Chapter 13, the VCR's refusing to accept or eject a cassette can be due to a dirty, slipping idler (which prevents the reels from being moved during the start-up test that the VCR system control runs when it is first turned on, causing it to shut down these functions).   But to replace this idler, you will need to remove the cassette front-loading assembly (although if there is no cassette in the VCR, you may be able to clean the rubber tire with a very long cotton swab stick moistened with alcohol). To find out how to identify and replace this idler, see Chapter 13.

7. Step Five: Remove the Complete Front-Loader Assembly from the VCR

It is time to take the big step of removing the complete front-loading assembly from your VCR so that it can be worked on (and also, if there is a cassette stuck in it, so that the stuck cassette can be removed). Consult Appendix III, for a detailed explanation of how to remove the front-loading assembly. Return to this point after you have finished removing it.

Look at your front-loader while you read the next section.

8. Design Similarities in Front-Loaders

When the VCR power is turned on and a cassette is inserted into a front-loader assembly, as the inserted cassette fills the basket, the gentle force of insertion is supposed to move the basket to the rear a short distance, which causes a change in the position of the contacts on one or more tiny electrical switches, as well as releasing one or more mechanical catches that held the basket latched in the fully unloaded position until the cassette was inserted.

A "leaf switch" consists of two or more thin, flexible strips of springy metal with electrical contact points that close together, or open up, in response to being pushed on mechanically. These important electrical switches may be "leaf switches" that are "normally closed" (that is, making contact until pushed), or they may be "normally open" (that is, not making contact) until pushed, as shown in the illustration.

Some leaf switches have more than two "leaves" stacked up in the same mounting. The leaves are attached to wires, and to plastic parts or circuit boards that are supposed to hold them in an exact, correct position relative to the rest of the front-loader assembly.

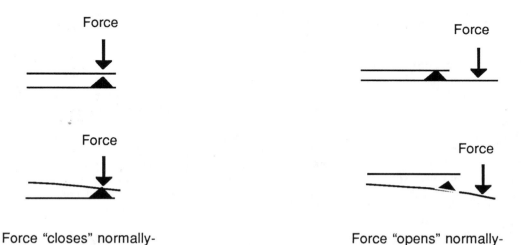

Force "closes" normally-open leaf switch.

Force "opens" normally-closed leaf switch.

Normal Operation of "Cassette In" Leaf Switches

Another common type of "cassette in" switch looks like a push-button switch:

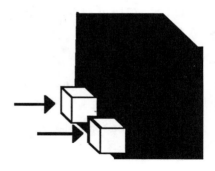

Typical Push-button Type "Cassette In" Switch

Depending on design, these switches may be normally pushed in when the VCR is fully ejected, and released when a cassette is inserted, or they may be normally out when the loading assembly is in the fully-ejected position, and pushed in by arms when a cassette is inserted.   Some Hitachi-category machines use this type of cassette-in switch.

Some models, such as some VCR's made by J V C, use an optical-type cassette-in switch.   A tiny, enclosed invisible infra-red light shines through a hole molded in a plastic piece hitting an optical sensor on the other side.   When the front-loading mechanism moves, the piece with the hole moves too, and cuts off the light, which causes a signal to be sent to the VCR's brain to tell it to send power to the cassette loading motor.

The cassette-in switches can be found on either side, or the top, of the front-loader, depending on the model.

You can already see one requirement that must be met by a properly aligned front-loader assembly: the cassette-in switches whose change of contact is supposed to activate the cassette loading motor must be positioned in such a way that normal insertion of a tape moves and changes them properly.   Generally you can tell whether the cassette-in switch is opening and closing properly by finding it and looking closely at it while gently inserting and withdrawing a cassette from the opening of the front-loader.

On many models, the switches are held in position by a screw that attaches their plastic base to the cassette loading assembly, and they can get knocked out of position by someone trying to insert a cassette without the power switch being on, or by rough handling, or sometimes just on their own.   Not only must the switches be correctly positioned relative to the rest of the mechanism, but also the contacts must make good electrical connection when they touch.   If the contacts are dirty or corroded, the switch may not activate the cassette loading motor correctly.

9.   Step Six: Clean the Switches Attached to the Front-Loader

If the switch contacts are not enclosed in an outside plastic housing, clean them by spraying them with head cleaner, or if that does not do the job, by rubbing both contact points with a piece of ordinary clean paper (like typewriter paper) that has been moistened with alcohol or head cleaner. If the contact points are inside a plastic enclosure where you cannot get at them directly, often they can be cleaned by spraying a lot of head cleaner into the plastic enclosure through a crack, and then pushing on whatever moves the contacts about fifty times.

When the multi-wired connector is attached to the front-loader and the VCR is powered up, you should be able to activate the cassette loading motor to start the loading process with no cassette in the basket by using your fingers to release any mechanical catches latching the basket in the fully-ejected position and then moving the switch or switches in the same way that a cassette should do.

Study the front-loading mechanism that you have removed from the VCR.   With your hand, try the experiment of inserting a "junk" cassette while watching the action of the switches.   As you push the cassette in slowly, you should see at least one of the switches change its state, either opening or closing.

If this does not happen, then this could be your problem. You will need to disassemble the mechanism and then reassemble it with the parts repositioned so that the switch gets switched when a cassette is inserted. Also, if you can, spray some head cleaner onto the switch contacts, or into the switch enclosure, and work the switch with your finger many times.

10. More on How the Front-loading Motor and Mechanism Work

The cassette-in switches are positioned on the front-loading assembly so as to be pushed on when a cassette is inserted, causing their contacts to change position. If everything else is right, this serves to send a signal to the little computer to send power to the cassette loading motor.

The cassette loading motor should then start up and rotate in a direction that turns the mechanism in such a way as to pull the right side of the cassette basket back to the rear, and then downward, in the L-shaped tracks, carrying the cassette inside it.

The illustration shows a simplified picture of one very common way in which the cassette motor accomplishes this. The motor side of a typical cassette loading assembly is shown with confusing frills and mechanical details omitted. In this design, the motor turns the worm gear, which turns the big gear, which moves the arm, which pulls the pin

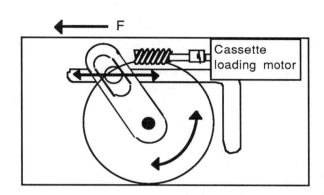

Simplified Picture of Typical Way that Motor Pulls Cassette Basket into VCR

attached to the cassette basket back along the L-shaped slot.

11. Step Seven: Check for Warped Plastic Motor Mount

Use a felt-tip pen to put an identifying mark on the outermost, or topmost, side of the loading motor (if it is attached to the cassette loading assembly) to help in positioning it later on reassembly.

One common problem with some front-loaders is that the motor overheats for one reason or another, and warps the plastic part to which it is attached, causing the motor now to be misaligned relative to the shaft that it is supposed to turn.

When this happens, the mechanism can bind and "freeze," sometimes preventing it from accepting or ejecting a cassette. If this has happened, the warped plastic part must be replaced, and it probably

would be a good idea to replace the motor too, while you have it apart, as a precaution against the new plastic piece getting melted too by the old motor in a recurrence of the same problem.

Also, any motor that got hot enough to melt a plastic part touching it probably has itself been internally damaged. Look very carefully at any plastic parts that touch the motor, studying them from both sides. Do you see the slightest warp or curve in the surface that does not look like it is supposed to be there? If so, you probably will need to replace that part to make the front-loader operate correctly.

12. How Gears Make the Two Sides of the Basket Move Together In Step

Additional gears are used to cause an arm on the other side to move in exactly the same way, so as to move the pin on the other side of the basket along the slot on the other side in step with the motor side. A typical set-up is to have a second, smaller gear mesh with the big one, and drive an axle that runs across the bottom of the front of the assembly, to drive an identical small gear on the other side, which drives a big gear and arm on the left side, that moves the left side of the cassette basket in time with the right side.

Thus, another requirement that a properly aligned cassette front-loading assembly must meet is that all the gears must be meshed in such a way that the turning of the motor causes both sides of the basket and cassette to move along the L-shaped slide tracks in parallel.

If one of these gears "skips a tooth," then one end of the basket moves ahead of the other, and can jam up in the mechanism. These plastic gears often jump a tooth or two, and when this happens, the two sides of the basket do not move together.

As a result, one side of the basket tries to pass around the right-angle bend in the L-shaped slot ahead of the other, often jamming the mechanism.

And even if both ends of the basket manage to get around the corner without jamming, one end or the other may fail to go all the way down to the end of its track in the fully-loaded position, leaving the cassette at a crooked angle. Or when the mechanism tries to eject a cassette, one side of the basket may be stopped at the end of the slot before the other side gets to the end of its track, preventing it from getting into the fully-ejected position. Then the mechanism may refuse to accept a cassette.

The two sides of the cassette basket either should move together in parallel, or, as in Fisher and some other makes, the nonmotor side should move so as to reach either end of its L-shaped track slightly before the pin on the motor side does so, in order to allow for the small amount of loose play that is unavoidable with plastic gears.

The "timing" of the one side relative to the other is generally most easily adjusted by removing the fastener, if any, holding on the outermost gear on the nonmotor side, sliding this gear off its axle enough to rotate it a little so that different teeth mesh on the nonmotor side, and then testing the operation of the front-loader by rotating the driving shaft like the motor does. It should not take very many trials to find the correct position by trial and error, because the gear teeth are relatively coarse, so the number of different possible meshing positions to be tested is small.

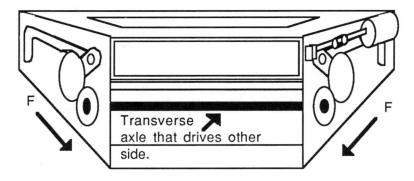

*Front View of Loading Assembly Drawn in Exaggerated
Perspective to Show Transverse Axle Connecting Two Sides*

Some models have little marks molded into the

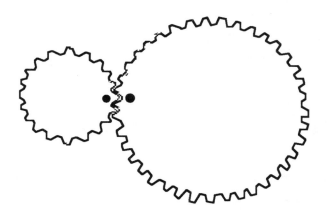

Two Gears with "Timing Marks" to Show How They Should Go Together

plastic gears that you can use to put the gears together with the right teeth meshed together. These are called "timing marks."

You cannot change the teeth on the gears, but you can change the way in which the teeth on one gear mesh with the teeth on another gear.

To "retime" gears means to take them apart and then reassemble them changing which tooth in the one meshes with which tooth in the other. To retime these gears, it is almost always necessary to slide one gear off the shaft on which it turns, rotate it a little, and then put it back on with a change in the way the teeth mesh. The way the teeth mesh should be changed in a way that makes the two sides of the cassette basket move in synchronization with each other.

These gears usually are held on either with a metal C-ring or with a catch or tooth molded into the center of the gear that latches into a slot cut around the circumference of the shaft on which it turns, near the end.

One way to remove metal C-rings is to get a very small screwdriver whose blade will fit into the little space between the C-ring and the shaft, insert the screwdriver blade, cover the C-ring and tool with a clean cloth to stop the C-ring and catch it if it decides that it wants to fly off, and twist or pry gently with the screwdriver to work the C-ring free.

To reinstall, line up the C-ring with the groove in the shaft that it goes onto, hold in place with needle-nosed pliers with one hand while covering with a cloth, then squeeze gently to force the C-ring to spread and go into the groove.

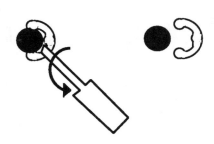

To remove C-ring, insert blade of small screwdriver in space between C-ring and shaft, cover with cloth (not shown) and twist.

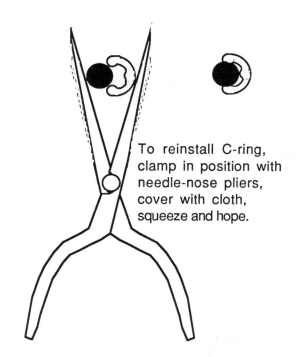

To reinstall C-ring, clamp in position with needle-nose pliers, cover with cloth, squeeze and hope.

Removing and Reinstalling Metal C Rings

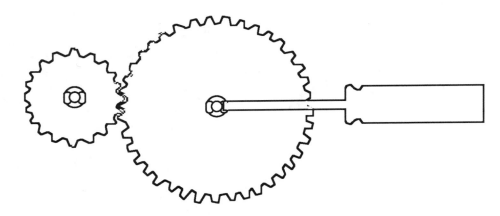

Using Screwdriver to Remove Gear Held on by Plastic Tooth

On other models, the gears have one or two spreadable teeth molded at the edge of the hole in the center. To remove one of these gears from the shaft, use a small screwdriver to spread back the plastic tooth while pulling the gear firmly off the shaft.

This design is easy to reinstall: just press the gear back onto the shaft until the plastic tooth catches in the groove in the shaft.

Adjustment Precautions

In adjusting the timing of the front-loader gears, observe these precautions:

Do Not Force a Stuck Mechanism to Move

1. If the mechanism seems to jam or stick at one point, never attempt to force it past this point, for doing so may break a plastic part.

Remember, too, that most front loaders mechanically latch themselves in the "fully-ejected" position once they get there, so that in order to manually turn the motor and run the mechanism through its cycle, you must unlatch these catches, either by inserting a cassette or directly, with your fingers.

Do Not Let C-Rings have a Chance to Escape

2. Never attempt to remove metal C-rings or snap rings in an environment where they would be difficult to find if they sprang away from you, as they like to do.

Do not work over a rug, or in a workshop with lots of cans, bottle, or junk around, because it might take you hours to locate a spring fastener that sprang away from you in such a location, and you might have great difficulty obtaining a replacement from anyone but the manufacturer (after a long wait). Instead, work on a large clean, bare floor, and cover the fastener and tool with a soft tissue or cloth to catch the fastener if it tries to spring away from you when removing or reattaching a springy fastener.

Make Alignment Marks on Parts Before Disassembling

3. Use a felt-tip pen that is not water-soluble to put your own "timing marks" on each moving part (except a worm gear) to show their positions in relation to each other before you do any disassembling of gears, preferably when the basket is at extreme end of its slide (if possible, the fully-ejected position).

This will give you a starting reference point for timing adjustments, and also help you get the mechanism reassembled back to the point at which you started, in case you become confused and want to restart.

Watch Out for Parts Inside the Universal Joint

4. On some models, an unconnected, free part inside the universal joint, on the shaft between the work gear and the motor, will simply fall out if you are not careful when the mechanism is disconnected. It is fine to remove this part and reinstall it again later, but carefully note and mark with a felt-tip pen which side went forward, and make a drawing of it in your disassembly notes.

Insure Correct Operation of Cassette Assembly Switches

5. Before, or at the same time as, adjusting the relative positions of the two sides of the basket as the mechanism slides them back and forth, check to make certain the switch or switches that signal the motor when to stop at the end of the track are correctly positioned to do so.

These switches usually touch some part of the gear assemblies that move the arms, and the position of the contacts of at least one of the switches should switch or change at each extreme endpoint of the excursion of the basket.

13. Step Eight: Test the Operation of the Mechanism Manually

If there is no cassette stuck in the front-loader currently, insert a "junk" cassette in the front entrance, push it in gently, and with your fingers or a small tool, rotate the shaft from the motor in whatever direction makes it try to pull the cassette in. Try one direction for a turn or so, then try the other direction if that does not work.

If it runs smoothly, watch the cassette while you turn the mechanism to move it all the way to the end of the L-shaped slot. If it jams up, or goes crooked, or starts turning hard, stop and try to eject it by running the mechanism the other way.

14. Step Nine: If the Mechanism Does Not Move Properly, Disassemble and Reassemble with Parts Properly Aligned for Smooth Movement

If a cassette sticks in the mechanism, or if there is already a cassette stuck in it, before removing the motor or driving side of the front-loader, sketch a drawing showing the position of each screw. Assign each screw a number, and put them one-by-one in containers with the same numbers, or nestled in numerical order, as you remove them, so that you will be able to get each screw back in the correct place on reassembly. (Do not count on your memory for this.)

Check to see whether additional disassembly hints are included for your model in Appendix VII.

If a cassette is stuck in the front-loader, disassemble to the point where you can remove it. If one side of the basket has been running ahead of the other side, disassemble to the point where you can change the way one of the driving gears meshes, so as to move it ahead or behind a tooth. (Make your own timing marks with a felt-tip pen, showing how all the gears meshed before disassembling them.) If the cassette in switch(es) was (were) not being activated correctly, disassemble down to the part that presses on them.

When the front-loader has been disassembled and reassembled so that the empty basket will move smoothly back and forth through its full excursion when you rotate the motor shaft manually, insert a cassette and make certain that it still moves properly when a cassette is in the basket.

If so, try powering up the cassette loading motor with a battery, as described in Chapter 8, and in the following section of this chapter.

If you cannot make the battery test for any reason, read the next section, then go directly to Step Eleven, reinstall the front-loading assembly in the VCR, and see whether it will work now on power from the VCR.

15.   Testing the Cassette Loading Motor

For this test, you will need a small 9 volt battery and two alligator clip leads, as described in Chapter 8 (or a comparable direct current power supply).

If you cannot obtain a 9 volt source (or do not wish to run this procedure for any other reason), then you have three options: (i) you can get another VCR that has the same cassette loading assembly as the one you are working on, unplug it and remove the entire cassette-loading assembly from the working VCR as well, and try installing it in your VCR. If the other cassette loading assembly does not work in your VCR, but did work in the VCR you took it out of, then obviously the trouble is in some other part of your VCR, and you will need to take it to a shop.

But if another loading assembly works in your VCR, then the trouble is definitely in your loading assembly. Assuming that you have got the mechanical parts so that they are all correctly aligned, then either your cassette loading motor is bad, or the contacts in the "cassette load" switch(es) are making bad electrical connections. (You could even determine which of these is your trouble by trying the experiment of removing the cassette loading motor from the known good assembly, temporarily installing it in your assembly, reconnecting and retesting. This test would require unsoldering and resoldering both motors.)

(ii) The second option if you do not use a battery is more expensive. You could simply order a new replacement cassette loading motor, install it, reconnect the assembly, and retest it with the new motor installed.

If it now works, you have solved the problem. If it still does not work, you know that you have a trouble elsewhere in the VCR, and it will need to go to the shop. Without a battery or other appropriate power supply, you will not be able to determine whether the old motor was good, or whether it got damaged as part of whatever happened to damage the circuits elsewhere in the VCR.

If the shop to which you take it does not object to working on a machine that you have already worked on, you could mention that the front-loader has a new motor in it, so that they do not install yet another good motor.

(iii) The third option is to reinstall the whole front-loading assembly in the VCR and see whether it will now work under power from the VCR. If it does work, hooray!, it's fixed. But if it does not work, without a battery to test it, you will not know whether the problem is with the motor, or something else.

---

16.   Step Ten: How to Test the Motor

Begin by reading the section of Chapter 8 on using a battery and clip leads to test the cassette loading motor, if you have not already done so. It is entitled, "Special Tools and Techniques for Front-Loader Troubleshooting: Using a Battery to Run the Front-Loader." Then return to this point.

Insert a "junk" test cassette into the front of the loading assembly, push it gently in, and with your other hand, manually turn the shaft or pulley that the motor is supposed to drive in a direction that will make it pull the cassette in. Stop when the cassette is about half-way to the point of being fully loaded. Support the front-loading assembly containing the

cassette in a normal upright position on the work-bench, and touch the probes from the battery to the two terminals on the motor in either direction or polarity.

The motor should start turning and move the cassette in one direction or another. Remove the battery probes before the cassette gets to the end of the track, and reconnect them in the reverse direction. The cassette should now move in the opposite direction.

If the cassette moves when you connect the battery, then the cassette loading motor is good. This is actually bad news if the loading assembly does not work when you put it in the VCR and connect it up, because assuming that you have correctly aligned it and the switches are OK, it means that the remaining trouble is outside the front-loading assembly and elsewhere in the VCR, so that you will need to take it to a shop.

On the other hand, if the cassette does not move when you connect the battery, and the mechanism is not jammed or warped, then the indications are that the motor is bad and will need to be replaced. This does not eliminate the possibility that there is a second trouble elsewhere in the VCR as well, but probably the only trouble is in the motor. If we replace the motor and it still does not accept a cassette, but a battery will now move it, then we know that we have fixed part of the trouble, but there is still a problem in the circuitry that probably will require professional service.

17. Replacing the Cassette Loading Motor

If the problem is that the cassette loading motor has a bad spot on it, and whenever it lands on this bad spot, it has trouble starting up again, there is a stop-gap solution that works in some cases. Sometimes the motor mounts are such that they will permit the whole motor to be rotated from its initial position and reinstalled in a new configuration that puts the burned spot someplace other than where the motor normally stops, so that it will start up again each time, and rotate past the burned spot by inertia, saving you from buying a new motor, as shown in the illustration.

Check to see whether you can reinstall the motor in a position that is rotated part way from the position in which it was previously mounted. If so, you can put it back in this new position, and test to see whether it now functions satisfactorily.

To replace a cassette loading motor, obtain the proper part number for the motor from your manufacturer, or from a cooperative parts supplier, starting with the sources listed in Appendix II. Parts numbers for some of the most common cassette loading motors are given in Appendix VII.

Replacing a cassette loading motor requires that you unsolder and resolder the connections between the wires and the motor. You'll need a soldering iron, solder, solder wick, and a little practice at desoldering and resoldering before you can do this job. If the use of a soldering iron makes this job too difficult for you, then you will need to turn the job over to a shop.

Wait until you have the replacement motor in hand before removing the old motor, so that you can double-check that it has the same mounting and wiring connections, and most importantly of all, so that your memory of the correct way to mount it will still be fresh when you go to install the new motor. Be sure to draw a picture showing which colored wire goes to which terminal on the motor before unsoldering the

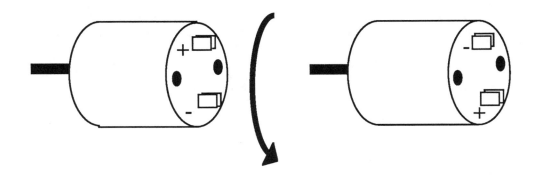

*Rotating this Loading Motor a Half-circle, 180 Degrees,
Leaves the Mounting Holes in Similar Relative Positions*

wires, so that you can get the wires connected to the new motor correctly. If you accidentally reverse the connections when installing the new motor, it will never work.

Try to avoid melting anything other than the solder when you use the soldering iron and solder-wick to remove as much of the old solder as possible before trying to free the wires from the terminals on the old motor. You can unfasten the motor from the assembly that holds it either before or after unsoldering the two wires, whichever you find more convenient.

When you remove the cassette loading motor on Fisher VCR's, and some other models, watch out for a little plastic disk suspended between the two halves of the universal joint, which falls out as you remove the motor. Watch for it as you remove the motor, and mark which side goes forward, so that you can put it back in the same direction when you install the new motor.

On some Panasonic-category VCR's, the

manufacturer recommends using a small wedge as mentioned in Appendix VII to hold the parts in preliminary position during reassembly, but this is not essential.

After installing the new motor, a battery should now make the cassette loading assembly go through its proper motions, when tested as described previously. (If it does not, then there is something still wrong in the alignment of the cassette loading assembly.)

If the VCR still will not accept a cassette after you reinstall the loading assembly in the machine and double-check to make certain that all multi-wire, multi-pin connectors have been plugged back together correctly, then there must be a second fault somewhere in the support circuits elsewhere in the VCR, and you'll need to take it to a shop. A likely suspect is the loading motor drive "chip" (integrated circuit), but before replacing this component, most technicians would use instruments to verify that system control (the VCR's brain) is sending the correct signal(s) to this drive circuit "chip" to tell it to power up the cassette loading motor.

---

18. Step Eleven: Reinstall the Front-Loader in the VCR and Test Whether it Works

After checking to make certain that the VCR is still unplugged from the wall, reconnect every multi-wire connector that you disconnected when you removed the front-loader assembly from the VCR. Make certain that you are not putting the multi-wire plug back upside-down when reconnecting.

Reinstall the front-loading assembly in the VCR, making certain that any little feet at the bottom of it that need to go into slots are in their proper position. Put the front cabinet piece back on, prop up any circuit boards over the front loader with a wooden stick, power up, and try to insert a cassette.

If the unit now works correctly, return to the

universal check-out procedure in Chapter 4 .

If it still does not work correctly, go back to the step of testing the motor with a battery, if you skipped that step. If you earlier found that the mechanism was misaligned or jammed, maybe the cassette loading motor got destroyed trying to move the stuck mechanism. Let us hope so, for that is probably the last remaining possible cause of the problem that you can repair yourself, using the procedures in this book. If the trouble is in the electronic control-and-drive circuitry for the cassette loading motor, you'll probably have to pay a shop more than the relatively inexpensive cost of a cassette loading motor.

Chapter 11

"Cassette In" Indicator Does not Light when Cassette Loads, or Indicator Lights When No Cassette is in VCR, or VCR Loads Cassette but Ejects Automatically Right After Loading

A lot can be learned about a VCR's troubles by reading what it says on the front panel sometimes.

1. "Tape In" or "Cassette In" Indicator Lights When No Cassette is in VCR, or Cassette Loads but "Cassette In" Indicator Does Not Light

Most VCR's have some sort of "Cassette In" light or indicator somewhere on the front panel, often as part of the clock display. If you are having problems with the machine not accepting or ejecting cassettes correctly, look at this indicator while the power is on. (If you do not know whether your VCR has such an indicator, or if you cannot find it, consult an owner's manual for your model, which, if you've lost yours, can be obtained from the headquarters for your brand listed in Appendix II.)

If you turn on the power and the "Cassette In" indicator comes on when no cassette is in the VCR, then you know that you definitely have problems. The little computer ("microcomputer") inside your VCR mistakenly "thinks" that there is already a cassette inside the machine when actually it is empty.

On some models, this will produce the result that the take-up reel will rotate continuously in the rewind direction whenever the machine is plugged in again and powered up, and the machine will not do anything else.

The machine's mistakenly thinking that it has a cassette inside when actually it does not also can cause it to refuse to accept a cassette when you try to insert one, because it thinks that it is protecting you against accidentally putting a second cassette in after the first.

Or, if you can see or feel with your fingers that a cassette already is inside the VCR, but the "Cassette In" indicator fails to come on when you power up, evidently the internal computer mistakenly thinks that there is no cassette inside the machine, when actually there is. And this could explain why it will neither eject nor play the cassette inside.

Erroneous front panel indications could, of course, be due to a problem with the indicator lamp or its supporting circuitry, but when these false read-outs go along with other improper behavior of the VCR that fits with them, then usually they indicate something wrong in the VCR's "thinking" somewhere.

It could be trouble with the electronic circuits by which the little inner computer does its logic, or it could be a partial failure in the internal power supply circuits, but the most common cause of problems such as these is a mechanical misalignment of the front-loading assembly.

Perhaps someone tried to insert a cassette backwards, or tried to force it in when the power was off, or any of several other things may have happened to cause some little electric switch that acts as one of the VCR's "sense organs" to get stuck in the wrong position, sending wrong signals to the little computer, confusing it.

The switches that are supposed to signal the presence or absence of a cassette, and also sense the position of the basket in the front-loader assembly,

usually are located on the side (usually, the right side) of the front-loader assembly.

Sometimes the problem is that the electrical contacts in the switch are dirty or corroded, but more often, some plastic gear(s) and/or other parts of the front-loader's mechanism have accidentally gotten knocked out of alignment, or caused to "skip a tooth" the last time it was used. Most of these front-loading assemblies are too lightly constructed by the manufacturers, to tell the truth. Sometimes the plastic or nylon parts simply warp, crack, or bend, just on their own, under normal careful gentle use.

To troubleshoot and repair such a problem, you will need to unplug the VCR, go to Chapter 10, and follow the procedures given there for testing, removing, and retiming the front-loading assembly.

However, if you happen to have access to a second VCR with an identical front-loader mechanism that is working correctly, you can quickly first determine whether the problem is indeed in the front-loading assembly, by substituting the known good front-loader for the one under suspicion in the malfunctioning VCR.

If the VCR with the problem works OK with the other, known good front-loader, then you know that the problem is in the front-loading assembly. But if the problem remains the same after you install the known good front-loader assembly, then you know that you have a problem in the computer or logic circuitry beyond the scope of this book.

J V C-category VCR's: If you are experiencing a problem with the "cassette-in" indicator in one of the VCR's made by J V C that uses a (hidden) light source that shines through holes molded in a round plastic part at the front of the right side of the front-loading assembly in order to signal to the microcomputer the position of the cassette basket, then your problem is almost certainly beyond the scope of this book. The problem is either a partial failure in the VCR's power supply, a defective sensor lamp, or trouble in the logic circuitry, and you will need to take it to a shop.

2. Cassette Basket Moves Back and Forth, on its Own, In and Out Short Distances, or Chattering

On Hitachi-category machines, this problem is usually caused by bad, marginal, or intermittently bad, "tape end-sensors" inside one of the two little

black plastic housings attached close to the back edges of the two sides, about half-way up from the bottom, as shown in the illustration.

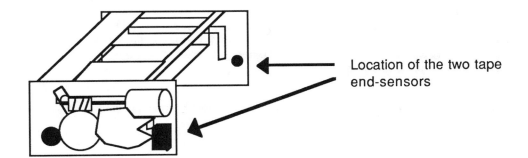

Location of Hitachi Tape End-Sensors

Inside each little black plastic enclosure is a photo-transistor soldered to the little circuit board:

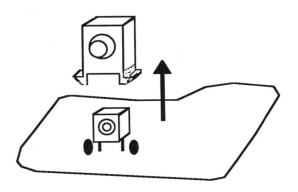

Hitachi Type End Sensor with Plastic Enclosure Removed

These parts are inexpensive, but replacing them requires desoldering the old parts after removing the loading assembly, taking off the little circuit boards on which they are mounted, pulling off their black plastic covers, carefully installing new photo-transistors so that they face in the same direction, and soldering them back in.

The need for desoldering and resoldering puts this repair beyond the scope of this introductory manual, but if you already know how to solder, or feel adventurous, go ahead. VCR technicians usually replace both end sensors, even though only one (which one?) may be bad.

These parts were used in lots of machines sold through Hitachi, RCA, Sears, and other brands, and they also seem to have found their way into VCR's manufactured by other companies too. The Hitachi part number is #5381681; the RCA number for the same part is 161757. If the parts arrive with a 90-degree bend already formed in the two little legs, just take some round-nosed pliers and straighten them out, or clip them off. You'll need to clip off the excess, anyway, after soldering them in.

3. Dew-Lamp Blinks on a Fisher and Machine will not Operate

If there is too much humidity inside a VCR, the tape might stick by surface adhesion to the spinning heads and cause serious damage. So, most VCR's are designed so that the little computer inside will shut them down if a special dew-sensor detects too much humidity inside the machine, and it will turn on a light on the front panel to tell you why it has shut down. This light will stay on until the danger from the humidity inside the machine has passed.

On some Fisher VCR's, however, this light serves a dual function. The little computer is also able to detect any failure in the "infra-red light-emitting diode" that is used to tell when the tape has come to an end, as explained in Chapter 2. This is the little light in the plastic enclosure on the reel table that goes up into the hole for it in the tape cassette, when it lowers into the machine.

If this light is not working in a Fisher for some reason, the little computer will shut down most systems in the VCR, and it signals that this has happened by making the Dew Indicator lamp BLINK, as opposed to staying on constantly. If you observe a blinking dew indicator on a Fisher VCR, try unplugging and replugging the little plug attached to the little wires that lead from the dew indicator to some circuit board before you order a new dew indicator. Often the problem is only that the contacts in the little plug have gotten slightly corroded, and unplugging and replugging cleans them enough to make it work again.

If that does not solve the problem, you can try replacing the sensor light (call a supplier for the part number), or take it to a shop.

4. VCR Loads, but Ejects Automatically Right After Loading

On Hitachi-category VCR's, this problem can be caused by marginal end sensors, the same problem described in Section 2 of this chapter.

On all machines, it can be caused by failure of the cassette-in switch to make contact to register the completion of the loading of the cassette. Clean the switch contacts, and if necessary disassemble the front-loader and retime it, as explained in Chapter 10.

"...Yes, I'm the salesman who sold you the $124,000 'ultimate home video system' last December. . . . I was wondering if you might be interested in upgrading your system? . . . "

Important: Use the check-out procedure
in Chapter 4 before using this chapter.

Chapter 12

Cassette Does Not Sit in VCR Correctly

When a VCR correctly loads and plays a cassette, the plastic cassette box is not resting on the bottom of the cassette basket, as it sits down in the machine.

Instead, the cassette box sits on the tops of several metal pins or pegs that are permanently fastened in a standard pattern into the base metal of the reel table. Holes in the same pattern are cut through the sheet metal bottom of the cassette basket, and during the very last part of its descent into the VCR, these pins poke up through these holes, and lift the cassette box slightly from the bottom of the cassette basket. Then the cassette box rests on these pins while the tape is being played.

VCR's are designed this way because the cassette must be positioned at one precise height above the reel table in order to play properly, and the position of the movable parts of the loading basket and loading mechanism cannot be controlled accurately enough to guarantee this.

If anything prevents the cassette basket from descending far enough into the VCR to leave the cassette box resting high and dry on the tops of these pins, then the machine may do any of the following: reverse its gears and eject the cassette, jam, stop and refuse to go into the Play mode, start playing but give a bad picture, quickly unload and stop, or damage the tape.

Failure to load the cassette fully so that the box is left resting on the pegs usually is due either to the presence of a foreign object on the reel table interfering with the basket's descent, or to some trouble in the cassette-loading assembly mechanism.

Take time to scout around very carefully in the region of the reel table underneath the cassette basket and loading assembly to see whether perhaps a child may have put a small toy or piece of plastic or paper into the cassette slot, or perhaps a coin, screw, or nail somehow fell into the machine. The mechanism on

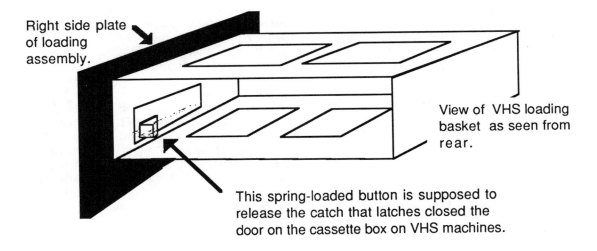

Right side plate of loading assembly.

View of VHS loading basket as seen from rear.

This spring-loaded button is supposed to release the catch that latches closed the door on the cassette box on VHS machines.

Location of Door Release Button on VHS Cassette-Loading Assembly

the reel table is so rich and complicated that it is easy to overlook such a foreign object on a quick glance, so mentally divide the area into a grid, and study it carefully, square by square.

If the cassette-loading mechanism moves squarely and evenly, but as it starts down, jams when the edge of the cassette door hits against the part of the mechanism that is supposed to lift the cassette door, then the trouble probably is associated with the little part in the cassette basket of VHS type VCR's that is supposed to push the button or lever in the cassette that releases the latch that holds the cassette door flap closed when the cassette is not in the VCR.

In Beta-type VCR's, the door to the cassette is released by a little sheet metal tab that extends up from the left rear edge of the bottom of the cassette basket, which being fixed, almost never fails.

Possibly the cassette door latch is not getting released due to the button not pushing it in far enough, perhaps because of metal fatigue in the spring behind it. You will need to figure out exactly why it is failing to release the door catch, and repair it by bending it back to its old shape, if this is possible, or else by replacing it, whichever is best.   There is no single answer or procedure that I can give you for all models, but you should be able to figure out what the problem is by examination and trial and error.   If it is a front-loader, you probably will need to remove and partly disassemble the front-loader assembly to accomplish this. See Appendix III for how.

Another possible cause of the cassette jamming midway on its descent on some VHS Panasonic-category models, and other models made by Sharp, is the central plastic part that holds the sensor lamp getting out of position so that the top of the lamp, instead of going neatly up into the hole made for it in the bottom of the cassette, misses and the cassette bottom hits on it, stopping the cassette's movement. If this is the problem, put the sensor lamp back into its correct location so that the cassette will go down over it correctly.

1. Front Loaders

If a front-loader's mechanism is out-of-time, the cassette may hang up in a half-loaded position, as explained in Chapters 8 and 10. (If you ever observe this happen while you are watching a front-loading VCR with the cabinet top off, immediately pull the power plug, in order to prevent the stalled little motor from burning itself up and self-destructing.)

When a cassette box hangs up midway on its route descending into a VCR, you can usually observe this visually. If this happens, after unplugging the VCR, look around and below the cassette carefully to try to see what is causing the cassette to get stuck.

Look to see whether perhaps the door flap on the cassette box is jammed onto the part of your VCR that is supposed to open the door. Try to manually unload the cassette by using your scribe or finger to rotate the loading motor that drives the front-load mechanism, or by turning the gear or pulley that it is supposed to drive, as explained in Chapter 8.

If no foreign object is found, then the next most likely possibility is that something is wrong with the front-loading mechanism.

One way to verify this is by inserting a cassette into the front slot when the VCR is unplugged, and trying experimentally to slowly turn by hand the cassette loading motor or belt drives in the direction that makes it load a cassette.

Watch carefully to verify that the latches that hold the cassette basket in its fully-unloaded, eject position come unlatched, as they should, when the cassette is pushed in, so that the basket can be pulled in by rotating the motor manually.

Check to see that one end of the cassette does not get ahead of the other end of the cassette as the loader carries it into the VCR. The cassette should go into the machine straight and square, with neither end ahead of the other. After the cassette has moved straight back into the machine for a distance slightly greater than the width of the cassette, both ends of the cassette basket with cassette should go "round the corner" at the same time, and start to descend vertically at the same time.

If one end of the cassette gets ahead of the other in this process, either the cassette is not being held tightly enough by the basket, or more likely, a loader gear has gotten out of time and one end of the basket is moving ahead of the other end. In this event, the entire front-loader mechanism must be removed and retimed, as explained in Chapters 8 and 10.

As explained earlier, if the cassette moves squarely and evenly, but as it starts down, jams when the edge of the cassette door hits against the part of the mechanism that is supposed to lift the cassette door, then the trouble probably is associated with the little part in the cassette basket of VHS type VCR's that is supposed to push the button or lever in the cassette that releases the latch that holds closed the cassette door flap when the cassette is not in the VCR.

Springy Cassette Box

Sometimes the cassette door opens OK, but as you manually turn the drive motor for the front-loader, the cassette and mechanism stop short slightly before coming to rest on the pegs, and will go no further. It may look as though the cassette is fully loaded, but you can detect a problem because the cassette box goes down a tiny bit more, or "gives," when you reach in and press on it firmly with your fingers, perhaps springing up slightly when you lighten your touch. Then either there is a foreign object in the VCR that you overlooked, and which still needs to

be found and removed, or else the front-loader assembly itself is out of alignment and needs to be "retimed" as explained in Chapters 8 and 10.

Sometimes the cassette and cassette basket repeatedly fail to go all the way down when the VCR is powered up and trying to load a cassette under its own power, but when it is unplugged and you turn the motor manually, the cassette will go down all the way.

In such a case, either (a) the front-loader mechanism has too much friction at the very end of its journey, or (b) the electrical switch that is supposed to sense when the cassette is fully loaded has gotten mispositioned so that it is telling the VCR's little brain to turn off the cassette loading motor too soon.

In case (a), the solution is to remove the front-loader and make certain that it is moving smoothly and effortlessly at the end of its journey. Spraying oil or lubricant on this mechanism is NEVER the right thing to do, not even in this case. More likely, one of the parts in the front-loading assembly is bent, broken, or out of alignment.

In case (b), insure that whatever is supposed to signal that the cassette is fully loaded, all the way down, is in the correct position and not sending the "cassette fully loaded" signal too soon.

2. Top Loaders

If a top-loader seems to go down OK, but the mechanism is forced to stop right before it goes all the way down, then probably either (a) there is a foreign object in the VCR that you have overlooked, or (b) the latch that is supposed to hold down the basket and cassette is not opening correctly to receive the arm or part of the top loader that it is supposed to catch and hold, or (c) the mechanism that is supposed to release the catch holding the cassette box door shut has failed to do so, or (d) the part that is supposed to press against the top of the cassette when it goes into the front-loader is bent or broken, or (e) in VHS machines, the sensor lamp has gotten out of place and is hitting against the bottom of the cassette box instead of going into the hole made for it in the box.

The solution to problem (a) obviously is to find and remove the offending foreign object. Take another careful look around.

The solution to (b) takes longer to explain, so it appears after (e).

The solution to problem (c) is to fix the button so that it pushes in far enough to release the cassette door, as described earlier in this chapter.

The solution to (d) is to replace the plastic cassette hold-down, if it is broken, or if it is metal and bent from use, rebend it so that it will press the cassette down firmly against the bottom of the cassette basket.

The solution to (e) is to reposition the sensor lamp correctly, as mentioned earlier.

The remaining possibility is that (b) the latch is already shut, rather than open, so that it will not accept the part of the top-loader assemby that it is supposed to grip, or else, the arm or part of the top-loader assembly that is supposed to be caught and held by the latch has gotten bent out of its correct position.

To take care of the possibility that the hold-down latch is not open to receive the part from the top loader that it is supposed to catch, start by powering up again and pushing the "Eject" button, and then trying again to load a cassette.

If the cassette still will not go down all the way, unplug, and then manually put the entire tape-loading mechanism into the "Eject" position (See Appendix VI for help), and manually trip the latch that is supposed to hold the basket down so that it opens. If this mechanism will not move in such a way as to permit this catch to be unlatched, then study the mechanism to try to determine why. I could not tell you the exact cause in your particular case without examining your VCR myself, because each case is different. If you cannot determine the cause, you will need to take it to a shop.

If the cassette latch mechanism is open and receptive as it should be, then the problem may be that something is wrong with the part of the top-loading assembly that the latch is supposed to catch and hold.

Find this part and examine it carefully. Usually it is a little arm with a horizontal rod or pin attached, extending below the cassette basket, but on a few VCR's, it is part of the folding arm assembly at the right or left side of the cassette basket. If it is a pin or rod attached to an arm at, or near, the bottom of the basket, check to make sure that it has not gotten bent in a way that prevents it from going into the latch mechanism correctly. If necessary, use a tool like needle-nosed pliers to adjust its angle or position slightly so that it will engage.

If a top-loader latches down, but when you press with your hand on the plastic lid over the cassette as it is playing, the picture improves, then possibly the problem is that the mechanism is bent or misadjusted so that the basket is not being held down quite as far as it should be in order to place the cassette box on the pegs.

Important: Use the check-out procedure
in Chapter 4 before using this chapter.

Chapter 13

VCR Will Not Fast-forward or Rewind

1. Preliminaries and Special Model Notes

Refusal to fast-forward or rewind can be caused by a foreign object, or broken piece of the VCR's own plastic, lodged in the loading gears or loading mechanism, so keep your eyes open as you look around in the machine performing the procedures below.

If the VCR does not move the tape when Fast-forward or Rewind buttons are pushed, first observe whether the VCR undergoes any change of state (that is, does anything happen?) when you press the fast-forward or rewind buttons.   For example, does the Rewind or Fast-forward light above the push-button, or on the button, come on when you press it?

Or, if the unit has no lights to indicate this, can you hear any motor sound, or noise, as if something is moving, or trying to move, when you push fast-forward or rewind? If there is any such indication of a change of state, but no tape motion, then go to Section 4 below. If there is no such indication of a change of state, go to Section 2 if it is a VHS machine, or go directly to Section 3 if it is a Beta.

ZENITH OR J V C: If the VCR is any of the models of Zenith or JVC with the little light that appears through a tiny window in the Fast-forward, Rewind, or Play mode when any of these buttons is pushed and if when you push FF or Rewind, the VCR actually starts moving the tape and then stops moving the tape and shuts down shortly thereafter, then you have a problem beyond the scope of this book. However, if there is some indication of mechanical action when you push the Fast-forward and Rewind buttons, but the VCR does not move the tape, then turn directly to the Zenith-JVC repair procedure at the end of Chapter 14, which also covers problems with the play function requiring similar disassembly.

J.C. PENNY or other GOLDSTAR-category VCR's: If the machine starts to rewind and then shuts down, and cleaning or replacing the idler does not fix the problem, go to the procedure in Chapter 14 for "shuts down after it starts to play" for these models.

SAMSUNG: If the machine starts to rewind and then shuts down, go to the procedure in Chapter 14 for "shuts down after it starts to play" under "Samsung."

2. Sensor Lamp Replacement

If there is no indication of a change of state at all on an older VHS with an incandescent-type sensor lamp, we first should check to see whether the sensor light might be burned out. The sensor light is a small light used on VHS machines to enable the VCR to tell when you have reached the end of the tape in either direction as it runs.

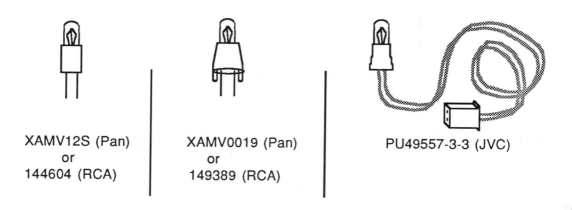

XAMV12S (Pan)   XAMV0019 (Pan)   PU49557-3-3 (JVC)
or   or
144604 (RCA)   149389 (RCA)

The Three VCR Incandescent Sensor Lamps and Their Part Numbers

The sensor light is located slightly to the rear of the center point between the two reels. The cassette fits down over it when it is put into the machine, and the light shines inside the cassette and out through two little square holes on the edge of the cassette to strike two tiny light sensors mounted at the ends of the cassette basket.

As explained in Chapter 2, VHS tapes are made with a clear plastic leader spliced to both ends of the tape. When the VCR reaches the end of the tape running in either direction, light previously blocked by the opaque tape is suddenly able to pass through the clear plastic leader spliced at the end of the tape. This light strikes one of the light-detecting "end-sensors," which sends a signal to the circuitry to stop the motors so that the tape will not be pulled out of the cassette at the end.

Likewise, when you rewind tape all the way back, at the other end, light coming through the leader at the other end of the tape causes the machine to stop rewinding and go into the Stopped state.

Auto Stop

But if the little light bulb itself burns out, another special safety circuit detects this and prevents the machine from operating at all. It puts the VCR into what is called "Auto Stop." Most VHS machines have infra-red sensor lamps, which rarely fail, but some machines, especially early vintage machines, have an incandescent lamp which burns out and needs to be replaced every couple of years.

Infra Red Sensor Lights

If your machine has the infra-red type, skip the rest of this section and go to Section 2. You can

determine which type of sensor your VCR has by looking at its physical appearance. If the top part looks like a little piece of clear plastic with two little domes molded on either side, then it is the infra-red type.

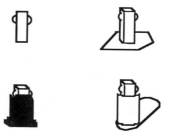

Typical Appearance of Some Infra-red Sensor Lamps

Incandescent Sensor Lamps

But if it looks like a tiny miniature light bulb, with a clear glass envelope and tiny wires inside that are so small that many of us cannot see them with the naked eye, then it is the incandescent type.

Two types of incandescent sensor lamps are made by Panasonic, and one type is made by JVC, and used in a variety of brands of VCR's. They look as shown in the illustration.

You can order the lamps as Panasonic parts with this part number, and use them in any Panasonic category machine that has a lamp with a similar shape in it.

The JVC lamp is often placed in a plastic housing that hides most of it and sometimes makes it difficult to tell whether the machine is using this type, or an infrared type, but the telltale sign is that the bulb will push out of the bottom of the housing and two wires

Typical Housing for JVC-type Sensor Lamp

come attached to it. The most frequently used housing for the JVC lamp looks like a little wishing well, as illustrated.

Use Exact Replacement Part Only

If the sensor lamp in your VCR resembles one of these three incandescent types, then you can use the given part numbers to order an exact replacement, whatever your brand happens to be. But you CANNOT SUBSTITUTE some other small light bulb for it. The little computer can tell if it is not the right lamp, and will keep everything shut down.

PANASONIC-CATEGORY TANK:

If the mechanical push-buttons will not stay down on an early Panasonic-category tank, the problem may be a burned out sensor lamp.

Testing the Sensor Lamp

On some VCR models, an incandescent sensor light comes on and burns brightly the moment that the power is switched on, but on other models, the lamp does not come on until a cassette is loaded into the machine. On a few models, the light does not stay on, but only flashes momentarily.

A technician might tell whether an incandescent sensor lamp is burned out by measuring across it with instruments, but we can discover the same thing by applying power to it from a small battery and observing whether the light comes on.

Unplug the VCR. Remove the screw holding down the little circuit board to which the base of the lamp is attached, taking extreme care not to drop the screw down into the mechanism. Hook up a 9-volt battery with clip leads as described in Section 19 of Appendix I.

If it is one of the Panasonic lamps, turn the lamp and little circuit board over right where they are in the machine to get access to the soldered connections on the bottom. If it is the J V C type, either follow the wires back to where you can get access to the metal pins to which they are attached, or penetrate

the insulation of the two attached wires with the pointed tips of your scribe and other sharp-pointed tool attached to the two clip leads, attached to the battery, whichever is easier. If the lamp comes on, it is good, and you can go to Section 2, but if it does not come on, the lamp is burned out and must be replaced.

Replacing a Bad Sensor Lamp

To fix the problem, you will need to obtain a replacement of exactly the same type, using the part numbers above and the sources in Appendix II.

If it is of the JVC type, which was also used in some Fisher and other brands of VCR's, then it comes connected to two long wires that just plug into a little plug on one of the circuit boards.

Once you have the new lamp in hand, use the procedures in Appendix III to disassemble the VCR only as far as necessary to find the place where it plugs in. Unplug the old lamp, lift up its wires sketching on your notepad the route they followed, and unfasten the old lamp. Fasten the new one down in the same position, and carefully lay the wires along the same path and plug it into the same receptacle.

If the sensor lamp is one of the Panasonic-type, then it is soldered to a little circuit board. Soldering is a slightly more advanced technique than most in this book, but it is still easy to learn and it is less expensive to buy soldering tools and do the job yourself than it is to have it done by a shop, especially if you will keep the VCR and will need to replace the lamp again the next time it burns out.

If you want to replace the solder-in type of lamp, see Section 20 of Appendix I for an explanation of the tools and techniques to do the job. For present purposes, I will assume that you have gotten from there or elsewhere the tools and general knowledge needed for soldering, and I will confine myself in the following paragraphs to the specific additional information you need for this special soldering job.

Double-check to make certain the VCR is unplugged. You have already removed the little screw holding down the tiny circuit board to which the lamp is attached, and turned it over right where it is. Find the two little solder blobs directly below the little lamp; do not unsolder the other two blobs attaching the incoming wires to the circuit board.

Unsolder only the two wires coming from the bulb and sticking through the circuit board right below the bulb, using solder wick to absorb the melted solder, as explained in Section 20 of Appendix I. Now at this point, slightly different procedures are used for removal and replacement of the two different types of solder-in bulbs. If the VCR has the XAMV12S type bulb, with its cylindrically-shaped base, skip the next paragraph.

If the sensor lamp has a sloping, conical-shaped base (Mat. part #XAMV0019), then you will also see two little plastic blobs that are the melted over plastic pins attached to the bottom of the base of the bulb, as shown in the illustration. Scrape these away with the hot tip of the iron, and gently pull the bulb free along with its conical base.

You can use fingernail clippers to trim the two wires on the new replacement bulb to approximately the same shortness as you can now see the old wires had. Feed them through the holes, and resolder the new wires. Next, use the hot tip of the iron to splay the little plastic pins on the new base like the ones you saw on the old base before you removed that lamp. Turn over the circuit board and lamp, carefully reattach with the screw, and test the VCR.

If the sensor bulb is of the type with the cylindrical, nonsloping, straight up-and-down base, then you remove and replace ONLY THE GLASS BULB AND ITS BASE, AND NOT THE OUTER PLASTIC BASE.

As before, unplug the VCR, remove the screw holding down the little circuit board to which the lamp is attached, turn the little board over where it is, and find the point where the two wires coming from the base of the bulb are soldered to the board. Use solder wick to remove all the solder from these two wires, but do not unsolder the two wires coming to the circuit board from elsewhere.

When you unsolder the two wires from the bulb, the bulb probably will not come out freely, because its bottom is probably glued inside the outer cylindrical plastic base at its bottom. But if you hold the black-plastic-base-plus-circuit-board assembly in one hand, and gently wiggle the little light bulb and its white plastic base with your other hand, it eventually will come loose and pull out. Be very patient and eventually it will come free. Take care not to break the glass, as that would make removing the rest of it much more difficult.

When it does come loose, remove it, and use fingernail clippers to cut the two wires on the new bulb to approximately the same length as the wires on the burned-out bulb. Put the new bulb in the black plastic base, making certain that the two wires go all the way through the two holes in the bottom and in the circuit board. It is not necessary to glue it back in — they did that at the factory so that they could do the soldering later. Solder the two wires back to the places for them on the circuit board and that will hold the bulb in. Put the circuit board back in place with the screw.

Plug the VCR back in, and test to see if it now will fast-forward and rewind. If so, return to the check-out procedure in Chapter 4. If not, go to Section 4 if there is now some movement or noise when you press "Fast-forward" or "Rewind," but if there still is no response, go to Section 3.

3. Fixing Reel Motor Problems

If absolutely nothing happens when you push "Fast-forward" or "Rewind," the problem may be that the motor that is supposed to move the tape is locked up or burned out. Different models of VCR's move the tape in the fast-forward and rewind modes differently. In all cases, of course, a motor is involved, but different types of motors are used, some of which you can test with a 9-volt battery and clip leads as described in Section 19 of Appendix I.

Some VCR's have a special "reel motor" that is located right underneath the reel table between the two reels. If you find your reel table in Appendix VI, the drawing may tell you whether it has a separate reel motor or not. Or, using the procedures in Appendix III, open up the bottom of your VCR and observe whether you see a reel motor located between the two reels underneath. A reel motor will be housed in a metal enclosure, from which you can probably see wires coming out. If what you see instead is a pulley with a belt attached to it coming from elsewhere, then it does not have a reel motor.

If you find a reel motor, you can test whether it is OK using a 9-volt battery. You probably will need to attach the clip leads to two sharp-pointed tools to use to penetrate the insulation on the wires going into the reel motor, although on some models, you can find points where the wires, or terminals going to them, are already bare. After double-checking to make certain that the VCR is unplugged, connect the wires and battery to the motor. You can connect them either way. If it does not run, double-check your connections; try reversing them. If it still does not run, then the reel motor is bad and must be replaced. Get the part number from Appendix VI (or call the company for the part number, using the information in Appendix I), order up a replacement reel motor, and install it.

If the reel motor does not run when the machine is powered up and you push FF or Rewind but it does run with a battery, or if there is no reel motor and nothing still happens when you push Fast-forward or Rewind, then we have about come to the end of what can be done for your problem within the elementary scope of this book.

A possibility remains that the cassette loading mechanism has a problem that is causing a crucial switch to fail to make contact when a cassette is loaded, so you may wish to consult Chapter 9 (if it is a top-loader) or Chapter 10 (if it is a front-loader), and use the procedures explained there to check out the "cassette in" switch(es) before taking it to a shop.

You can also try removing the cabinet front panel, using the procedures in Section 8 of Appendix III, and pushing directly on the buttons under the front panel. Sometimes the panel gets dislodged so that the motion of pushing on the outside plastic cosmetic buttons does not get transferred to the actual push-buttons underneath.

4. Idlers, and Clutches

If you heard or saw something happen when pushing the fast-forward or rewind buttons, that's good, because in this case, the problem is most likely due to either a slipping or broken belt, a slipping idler, or a weak clutch drive (although it is still possible that there is a trouble in the electronics).

Turn off power switch and unplug machine.

With the cassette out of the machine and the cassette basket up or in the unloaded position, observe the two reels and determine what is supposed to contact and move the reels when rewinding or fast-forwarding. In many cases, it will be a centrally positioned plastic wheel with a little rubber tire around it, about 1 to 2 inches in diameter. This is called an "idler."

In most cases, this idler will be attached to a mechanism that permits it to "swing" from side to side, touching either the edge of the supply reel, or the edge of the take-up reel. Except for some models of Fisher-category machines, the idlers on most VCR's are located on the top of the reel table. If your machine has an idler with a rubber tire, proceed as described in Section 5. Otherwise, go to Section 7, "Is the Problem a Broken or Weak Belt?"

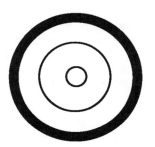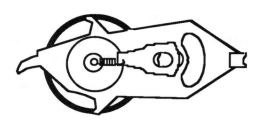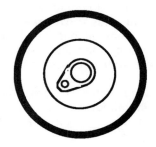

Some Typical Idlers

5. General Procedure on Machines with a Fast-forward & Rewind Idler

Move the idler back and forth, observing the different surfaces that it is supposed to contact. In some cases, you will find that it will contact one of the two reels, but stops swinging a little short of contacting the other reel. This does not necessarily indicate a problem, as it may be that movements that should occur in the mechanism underneath normally permit the idler to swing the rest of the way to contact the other reel under normal operation.

SPECIAL MODEL NOTES

If the VCR is a Sharp or Montgomery Ward, go directly to the replacement procedure for Sharp and Montgomery Ward idlers.

If the VCR is an Emerson 870, it uses a Sharp idler. Turn directly to the paragraphs under Sharp later in this section for part number and procedure.

If the VCR is a Fisher, or Sears of the models manufactured by Sanyo, using a Fisher-type idler, turn directly to the replacement procedure for Fisher idlers.

Some Lloyds, such as the L845, use a Sharp idler. Turn directly to the paragraphs under Sharp later in this section for part number and procedure.

If the VCR is a Samsung VT210 TB, VT215T, VT221T, VT225T, VT311TA, or VR 2310, it uses a Sharp idler. Turn directly to the paragraphs under Sharp later in this section for part number and procedure.

General   Procedure

Moisten a cotton swab stick with rubbing alcohol and thoroughly clean off the surface of the rubber tire that is supposed to contact the edges of the two reels, the edges of these two reels where the idler is supposed to contact them, and most important of all, clean the axle or shaft that meets the edge of the idler and drives it from underneath.

This very important part may be hidden from view by being under a cover of some sort, so you may

only be able to see it by peeking at an angle from the side. In fact, you might not be able to see it at all, but be assured that it is present, for it directly drives the idler wheel that is supposed to make one of the reels turn during rewind, and make the other reel turn during fast forward.

A common cause of no FF or no rewind is slipping occurring between the idler and this drive axle, so be sure to clean it very thoroughly. Even though this drive shaft comes up from underneath the reel table, you can clean it from the top side. Use several cotton swab sticks. Be careful that you do not accidentally touch the cotton swab stick to some other part of the mechanism that is normally covered with oil, contaminating the cotton swab with oil that you might accidentally carry back to the idler!

On the Hitachi VT 8500A and some other VCR's, the rubber is around the reels against which a moving plastic drive swings and turns. Clean these rubber and plastic parts in the same manner as for machines where the rubber is on the swinging part itself.

While cleaning the rubber, look at it closely and carefully. Are there fine cracks in the rubber? Is the edge of the rubber shiny, instead of being a dull black like it should be? Look around the edge of the reels for small black grains of worn-off rubber.

If any of these conditions is observed, the idler is due for replacement. It may work OK for a week or a month after you clean it, but sooner or later you will need to replace it, so you may as well make a note to order a replacement idler now. In fact, one good way to prove that the problem is due to a bad idler is to clean the idler and observe that the fast-forward/rewind functions have been temporaily restored: this indicates that it is time to replace the idler.

Part numbers for many idlers are given in Appendices IV and VI, and procedures for replacing them are given in the next section. In most cases, the problem can be fixed temporarily, at least, by simply cleaning the idler, as explained in the cleaning procedures, but eventually you will need to replace any slipping idler, and you may wish to do so now to save time in the future.

6. Special Idler Replacement Procedures

If the idler is held in place by a C-ring, or a compression ring, it can be removed using the procedure explained in Sections 9 and 10 of Appendix I. Be sure to put tissue paper down in the spaces and holes around the idler, and cover the axle and screwdriver tip with a cloth before springing the stuck to the idler, it may still be on the shaft. Do not forget and turn the machine over without removing any washer first, to make certain that it does not fall off and get lost. Put the washer back on the shaft, install the replacement idler with the same side up as you observed on the idler you removed.

fastener loose, so as not to loose it.

When the fastener is off, slide the idler up the shaft very slowly and turn it over slowly the second it comes off so as not to loose any fiber washer that may be sticking to the bottom of the idler. If it is not

Replace the metal fastener using the procedures in Appendix I and again taking care to protect against it springing off and getting lost. Clean off any oil on the new idler, plug in, power up, and test the VCR.

Fisher Idler Replacement

Fisher A

If you have fast-forward/rewind problems in a Fisher-category VCR with a reel table of the type called "Fisher A" in Appendix VI, then the problem probably can be fixed by replacing the little idler drive, a part that looks like a little metal shaft with a small rubber tire part way around it.

Idler drive: 143-0-4804-00100
(SFC)

Small Fisher Idler Drive

You can try first just cleaning the idler drive while it is in the machine, but probably if that works at all, it will only work for a few weeks, and then you will need to replace it to produce a more durable repair.

If you are able to work in close quarters, you can remove and replace it without removing the cassette basket loading assembly. Note carefully how high it is mounted relative to the plastic wheel it rubs against. Then use a 1.5mm hex key (see Section 11 of Appendix I) to loosen the little set screw in the side of

it just enough to be able to slide it off. Slide on the new idler drive and tighten the set screw so that the new part rides at the same height as the old one did.

Fisher B

The Fisher B type idler replacement is covered at length after the Fisher C idler replacement procedure covered next.

Fisher C

To remove the '00700 idler, use needle nosed pliers to grip the plastic lock washer near the slit, and unwrap it from the axle. See Section 10 of Appendix I for explanation of how to do this, if necessary.with the idler removed, use an old toothbrush and alcohol to scrub the plastic knurled wheel that contacts the rubber surface of the idler and drives it. You'll see it clearly once the idler is removed. Turn it around with your finger and scrub it vigorously, up and down, to get it absolutely clean.If you fail to get it clean, the new idler will fail.

Put the new idler down over the axle, and use needle-nosed pliers to put the plastic lock washer back into the grove in the axle. Test out.

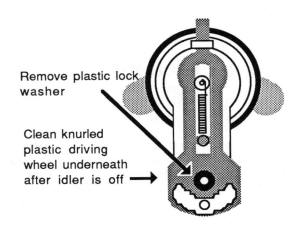

Remove plastic lock washer

Clean knurled plastic driving wheel underneath after idler is off →

Fisher 143-04204-00700 (SFC) Idler

Fisher B

A common idler used in many Fisher-category machines sold under the names Fisher, Sears, and other names looks like this when it is out of the unit.

Fisher idler:
143-0-4204-00400 (SFC)

Appearance of Common Fisher Idler —Fisher #143-0-4204-00400 and #143-0-4204-00300

The rubber tire on this idler frequently wears out and needs replacement. There were two versions of these idlers used in early models, both identical in appearance, except one used a slightly wider rubber tire. Both can be replaced with the same 143-2-4204-00400 idler, however, which is the part number you should use. In later models, the company replaced this idler with a design that looks similar except that it has geared plastic teeth in place of the rubber tire around the idler wheel.

You cannot see this idler easily inside the VCR, because, unfortunately, it is hidden under layers of other mechanism. But you can determine whether your Fisher, Sears, or Realistic VCR uses one of these idlers by removing the top of the cabinet and looking down from the top into the area between the two reels. If what you see looks like the following illustration, then you may have one of these idlers.

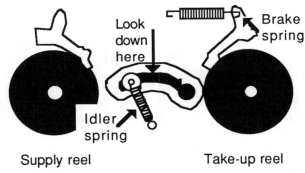

Top View of Reel Table Using this Idler

To determine whether you have the version with the rubber tire that regularly needs to be replaced, or the version with gear teeth around the outside, get a flashlight and peek down into the slot at a point where you can see the edge of the wheel (as indicated by the arrow in the diagram), and observe whether the outer edge of the wheel is a rubber tire or a plastic gear. If it is a rubber tire, it's a good bet that you need to replace this part to fix the fast-forward/rewind problem.

You can order one of these idlers from a Fisher supplier; or you can order a generic substitute for just the rubber part that wears out, the rubber tire around the wheel, as Tenma part #32-705 from M C M; or you can order a complete substitute idler from MCM as part 32-830 (addresses in Appendix II). The complete idler from Fisher tends to be rather expensive and slow to be shipped, and the Tenma substitute tire sometimes does not quite fit, so I would recommend replacing the whole idler.

This idler is actually a precision part, and some companies are marketing inferior cut-rate imitations that often do not work, or only work for a short time, so I would recommend using either the Fisher original part, obtained from a Fisher parts supplier, or the MCM replacement idler. Anything else is a risky substitution, I've found. Might as well order two or three while you are at it, because they need to be replaced regularly.

Do not disassemble further until you have the part in hand, because you may forget how things came apart while waiting for the part, or lose some small screws, clip rings, etc. In fact, it would be a good idea to put the cabinet back together while you are waiting, so as not to loose the screws, etc.

It would be wise also to order up replacements for the two belts that go underneath, because they are inexpensive, probably tired out too, and this would be a convenient time to replace them.

The larger reel belt is Fisher #143-2-7504-00400 (SFC) or you can use #143-2-7504-00600 (SFC); but I prefer to substitute a Hitachi #6355471

(Hit), which is quicker to get and actually seems to work better.

The smaller loading belt is Fisher #143-2-7504-00300 (SFC) or equivalently #143-2-7504-00700, but I have found that a Panasonic VDVS0122 (Pan) or RCA #148664 (RCA) works better and is quicker to get. When you have the part(s) in hand, proceed as follows.

Plug in the machine and push "eject" to put the basket in the cassette-ejected position. Then unplug the machine.

Remove the top cabinet part, if you have not already done so. Put the cabinet screws in a container, number it, and record location in your disassembly notes.

It is not absolutely necessary to remove the cassette basket to replace the idler. If you have small fingers or do not mind working around the minor obstacle it presents, you can do the following procedure without removing the basket. If you are uncertain, you can start the procedure, and then remove the basket if you find that you must in order to complete the following steps. For clarity, all the illustrations in this section will picture the situation as it looks with the basket assembly removed. If you are not going to remove the cassette loading assembly, skip the next paragraph.

If you decide to remove the cassette loading assembly, first check to see whether there are any plastic knobs extending through the front cabinet piece that need to be removed before the front panel can be removed. If so, remove them. Then remove the cabinet front panel by unscrewing any screws at the bottom edge hold-downs, and gently prying free the plastic tabs at the top, or rotating the front panel up. See Appendix III for further help. As I mentioned, on many of these Fisher-category machines, it is unnecessary to remove the front cabinet or cassette loading assembly to replace the idler. But if you do decide to remove the cassette loading assembly, mark with a felt-tip pen the position of any ground wires or pieces of springy metal connected under any of the screws holding the basket in, then remove the screws holding the cassette basket mechanism and lift the basket assembly out. Pull any wires connected to it by a multi-wired plug loose by wiggling the plug by gripping the wires and pulling them from side to side until the plug comes loose. Set the loading assembly aside.

Look at the spring attached to the part of the idler that rides from side to side in the crescent-shaped hole that you looked down earlier with a flashlight. Before you remove the spring attached to the center of the idler, notice carefully how it is attached at both ends. Now remove it by stretching it slightly with needle-nosed pliers, or the bent end of a scribe, and lifting it free. Grip one end of the spring, pull just

enough to free the end, and lift clear. Put it in a container, number it, and record this number with the notation "top idler spring," to prevent confusing it with other springs removed later.

If this spring is missing, this could be the whole problem. If you cannot find it lying around inside the VCR, you will need to order an new one, Fisher part number 143-2-6704-02600 (SFC). This spring fits all makes using this design.

Remove the spring attached to the plastic brake arm that pushes against the rear edge of the take-up reel (the reel on the right). Put in separate numbered container, so as not to get the two springs confused with each other. Next, remove the plastic brake by pushing back the little plastic catch that holds it down, and lift it free. Put it in a separate container.

Removing Take-up Reel Brake Spring and Brake

Next, we need to remove the take-up reel. It is held in place by a split plastic washer. You can remove this washer by gripping it tightly with a very sharp pair of needle-nosed pliers close to the place where it is split, spreading it slightly and twisting it off. Use needle-nosed pliers with a good grip at the end to grab hold of the split plastic washer holding down the center of the take-up reel, pull it out slightly, and off, as described in Section 10 of Appendix I. Concentrate all your attention on what you are doing at this time so as not to drop this lock washer. Put it in its own numbered container, and record "take-up reel hold-down washer" in your disassembly notes, so as not to confuse it with the other lock washers that you will remove from below later on. (In case you do lose this washer, its part number is 143-2-6304-01800 (SFC).) Do not remove the plastic reel part until you read the next paragraph.

One of the easiest parts to lose is the spacer washer that is usually under the plastic take-up reel, because it often sticks to the bottom of the plastic part as you remove it, making it appear as if there is

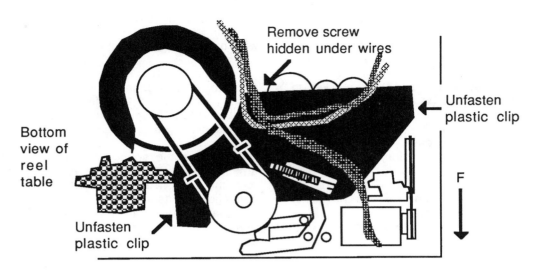

Underneath View of Fisher Idler Drive Mechanism

no washer there, and then quietly falling off later when you are not looking. While sliding the take-up reel up the axle, locate the crucial plastic spacer washer underneath, remove it, and put it in its own container, recording its container number and "reel spacer washer" in your notes. So, when you slide the plastic reel up, watch very carefully for a washer stuck to the bottom of it — or perhaps it will remain stuck to the metal below.   Put in its own container and note the number in your notes. Do not leave it on the axle because it will fall off and get lost when you turn over the VCR in later steps. This is a precision spacer that is necessary to hold the reel at the right height so that the VCR will play correctly, and different spacers are inserted in machines as they come down the assembly line to correct for minor manufacturing differences, so it is not easy to replace it.

Swing the little part of the idler that shows above the crescent slot in the reel table over toward the wide space on the right-hand side of the arched track in which it runs, so that later in the procedure, it can be pulled free from below through this wider space.

This is the time to use alcohol to clean the sides of both reels that contacted the idler rubber tire. Use a cotton swab stick moistened with alcohol to reach down through the places where the opening in the metal of the reel table gets wider around the plastic reel part to clean its sides below the level of the metal.

Turn the VCR over so that it is upside down, being careful not to bend or break any of the circuit boards on the sides or on the top in doing so. After using a felt-tip pen to mark the positions of screw heads holding on the bottom metal cabinet piece, remove these screws, looking at each carefully as you remove

it. If any is shorter or longer than the other screws, put a special mark by it, and note this fact in your disassembly notes. Remove the bottom cover.

If a circuit board covers the machinery under the reel table on the model you are fixing, remove any screws that hold down this circuit board . Some screws should not be removed. Usually there will be a white arrow pointing to the screws that need to be removed, or their heads will have a slight redish tint. Put these screws in a numbered container and record their location.

Release any plastic locking tabs that hold down the edge of any circuit board covering the mechanism, and swing this circuit board out gently on its hinges. Let it rest in this position. Be very careful not to break it in subsequent steps. If it appears to be bending under its own weight, put something (like a stack of books) under it where it hangs out, to support its weight. Some technicians prefer to work on the bottom of the VCR with the machine balanced on one end, but I always have trouble with the disassembly steps below when the machine is in that position.

Most models have a mechanism that looks approximately as shown in the illustration. On some models, instead of the large plastic piece shown in black, a smaller piece made from sheet metal was used to do the same job, as in the next illustration. The following procedure works for both.

Remove the rubber belt that goes between the capstan pully and the plastic pulley that drives the idler ("idler drive pulley assembly").   Put in numbered container. If you found the reel belt lying loose in the machine when you opened up the bottom, you've found the immediate problem.   You can just replace it and put everything back together without taking out the idler, but you should put all new rubber in now, while you've got it open, since the idler will need to be replaced sooner or later. In fact, a bad idler may cause a new belt to pop off too.

When you remove the idler drive pulley assembly later, it will look as shown in the next illustration:

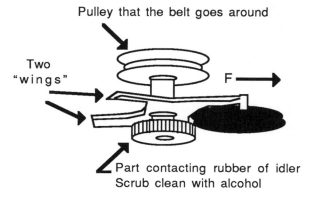

Pulley that the belt goes around

Two "wings"

F

Part contacting rubber of idler
Scrub clean with alcohol

Common Fisher-type Idler Drive Assembly

You cannot see it from this angle until it has been removed, but I show it to you now to help you understand what is done in the following steps. This is not the part that is bad (probably), but it is the part we must remove to get at the part that is bad.

Look at the long, thin plastic arm, with a bend in the middle that starts from next to the plastic pulley that drives the idler and runs up under the metal or

plastic cover. On some models, this arm was shaped in such a way that you can slide the idler drive pulley assembly straight up and out without having to remove the plastic cam arm, as in the illustration, while on other models, the arm is shaped so that part of it stands between you and a "wing" on the idler drive pulley assembly, in which case the plastic cam arm needs to be removed to get out the idler drive pulley assembly and also the idler itself. Study the situation in your machine to judge whether the idler drive can clear the arm and the black plastic, or big metal, cover, and come up freely, in which case you can skip down to the first paragraph under the heading "Removing the Idler Drive Assembly" where the slotted plastic washer holding the idler drive pulley is removed.

Removing the Cam Arm

Remove the spring attached between the plastic cam arm and the plastic or metal cover over themechanism.

Remove the screw(s) holding down black plastic cover piece (or sheet The screw(s) may be hidden beneath some wires. Notice that in addition to the screw(s), the cover piece is held in place by one or two plastic clips going through the metal beneath at the end closer to the center of the VCR. Probably it will be necessary first to work loose the wires running across it. Mark their route with a felt-tip pen before removing them, so that it will be easier to get

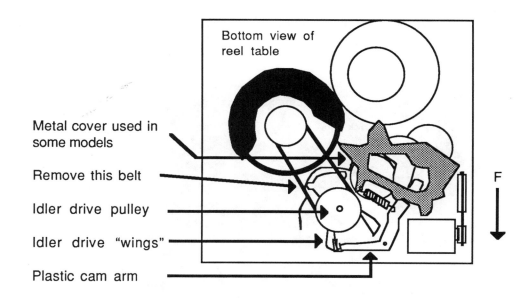

Bottom view of reel table

Metal cover used in some models

Remove this belt

Idler drive pulley

Idler drive "wings"

Plastic cam arm

F

Fisher-category Reel Table with Metal Cover and Idler Drive Pulley that Can be Removed Without Removing the Plastic Cam Arm

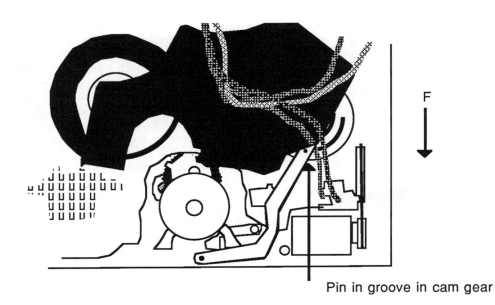

F

Pin in groove in cam gear

Fisher Idler Mechanism Exposed: Cam arm pin

them back into the same route later when reassembling. Free up any plastic hook(s) holding the plastic cover down to the sheet metal of the table. Free enough of the wires clipped down on the black plastic cover to permit swinging it a little to the side to expose the idler mechanism underneath, as shown in the illustration.

Look again carefully at the long white plastic cam arm. Notice the exact groove in which the metal pin at one end (to which the arrow points in the picture)

rides in the large, round white plastic "cam gear." It is somewhat hidden under the plastic part, wires, etc. Draw a picture of this in your disassembly notes — later, you will be glad that you did so.

Now look at the other end of this rod. It also has a metal pin in it. Do you see the two plastic wings nearby that move easily from side to side? Notice how, at the furthest extent of its motion, one of them hits against the metal part of this pin, and not against the plastic of the rod. Use the same sharp-pointed

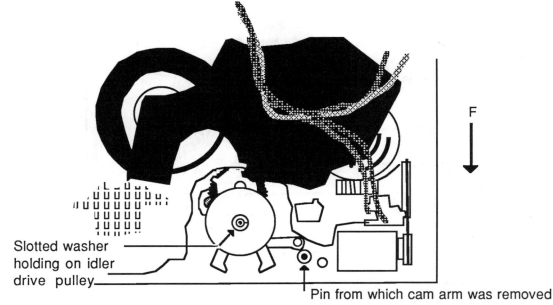

F

Slotted washer holding on idler drive pulley

Pin from which cam arm was removed

Appearance of Assembly after Removal of Plastic Cam Arm

Arc or arch cut in sheet metal to remove idler drive

F

Later Fisher Table Configuration

needle-nosed pliers trick used earlier to grip the plastic lock washer holding down this arm, gripping it near the cut in it, and twist it off gently, as you did with the one above holding down the take-up reel. Be super-careful not to let it drop into the mechanism. Put this slotted washer in a numbered container and record the number and location in your disassembly notes. (Do not lose it, but if you do, the part number is 143-2-6304-01500 (SFC).)

Underneath this plastic arm may, or may not, be another washer that is easily lost when you remove the arm, so watch carefully as you slide this arm up the axle it rotates on. Find this spacer washer, if one is present, and put it into a separate numbered container. It may be sticking to the underside of the rod, as you remove it, so watch out! Wherever it is, if there is one, get it off and into a separate numbered container, to maintain order. Most machines do not have this washer.

Removing the Idler Drive Pulley Assembly

Use needle-nosed pliers to grip and remove the plastic lock washer holding down the idler drive pulley (the plastic pulley from which you removed the belt earlier). Same part number as the washer attaching the cam arm. You may need to remove, or move slightly, the large flat black plastic part before pulling the idler drive pulley up and out.

Now we come to another parting of the ways, a fork in the procedure, where what you do next depends on whether you have one of the earlier models or a later model. On some early models, the designers neglected to make a space in the sheet metal to pull the idler drive pulley assembly directly up and off, necessitating a longer procedure to replace the idler, while in later models, they thought of this and cut a small arc in the sheet metal, as shown. The difference is in the small arc cut in the sheet metal in the later models, as shown here.

If you have one of the later models, you can remove the idler drive pully assembly by moving the geared part of the swinging mechanism to a position under the access cut and pulling it straight off the shaft, up and towards you as you look down on the bottom of the table. As you remove the drive, watch for a washer on the shaft at the bottom, or perhaps sticking to the underside of the drive assembly when you remove it. Take this crucial washer off, and put it in its own numbered container, and record its location in your notes. Part number: 143-2-6304-00600 (SFC).

If you have one of the earlier models, you will note that the idler drive pully assembly can only be slid up the shaft a short distance before the part with the toothed black plastic gear or wheel hits against the sheet metal. We need to get it free before we can replace the idler, obviously. There is a long way to do this, and a short-cut method. Here is the short method.

Remove the screw at the left corner of the sheet metal to make space to rotate the idler mechanism out through it. Put in numbered container.

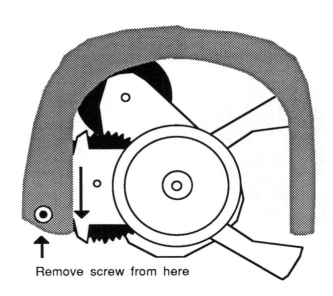

↑ Remove screw from here

Short Cut Method of Removing Idler in Early Models

Now pull the edge of this sheet metal up slightly, just enough to rotate the geared end of the idler drive pulley assembly through and past. Push the whole head of this swinging gear-idler mechanism counter-clockwise and push it down beneath the sheet metal while turning and bending one plastic wing up over the post attached to the idler assembly that it catches against, which limits its rotation. Some bending of the sheet metal and plastic wing, and some forcing, are necessary to rotate the part CCW enough that you can remove it. It is OK to bend the wings on the idler drive pulley assembly, and to bend the sheet metal to the left of this assembly to squeeze it under and

around — the only part that you must take care NOT to bend is the metal of the idler assembly to which the post is attached that the wing is hitting (if you want to reuse it). Work the idler drive assembly free as needed, and pull it straight up, watching carefully to catch the crucial spacer washer around the shaft underneath, or perhaps stuck to the bottom of the idler drive piece you are removing.

Idler Assembly Removal

Use the same needle-nosed-pliers method to remove the plastic lock washer that is holding down the short white plastic arm connected to the idler assembly itself. Put it into a separate, numbered container.

If the idler has happened to swing to a point on its arc of travel in the track on the topside back to where it is caught in place, then you will need to swing it back into the position in which we put it as the last step of the procedure earlier when we were working on the top side of the machine, in order to slide it free.

Remove the old idler assembly.

Now, at last, the worn idler assembly is out of the machine and can be given a new tire, or preferably, the whole idler assembly replaced with a new one. If you ordered up the whole idler assembly, you can skip over the next two paragraphs.

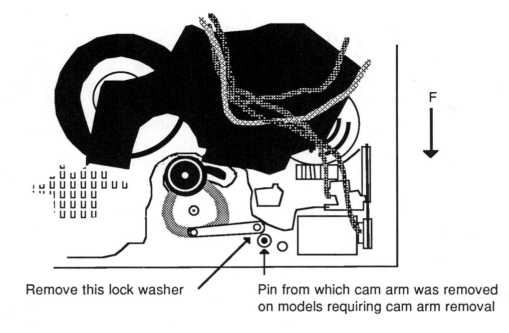

Remove this lock washer ╱   Pin from which cam arm was removed
on models requiring cam arm removal

Exposed Fisher Idler

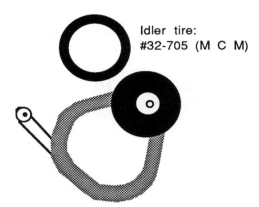

Idler tire:
#32-705 (M C M)

Complete idler assembly:
143-0-4204-00400 (S F C)
or # 32-830 (M C M)

Fisher Idler Assembly with Tire Removed

Replacing Only the Tire

If you are putting a new tire on the old wheel, you must remove the old tire carefully, with minimal bending of the sheet metal loop, clean the old wheel with a cotton swab and alcohol, and stretch the new tire onto it.

If the new tire is slightly wider than the old tire, and will not go down into the grooves on the wheel, then you should know that this tire is not going to work on this wheel for very long. Probably your machine had the 143-0-4204-00300 idler instead of the '00400 idler, so the tire does not quite fit. You can use your thumbnail to force the sides of the new tire down into the groove on the wheel, but you really should order up a whole replacement idler as I advised earlier.

Make certain that the new tire is on the old wheel in such a way that it is completely round, with no bumps or high spots. Holding it by the little pin through the center of the wheel as an axle, run it back and forth a few times on a clean surface to make certain that the little tire is roundly mounted on the plastic wheel. With alcohol, clean the tire after mounting, and straighten out any bends that may have gotten into the metal loop.

Reinstallation of Idler

Make certain that the shaft on the idler to which the plastic arm is attached is perpendicular to the surface of the sheet metal to which it is attached, and not bent to either side. Hold it up and look at it edgewise as shown in the illustration, and swing the plastic arm from one side to the other, making sure that it does not come closer to the sheet metal on one side than on the other. If it does, try to straighten it,

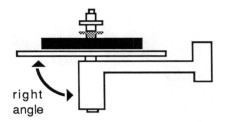

right angle

or return it and get another. It will not work if it is not at right angles to the sheet metal piece.

Put in the good idler, simultaneously sliding the plastic arm attached to it down on the pin it came off. Make certain that the little part at the center of the wheel of the idler gets into the wide part at the end of the topside crescent-shaped slot. Make certain that it tracks correctly in the crescent-shaped slot. Do not yet replace its lock washer.

If you stand the VCR on end, and lift up the idler to the high end of the crescent-shaped slot in which it tracks, and then let go, the idler should fall down the track freely of its own weight. If it does not and binds, then find out why and get it so it does fall freely before continuing.

Reinstalling the Idler Drive Pulley Assembly

Clean off the round plastic driving surface of the pulley drive assembly with alcohol.

Put the spacer washer for the pulley assembly back onto the shaft from which you removed it. If it seems to have disappeared from the container you put it in, check the bottom of any container that you nested inside it to see whether it is hiding stuck to the bottom of it. Slide the swinging gear mechanism with pulley down onto its axle, making certain that the round part at the bottom goes into the open area in the center of the sheet metal loop of the idler, and that the tire of the idler is not pinched under this round part.

If you have one of the later models, the idler drive assembly will easily slide all the way on. In that case, you can skip down to the paragraph after next. But if you have an earlier model, then you will have to go through a reversal of the bending procedure you used earlier to remove it. Position the part with the toothed gear in about a seven o'clock position, pull up the corner of the sheet metal with the hooked part of your scribe or other suitable tool, TAKING CARE TO KEEP YOUR FACE TO THE SIDE, AWAY FROM THE DIRECTION YOU ARE PULLING, IN CASE THE TOOL SLIPS LOOSE AND SUDDENLY SPRINGS IN THE DIRECTION IN WHICH YOU ARE PULLING IT, and slide

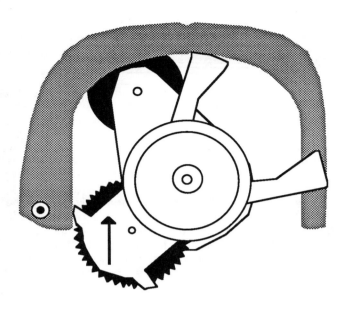

Preinstallation Position of Early Fisher-Sears Idler

the whole geared head under the edge of the sheet metal.

You may need to twist the part with the swinging gear to get it started under the metal corner that you loosened up. Get the teeth of the black gear under the metal, and then the tabs of the white plastic housing. Then rotate it all back into the position in which it was originally, bending one "wing" up over the end of any posts in the way as you swing it into position. As you start to rotate this drive assembly clockwise into the correct position, you will find that one of the plastic wings that you had to bend up earlier to get the part out now must again be bent up to get around the plastic piece attached to the sheet metal loop on the idler. Bend it up enough to clear this pin, and continue to force these parts to rotate around until they get into correct position. As you slide this part the last little distance down on its shaft (with washer already in place), make certain that the rubber tire of the idler runs against the plastic driving surface of the swinging gear piece, and is not caught under it — this is very important.   The parts should drop into position then. You may wish to express a few choice words for the designers of this configuration in the process of getting these parts in place.

Make certain that the idler assembly still swings freely across the entire arc of the crescent-shaped track.

Reinstall the slit washer holding down the idler drive pulley assembly. Grip the washer in needle-nosed pliers close to the cut and simultaneously push it on over the shaft and down as far as it will go, pushing with your finger too. Concentrate carefully on what you are doing so as not to drop or lose this washer!

Before completing the reassembly on the bottom side of the reel table, it can save time first to turn the VCR back over, reinstall the crucial take-up reel spacer washer, the take up reel, the slotted washer holding on the take-up reel, and the idler spring. After doing this, lift the machine up slightly and spin the idler drive pulley underneath with a finger while watching the two reels.   When you spin it one direction, one reel should turn; when you spin it in the other direction, the other reel should turn. If it fails this test, determine why, disassemble and reassemble until it works right. (If it is not working right, it may be that the plastic of the idler drive assembly is bent so that the black part of it is brushing against the side of the new idler that it is supposed to pass freely over. If so, bend it away, or replace it.) If it passes this test, turn it over and continue with the reassembly.

Put the spacer washer (if there was one) onto the axle for the plastic cam arm. If there was one but it seems to have disappeared, try looking for it stuck to the bottom of the container that was nested inside this container.

Install Split Washer Holding Idler On

Reinstall the lock washer holding in place, on its rod, the plastic pin that is attached to the idler. If a lock washer seems to have mysteriously disappeared, try looking for it stuck onto the bottom of the previous container.

Install Cam Arm Split Washer

Put the white plastic cam arm rod back in place, remembering which end went where in the cam gear, fitting it on its axle while putting the pin on this farther end into the same track in the cam gear that it came out of, which you looked at carefully earlier in the procedure before disassembly. Put down the split washer that holds it in place.

Check to make certain that when the idler drive swings from side to side, the wing of the idler drive assembly strikes against the metal pin attached to the other end of this rod, and not against the plastic. If you have an early model, you bent the wing(s) before, during disassembly and reassembly, so they may now be bent out of position. Make certain that they hit the pin in its middle, and not close to the end attached to the plastic, because when you turn the machine over into the rightside-up position, the part with the wing will slide a tiny bit down, in the direction of the bottom of the machine. If it hits the plastic part instead of the metal, the take-up reel will not turn properly to reel in the tape during play, and the VCR will keep shutting down.

Complete the reassembly by following in reverse order the earlier steps of the disassembly procedure. If you had to remove a screw from the sheet metal to get out the idler drive pulley assembly on an early Fisher, put it back in. Attach the plastic or metal cover. Hook up the spring. Put the wires back into place. Reinstall the fast-forward belt — install a new one if you've got one. And install a new loading belt while you're at it, if you have one. Fasten down the bottom board with screws and plastic catches. Reattach the front cabinet piece if you had to remove it to get the bottom open. Turn the machine over, and reinstall the take-up reel spacer washer, the take-up reel, the split washer locking it down, if you have not already done so. Reinstall the plastic brake and brake spring, idler, idler spring, and cassette loading assembly (if you removed it), and test.

It should now fast-forward and rewind. If the fast-forward and rewind work, but the take-up reel does not turn in play, go back and check to make certain that the plastic wing on the pulley assembly hits the metal pin, and not the plastic, on the plastic cam arm rod, as explained earlier.

Goldstar-category Idlers

If you have a Goldstar, Samsung, or Emerson that has the idler with the four little raised dots at the four points of the compass near the edge, that looks like this, you may run into the problem that all the parts suppliers want to sell you a whole assembly with three motors attached to a plate with the idler attached, just to replace the idler. On many of these

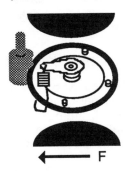

← F

machines the idler is the same as the Sharp NIDL0006GEZZ (Shp) idler. This same idler goes into units made by completely different manufacturers, oddly enough.Compare it with the picture of this idler listed under "Sharp" in Appendix IV, and if it is the same, then order up one of these from a Sharp supplier, and follow the procedure for replacing the Sharp '0006 idler given later in this section. If it does not have the same body as the Sharp idler, probably the best bet is to go ahead and order up a Sharp idler, take the rubber tire off of it, and just replace the rubber tire reusing the old idler plastic parts. The only supplier I have found that can provide just the wheel and tire alone is Electrodynamics, part number is 08-4290 (Edc), but in my experience, the rubber on this generic substitute idler only lasts about six months.

Hitachi-category Idlers

HITACHI 1

If the idler in your RCA, Sears, or other Hitachi-category machine looks like the one on the following page, then it is one of the easiest idlers of all to replace. Your dealer's parts catalog may call this an "arm," rather than an "idler," because it swings like an arm, but it is the same part. Its part number is

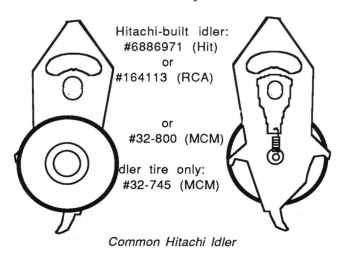

Hitachi-built idler:
#6886971 (Hit)
or
#164113 (RCA)

or
#32-800 (MCM)

dler tire only:
#32-745 (MCM)

Common Hitachi Idler

is 6886971.

If your machine uses this idler, you should always keep a spare on hand, because they do wear out. (They also come in three-packs as #X480041 (Hit)).

To replace this type idler, proceed as follows.

Put cassette basket in the ejected position and unplug machine.

Look carefully at the center of the top of the idler that you can see down below the raised basket. Try moving it from side to side.   Do you see how it swings?   On the left it will probably swing over to contact the edge of the supply reel, while on the right, depending on the particular model, it may be prevented from quite reaching the edge of the take-up reel by a little plastic pin coming up from the mechanism below, but it still swings over a little way.

Look at the round shaft protruding up through the idler from below, the shaft around which it swings. Notice that the idler is prevented from riding up this shaft by a thin, movable piece of sheet plastic that hooks into a groove scored in the shaft at one end, and that is connected to a tiny spring that goes to the

metal piece at the center of the idler's wheel on the other end.

To remove the idler, use a scribe or other pointed tool to slide this thin plastic catch in the direction away from the center of the idler wheel — that is, in the direction that stretches the spring —and at the same time, slide the idler up the shaft in a vertical direction. It will lift right up and out.

Slightly to the left of the idler, and to the rear of the take-up reel, there is a hole in the reel table through which, among other things, a plastic part can be seen that has a short vertical plastic pin coming up. Look to see whether the metal piece is behind it, as it should be, as shown in the illustration, or stuck in front of it, as it should not be.

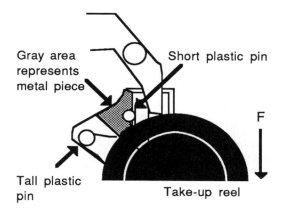

Gray area represents metal piece

Short plastic pin

F

Tall plastic pin

Take-up reel

Correct Position of Pin from Idler Drive

If the metal piece is behind as it should be, go directly to the next paragraph. If the metal piece is sticking out in front, then you need to retime the mechanism as well as replace the idler. To do this, turn the VCR over and open out the bottom circuit board (see Appendix III for help doing this). Remove the plastic idler drive mechanism by removing the reel belt and two screws (notice that the screws are different).

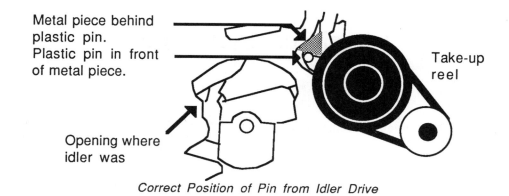

Metal piece behind plastic pin.
Plastic pin in front of metal piece.

Take-up reel

Opening where idler was

Correct Position of Pin from Idler Drive

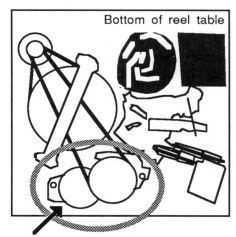

Bottom of reel table

Remove idler drive assembly

Plug in the VCR with the plastic idler drive mechanism removed. The motor will take it back to its normal stop position. Replace the two screws (the one with the fine, machine thread goes on the supply side (left side, as you view the bottom of the upside-down machine).

It would be a good idea to take this opportunity to replace the two loading belts on the little pulleys. See Chapter 14 for help doing this, if you need it.

Use a cotton swab moistened with alcohol to clean all the way around the edges of both plastic reels. Also, clean the plastic shaft that the rubber tire of the idler pressed against when it was in the machine. You can make this plastic drive shaft rotate, so as to clean all sides. On many models, you can make this drive rotate by turning an exposed flywheel on the top of the capstan motor located at the right rear corner of the reel table (but on later models, this flywheel is inside the motor enclosure, so you cannot reach it).

Install the new idler by reversing the procedure. Slide it down the shaft, use the sharp scribe point to move the thin plastic piece in the direction that stretches the spring slightly, and slide it down to catch in the groove on the shaft. This would also be a good time to clean the rest of the machine using the procedures in Chapter 3. You may as well clean off the new idler as well, since you may have gotten some oil onto it in the reassembly process.

Plug in the machine, insert a cassette, and first check to make certain that it now Fast-forwards and Rewinds. If it does, return to the basic universal check out procedure.

If the machine will rewind, but will not fast-forward, trick the machine into operating with no cassette inside it, following the procedures in Appendix V. Then watch what happens when you hit fast-forward. Watch the plastic pin coming up through the opening in the reel table beside the idler that we looked at earlier right after removing the idler.

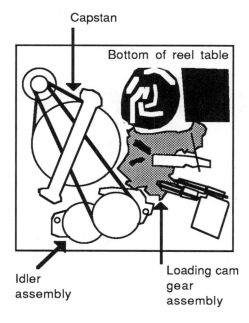

Capstan

Bottom of reel table

Idler assembly

Loading cam gear assembly

Location of Loading Cam Gear Assembly

If it is preventing the new idler from swinging all the way over to the right to contact the take-up reel, you probably have a broken lever in the loading cam gear assembly underneath the reel table. This assembly is located behind the loading motor and on the supply-reel side of the capstan.

It is shown as gray in the illustration. It cannot be repaired, and replacing it is really an advanced procedure beyond the elementary level of this book, but if you want to be adventurous, here is what you do. On top of the reel table, remove the capstan oil retainer; make sure the mechanism is in the fully unloaded position and put rubber bands on the P-guides to hold them at the front of their tracks; locate the two screws through the sheet metal in front of the heads that hold on the loading cam gear assembly that is below, and remove them. Turn over the machine and remove the idler assembly, capstan brace and capstan, and finally the loading cam gear assembly. Replace it with a new one, taking the time to get the parts from the loading mechanism into the right slots in the cam gear assembly. Reassemble.

Hitachi 2

A less common Hitachi-category reel table looks as shown in the next illustration.

Here there are two idlers to replace, the one in the center, between the two reels is responsible for the fast-forward and rewind functions. Its part number is 150614 (RCA). The other idler, located to the right of the take-up reel, called the "take-up

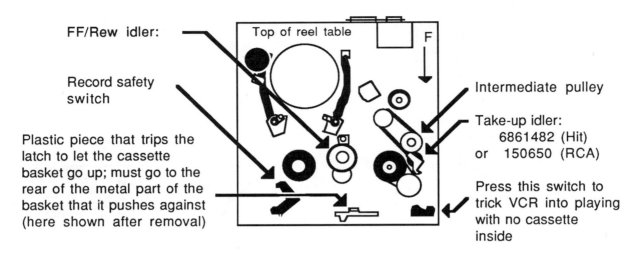

FF/Rew idler:

Record safety switch

Plastic piece that trips the latch to let the cassette basket go up; must go to the rear of the metal part of the basket that it pushes against (here shown after removal)

Top of reel table

F

Intermediate pulley

Take-up idler:
6861482 (Hit)
or   150650 (RCA)

Press this switch to trick VCR into playing with no cassette inside

Hitachi-category Reel Table with Two Idlers

idler" isresponsible for making the take-up reel wind in the tape after it has passed the heads during play and record. If it slips, the machine may shut down shortly after it goes into play, or it may damage the tape. Its part number is 6861482 (Hit) or 150650 (RCA).

To replace the FF/Rewind idler, first remove the record safety switch relay arm, located forward of the supply reel. It needs to be removed in order to remove the basket assembly. Note how the little plastic piece near the take-up reel that pushes against the metal part of the basket assembly to make it eject is positioned rearward of the metal part it pushes against, so that you will remember this correct position of the parts when you reinstall the basket assembly.

Remove the loading basket assembly by removing the two screws that hold it down.

Note the height of the FF/Rew idler relative to the pieces it contacts. Use a 1.5 mm hex key to remove the allen-head fastener holding the brass piece on the motor shaft. Slide the whole assembly up the motor shaft, slide on the new one, and lock it down with a 1.5 mm hex key.

To replace the take-up idler, first remove the fastener locking down the intermediate pulley, the pulley, and the belt to it. (See Sections 9 and 10 of Appendix I if you need help removing these fasteners.)

Remove the fastener holding on the take-up reel. As you slide the take-up reel up and off the shaft on which it is mounted, when it gets part way up the shaft, be careful to push the little bearings on the shaft back down. Do not let them come off with the reel.

Look carefully at how the take-up idler is positioned in the machine, and bend the flexible part of the new one into the same position. Remove the fastener holding down the take-up idler and slide it off its shaft. Put on the new one before you forget how it goes, and replace its fastener. Reinstall the take-up reel, its fastener, the intermediate pulley, its fastener, and the belt.

When replacing the cassette basket, make certain to put it in with the little plastic piece near the take-up reel on the reel table to the rear of the metal part of the basket assembly that it pushes against, as you observed earlier.

MITSUBISHI

The Mitsubishi HS 329 UR has the same reel table as a Hitachi using the 6886971 (Hit) idler. Use replacement procedure under Hitachi.

SEARS

Some Sears VCR's, including the Sears model 934.53323550 use the #6886971 (Hit.) idler. For this idler, use the replacement procedure under Hitachi.

Panasonic-category Idlers

Idler replacement on most Panasonic-category VCR's is simply a matter of removing a C-ring and sliding the old idler up and off the axle, cleaning the surrounding parts, sliding the new idler on with the same side up, and replacing the C-ring. Consult Sections 9 and 10 of Appendix I to learn how to remove and replace the C-rings, and be sure to put a cloth down over the tool and C-ring before snapping it off or on, because they do love to fly away and get lost. The cloth will catch and hold them right where you're working.

A small amount of disassembly is required to replace the VXPO521 idler, which looks as shown in the picture. A spring stretched above the idler can be unhooked at the end on the left to move it out of the way before removing the plastic lock fastener and sliding the idler up and off. You can get just the idler wheel and replace it in the assembly by removing a metal C-ring, spacer, and spring — but for my money, it's easier just to order up a whole new VXPO521 assembly.

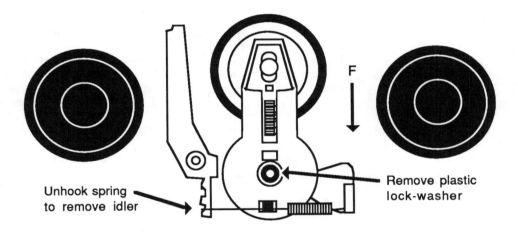

Unhook spring to remove idler

Remove plastic lock-washer

Removing Spring to Remove VXPO521 (Pan) Idler

Samsung-category Idlers

If you have a Goldstar, Samsung, or Emerson that has the idler with the four little raised dots at the four points of the compass near the edge, that looks like this, you may run into the problem that all the parts suppliers want to sell you a whole assembly with three motors attached to a plate with the idler attached, just to replace the idler. On many of these machines the idler is the same as the Sharp NIDL0006GEZZ (Shp) idler. This same idler goes into units made by completely different manufacturers, oddly enough.Compare it with the picture of this idler

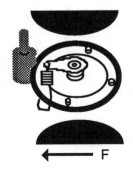

listed under "Sharp" in Appendix IV, and if it is the same, then order up one of these from a Sharp supplier, and follow the procedure for replacing the Sharp '0006 idler given later in this section. If it does not have the same body as the Sharp idler, probably the best bet is to go ahead and order up a Sharp idler, take the rubber tire off of it, and just replace the rubber tire reusing the old idler plastic parts. The only supplier I have found that can provide just the wheel and tire alone is Electrodynamics, part number is 08-4290 (Edc), but in my experience, the rubber on this generic substitute idler only lasts about six months.

Sharp-category Idlers

Two of the most common configurations for idlers in machines in the Sharp category, including many Montgomery Ward VCR's, are shown in the next illustration, along with Sharp and alternative part numbers. (For short, let's call the one on the left the "'0005" and the one on the right the "'0006" idlers in the following discussion.)

Before you can repair the machine, you will need to obtain a replacement idler, or at least a new tire for the idler. With the cover off and no cassette in the VCR, look down between the reels, and see whether your machine has one or the other of these two types of idler. If so, order up a replacement (see Appendix II for sources). If you need to get your machine running again, you probably can get by for three or four weeks just by cleaning the rubber tire on the idler and the surfaces it rolls against, but cleaning is only a temporary fix.

If you are ordering from a Sharp supplier, it might be a good idea also to order a replacement little spring that is supposed to hold the idler firmly against the shaft that drives it. The part is inexpensive, you might lose the one you've got already during disassembly, and the old spring may be weak and need replacement, if not now, then at some time in the future. The part number for this spring is MSPRT0181GEFJ.

If you are going to replace the idler and not just clean it, you should wait until you have the part(s) in hand before beginning the disassembly procedure, because if you disassemble it and then wait for the part, your memory of how the parts go together may become faint, or worse, you may lose some of the crucial little parts in the interval. In the meantime, while you are waiting, you probably can get more hours of use out of the VCR by cleaning the idler, reels, and drive using the procedures in chapter 3.

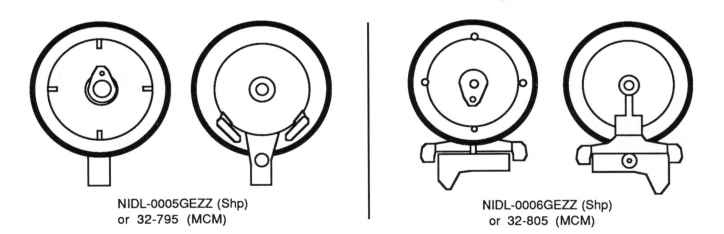

NIDL-0005GEZZ (Shp)
or 32-795 (MCM)

NIDL-0006GEZZ (Shp)
or 32-805 (MCM)

Fronts and Backs of Two Common SHARP-category Idlers

Removing Parts to Get at a '0005 Idler

When you have the needed new part(s), proceed as follows.

Unplug the machine, remove the bottom cover, put all the screws in a container, number it, and record the container number and screw location in your assembly notes.

Either tape shut any hinged plastic doors or lids on the front panel, or carefully remove them altogether, so that they will not get broken when you remove the front cabinet piece — which you will need to do in order to remove the cassette basket mechanism, so as to replace the idler. Follow the procedure in Appendix III to remove the front cabinet piece. This will involve removing some screws on top, and unfastening plastic catches on the sides and bottom of the front cabinet piece. When you first take the front off, remove any little black plastic or felt covers over the small slide switches on the front, so that they will not get lost.

If a top circuit board is attached to the rear of the cassette basket assembly housing by two screws, remove the screws and fold the circuit board open on its hinges.

Next remove the cassette basket assembly. Unplug the wires that connect to the basket assembly at a rear corner. Stuff some tissue paper in nearby holes to prevent losing screws down them. Remove the two to four red screws that hold the assembly down to the table. On some models, you will need also to remove two larger black screws that hold down a plastic strip that runs along the front of the front-loading assembly below where it opens out into the front opening for the cassettes to enter.

On some models, when the screws are removed, you will need to slide the whole basket assembly a short distance forward horizontally, toward the front of the machine, to unhook two metal catches that extend from it down below into the sheet metal table. Slide it about 6mm (1/4 inch), and then lift it up and out when it comes free. Put the different types of screws in different containers, number and record

their location in your notes.

A few models had a plastic cover over the whole area in which the idler is located. If this is the case on your machine, remove the counter belt from the take-up reel, and remove this cover by sliding off the grommet holding it down.

Some models have a black plastic piece with a "cassette down" switch attached that stands between you and the parts you need to remove to get out the idler. If yours does, remove the screws holding this piece down, and put them in a numbered, labeled container. You cannot remove this piece entirely if wires are attached, but it is enough just to push it to the side slightly, out of your way.

Carefully remove the little spring that is hooked to the piece of metal in the center of the idler. You can do this by gripping it with needle nosed pliers, and working it free. The idler shown in the illustration is of the '0005 type, but you do exactly the same thing with the '0006 type. Be very careful not to drop this spring down into the mechanism underneath, because it could be extremely difficult to remove.

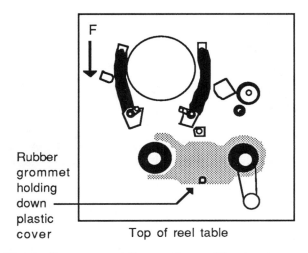

Rubber grommet holding down plastic cover

Top of reel table

Plastic Cover over Idler on Some Sharp-category VCR's

Removing Spring from Sharp-type Idler Assembly

Next put a mark with a felt-tip pen on the forward edge of the sheet metal piece that held the spring that you just finished removing, so that you will not be uncertain later which way it faces when you reinstall it. This mark will help you reinstall it correctly later. It is held down by the two little screws indicated by the two arrows in the illustration.

Before removing these two little screws, let me explain something. In addition to holding down the small formed sheet metal part to which one end of the spring you just removed was attached, these two little screws are also holding the outer enclosure of a small motor that drives the idler, which drives the reels. You cannot see it because the screws have it pulled up against the bottom side of the reel table underneath.

When you remove the second screw, the motor will drop free a short distance below, until it hits on the circuit board underneath. I think that the hardest thing about this Sharp-category idler replacement job will probably be getting both these screws screwed back into the motor enclosure underneath. There are basically three approaches to get the motor reattached, which I need to explain now, because they influence the next disassembly step.

(1) You can open up the bottom circuit board and with one hand push the motor back up into place from the bottom while simultaneously starting the first screw from the top. You need a magnetic screwdriver or three hands to do it this way.

(2) Without opening up the bottom circuit board, you can reach down from the top with an L-shaped tool

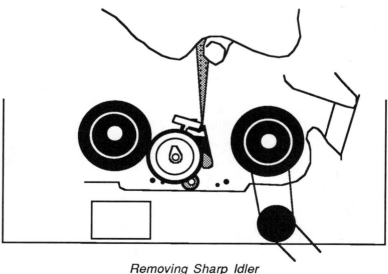

Removing Sharp Idler

(like a scribe) through holes in the reel table, hook around the bottom of the motor, pull it up with the L-shaped tool, fiddle with it until the screw holes in the motor enclosure align with the holes in the loose little formed sheet metal part and in the reel table, and get a screw threaded in to pull the motor up far enough to start the second screw. Again, you will probably want a magnetic screwdriver.

(3) You can remove one of the two screws completely, but only loosen the other screw enough to get the old idler out and the new one in, and then retighten the loose screw to lift the motor back up high enough to get the second screw threaded into it.

I always find the third approach fastest and least frustrating. It took me a number of jobs to figure this out, but you might want to benefit from this knowledge and start with the third approach.

Carefully remove one of the two screws, loosen the other part way, and try to swing the formed sheet metal part around out of your way enough to remove the idler in the next step.

Now you need to wrestle the old idler loose. See how there are two slots cut in the sheet metal, which are mostly hidden underneath the idler, in kind of a V-pattern? Swing the idler from side to side and you will see different corners of this notch. The trick I use on the '0006 idler is to push the center part of the idler off into one of these slots, and then pry the plastic assembly over and out through the slot in the sheet metal. A scribe or very small screwdriver is a good tool for doing this. On the '0005 type, just lift up the foremost edge or side of the round part of the idler enough to clear the little knurled shaft in front of it, and pull it forward, out and over this shaft.

The spring held the idler against a round plastic shaft coming up from the motor below. Clean the exposed plastic shaft, and the sides of the two reels with a cotton swab stick dipped in alcohol. Rotate each reel as you clean it, so that all sides will be cleaned free of oil (which could make the new idler fail too) and other residues.

If your machine's old idler is of the "number 6" type shown earlier, compare it with the new idler you've gotten. If the new idler has a little peaked point of plastic as shown in the picture above, but your old idler was flat across the top, you can cut off the peaked point to make the two completely similar.

A good tool for doing this is a pair of diagonal wire cutters. Just squeeze them down along the top line that you want to straighten. It is probably best not to try to use a knife, as you might accidentally cut your hand. Rather than that, it is better just to leave it on, uncut.

Wrestle the new idler down into the slots in just the opposite way than you removed the old idler. The

plastic is flexible to some extent, so you can bend the new idler a little in this process without breaking it. Don't push your luck, however. Stick one corner part under, and then wrestle the other part under. On the '0006 idler, make certain that the plastic arm connecting the round idler to the rectangular part at the top goes into the narrow top part of the slot.

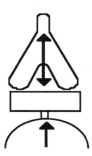

Narrow Part of Idler Goes into Narrow Part of Slot

Also, make certain that the rubber tire on the edge of the new idler is against the side of the cylindrical plastic drive wheel coming up from the motor underneath, and is not left caught wedged on top of it. Take a few minutes to get this right. The idler mechanism should now be swingable from contacting the supply reel on the left, to swinging over to contact the take-up reel on the right.

If you were able to do all this without removing both of the screws that held on the formed sheet metal piece to which one end of the spring attaches, you only need to bring the other side of the motor enclosure up to the other hole so that the screw holes through the formed piece, the reel table, and the motor enclosure all line up, insert the second screw, and get it started. One way to lift the other side of the motor enclosure is by reaching down through the V-slot in the reel table with an L-shaped tool, catch it under the motor, and lift it up. Or alternatively, you perhaps can grip the plastic idler drive from the motor with your fingers or needle-nosed pliers and pull it into place. Tighten the two screws to a gentle tightness.

If you removed both screws, then use the following procedures. Put the sheet metal bracket to which the spring connects, back in place with the side you marked facing forward in the VCR. Now comes the part that requires the most patience. You must get the two little screws back into their holes, through the reel table, and screwed into the two little threaded holes in the metal housing on top of the motor below.

The frustrating thing is that the motor moves when you take the two little screws out, and it is difficult to get it lined up with all the holes so that the screws will start threading into the holes in it. And even when it is lined up, it tends to swing down when you press on the screws to turn them.

As mentioned earlier, some technicians avoid this problem by setting the machine on edge, on one of its ends, removing the fasteners that hold the bottom circuit-board to the plastic chassis, gently prying loose all the little multi-pinned plug-in connectors between the front circuit board and the bottom circuit board on the side under the table, and reaching in with one hand to push the motor up against the underside of the reel table while simultaneously starting the screws from the other side, initially holding them with a magnetic screwdriver.

However, if you can stand a bit of frustration (think of it as a Zen exercise), it is usually just as fast, and much safer from the standpoint of not breaking bottom or front circuit boards, to fiddle around with the parts from the top until you get both screws started.

Here are some tricks. Get one screw already to go, by attaching it to the magnetic screwdriver if you have one. Or get someone else to help with this step. Move the motor around, by gripping the plastic drive gently with fine needle nosed pliers until you see the threaded holes in the motor case line up with the holes in the table. Then pick up the sheet metal spring holder, making certain that the front side is forward, and set it down with the holes lined up with the holes in the reel table and motor case below, while continuing to hold the motor in place with the needle nosed pliers. Now with your free hand, pick up the magnetic screwdriver with screw attached, and start the screw turning in the hole. Make certain that it

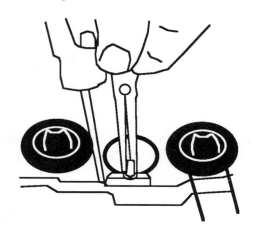

One Way to Get the Reel Motor Attaching Screws Started

tight, leaving the motor a little loose to maneuver it to align the other hole for the second screw. Repeat the procedure until you get the second screw started. You can move the motor a little with the pliers or with a scribe pushed down through one of the V slots. Get the second screw started, then screw both screws up to a gentle tightness.

Attach spring. The larger loop at one end of the spring goes around the little pole or catch on the sheet metal piece you just reattached. It fits down under a little collar or raised part of this pole down at the bottom where it goes into the sheet metal. The smaller hook at the other end should hook into the piece at the center of the idler with the open part of the loop pointing downward. After getting the end with the bigger loop connected, use needle-nosed pliers to stretch the spring out slightly and hook the other end through the little hole in the metal part connected to the center of the new idler. Check again to make certain that the idler still can swing from side to side, as before.

While the cassette basket is still out is a convenient time to clean the tape path and heads. See Chapter 3. You already cleaned the edges of the reels, of course, but you should now also clean the rubber around the new idler to remove any oil that may have gotten on it during the installation process.

Put the cassette basket in place. If your model has feet that catch down through the slots in the table, insert them, slide the assembly into place, and fasten it down with the screws you removed. If your model has two screws with large black heads attaching a plastic piece at the front of the cassette loading assembly, reinstall those parts. Plug the multi-wired connector back into the right side edge of the basket assembly. Make certain that the alignment guides on the one plug are aligned with the slots in the edge of the other plug as you push it on.

Replace any little felt or plastic black covers you removed from the small slide switches on the front of the VCR. Reattach the cabinet front, pushing it on gently directly from the front, and making certain that all plastic catches on the sides and bottom snap into place correctly. Put back any screw(s) you removed from the top catch(es). Reattach the bottom cover. Plug machine in, insert a tape, and test Fast-forward and Rewind. If these now work, hit Play and Record. Your problem should now be fixed. Return to the universal check-out procedure in Chapter 4.

Early Sony-category and Zenith Beta Models

If your VCR will not rewind the tape, and it is an early Sony-category or Zenith Beta model with the mechanical push-down switches that looks something like this,

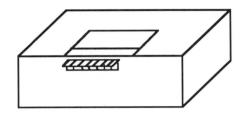

"Tank" Exterior

then you probably need to install the newer pulley kit and pulley, if this has not been done before. See Appendix III for help, if necessary, removing the top cover. The reel table you see should look something like that shown in the next diagram, if we are talking about the same machine.

You will need a pulley kit plus a pulley. The Zenith part numbers are #861-291 (Znt) and #861-519 (Znt). You may also want to replace the other pulley; its Zenith number is 861-490 (Znt) , or for the same part from Sony, X-3661-520-0 (Sny).

Do not be surprised if the new parts look a little different from the old parts. The kit is a factory modification improving on the old design, so if your VCR has never had this modification made on it, the

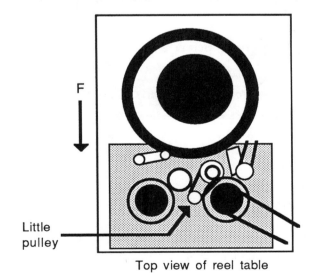

Little
pulley

Top view of reel table

Early Sony-Zenith Reel Table

new parts will look different. When you've got the parts, go ahead and remove the front panel piece too. Also, you will need to remove the cassette top-loading assembly and basket: once you've got the screws out, the removal trick is to lift up the right side of the basket assembly first, and then slide out the left side.

Remove the belt from the little pulley and remove the little pulley using two thumbnails to pull back the little plastic tabs at the top center that hold it to the shaft that comes up through its middle.

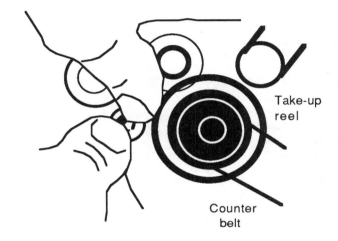

Take-up
reel

Counter
belt

Using Thumbnails to Unfasten and Remove Little Pulley

Look carefully at how the counter belt goes around the take-up reel. If we allow this belt to go slack for even a second as we remove the take-up reel, it will come off its pulley at the other end, which is difficult to reach to reinstall the belt. Try to do this as follows.

Remove the cross-slot screw in the center of the take-up reel. Get out the new take-up reel that comes with the kit. As mentioned, it will save a lot of time and frustration if you can slide the old take-up reel up the shaft without letting the counter belt attached to it go slack, and being careful to leave the crucial washer-spacer underneath still on the shaft; slide the old reel off while keeping it close by in a position that keeps a gentle tension on this belt; slide on the new reel, pushing it down part way to where it probably will be stopped by a little plastic brake whose spring has pushed it in to where it prevents the reel from going all the way down. Let it stop here for a moment. Then take the counter belt between your fingers still keeping tension on it, pull the old take-up reel free, and gently stretch the counter belt around the newly installed reel.

If you are replacing the other pulleys, now is the time. One of them can be replaced without moving the take up reel, but to get the other one off, you may need to slide the take-up reel part way up its shaft again, being careful not to let the counter belt come off. Consult Appendix I for how to get the fasteners off and back on again. When you've finished changing out other pulleys, use your fingers to push the little plastic brake out enough to let the take-up reel slide down past it.

To install the new little pulley, it is necessary to push from underneath on the other end of its axle as you press it on. You can save yourself the trouble of opening the bottom circuit board by inserting the L-shaped end of a scribe or similar tool through a hole in the reel table down below the midway area between this pulley and the take-up reel. Then rotate it so that it hooks under the plastic pulley underneath, pull up on it while aligning the flat side of the hole in the new pulley with the flat side of the axle, push the new pulley down on the axle, and spread the plastic clips so that they catch in the groove in the axle.

Replace the belt around the little pulley, the screw fastening on the take-up reel, the cassette basket, and test.

Toshiba

To replace the idler on Toshiba machines that look as shown under Toshiba B in Appendix VI and like this,

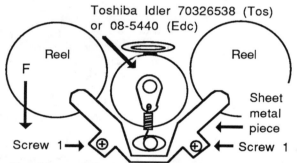

Top View of Typical Toshiba Idler Arrangement

remove the washer over the large end of the spring attached to the idler, free this end of the spring, and remove the two screws marked "1" in the picture (over). Remove the bottom cover from the VCR. From below, remove three screws marked "2." You can also unplug the connector marked "3," or just leave the motor hanging, whichever is more convenient. Now turn the machine back rightside up, and wrestle out the old idler, put in the new one, and reassemble in reverse order of disassembly.

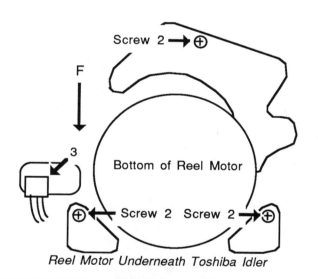

Reel Motor Underneath Toshiba Idler

7. Is the Problem a Broken or Weak Belt?

If you've heard or seen something happen when pushing the fast-forward or rewind buttons, that's good, because in this case, the problem is most likely due to either a broken belt, a slipping idler, or a weak clutch drive (although it is still possible that there is a trouble in the electronics).

Open up the bottom of the VCR (for help, see Appendix III).

Look inside of the top and bottom of the machine for any broken belts. If the machine starts to move the tape and then cuts off after a few seconds, look especially for a broken belt going to a counter on the front panel. If you see a broken belt lying inside the machine, remove it, obtain a replacement new belt (see Appendix VI for part numbers and Appendix II for parts sources), and replace the belt. Test to see whether this fixes the problem. Does the machine now rewind and fast-forward a tape at a good speed? If so, go to the last paragraph of this section. If there is no broken belt, or if replacing a broken belt is not enough to solve the problem, then continue as follows.

For Sanyo-category VCR's with the little belt on the idler assembly located midway between the two reels, if that little belt is broken, go to the procedure at the end of this section.

With the bottom open, stand the machine on its end (I mean, the right side or the left side). Make certain that there is no metalic object under machine on the bench.

Plug the unit back in, turn on power. Insert a spare cassette or a dummy cassette (made as described in Section 17 of Appendix I). Press "Rewind" or "Fast-forward" buttons. Check to see whether a belt that should be turning part of the mechanism is slipping. Sometimes this can be checked by using the eraser on a pencil to "help along" the pulley or other parts that the questionable belt is supposed to be driving.

If a belt is slipping, check for foreign objects in the mechanism that it is supposed to be driving — like a coin, a toothpick, a screw, that sort of thing. If you find one, remove it, of course, and see if the machine will now fast-forward and rewind. If so, return to basic check out procedure in Chapter 4 (although you may wish to take this opportunity to clean and lubricate the machine first, as explained in Chapter 3).

If a belt is slipping, but the problem is not due to a foreign object in the mechanism, carefully disassemble the mechanism only enough to remove the slipping belt(s), put the belt(s) briefly in a bottle of 90% rubbing alcohol, remove, carefully wipe dry with a paper towel or clean rag.

Before putting the cleaned belt(s) back onto the pulley(s), you must clean the groove in the pully(s) with cotton swab sticks dipped in the alcohol. NOTE: The pulley(s) must be dry before putting cleaned belt(s) back on. (Also, as emphasized elsewhere, do not use cotton swab sticks to clean the video heads.) If any belt has cracks or a shiny surface after cleaning, you should order a replacement, even if you can make the old belt work temporarily. You'll have to replace it someday soon, even if you can get by with the old one in the meantime.

Test to see whether this fixes the problem. Does the machine now rewind and fast-forward a tape at a good speed? If the belt still slips, then probably it is stretched or weak, and needs to be replaced. Remove it, obtain a replacement new belt (see Appendix VI for part numbers and Appendix II for parts sources), and replace the belt. If it now moves at a good speed, proceed as follows.

If you happen to have one of the special dummy cassettes with holes in the top made as described in Section 17 of Appendix I, you can check the rotary force ("torque") with which the reels are turned in the fast-forward/rewind modes. It should be strong enough that you can barely stop the reel from turning by gripping it between thumb and forefinger (400 gram-centimeters of torque, if you want to be exact and have a torque meter). If it is good, reassemble the machine carefully and return to the basic procedure in Chapter 4.

If FF and rewind still do not work properly, go to the next section, unless it is a Sanyo of the type described in the next paragraph.

Sanyo

Many Sanyo-category VCR's used an idler assembly located between the two reels that looked like one of the mechanisms in the illustration. For clarity, the illustration shows only the idler assemblies, as if out of the VCR.

A common problem is that the little belt breaks. The belt comes with a larger "roller assembly" that completely encloses it, which is somewhat expensive. You get a new rubber idler (part with rubber tire) with the complete assembly, but if the rubber on your old idler is not cracked or dried out, you might be able to get running again by just cleaning the old idler and installing a new belt, part #143-2-564T-03200 (SFC). The following procedure works for both courses of action, illustrated with the 143-0-662T-

Roller Assembly: Part #
143-0-662T-01202 (SFC)

Roller Assembly: Part #
143-0-662T-10350 (SFC

Sanyo-type Idler Assembly that Comes With Belt Attached

10350 assembly

The fact that the new belt comes attached to a complete idler assembly might make it appear that the whole idler assembly should be replaced, if you have a new one. You can do this, but the bottom sheet metal plate is difficult to get up. It is much easier to take the important parts with rubber on them off the old idler assembly and just replace them with the new part(s) you've got.

Starting with the new idler assembly, remove the locking fastener that holds down the top swinging part. (How to remove these fasteners is explained in Appendix I.) Slide it off.

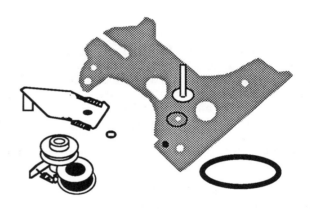

Disassembly of Sanyo-type Belt /Idler Assembly

idler part, taking care to get out the washer between them. Put the new bottom part onto the shaft that the old part came off of in the VCR, put the washer on the shaft, and then slide on the top part.

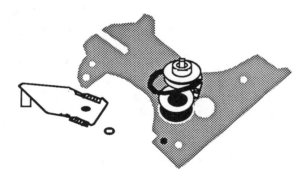

Disassembly of Sanyo-type Idler Assembly

If you are only replacing the belt, skip down to the beginning of the next paragraph. If you are replacing the idler as well as the belt, next, slide off the part with the plastic pulley together with the idler with the little rubber tire around it. The idler is actually a part separate from the pulley that was together with it. It is OK to take these off separately, but take care not to lose the little washer that was between them around the shaft.

Remove all the same parts from the old assembly in the VCR. Now separate the pulley part from the

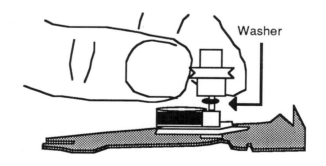

Reassembling Sanyo-type Belt and Idler Assembly

Slip the new belt over the pulley, then slide the swinging part onto the shaft, and put the fastener on. Pull the other end of the belt over the nearby shaft that drives it, and test.

8. Realigning Position of Top-loading Basket Assembly in Portables

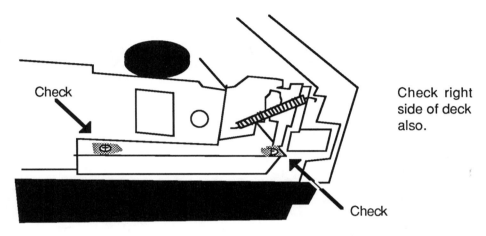

Check

Check right side of deck also.

Check

Location of Screws Attaching Top-loader to Deck

If a Panasonic-category portable deck is having trouble with fast-forward and rewind functions, the problem may be that the top-loading assembly containing the cassette basket has gotten knocked out of alignment relative to the reel table. This is especially likely if the deck has been dropped.

With the top of the cabinet removed, look along the right and left sides at the points where the sheet metal sides of the top-loading assembly rest on the metal reel table. The screws that attach the loading assembly to the reel table will have dabs of colored (usually red) lock-paint on them.

Look at them carefully to determine whether any has had the metal from the top loader under it moved.

You can tell because there will be a circular impression in the dried paint where the round head of the screw used to be, and then the head of the screw will be seen to be displaced a short distance forward or to the rear from this impression in the paint.

If this situation is observed to exist, loosen slightly the screws holding down the assembly, remembering to push down hard on the screwdriver as you start to turn each screw, so that you will not strip the cross slot in the head. When the screws have been loosened slightly, slide the top-loading assembly a short distance to the point where the old paint-impressions are exactly under the screw heads, and then tighten them down again. Now test and see whether this solves the problem.

9. If All Else Fails

If you cannot determine why the motor that runs when you push FF/Rewind does not cause the tape to be wound, try "tricking" the VCR into going through its fast-forward/rewind motions with no cassette in the VCR. Appendix V tells you how to set the machine up to run with no cassette in it. Turn to Appendix V to

do this, and return to this point when you've got it set up.

Now push the fast-forward/rewind buttons and observe what is happening. You have got a motor that runs someplace, and you have got one or the other reel that is supposed to be turning, but is either not turning, or not turning with sufficient rotary force. The trouble lies somewhere in the mechanism between these two points.   Figure out how the motion is supposed to be communicated from the driving motor to the reel, and then find where the slippage occurs, or the motion fails to get communicated.   This point is the source of your trouble.   Replace, clean, or repair the part that is failing to transfer the motion.

". . . It says here that you're not supposed to clean a VCR that way. . . "

Chapter 14

VCR Will Not Move Tape Properly in "Play" Mode

This chapter concerns only problems that the machine may have just moving the tape in the play mode. For example, if you inserted a cassette and hit "Play" and the VCR did not start moving the tape properly, or it started, ran for a moment, and then turned itself off, then this is the right chapter for you to be reading. Later chapters cover problems in the picture or sound that show up on the TV.

If the problem is that the machine starts to play, and actually plays for a while, and then keeps shutting down, sometimes after playing for more than half an hour or so, then you may be able to save time by turning directly to Section 7 at the end of this chapter, especially if the VCR is manufactured by Samsung. Otherwise, continue reading below

SPECIAL MODEL NOTES

FISHER, HITACHI, AND R C A

If the machine is in either the Hitachi or Fisher category, and it makes a squealing or screetching sound when you plug in, or power up, or when you hit "Play," maybe then shutting itself off, turn directly to the procedures for replacing the loading belt(s) in Section 3. If the VCR makes a squealing noise and then plays, go also to Section 3. On a Fisher-category machine, if replacing the loading belt does not solve the problem, check for a broken plastic pin on the bottom of the motor mount bracket.

HITACHI AND R C A

If the VCR is a Hitachi or R C A that eats tapes, or just won't play right, and the top of the reel table looks like the next illustration,

and the idler (where the arrow points) is of the type with a toothed plastic gear that engages the reels (rather than a rubber wheel), turn to Section 1 of Chapter 15 to determine whether the problem is a broken idler-drive assembly plate. If not, return here.

ZENITH AND J V C

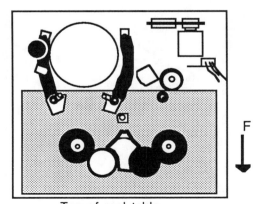

Top of reel table

A Common Zenith-JVC Loading Motor Configuration

Zenith and J V C: If the machine is a Zenith or J V C table model in which little lights show through square holes in the push buttons on the front panel and it has a little motor and belt back behind the heads in the right rear corner of the table as shown in the illustration, turn directly to the special Zenith-JVC loading gear repair procedure in Section 4 of this chapter. (If all this happens except that the machine plays for a short while before it shuts down, then you have an electronics problem beyond the scope of this book.) If all this happens except that the take up reel does not move at all, or just twitches, then run the Zenith-JVC idler-clutch assembly cleaning or replacement procedure also in Section 4.

1. Preliminary Diagnosis of the Problem

It is assumed that you have unplugged the VCR, opened the cabinet top and circuit boards above the reel table, powered up again, loaded a cassette, and pushed "Play," and observed the machine fail to load the tape properly or fail to move the tape properly in play fashion.

In particular, the following events should happen: The video heads rotate rapidly. The tape loading mechanism pulls a loop of tape out of the cassette and wraps it around the video heads. The capstan starts turning. The pinch roller moves over to press the tape tightly against the capstan. The take-up reel starts rotating clockwise. The counter starts counting numbers. And the mechanism should continue moving the tape slowly past the heads until the end of the tape is reached or until someone pushes another control button on the front of the VCR.

The first step is to watch carefully exactly what does — and what does not? — happen in this sequence of events. Do the heads spin? Do the guides pull some tape out of the cassette? Does the capstan start rotating? Does the pinch roller move all the way over to press the tape against the capstan? Does the take-up reel start turning and continue to turn? Does the counter move? At what point does abnormal behavior first appear?

Observe whether the cylinder (also called "upper cylinder" or "video heads") starts to rotate. If not, go to Section 2. If it does, proceed to the following paragraph.

Observe whether the machine pulls the tape out of the cassette and loads it fully. If the machine is a Beta, skip over the next paragraph.

If the machine is a VHS, observe the movement of the P-guides (also called "loading posts" or "tape guides" — See Chapter 2 for terminology and location). On most VHS machines these should move from the opened plastic door of the cassette in towards the rear of the machine, running on tracks in the table, and stopping against the "V-blocks." These are the stoppers at the rear end of the tracks. Most

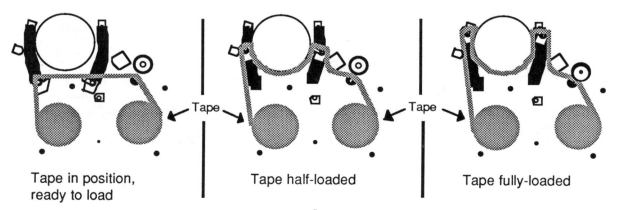

Tape in position,   Tape half-loaded   Tape fully-loaded
ready to load

Tape Loading in a VHS Machine

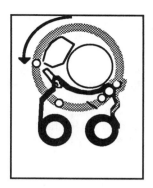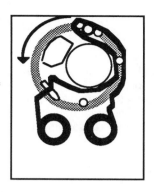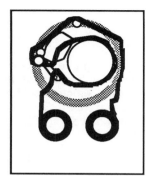

Tape Loading in One Type of Beta Machine

V-blocks have a notch that the P-guides are supposed to come to rest against, for which reason they are also called "V stoppers." (Incidentally, NEVER try to move or adjust the position of the V-stoppers; they are set in place with an elaborate jig at the factory, and if they are ever moved again, it is extremely unlikely that even the best technician could ever get them back into their exact position again correctly.) This operation of the P-guides pulling the tape out of the cassette and running along the tracks until the P-guides, with tape looped around them, stop against the V-blocks is called "loading," or more precisely, "tape loading," as explained in Chapter 2. After hitting "Play," look to see whether your machine completes this loading process correctly Do the P-guides MOVE ALL THE WAY to the V blocks, or does one or both stop short?

If the P-guides do not move, or if one (or both) stops before reaching the V-blocks (perhaps then reversing the direction of its motion after a moment and returning to the starting unloaded position), go to Section 3 below. If they move all the way forward, then go to the paragraph after next and check what happens to the pinch roller.

If the machine is a Beta, there are more possible variations on the loading process used in different models, but a typical pattern on early Beta's involved a single loading ring using a circular motion to wrap tape most of the way around the stationary cylinder with video heads. If the tape loading ring does not move, or starts to move but stalls before getting all the way around to the position where the little catch latches it in the fully-loaded position, go to Section 3. If not, and it goes all the way around, then continue reading here.

Sometimes people do not realize that it is the capstan, and not the take-up reel, whose rotation is supposed to pull the tape through the mechanism and past the various heads at the correct speed. Check to make certain that the rubber pinch roller moves into a

position firmly contacting the capstan shaft after the tape-loading mechanism reaches its end point, and that the capstan starts rotating to pull the tape along the tape path. If you cannot tell for sure visually whether the capstan is rotating and making the pinch roller turn, take a screwdriver with an insulated handle and keep it on the top of the rubber pinch roller near the edge, as you push "Play" and watch the sequence of events. You should feel the pinch roller start to rotate after it moves in and presses the tape against the capstan. If not, go to Section 5.

Is the unit stuck in the "Pause" mode? It won't play if it is in pause.

If the tape loads, the video heads turn, and the pinch roller contacts the capstan which starts to rotate, but the take-up reel fails to turn, then the safety system will shut everything down on most machines. If so, go to Section 6. If you see a loose loop of tape spilling into the machine from between the capstan and the place where the tape goes back into the cassette box, then you know that this is the trouble. But on most machines, the safety system will detect the problem with the take-up reel, shut everything down, and unload the tape even before it spills, so the best way to tell whether the take-up reel is turning on a VHS machine is to look at it through the window in the top of the cassette.

If the loading appears to be completed, and the tape starts to play, with a picture coming up momentarily on the TV screen, but the VCR makes a squealing noise and then shuts down, a likely cause of the trouble is that a loading belt is slipping. To check for this, you need to locate the loading motor and check to see whether it has a loading belt that is slipping so that loading is never quite completed all the way. Go to Section 3. But if the machine plays for longer than a minute, with a picture on the screen and the take-up reel winding in the tape, then go to Section 7.

2. Video Heads Not Turning

If the video heads or cylinder fail to start rotating, the usual cause of this problem is a trouble in the motor or electronics beyond the scope of this book. However, there are a few possible causes of this problem that you should check before sending the machine to a shop.

With the machine unplugged, try rotating the video head assembly with your finger to see whether the bearings might be locked-up, frozen, or stuck. If so, try rotating the video head assembly by hand to see if you can unstick it. I have seen machines with this problem that were completely restored to health by just turning stuck heads so that they came unstuck. Sometimes a shock or bump from someone setting the machine down too hard will cause the bearings to jam so that the little motor cannot start the heads turning, but once they are freed again, the machine works fine.

A few very early Sony and other models used a belt to drive the heads; on those models, the breaking or slipping of this belt can also cause the heads not to rotate.

In a few cases, a video tape will adhere to a slightly damp cylinder, wrap around it, and jam it — but you probably would have noticed if this was the problem, and removed the tape.

If none of these is the problem, you will need to take the machine to a shop.

3. Mechanism Not Fully Loading the Tape

If the P-guides on a VHS machine, or the loading ring on a Beta, do not go all the way into the fully loaded position, or the pinch roller does not get pushed over against the capstan, then there is a problem in the tape-loading mechanism.

Also, as explained earlier in this chapter, if the machine appears to complete the tape-loading process, and the tape starts to play with a picture coming up momentarily on the TV screen, but the VCR makes a squealing noise and then shuts down, a likely cause of the trouble is that a loading belt is slipping.

To check for this, you need to locate the loading motor and check to see whether a loading belt is broken or slipping so that tape-loading is never quite completed all the way. Watch the pully or gear attached to the loading motor very carefully as you push "Play" with the cassette in the VCR. If the loading motor never stops and comes to rest, but is always running, first in one direction, then suddenly in the other, this usually means that the loading operation is never being completed. (If the sensors indicated that it was completed, the little computer would have shut off the motor. After a few seconds of trying to load, the little computer decides that it had better unload and shut down, which is why it

reverses the direction of rotation of the loading motor suddenly.)

The most likely cause of failure to load fully is a worn or oily belt, probably the smaller belt, if there are two. This can be tested conclusively by removing the old loading belts, cleaning the pulleys, and replacing the old belts with new belts. Or there may be excessive friction in the loading mechanism which requires relubrication.

If you do not have ready access to a source for these belts, you often can get a few weeks of additional use out of the VCR while you are waiting for belts to be sent to you by removing the old belts, soaking them in rubbing alcohol for a few minutes, cleaning the pulleys with alcohol and reassembling. But you should order up some new belts at the same time, because cleaning alone is only a temporary solution.

The pictures in Appendix VI should help you locate the loading motor and loading belt(s), if any, on your model.

The following is additional information on servicing the loading mechanism on common models.

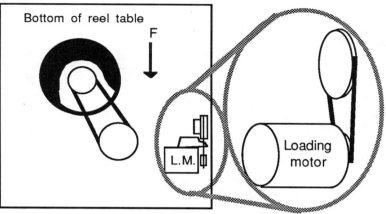

One Common Configuration of Fisher-category VCR's

FISHER-CATEGORY VCR'S

A slipping loading belt is a common cause of failure to start up correctly in Fisher VCR's, especially all those that on top had a reel table like that shown in the left picture, and underneath had a loading motor on the lower front corner of the reel table connected by a belt to the rest of the loading mechanism, as shown in the center picture. See Appendix III for help in opening up the bottom of the VCR to gain access to the loading motor and loading belt.

You should be able to get the old belt off without removing the loading motor, if you use your fingers to pull it off the sides of the pulleys closest to you and then use a scribe or other long, thin tool to work it out free from the bottom edges.

On these machines, it is always better to replace the old belt with a new belt than to try to clean and reuse the old belt. The Fisher part number for the replacement belt is 143-2-7404-00300 (SFC) or equivalently 143-2-7504-00700 (SFC). However, I have found that a VDVS0122 (Pan) belt made by Matsushita for its machines is generally easier to get, works better, and lasts longer, so I recommend substituting the Panasonic belt, or its RCA equivalent, #148664 (RCA).

While you're down under the reel table, it would be a good idea also to replace the reel belt, which is the other, larger belt shown in the picture. The Fisher part number is 143-2-7504-00400 (SFC) or equivalently, 143-2-7504-00600 (SFC), but here I have found that substituting a Hitachi 6355471 (Hit) works better. Take your choice.

Once you've got the loading belt off, take a careful look at the small pulley attached to the loading

motor. Look in the groove as you rotate it to see whether it has separated. If it is OK, skip down to the section below entitled, "Replacing a Fisher Loading Belt." But if the pulley has been partly split, or cut through, go to the next section entitled, "Replacing a Fisher Loading Motor."

Replacing a Fisher Loading Motor

If one of these machines is allowed to go too long with a slipping loading belt, the heat and friction generated by the slipping of the belt around the little pulley can melt or burn through the plastic that it is made from.

When this happens, you used to have to buy a whole new motor, just to replace the little pulley. The part number of the motor is 4-5254-00250 (SFC) or equivalently, 4-5254-00251 (SFC), and it is expensive. But now just the pulley alone is available from Electrodynamics as part # 08-4295 (Edc).

The only alternative is a long shot: if you can find a shop or someone who has a burned-out little Hitachi loading motor in their junk pile, you can pull the pulley off it and push it onto the axle of a good Fisher motor after pulling off the bad pulley. To replace the Fisher motor (or to replace its pulley, if you found one), remove the other, larger belt (a reel belt), and remove the spring attached to the black-plastic, or sheet-metal, cover.

If the model has the black plastic cover, remove the one screw and unfasten the two plastic clips extending from the cover into the sheet metal underneath. If the model has the sheet metal cover, unfasten it by removing two screws. Follow the brown-covered wire(s) from the loading motor back to their other end at a little plug, and unplug it.

Loosen the plastic-covered metal piece holding these wires down on their route, and cut the plastic wire-tie with fingernail clippers, taking care not to cut the wires. Look carefully at how the metal pin on one end of the long plastic arm runs in the track for it in the cam gear, and then use needle-nosed pliers as explained in Section 10 of Appendix I to remove the split plastic washer holding down the long plastic arm and remove the arm too.

Remove the three screws holding down the motor bracket, and tip the bracket and motor up at the front, work it free, and slide it out. Remove and replace pulley, if you found one, or remove and replace the whole loading motor, if you were not so lucky, by removing the two screws going through the plastic into the front of the motor and install the new motor.

Make certain that the square plastic pin holding on the sliding rod attached to the bottom of the plastic motor housing is not broken off; if it is broken, you will need to order a new plastic motor bracket (part #143-2-3404-06800 (SFC)) and replace it; you also will need to find a hardware store with very tiny (like size 0/80) nuts and bolts to attach the motor to the different type of brackets they send as replacements.

Rotate the large plastic cam gear CW as far as it will go and hold it there while reinserting the motor and bracket, so that the little plastic pin on the motor-side of the sliding rod attached to the motor bracket will go in on the left side (when looking at the upside down VCR) of the stopper on the sliding metal rod. The teeth on the worm gear must mesh with the teeth on the cam gear.

After replacing the motor and bracket, make certain that the extended pin on the sliding rod attached to the bottom of the motor housing goes through the hole in the reel table to come out on the front side of the brake lever as shown in the first Fisher illustration earlier.

Plug in the wire from the motor and fasten down the plastic-covered metal wire-holder. Don't worry about any plastic wire-ties that you clipped; just discard them. When you reinstall the long, white plastic lever, make certain to get the metal pin on one end back into the same track that it came out of in the white plastic cam gear. Fasten it down with the split plastic washer (part #143-2-6304-01500 (SFC), in case you lost it). Reinstall plastic or metal cover, put the wires back under the clips that hold them, replace the reel belt, and reinstall the little spring.

Replacing a Fisher Loading Belt

To replace the loading belt with the motor in the VCR takes a little patience, because they don't give you much room to work. Try catching the belt in the grooves of the two pulleys on the side opposite to you, and then stretching the belt over the edges of the pulleys closest to you. You may want to use the bent end of a scribe or similar tool to help get the belts caught properly in the far sides of the pulleys. It usually takes several attempts. When you finally get the belt on both pulleys, rotate the little pulley attached to the motor a few turns while checking to make certain that the belt is not twisted (if it is, untwist it).

Put the front back on, if you took it off. Plug in, power up, and test to see whether it will play a tape. If it still makes a screeching noise and then shuts down, and if you have not removed the motor, then unplug, read the section above that you skipped, and check to see whether the end of the little square plastic pin that holds the sliding rod fastened to the bottom of the plastic motor mount has broken off. If you have sharp eyes, you can tell by peering down while poking the rod with a long thin tool, or if you are like many of us, you will need to run the motor removal procedure above to check for this.

When done, reinstall bottom plate and everything you took off on the top.

HITACHI-CATEGORY AND R C A TABLE MODELS

This subsection applies to Hitachi-category table models with reel tables that look like this on top:

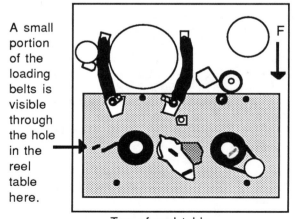

A small portion of the loading belts is visible through the hole in the reel table here.

Top of reel table

Common Hitachi-category Reel Table Requiring Replacement of Loading Motor Belts

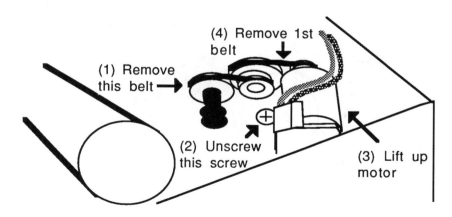

(1) Remove this belt →

(4) Remove 1st belt

(2) Unscrew this screw

(3) Lift up motor

Common Hitachi-category VCR Loading Belt Arrangement

See Appendix III for help removing the bottom cover to gain access to the bottom circuit board and the loading motor on the other side of it, and for help folding out the bottom circuit board. On some Hitachi-category machines, the bottom circuit board can be opened on its hinges without removing the front cabinet of the VCR, but on other models, it is necessary to remove the front cabinet to open up the bottom circuit board.

After you have opened the bottom circuit board on its hinges, you can now see the mechanism underneath. Find the loading belts. They look like the illustration above:

The problem, including the squealing if it is making a noise, may come from one of these belts slipping. This can be tested conclusively by removing the old loading belts, cleaning the pulleys, and replacing the old belts with new belts. There are at least four different reel table designs that look like the picture, but use slightly different sized belts, listed one after another in Appendix VI. Take the time to find which one most closely matches your VCR.

If you do not have access to a quick source for these belts, you can often get a few weeks of additional use out of the VCR while you are waiting for belts to be sent to you by removing the loading motor and old belts, soaking the old belts in alcohol for a few minutes, cleaning the pulleys with alcohol, and reassembling everything. To do this, follow the procedure described below, except putting old cleaned belts back on instead of new belts. But if you can possibly get new replacement belts, putting new belts on is the preferred course of action. (And an emergency, you can get by temporarily by using a 6355601 (Hit) or 157061 (RCA) to replace the belt attached to the motor, and a 6355591 (Hit) or 157062 (RCA) on all models except those in Appendix VI that call for a 6356111 (Hit).)

Replacing Hitachi-category Loading Belts

Unplug VCR. Remove the smaller of the two belts (the one not directly connected to the motor). You will need to work it out around the long mechanism attached to the outer pulley (the one on the left in the illustration).

Unplug the multi-wired connector attached to the motor. Remove the screw that is holding the motor down, being super-careful not to drop the screw. The motor mount has a small metal tooth that catches underneath the sheet metal on the side opposite to the side with a screw. After unscrewing the screw, remove the motor by lifting up on the side with the screw, and pulling the motor up and out toward the side that was held down with the screw. Pull the motor just far enough back to get the larger belt attached to the motor pulley off it.

Remove this belt. Clean the motor pulley, and the two other little pulleys, with alcohol and cotton swabs or an old toothbrush (to get all the way down into the grooves of the pulley).

Install the larger new belt around the motor pulley and adjacent pulley, put the motor back in place remembering to catch the tooth under the sheet metal, and line up the motor so that the new belt runs in a straight line parallel to the line that the second belt will follow between the other two pulleys. Tighten down the screw.

Next, slip the other, smaller but thicker belt over the funny little black plastic end coming from the third pulley (on the left in the illustration), being very careful not to touch the clean new belt to the greasy axle of the pulley. If you do accidentally get grease on it, take it back off, clean it with alchohol, and try

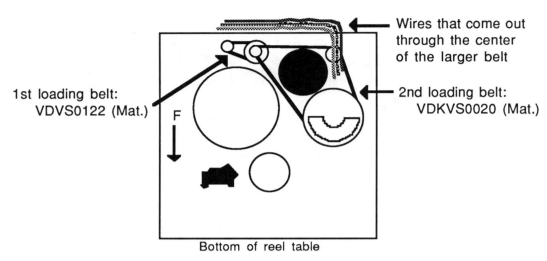

1st loading belt:
VDVS0122 (Mat.)

Wires that come out
through the center
of the larger belt

2nd loading belt:
VDKVS0020 (Mat.)

F

Bottom of reel table

Panasonic-category Reel Table with Wires in the Way of Removing One of the Loading Belts

again. Or, you can use a cotton swab to clean the belt again after it is installed.

You may need to use a scribe, or other long thin tool, to help you get the belt around the bottoms of the two pulleys. When the second belt is installed, the two belts should lie in lines that are straight or parallel with each other (with an offset where the one belt ends and the other begins). If they are not straight, loosen the screw holding the motor down and adjust the position of the motor. Put fingernail paint or lock paint on the screw if you've got some.

Reattach any wires to the motor that you removed.

Plug VCR back in and push "Play." If the unit now goes into the play mode and continues to play without shutting down, you have fixed the problem. Unplug and reassemble.

PANASONIC-CATEGORY TABLE MODELS (INCLUDING SOME PANASONIC, SYLVANIA, MAGNAVOX, G.E., CURTIS MATHES, J.C.PENNY, QUASAR, PHILCO, AND CANON)

On many of these brands of machines, replacing the loading belts is only a matter of opening up the bottom of the VCR, pulling off the old belts, and putting on new belts. But on some models that underneath look as shown in the illustration, a few words of additional explanation may be needed, because when you go to remove the larger loading belt, you may find that you cannot remove it because a group of wires that run out through the space in the middle of the belt, and possibly also a clear plastic part with a hole in it through which they run, prevent it.

If you wish, you can replace only the smaller belt with a new belt, leaving the larger belt unchanged, and in most cases this will solve the problem with less work. You could, if you wish, start by replacing only the smaller belt to see whether this is enough. If it is not, then you can run the longer procedure to replace the larger belt too. Or you may agree with me that you might as well replace both now, while you've got the machine opened up, so that you will not have to do this again for a long time.

To replace only the smaller belt, make certain that the VCR is unplugged. Pull the larger belt free from the intermediate pulley (that is, the pulley around which both belts go). Then remove the smaller belt from that pulley and the first pulley (on the motor). Clean the grooves of the pulley on the motor and the two sets of grooves on the middle pulley with alcohol and cotton swabs or an old toothbrush, to get down deep into the grooves. Put a new smaller belt on, reassemble power up, and test.

Replacing the Larger Loading Belt

If you are going to replace the larger belt too, then in order to get the larger belt off, it will be necessary for you to unfasten the other end of those wires coming up through the center opening of the larger belt. The procedure for doing this is described below.

Double-check to make certain that the VCR is unplugged. Loosen the screw holding down the clear plastic piece that covers part of the larger belt, and remove the clear plastic part.

Follow along the wires coming out of the middle of the larger belt, freeing them from the plastic-covered

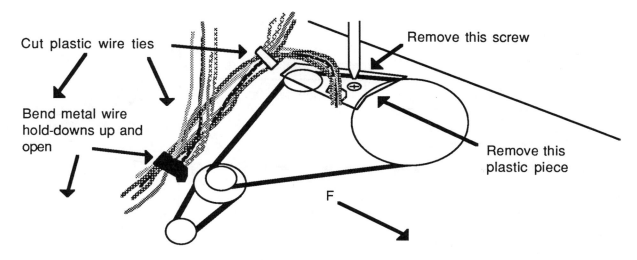

Cut plastic wire ties

Remove this screw

Bend metal wire hold-downs up and open

Remove this plastic piece

F

First Step in Removing Panasonic, etc., Loading Belts

metal clips bent over and around them, and clipping the plastic wire ties as you go. Use fingernail clippers or scissors to clip the plastic wire ties, removing them and throwing them away as you go along. Be very careful not to cut the wires. Work the wires free as you follow them along, and free them as they run up onto the circuit board that you have folded up and back.

Work up to the point on the opened-up bottom circuit board where all these wires attach to one connector and unplug it by gripping all the wires firmly and gently wiggling them back and forth until

the connector comes loose. Pull it straight off so as not to bend the little pins. If you do bend a little pin, straighten it back up with needle-nosed pliers.

Clean all the pulleys. Change (or clean and replace, if you have no new ones) both belts. Put on the small belt, then the larger belt. You may need to

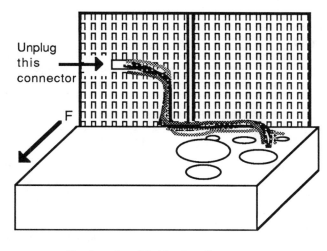

Unplug this connector

F

Unplugging Multi-wire Connector

use a scribe, or other long, thin tool to get the larger belt around all three of its pulleys.

Reassemble, going back in reverse order through the disassembly steps. The plastic wire ties that you cut were mainly there to aid in keeping wires straight during mass production of the VCR, and they do not need to be replaced. The plastic-covered metal wire holders are enough to hold the wires in place, if you will just remember to bend each of them back down around the passing bundle of wires as you go along. But if you feel compulsive about it, you can obtain replacement plastic wire ties from an electronics parts supplier to replace the ones you clipped, or just tie them back with short lengths of string, but it is not really necessary, as long as you put the wires back along the path from which you loosened them, and bend the plastic-covered metal clips back down over them. That's enough.

Replace the circuit board, turn the VCR over so that it is rightside-up, plug in, and test. Hopefully this cures your problem.

EARLY PANASONIC-CATEGORY "TANKS"

The following procedure is for early Panasonic-category VCR's with the mechanical push-down switches, that look approximately like this:

A common cause of failure to load the tape on these machines is breakage of a loading gear. To check for this possiblity, follow the special procedures in Appendix III for opening up the bottom of one of these machines without damaging the power switch on the top.

The loading gears are located near the loading motor in approximately the location indicated in the

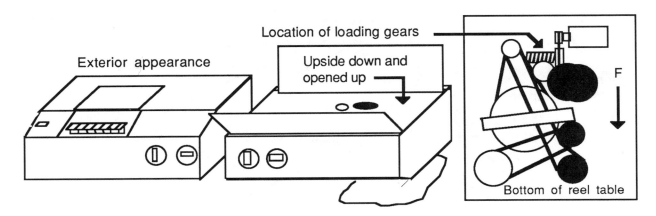

Opening Bottom of Early Panasonic-category "Tank" to Get to Loading Gears

Illustration. Look carefully at all the gears to see whether one of them has a crack splitting it. You must look carefully, because sometimes it is hard to see. Check especially the one down on the other end of the shaft, hidden on the other side of the metal plate, barely visible through a slot in the reel table.

If in doubt, you may be able to tell by rotating the loading motor by hand and watching to see whether the tape-loading mechanism on top moves between the unloaded position and the loaded position. Or you can go ahead and remove it, using the procedure that follows and visually check the lower gear for a split. You can also tell by putting the lid back on the cassette basket, inserting a tape, balancing the machine on one end so that you can see what the loading motor does, powering up, and hitting play. Watch the gears as the loading motor turns to see whether one of them is failing to engage properly, due to being broken.

If this is the problem, you will need to obtain the kit for replacing these gears. The part number for the complete loading gear assembly is VXP0225 (Mat) or 144207 (RCA). The new gears are better made, so they should have a longer life than the original gears. You can also save money by ordering just a replacement kit for the two gears that fail, part #32-435 from MCM. If you use this replacement gear kit, note that the instructions that come with it may incorrectly show the smaller of the two gears mounted backwards on the axle. The portion with the set screw goes outside, not inside, to make this gear and the old small gear with which it meshes offset relative to each other, as they were in the original assembly you removed.

To remove and replace these gears, it is necessary to detach both the old loading motor bracket and the video head and cylinder motor, in order to get out one of the screws holding in the old loading gear assembly. And you will probably need to use a good magnetic screwdriver or parts picker-upper to get

some of the screws back in. Wait until you have the new parts in hand before beginning (unless you are opening it up to check whether the hidden gear is indeed cracked).

Make sure the VCR is unplugged, and try to get the P-guides back into their fully unloaded position, if possible, by gently pushing on them and/or rotating the loading motor clockwise.

Unplug the multi-wire plug from the little circuit board near the worm gear.

Make a mark with a felt-tip pen near one of the two screws holding the dark metal weight to the bottom of the cylinder motor, remove both screws, put in a numbered cup, and before it moves, make another mark on the corresponding outer edge of the metal underneath, so that you will know which way to put it back on later.

Remove the little belt from the loading motor, and remove any of the screws that you can reach that fastens down the sheet metal bracket to which the loading motor is attached. There will be one screw you cannot get out because the video cylinder drive motor is in the way.

Look at the wire running through the hole in the reel table alongside the cylinder motor, so that you will be able to put it back in the same place.

Still working from below, remove two of the three screws attaching the cylinder assembly that are farthest away from the last screw attaching the loading motor bracket that you could not reach earlier. Now loosen the third screw a few turns — not all the way, but only enough to allow the cylinder assembly to hang loosely from it and be pushed out of the way enough to remove the last loading-motor-bracket screw.

If You Must Remove the Cylinder Motor

If you cannot get the last screw out this way, stand the VCR on one of its ends, so that you can have access to both sides of the reel table. Have one person grip the upper cylinder through a paper towel at the top edge and as far away from the video heads as possible, while the other person removes the third and final screw attaching the cylinder-motor assembly. It is not necessary to remove the cylinder-motor assembly from the VCR, but only to detach it for a moment, so that you can push it a little out of the way, in order to get out the final screw holding in the loading motor and gear assembly.

While your friend holds the cylinder-motor assembly, remove the old loading motor and gear assembly. Remove the loading motor from the old bracket and bring in the new bracket, and attach the motor to it, putting the loading belt in its proper place. Attach the new assembly with the motor and belt to the underside of the reel table. Have your friend put the cylinder-motor assembly back where it goes, line up the holes, and use a magnetic screwdriver or parts picker-upper to start the screws. Make sure that the cylinder-motor assembly seats back down all the way on the clean metal that it was mounted on, and take care not to bump the video heads while you are doing this.

Remove Loading Mechanism

Push the belts out of the way enough to remove the whole loading gear assembly with motor attached. If you are replacing only the loading gears, you can do that now. See Appendix I for help removing the C-ring and the plastic catch holding on one of the gears, if you have trouble. Install new gears. Before removing the old gears from the axle, note the way that the two gears that go down into the cavity are offset. One sticks out more than the other. When installing the smaller of the two new gears (which is one of this pair), be sure to put it on in the direction that gives it the same offset. (On some gear kits, the picture in the assembly instructions shows this gear installed with the wrong side out.)

If you are replacing the whole assembly, remove the screw and lock-washer attaching the small circuit board with wires running to the loading motor. And use a 2.5mm hex key (allen wrench) to remove the three little bolts holding the loading motor to the bracket. Attach the old motor and circuit board to the new bracket.

Go back through the preceding steps in reverse order to reinstall. Push the P-guides up against the V-stoppers (in the "fully-loaded" position), and then put back into position the assembly that you removed. When you've tightened down the three screws attaching the cylinder motor, wiggle the wire running up through the hole alongside it to make certain that it is free and not pinched. The wide belt (capstan belt) goes back on with the stripe marks on the outside, and oriented as shown in Section 5 of this chapter.

Remember to plug the little plug back in. Power up and test to see whether the machine now loads properly.

4. Special Procedures

J V C and ZENITH

The following procedure is for people who have one of the models of Zenith or JVC machines with the little lights that show through windows in the push buttons. If the fast-forward and rewind operations are not working correctly, or it is not loading and moving the tape properly when you push "Play," but you can hear a motor start up when you push the buttons, you probably have a problem in the idler-clutch assembly or in the tape-loading mechanism.

You could have a bad idler-clutch assembly, or a bad loading gear assembly, or mistimed loading gears, but in most cases, just a proper cleaning and relubrication of crucial points will fix the problem. For this job, most technicians would use the special "Molytone" grease described in Section 14 of Appendix I, but if you don't want to go to the expense and bother of ordering up some of this special grease, you probably can get by with a light white lithium general purpose grease.

If there is a cassette in the VCR, eject it. Unplug the unit. Remove the two little brass screws holding down the circuit board (the board with parts soldered to it) that is fastened down over the reel table above the video heads on most models. Put the screws in a

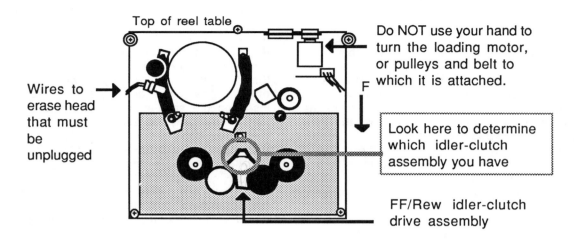

Top Side of JVC/Zenith Idler-Clutch Drive Assembly and Reel Table As Seen in the VCR

numbered paper container, and record their location in your disassembly notes. Prop the circuit board in an up, open position with a wooden stick caught under a securely fastened, larger component on the board.

Before we get started, I should mention that this is one of the few VCR models on which you should NEVER TURN THE LITTLE LOADING MOTOR OR ITS PULLEY BY HAND. The reason is that this can damage the teeth on a gear attached to the loading motor underneath, and it is a lot of trouble to replace this part. The little motor does not have enough power to damage this gear, but your finger does have enough power to do so. This is the little motor, pulleys, and belt located at the right-rear corner of the reel table.

We are going to partially disassemble the VCR, clean several crucial points deep in the mechanism, relubricate three of them, and reassemble. Since much of the same disassembly is used when fixing fast-forward/rewind problems in these machines, the procedure for doing that is also included here for readers who have been referred to this chapter by Chapter 13. If you are not having FF/Rew problems, you can skip the paragraphs included under that heading here, and go ahead to the heading "Play Problems."

Fast-forward/Rewind Problems?

If you are having trouble with the fast-forward/ rewind functions, another crucial point to be cleaned is a little rubber idler tire deep inside an idler-clutch assembly. For fast-forward/rewind problems, many technicians would simply replace this part with a brand-new idler-clutch assembly, because they do not want to have to do the disassembly twice in case cleaning is not enough. They also do not want the customer to have to bring the machine back in a second time a month or so later, in case cleaning only

temporarily fixes the problem for that long. It is no more trouble to install a new idler-clutch assembly, than it is to clean the old one, because the old one must come out to be cleaned, and then reinstalled.

But the do-it-yourselfer is in a somewhat different situation. I will tell you how to get the part number, from which you can call and get the current price of a new replacement idler-drive assembly, from which you can decide whether to try to get by with a cleaning, or to replace it.

Another factor, of course, is the amount of time it will take to obtain a replacement part. Even if merely cleaning it only makes it work for a month, some readers may want to get their VCR running again immediately, so that they can have the use of it during the time they are waiting for a new idler-clutch assembly to arrive.

If you are only having problems with the play function, and not the fast-forward/rewind function, you have the option of not bothering with the idler-clutch assembly, or cleaning it either, if you want, but only cleaning and lubricating the other points mentioned above and detailed below.

To get the idler-clutch part number, take a look at the idler-clutch assembly. As the top side appears through a opening in the reel table between the two reels, it looks as shown in the illustration:

Two slightly different idler assemblies are used in different models. You can tell which assembly you have by looking at the part of the idler where the circle is in the illustration. The immediately visible difference between the two assemblies is that the #861-646-04 (Znt) has one little plastic pin at the center point indicated, while the #861-651 (Znt) has two little plastic pins, one in the center and another off to the side. For clarity the illustration shows the top side of the idler-clutch assembly alone, without the parts on the surrounding reel table, and as viewed from the rear, looking forward.

If you have a JVC that uses parts that look like this, you can buy your replacement parts from a Zenith supplier, because they are exactly the same. At this point, you can stop and wait to order up a new idler-clutch assembly, or you can go ahead and see whether just cleaning and relubricating the crucial parts will solve the problem. Continue as described in the following subsection, "Play Problems."

Play Problems

To clean and lubricate the essential parts, we must disassemble the VCR far enough to gain access to the under side of the reel table. To do this, we must remove the metal bottom cover, the front cabinet panel, the cassette basket assembly, and the reel table itself.

Before you can remove the cassette basket assembly, you will need to remove the front cabinet part, in order to do which you will first need to remove the metal bottom cover.

Make certain that the VCR is unplugged. With a felt-tip pen, first mark on the cover the positions of the heads of visible screws so that you can get these screws back into the correct holes (some holes have no screws). Notice that some of the screws, especially the one at the rear center, may be shorter than others.

Look at the screws as you remove them one-by-one. If any is shorter, longer, or has different threads, put in in a separate container and mark the hole from which it came on the bottom cover with a felt tip pen and note. Put a distinctive mark by holes with shorter screws as you remove the screws, so that you can get them back into the right holes again, and put these screws in a separate container with a

notation as to their location.

Follow the procedure in Section 8 of Appendix III to remove the front cabinet panel, and then the cassette front-loading assembly, and then return to this point.

Check to make certain that the P-guides are in the "fully unloaded" position — which is, at the end of their tracks closest to the cassette reels and farthest from the heads. If they are, go to the next paragraph. If they are not in the fully unloaded position, attach a 9-volt battery and clip leads (Section 19 of Appendix I) to the terminals of the loading motor in such a way as to make the pulleys on the backside of the loading gear assembly rotate in a clockwise direction (as viewed from the front of the VCR) until the clutch between the larger pulley and the worm gear to which it is connected starts to slip. (For more on this, see the JVC section of Appendix VI.) If you are not having fast-forward/rewind problems, skip the next two paragraphs.

If you are having problems with the fast-forward/rewind functions, you will need to remove the clutch-idler assembly, either to clean it or to replace it. It has to be removed from underneath the reel table, but before that can be done, the little springs attached to it on top must first be removed.

If you have to remove the clutch assembly, make a careful sketch of the way the springs attach to the idler-clutch assembly on your VCR, number them as you remove them, and put them in separate containers along with their fasteners. When you unfasten any little springs attached to the idler drive assembly on the top side, leave them attached to the reel table (unless they look like they will fall free, in which case remove the one on the left, put it in a numbered container and record "left top spring," and then remove the right-hand one, put it in a differently numbered container, and record its location. Section 10 of Appendix I explains how to remove the plastic split-washer fasteners used in these machines.

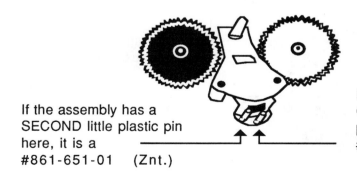

If the assembly has a SECOND little plastic pin here, it is a #861-651-01 (Znt.)

If the assembly has only ONE little vertical plastic pin here, it is a #861-646-04 (Znt.)

How to Determine Which Type Idler-Clutch Assembly You Have

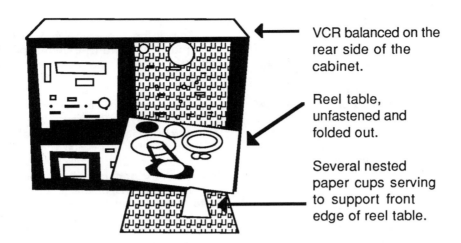

VCR balanced on the rear side of the cabinet.

Reel table, unfastened and folded out.

Several nested paper cups serving to support front edge of reel table.

Zenith - J V C Reel Table Folded Out to Expose Underside

Early and Later Mechanisms

Two different designs were used, and the cure for a failure to play problem is different for the two different designs. One way to tell which you have is to look at the white plastic gear on the underside of the mechanism. On the early version, there is a spring visible on the side of the gear as shown in the illustration. On the later version, the spring is on the other side of the gear and not directly visible.

On many of the later models, you can get access to the bottomside of the mechanism by just taking off the bottom cover of the cabinet. If this is your situation, and there is no spring visible on the side of the white plastic gear that faces you as you view it from the bottom, then skip to the later section under the heading "Later Mechanism Play Problems." If a circuit board stands between you and the bottom of

the mechanism when you remove the bottom cover, then proceed as follows to get access to the bottom of the mechanism. Do NOT try to lift up the bottom circuit board. Instead, proceed as follows.

Removing the Reel Table Mechanism

Unplug the little two-wire connector that goes to the erase head, whose position is indicated in the first illustration.

To fold out the reel table, begin by taking out, or loosening, the five screws that hold it down around the edges. Their approximate locations are indicated in the first picture of the reel table earlier. Notice that there may be a lock washer under the middle screw in the back, and large disk washers with the

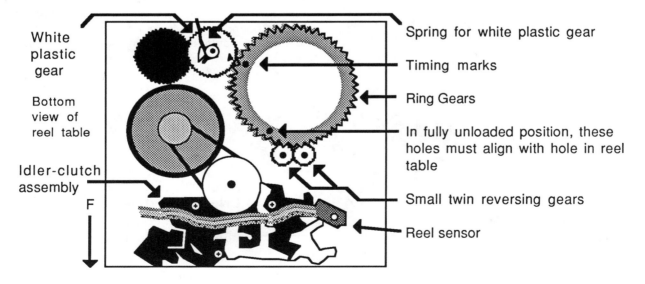

White plastic gear

Bottom view of reel table

Idler-clutch assembly

F

Spring for white plastic gear

Timing marks

Ring Gears

In fully unloaded position, these holes must align with hole in reel table

Small twin reversing gears

Reel sensor

View of Underside of this Zenith-JVC Reel Table

two screws at the back corners.   Four screws are at the four corners, approximately, and the fifth screw is in the center of the back edge, probably hidden under some wires. This middle screw probably has a lock washer around it — Be careful not to drop this into the stuff below, for it might be hard to find. Maybe put a paper towel below it to catch it if it falls. The two screws at the back corners probably have large brass washers molded to the screws.   You do not actually have to remove them completely — it will probably be enough to unloosen them most of the way. In the next step, you then should be able to pull the reel table out and fold it down toward the back of the VCR without removing any of the wires attached to it.

Get ready several nested small paper cups, or other soft object about 3 to 4 inches thick, to use as a support for the reel table.   Stand the VCR on its rear edge, with the circuit board that used to be over the heads hanging out.   Put several nested small paper cups, or other soft object about 3 to 4 inches thick, on top of the rear circuit board.   Fold the reel table out extremely carefully and gently, so as not to break any of the little wires in back of the cylinder assembly. Handle it slowly and gently, as you would an expensive delicate, fragile, glass figurine.

On the underside, you will see two round metal ring gears, which mesh with a smaller white plastic gear, which overlaps (and meshes) with a black plastic gear that only has teeth part way around it, and other parts that look as shown in the illustration. You should see a spring held down against the side of the gear as pictured — if there is no spring here, skip on ahead to the subsection entitled "Later Mechanism Play Problems."   If you are only fixing a fast-forward/rewind problem, skip ahead to the subsection entitled "Servicing the Idler-Clutch Assembly."

Look at the white plastic gear that has the spring. This gear has a little triangular-shaped timing mark molded near the edge at one point and meshes with the big metal ring gear in such a way that this timing

mark is in step with a round hole drilled through the top ring gear.   You may find the gears turned so that at the present moment the teeth near the marks are not touching, but they should be aligned in such a way that you can see that as the gears turn, the teeth by the marks would touch each other when the gears turned around.

Since there are also other round holes through the ring gears, use a felt-tip pen to put a mark by the hole beside the teeth that would mesh with the teeth beside the triangular mark in the white gear, so that you can put it back together correctly.

Notice another hole drilled through the two ring gears in the place where a cotton swab stick is shown stuck through in the illustration.   These are also timing marks.   Push a thin wooden stick through the two holes in the ring gears and through a third alignment hole drilled in the reel table underneath. This aligns all three holes.   If you cannot get a stick through because the holes are not aligned, then wait and do this immediately after removing the white plastic gear.

The long end of the spring on the white gear should go through a hole cut in the side of the reel table, while the other end should be tensed and caught against a little step, or raised edge, molded in the plastic surface of the gear.   If the spring is not caught against this step, but has slipped off it, this is part of your problem, but even if it is in the correct position, remove the plastic washer at the center, make a little sketch of which way the spring is positioned before removing it, then free up and remove the spring too.

Slide the white plastic gear off its axle.   If you could not get a wooden stick through the three holes earlier because they were out of alignment, move them into alignment now and put a stick through all three.   You should not need to move the ring gears very much in order to do this.   Use alcohol and cotton swab sticks to clean every bit of old sticky grease off

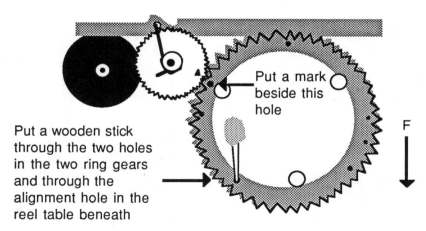

Put a wooden stick through the two holes in the two ring gears and through the alignment hole in the reel table beneath

Put a mark beside this hole

F

Timing Marks and Positioning of Two Gears

Spring slips off
rounded edge

Using a knife to under-
cut edge slightly

Spring catches OK
on undercut edge

Exaggerated Side View of Undercutting Edge that Catches Spring

the axle and also out of the hole in the center of the white plastic gear. This old sticky grease is usually the whole problem

If when you opened up the reel table and looked earlier, you found that the spring on the white plastic gear was not caught on the little step or edge molded in the top of the gear, take a sharp tool and undercut this edge slightly on the gear. Use little strokes, shaving off a tiny bit on each stroke, until you have reshaped the edge so that the spring will not slip off it again. Take care not to cut yourself. In case you need them, here are replacement part numbers: white plastic gear #834-168 (Znt), spring #880-1796 (Znt), & washer 893-1430 (Znt).

Put new Molytone grease or a white lithium grease substitute on the axle and hole in the center of the white plastic gear, and slide it back on the axle, making certain to mesh the teeth with the ring gear in such a way that the triangular timing mark is mated with the timing hole that you first marked earlier.

Look at your sketch of how the spring was positioned and put the spring back on with the correct side up and the long part through the hole in the rear edge of the reel table. Put the plastic locking washer back on. Then use needle-nosed pliers, or a scribe or

similar tool, to rotate the short arm of the spring back around to where it catches on the step or edge molded for it in the surface of the gear.

Next, our attention shifts to the twin little white gears that reverse the movement between the two ring gears. Remove these, one at a time, clean every speck of sticky goo off its axle and out of the hole through the gear, put on fresh lubricant, slide it back on the shaft, and reattach with the plastic locking washers. Clean and relubricate both gears.

Later Mechanism Play Problems

On later models, the manufacturer solved the problem of sticky grease that plagued earlier models, but a new problem got introduced. The plastic or nylon tracks on which the P-guides slide were held down at the end with a screw through a hole molded in them. As the material aged, it shrank, splitting the hole and causing the material to be pushed slightly in toward the open part of the track, narrowing it enough that the P-guides jam up shortly before they reach the V-stoppers.

This problem often can be solved simply by removing the screws, and leaving them out. Or, you

Wooden stick through the two holes in the two ring gears and through the alignment hole in the reel table beneath →

F

Twin Plastic Reversing Gears

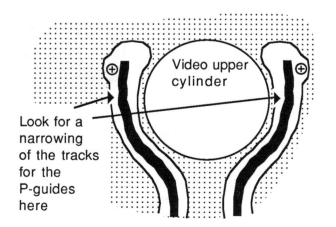

Look for a narrowing of the tracks for the P-guides here

Video upper cylinder

Area that Sometimes Binds on JVC-category VCR's

can take a fingernail file and file away a little bit of the plastic or nylon material that faces into the open area, making the open area a tiny bit wider.

After Fixing Play Problems

If the VCR has only a problem with the play function, and no problem with the fast-forward/rewind functions, you have the option of skipping over the steps of removing and cleaning the idler-clutch assembly and going directly to the reassembly procedure below. But if you are going to clean or replace the idler-clutch assembly, proceed next as follows.

Servicing the Idler-Clutch Assembly

Remove the rubber belt running from the capstan to the pulley on the bottom side of idler clutch drive assembly. Go to the underside of the idler drive assembly, remove the screw holding the black plastic idler motion detector, and lift up the wires running across the plastic assembly. Let the part with the wires attached hang off to the side, put the screw in a numbered container, and record where it came from in your notes.

Using needle-nosed pliers, or a scribe, unfasten at the plastic assembly end the springs whose other ends are attached to other parts fastened to the reel table. Leave the springs hanging from their other ends attached to the reel table.

Remove the three brass screws holding down the idler drive clutch assembly. Put in a numbered container and record their location.

This next step is optional, but it makes the following step easier. Remove one of the plastic overhanging pieces, probably the one closest to the take-up reel side of the assembly (on the left when viewed from underneath, as now), by removing the split plastic washer holding it on. Then slide the plastic part off the axle on which it rotates. Put the part and washer in another numbered container.

Gently remove the whole plastic idler clutch drive assembly mechanism, pulling it up off the two locating pins and working it around the various movable plastic parts that overhang it.

If you are going to try just cleaning the old assembly and replacing it, now is the time to do that. Be sure to clean the little movable rubber wheel that you can just barely see inside the mechanism. That is crucial. Otherwise, replace the assembly.

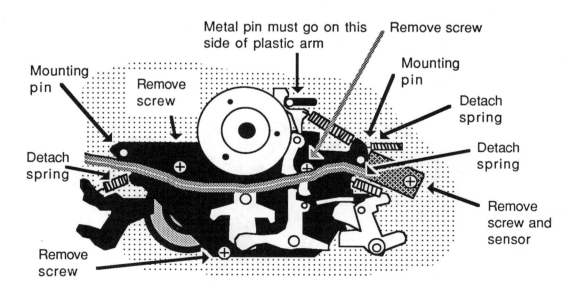

Underside of JVC/Zenith Idler-Clutch Assembly

Install the new, or old cleaned, idler-clutch drive assembly in machine, working it down over the locating pins and around the various overhanging movable plastic parts. Make certain that the plastic leg is on the correct side of the movable metal pin, as shown in the preceding illustration.

Fasten down the assembly with the three screws, put the plastic lever and lock washer back in place (if you removed these), reattach the springs, making certain that they run the shortest routes and are not tangled at their other end, put the plastic reel motion sensor back into place with the plastic pin inside the mounting hole and fastened down with the screw. Run the wires back through the plastic guides made for them in the assembly, and reinstall the rubber belt.

Putting the Reel Table Back into Place

With one hand holding the reel table, carefully rotate the VCR off its back and into its normal position, resting on its bottom.

Locate the multi-wired plug that you removed from the cassette-basket assembly earlier, and the wires unplugged from the erase head, and make certain that they are out on top somewhere where you can see them, and not caught under the reel table underneath, because later you are going to need to plug these connectors back into the cassette basket assembly and erase head.

Gently work the reel table back down into its original position, being careful that none of the wires attached to it, or around it, catch or hang up around the edges of the table or the surrounding plastic enclosure as you put it down. Be prepared to take a

little time to make certain that all four corners are all the way down before putting in the screws. Make certain that the little mounting pins molded into the plastic at the front of the plastic chassis assembly go into the holes in the metal reel table alongside the screw holes.

Replace the screws holding down the reel table. If you completely remove the screws at the back corners, you may need to use needle-nosed pliers to get them back into starting position. Put the central screw in the back, making certain that no wire is pinched under the tab into which it goes.

Reattach the little springs to the topside of the clutch-idler assembly, making certain that everything corresponds to your sketch.

While the cassette basket is out is a good opportunity to clean the heads and tape path (optional).

Pick up the cassette basket assembly, bring to the VCR, and plug in the multi-wired connector. Reinstall the cassette basket, taking care that the plastic feet at the bottom fit down into the slotted holes cut for them in the reel table, and that no wires are caught or pinched anywhere. Make certain that the top metal piece with the screw holes is all the way down onto the plastic posts before putting in the screws. Remember to put the springy brass piece in place (with arrow facing toward the front of VCR) and the wire with the terminal on top of it, before putting in the right rear screw.

Replace the front cabinet piece, reattaching all multi-wire connectors that were disconnected, running these through any notches provided for them in the plastic parts or circuit boards. If a small sheet metal piece inside the front cabinet part was pressed against the piece underneath, put it back in the position indicated by your disassembly notes.

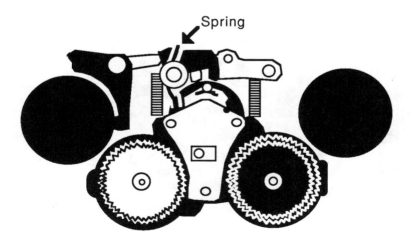

Top View of 861-646-04 (Znt) Idler-Clutch Assembly Installed

If there were little black pieces of thin plastic or felt with square holes around some of the slide switches, put them back on. If you had to remove a ribbon cable and grounding foil underneath to get the front cabinet piece off, reattach these. The ribbon connector is attached by inserting it into the slot in the plastic connector, and holding it there while you push the hinged part back down.

If you had to disconnect a foil strip, reconnect it to the binding post to which it was attached with the screw and under the springy brass piece, if there was one. Catch the plastic cabinet front on the little hinge-like protrusions below, and rotate it back into place, helping the plastic tabs on the top into their latching position with a scribe or small screwdriver, if necessary. If there was a metal piece over the center tab, replace it, and replace any top screw or screws. Replace any knobs you removed.

Reconnect the bottom sheet metal piece, putting the correct screws back into the correct holes, as indicated by your notes.

Plug in and test to see if it now will fast-forward and reverse.

---

5. Capstan and Pinch Roller Not Moving the Tape

If the capstan and pinch roller are not moving the tape as they should, the first thing to determine is whether the problem is that the capstan shaft is not rotating, or that the pinch roller is not coming all the way over to press the tape firmly against the pinch roller.

Never grip the capstan shaft with pliers or any other tool, because you will put a scratch in it that will ruin it. To determine whether it is rotating, put the machine in good light and watch the capstan shaft very carefully as you push "Play." If you cannot tell by looking at the capstan, touch the edge of the pinch roller with a scribe or screwdriver as it comes into contact with the capstan to see whether it starts rotating as it should.

If the capstan shaft is not turning, go to the first section, "Capstan not Turning," below. If the capstan shaft is turning but the pinch roller is not turning, then the problem is that the pinch roller is not being moved all the way over into contact with the capstan, and you should go to the section entitled, "Pinch Roller Not Moving All the Way Over," below.

Capstan not Turning

If the capstan is not turning, this could be due to a capstan bearing that has seized, or to a broken or slipping capstan drive belt on the underneath side in models with a capstan belt, or it could be due to a burned out capstan motor or other electronic problems beyond the scope of this elementary book.

Unplug the VCR. Check to see whether the capstan is belt-driven. If you cannot determine this from Appendix VI, follow the procedures in Appendix III to open up the bottom of your machine and look.

If it is not belt-driven and the capstan is not turning, you will need to take it to a shop. Exception: If it is one of the Fisher models with its loading motor underneath attached to a pulley driving a belt, as pictured under "Fisher-category VCR's" in Section 3 of this chapter, try replacing the loading belt following the procedure in Section 3. It could be that the belt is stretched, slipping, and not moving the loading mechanism quite far enough to cause the little computer to start up the capstan motor in   Play.

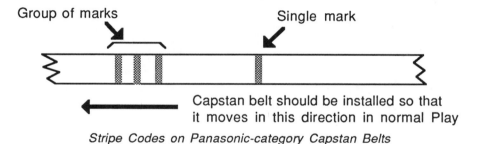

Stripe Codes on Panasonic-category Capstan Belts

If it is belt-driven, open up the bottom of the VCR and check the capstan belt. If it is broken, obtain the part number from Appendix VI, or from a parts supplier, and order a new one.

On some Panasonic-category VCR's there were three different belts, with three different part numbers, for a given model. They were of slightly different sizes, designed to compensate for small speed differences due to variations in manufacture, and in truth, any will work, but the exact replacement is determined by the number of stripes painted on the outside surface of the belt.

There will be a single thinner mark by itself, or a pair of marks, and then a second number of marks set apart in a group. The number of marks in the second group of marks determines the exact replacement, as specified in Appendix VI, and the other, smaller group of marks shows in which of the two possible ways you should put on the new belt. The new belt should be put on so that it moves in a direction away from the single or smaller group of marks: In Appendix VI, the belt in the illustration would be designated "3 + 1" marks. To determine what direction to put on a coded capstan belt, remember that all VHS capstans are supposed to rotate CCW (as viewed from the top) during normal play. This means CW as viewed from the bottom.

In recent years, Matsushita (Panasonic) has phased out the three belts for some machines, and gives just a single belt to replace any of the three. If you are given a substitute part number when you call in to order a Matsushita capstan belt, just accept the belt they give you — it is all that is available, and to tell the truth, the difference between the three belts was unnoticeable anyway.

If there is no broken capstan belt, try turning the capstan flywheel with your finger. If it seems stiff and resistant to rotating, slip off the belt and try again. If it is definitely stiff, you can try turning the VCR back over, lifting the capstan oil seal, putting two drops of oil on the bearing, and rotating the capstan flywheel back and forth to see whether it comes free. But you really should remove the bracket holding the flywheel on the bottom, slide the capstan and flywheel out of the bearing from the bottom, clean off the capstan shaft, reinstall and lubricate. If the capstan still does not start up and rotate when you power up and try to play a tape after these measures, you will need to take it to a shop. Either the capstan motor is bad or there is a trouble in the circuitry.

Pinch Roller Not Moving All the Way Over

If the capstan turns, but the pinch roller fails to move over all the way to press the tape against the capstan, the first thing to determine is what, on your machine, is supposed to move the pinch roller over to press against the capstan.

On most VCR's the same motor that operates the mechanism that loads the tape, the tape-loading motor, is also supposed to move the pinch roller into contact with the tape. But on a few early VCR's, the pinch roller was moved by means of a "solenoid."

Does the VCR Use a Solenoid to Move the Pinch Roller?

For our purposes, you do not even need to know that a solenoid is an electromagnet constructed from a wire coil in an enclosure with a metal plunger that can slide in and out of the coil. You need only to be able to recognize a solenoid when you see one. Here's the configuration on one group of early Panasonic-category VCR's:

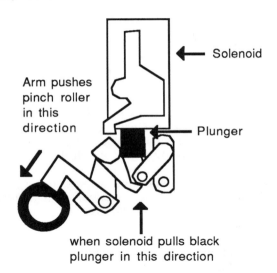

Typical Pinch-roller Solenoid Configuration

When electrical current runs through the coil, it produces a magnetic field that causes the plunger to be pulled into the coil with an audible snapping noise, pushing or pulling the pinch roller into contact with the capstan shaft, with the tape pinched in between.

There are two ways to tell whether your VCR has a solenoid. If a picture of its reel table is included in Appendix VI, you can see whether the presence of a solenoid is indicated by this symbol on the diagram.

Symbol for Pinch-roller Solenoid Used in Appendix VI

Or, you can determine whether the pinch roller in your VCR is activated by a solenoid by looking at the mechanical linkage connected to it, and seeing whether it traces back to an enclosed assembly with a plunger that goes into it.

Since most VCR's move the pinch roller by the action of the loading motor rather than a solenoid, probably your machine does not use a solenoid to move the pinch roller, and you can skip over the next paragraph.

If your VCR has a solenoid and the solenoid is being activated, you can tell by the snapping noise it makes right at the time it pulls the pinch roller over. If it is making this snapping noise, but the pinch roller is not moving into contact with the capstan as it should, you can try cleaning the moving parts connected to the pinch roller and solenoid, but probably you will need to take it to a service center to have electrical work done on the machine. If there is no sound or other indication that the pinch-roller solenoid is being activated, then you will also need to take the machine to a shop.

VCR's on which the Loading Motor Moves the Pinch Roller

If the VCR does not have a solenoid, and the capstan turns, but the pinch roller fails to move over all the way to press the tape against the capstan, this problem could be (1) the loading motor is not moving the tape-loading mechanism quite all the way into its fully-loaded position so that the last step of moving the pinch roller over is not quite completed, or (2) the arm to which the pinch roller is attached has bent slightly over the many times it has been used so that even when the arm is moved over into correct position, the pinch roller stops short of contacting the capstan.

Let's wait until we have eliminated the first possibility before taking the ultimate step of trying to bend the pinch roller arm back .

With the VCR unplugged, balance it securely on one end with the top and bottom opened up so that you can watch the behavior of the loading motor. Get a good light, like a strong flashlight, and look right at the first pulley or gear that is attached to the loading motor.

The big question is: When you power up, insert a cassette, and hit play, does the motor never stop turning, but only go straight from turning in one

direction to turning in the opposite direction (to unload the tape), or does this motor stop turning and stand still for a few seconds when the tape-loading mechanism gets to its fully-loaded position, and then an instant later, start up again and unload the tape? In other words, does the loading motor come to a dead stop for a brief time and then start up again (in the opposite direction)?

If the motor never comes to rest in a still position, this means that something is failing to tell the VCR's little computer that the process of loading the tape has been successfully completed. Either it is not loading the tape fully, or a sensor that is supposed to indicate completion of the loading process has failed, or there is a trouble in the electronics.

Unplug VCR, replace all loading belts, and spray head cleaner into the mode switch (see Appendix VI for its location). Also, on a VHS machine, clean and lubricate the tracks that the P-guides run on. Let everything dry and test again. If the loading motor still does not come to rest, you will need to take the machine to a shop.

Bent Pinch-roller Arm

If the loading motor does come to rest and stays stopped for a few seconds, and the machine does start to try to play the tape, but the capstan fails to move the tape because the pinch roller never comes quite all the way over to contact it, then the pinch roller arm may be bent.

Shut off the machine and unplug it. With the mechanism in the fully-unloaded position, first figure out what direction you need to tip the top of the pinch roller to make it tip slightly toward the capstan. At most, you need to bend the arm the smallest amount you can, an amount so small that it cannot even be detected by the naked eye.

Use you fingers to grip the pinch roller and apply a small gentle pressure in a direction that would tend to tip the top toward the capstan. Just bend it a little bit, a very tiny amount! If you overdo it and then try to bend the metal back the other way, it may break, so take care not to overdo it!

After bending it slightly, plug in, power up, and try playing a cassette. If the capstan still does not come over far enough, you can try bending it a tiny bit once more, but if that does not fix the problem, you will need to take it to a shop.

6. Take-Up Reel Not Turning

If the tape loads and the capstan moves the tape, but you see that the take-up reel is not turning, so that the tape is not being wound onto the take-up reel, but perhaps is spilling out into the machine between the pinch roller and the cassette box, hit "Stop," then hit "Rewind" to try to reel the loose tape back into the cassette.

Note: If the take-up reel is not turning, the VCR's safety system may detect this fact, unload the tape, and shut the machine off before you can even detect with your eye any loose loop of tape starting to form between the capstan and the take-up reel. Thus, "mysterious" shutting-down behavior on the part of the machine may actually be due to the take-up reel not rotating to wind in the tape.

If the machine fails to reel in the tape and leaves a loop of tape hanging out even after it has come to a stop, go to Chapter 7 or 8 for procedures to remove the cassette without damaging the tape, and then return to this point.

Failure of the take-up reel to turn, or its failure to turn fast enough, may be caused by failure of the loading mechanism to fully load the tape all the way, in which case, you should go back to Section 3, or it may be caused by a problem in the mechanism that is supposed to cause the take-up reel to rotate in play, such as a play belt or take-up reel idler.

If there is any question about this, use the procedure of watching whether the reel motor ever comes to a dead stop, explained in Section 5, to determine whether the loading motor is fully loading the tape. If the loading motor does come to a full stop before reversing and unloading again, but the take-up reel never starts turning, then go on to the next paragraph.

If the take-up reel fails to rotate to wind in the tape, the safety circuits should automatically stop the VCR before any tape spills out. If that is happening in your machine, eject the cassette and look to see what is supposed to make the take-up reel rotate.

If there is an idler (a rubber tire on a little plastic wheel) mounted near the take-up reel, clean it (as described in Chapter 3) or replace it, using part numbers from Appendix VI.

Some models have a special "play idler," while other models use the same idler that is used for fast-forward/rewind operations. In machines with a separate play idler, this idler may be driven by a belt (located either above or below the table), or it may be driven directly off a small knurled enlargement of the capstan shaft just below the oil seal, or the play idler may be driven indirectly off the capstan by a belt.

If the VCR does not have a separate idler for "Play," probably the same fast-forward/rewind idler is also supposed to drive the take-up reel in the Play mode or it is directly driven by gears. Replace any broken belts, clean or replace all idler rubber tires, and clean the surfaces they are supposed to contact.

More information on how to replace idlers, including procedures for removing and replacing metal C-rings and plastic locking washers is found in Chapter 13 and in Sections 9 and 10 of Appendix I. If the VCR is one of the Fisher-, or Hitachi-, or Panasonic-category models described in Chapter 13, follow the detailed procedures for idler replacement given there. If you do not have a replacement idler on hand and just cleaning the idler makes the take-up reel work again, you'd better order up a replacement idler, because the old one is on its last legs.

If cleaning or replacing the idler that is supposed to drive the take-up reel still does not make it wind in the tape, then check out what is supposed to drive this idler. For example, some Panasonic-category VCR's have their play idler driven by a "play belt" coming from the reel motor; if the belt is broken, the take-up reel will not operate correctly. Other Panasonic-category VCR's have the play idler driven by another idler that contacts a metal wheel which is attached to the capstan and peeks out from the metal housing right below the capstan oil seal; clean this idler, and if that does not make the machine play properly, remove the capstan (from below) and clean the knurled part that contacts this idler.

If after cleaning, a play idler still fails to move the take-up reel, or if you still cannot find the cause of the failure of the take-up reel to rotate during play, you will need to check to make certain that the idler is moving all the way into contact with the reel or part that it is supposed to drive.

Since this assembly is probably out of sight under the cassette when the cassette is in position to play, it may be necessary to remove the cassette basket or loader, and "trick" the machine into going through the motions of playing with no cassette in it. Use the procedures in Appendix V to do this, so that you can observe exactly where the failure occurs in the mechanism that is supposed to drive the take-up reel.

Determine whether the play idler is coming all the way into contact with the part it is supposed to drive,

by watching the machine go through the motions of playing with no cassette it it. What is stopping it? If it is being prevented from doing so by a stopper that is part of the mechanism, such as a movable pin sticking up from below the table, then the problem may be that it is not loading fully.

7. VCR Starts to Move Tape in Play, but Afterwards Shuts Down Automatically

If the heads start turning, the mechanism fully loads the tape, the pinch roller presses the tape against a rotating capstan, the loading motor comes to rest, and the take-up reel starts rotating to reel in the tape, but then the VCR mysteriously stops playing the tape, unloads it, and goes into the Stop mode, then probably the trouble is one of two things.

It may be that something is failing to tell the little computer that the take-up reel is moving properly, so the little computer is shutting down the system because it is afraid that your tape may get damaged if the take-up reel is not reeling it in properly. Or on a VHS machine, it may be that one of the tape end-sensors (explained in Chapter 2) is malfunctioning and telling the little computer that the end of the tape has been reached when actually it is still in the middle of the tape, so the little computer is shutting down the machine because it thinks you've come to the end of your tape.

Problems with the Sensor to Detect Take-up Reel Rotation

On early vintage machines, one or more belts run from the take-up reel out to a mechanical numerical counter at the front of the machine. The counter has a magnet in it that causes pulses to be sent to the VCR's logic circuits when the counter is turning. That's how it tells whether the take-up reel is turning.

But if one of these belts breaks, or if the counter jams because the plastic has warped, or because there is dirt in the little gears, then the VCR's little computer gets the message that the take-up reel is not turning even though it really is turning, and so the computer stops the machine. The solution, obviously, is to replace the broken or slipping belt(s), and/or spray out or replace the mechanical counter. If the

belts to the counter are good and you see it counting off numbers while the tape is playing before it shuts down, then you probably have a problem in the electronics beyond the scope of this book.

Optical Take-up Reel Sensor Problems

Later machines do not use belts that run to mechanical counters. Instead, they tell whether the take-up reel is moving by focusing a tiny invisible light on an alternating black-and-white pattern printed on a thin piece of metal placed on the bottom of the take-up reel. When the take-up reel is turning, a miniature electric eye sees the light alternate between bright and dark as the reel turns.

This system is susceptible to its own problems. Sometimes the pattern on the thin piece of metal gets covered with dust, and the eye does not see what it is supposed to see. The remedy in this case is to remove the take-up reel and dust off the piece of metal with the pattern.

Remove the reel by removing the fastener at the top of the axle or spindle on which it turns, and sliding the reel up. Just before it comes off its axle, make certain to slide any little washers or bearings back down the shaft so that they do not fall off and get lost. DO NOT PUT ANY LIQUIDS OR OTHER CLEANING CHEMICALS ON THE PIECE OF METAL WITH THE PRINTED PATTERN. Just dust it off, and see whether that corrects the problem.

SAMSUNG

To clean the take-up reel sensor, make certain the VCR is unplugged and remove the cabinet top, bottom, and front panel piece. See Appendix III for help, if necessary.

Unplug the multi-wire connector going to the cassette front-loading assembly. Remove the screws (probably three) holding the front-loader to the reel table, slide the assembly slightly toward the rear, lift up and away.

Remove the C-ring attaching the pinch roller.

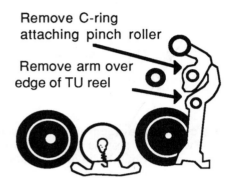

Remove C-ring attaching pinch roller

Remove arm over edge of TU reel

See Sections 9 and 10 of Appendix I if you need help with C-ring removal. Next, remove the C-ring holding down the arm that covers part of the edge of the take-up reel. Slide the arm up and off. Remove the C-ring and washers holding down the take-up reel. Slide the take up reel up its shaft, taking care not to lose the washer on the bottom.

Moisten a cotton swab with your breath only, by breathing on it with a hot, humid breath, no chemicals, and wipe off the shiny squares around the bottom side of the take-up reel you removed.

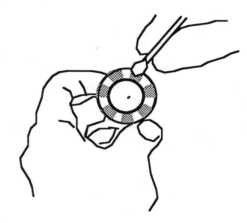

Reassemble by going through the above steps in reverse order. Make certain that the pin on the arm goes on the left side of the piece to which the spring hooks. Use needle-nosed pliers to squeeze C-rings back onto their shafts. And make certain that the pin on the pinch-roller arm goes into the oblong hole on the arm.

Replacing the Sensor

If cleaning the reflector with the pattern does not solve the problem, the next thing a VCR repair technician would try is replacing the little part that contains both the miniature light source and the electric eye, called an "opto-sensor" or "photointerupter."

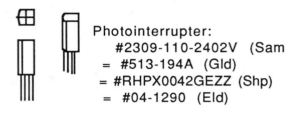

Photointerrupter:
#2309-110-2402V (Sam
= #513-194A (Gld)
= #RHPX0042GEZZ (Shp)
= #04-1290 (Eld)

This part is also available from Electrodynamics as #04-1290, and it is very inexpensive.

The replacement procedure is of intermediate difficulty, because it involves unsoldering several wires connected to the circuit board underneath on which the part is mounted, as well as unsoldering the old part and soldering in a new one. After writing down the points to which they connect, it is necessary to unsolder four wires running to the circuit board underneath, and remove the circuit board from underneath, in order to replace the photointerrupter.

Make sure that you install the new photo-interrupter with the correct orientation. It has two fine wires, and two thicker wires, and at the time that the old part is removed, you need to draw a little picture showing which type of wire went into which holes , so that the new part can be put in the same way.

GOLDSTAR and J.C. PENNY

If a Goldstar or J.C. Penny machine is stopping during play, look to see whether there is a circuit board under the take-up reel, and if so, what number is printed along the right edge.

If the number is 513-194, with no "D" after the last numeral, then you need to order up the new replacement board number 513-194D, with a letter "J" stamped on it, from Goldstar.

To replace the old circuit board (1) remove the cassette front-loading assembly, (2) remove the fastener attaching the plastic take-up reel, (3) slide take-up reel part way up its shaft, (4) remove screw attaching circuit board, (5) unplug connector to circuit board, (6) work old circuit board out around overhanging parts. (7) Work new circuit board into the same place, plug in, attach screw, slide reel

down, attach fastener, and reinstall cassette front-
loader.

"... *Sure, lady, $17,800 is a lot of money, but wouldn't you rather have him home at night watching a new Supervision X-100 than out tomcatting around? ...*"

Chapter 15

VCR Damages Tape
(or "Eats Tape")

Exactly what kind of damage is being done? Some of the different ways in which a VCR can damage tape are:

1. The machine may eject cassettes with a loose loop of tape hanging out, damaging the tape. This problem is usually caused by the take-up reel failing to reel in tape while playing, or the mechanism failing to reel all loose tape back into the cassette when stopping.

2. Tape may get wrapped and tangled around the cylinder with the video heads.

3. The machine may put frills or scallops, or other damage, along the top or bottom edge of the tape.

4. The tape may be scratched horizontally along as much of it as was run through the machine. These problems are treated separately in sections with the same number.

5. The tape may actually be broken apart, torn in half.

1. VCR Ejects Cassette With Tape Hanging Out

The most common causes of a machine ejecting cassettes with tape hanging out are the problems covered previously in Chapters 13 and 14. Usually, whatever is causing a VCR to do this will also cause it to fail to fast-forward or rewind a tape properly, and these problems normally show up in the earlier testing of these functions as part of the Universal VCR Check-out Procedure in Chapter 4, sending the reader to Chapter 13 or 14.. But these problems can be intermittent, or due to other causes, and appear first as the problem of "eating tapes."

Most VCR technicians will begin their repair of this problem by treating the machine as if it had failed the fast-forward/rewind and play tests, doing whatever would be done if were not performing those operations correctly.

This means replacing (or at least cleaning) all rubber drive parts, with special attention to any idlers with rubber tires that contact either reel and

checking all belts. So, turn to Chapter 13, and proceed as if the machine had failed the fast-forward/rewind test. This almost always fixes the problem, but if not, turn to Chapter 14 and perform the appropriate maintenance as if the machine were not playing the tape correctly. If not, return here.

SPECIAL MODEL NOTE: HITACHI and R C A

In some Hitachi and RCA machines, this problem can be caused by the breakage of the small clutch drive belt (belt #157053 (RCA) or 6355561 (Hit)on the under side of the chassis hidden by the two pulleys. Check for that. (Do not confuse it with the different belt having the same part number on top.)

It can also be caused by breakage of part of the clutch-idler plate assembly underneath the reel table. This problem was common on the models that use the idler with the toothed plastic gear (rather than a rubber tire) on the idler and that look like this on the top side of the reel table,

Top of reel table
F

and on the bottom side, look like this,

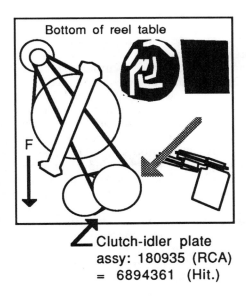

Bottom of reel table
F

Clutch-idler plate
assy: 180935 (RCA)
= 6894361 (Hit.)

and which use the version of the idler-clutch assembly that has a plastic base plate.

The problem is that, in time, a plastic pin that holds one end of a spring under tension breaks off from the force against it. Look at it from the angle indicated by the gray arrow in the illustration. Find the pin that holds the spring. It should look like this:

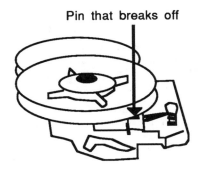

Pin that breaks off

For clarity, this illustration shows the plate in isolation from the rest of the VCR.

If this pin is broken off, that's the problem. The solution is to replace it with a new assembly. The replacement part (#6894361 (Hit) or equivalently, 180935 (RCA)) has been beefed up, so that you should not need to fix this problem again.

When you have the new part, remove the old assembly by first removing the idler from above, by sliding the little plastic clip that holds it on, and sliding the idler up and off the shaft.

Turn VCR over, remove belt, and remove two screws holding down the plate assembly,

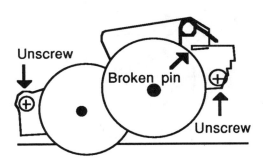

Unscrew
Broken pin
Unscrew

keeping track of which screw came from the left, and which from the right.

Plug in the VCR for a moment, then unplug it when the motors stop moving.

Make certain it is unplugged. Work the new assembly into place and attach with the screws. putting the screw with the coarser thread closer to the loading motor side. Put the belt back on, turn the VCR over and make certain that the plastic pin by the take-up reel is in front of the metal piece, as explained in the discussion of Hitachi-category VCR's in Section 6 of Chapter 13.

SHARP

On some early Sharp-category machines, like the Sharp VC 8400, a little rubber pad on the brake that presses against the take-up reel (at about 7 o'clock on the side) wears away or disintegrates, causing the reel to overshoot instead of stopping as it should at the end of a rewind operation. This causes tape to be left hanging out of the cassette when ejecting. The solution is to replace the brake pad with a new one. The part number is PGUMM0027GEZZ (Shp).

2. Tape Gets Wrapped and Tangled Around the Cylinder

This problem is generally caused by something like moisture or sticky tape particles making the tape adhere to the sides of the cylinder. Also, this accident frequently happens through no fault of the VCR when someone tries to play a cassette too soon after running a wet-type head-cleaning tape through the machine.

Hopefully, the video heads did not get broken when this happened.

If you open up a VCR and find that the tape is wrapped and tangled around the upper cylinder (where the video heads are), then gently unwrap the tape. Cut it with scissors, if necessary — that part of the

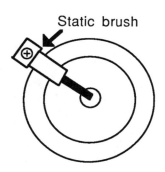

Static brush

tape is undoubtedly ruined, and it would not be a good idea to run it through the VCR again.

In order to get the tape unwrapped, you may need to take out a screw attaching a static brush that rides against the top of the cylinder, and temporarily remove the static brush.

If someone used a wet-type head cleaner and then put a cassette into the machine too soon before it dried, then you know how to prevent recurrent of the problem: Don't do that again.

Moving the machine from a cold place (like an air-conditioned room) to a warm humid environment and putting a tape in before the condensation evaporates can also cause this problem. The dew sensor is supposed to prevent the machine from starting up if moisture is condensed inside, but sometimes this safety system does not detect the problem in time.

If the problem was not due to moisture, then clean very carefully and thoroughly the surfaces of the upper and lower cylinder that the tape contacts, checking especially for any deposits stuck to the metal, and removing them with alcohol. If there was something sticky on the tape (like jam or jelly on a children's tape), the tape should be discarded, too.

3. VCR Puts Frills, Scallops, or other Damage Along Edge of Tape

If the machine seems to be putting frills or scallops along the top or bottom edge of the tape, something must be hitting the edge and bending it over as the tape runs through the tape path.

Check first to make certain that the capstan oil seal ring has not come up the capstan shaft to where it is hitting the tape. If it is, push it back down again.

Push oil retainer washer
back down as far as it
will go.

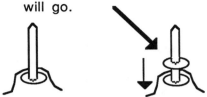

If the capstan oil seal is not frilling the tape, check to make certain that the cassette box is going all the way down onto the pins in the reel table. This is covered in Chapter 12.

If the cassette is sitting correctly in the VCR, then one of the guides or pins that the tape passes around may be too high or low.

Checking this possibility, and making the necessary readjustments, is really too advanced for this book. A technician would need to watch a fresh, undamaged tape carefully as it plays to see at what point it is getting damaged, and then correct the cause of the damage

If he changed the position of any tape guide, he would monitor the picture while doing so, and if the change did not solve the problem, he would put the guide back where it was before he changed it. It is extremely unlikely that more than one guide could be out of position (unless someone who does not know how to make adjustments has gotten into the VCR).

SAMSUNG

If the VCR is in the Samsung category, and it is damaging the edges of tapes, especially at the very beginning of tapes, the problem may be due to wear in the clutch for the take-up reel.

As this clutch wears, instead of slipping more easily, as you might expect, it grips with more force. As a result, the take-up reel, as it winds in the moving tape, pulls on it too hard, causing the edge to crease as it comes out of the cassette on the supply-reel side.

The problem is worst at the beginning of the tape because the take-up reel can apply the greatest rotary force then.

The solution is to dismantle the mechanism and replace the clutch on the take-up reel. Unfortunately, this is an extremely difficult major undertaking whose extent and complexity puts it beyond the scope of this book. You will have to take it to a shop and hope that they can do it.

4. Tape Gets Scratched as it Moves Through the VCR

If long, horizontal scratches appear on tapes after they have been run through the VCR, you first should carefully clean the entire tape path with cotton swabs and alcohol, as explained in Chapter 3. As you clean, look especially for a tiny particle of something stuck to one of the guides or rollers that the tape runs across.

After it dries, power up, insert a fresh new blank tape cassette, put the machine in play for a few minutes, push stop, eject, open the flap on the cassette box, and check to see whether the new tape is getting damaged.

If the machine still damages the tape, try cleaning it again.

If it still scratches the tape and you cannot find anything in the tape path that is scratching it, put the fresh blank back in, fast forward to a new, undamaged part, start it playing again, and while it is playing, pull out the plug, so that the machine stops with the tape still loaded.

Use a felt tip pen to mark the tape near where it is touching the capstan, put the plug back in, let the machine unload and eject the tape.

Find the point where the scratching begins on the new stretch of tape, and compare the beginning point of the scratch with the position of the mark you made.

From the distance, you can figure out at what point in the tape path the scratching is beginning. Go to that point, and hopefully find and fix the problem by smoothing off whatever is putting the scratch in the tape. The cause could be anything at that point that has a small sharp protrusion. There is no way that I can tell what that is on your machine without looking at it. You will need to find it and take the appropriate measures to correct it on your own.

5. VCR Breaks Tapes

There are two different manners in which a VCR may break tapes. It may break them by ejecting the cassette with loose tape hanging out, or it may break them by failing to stop at the end of the tape in play, rewind, or fast-forward.   These are completely different problems.

If the VCR breaks tapes by leaving them hanging out of the cassette when the cassette is ejected, so that they catch on the mechanical parts and get torn, then the problem is that the VCR is failing to wind all loose tape back into the cassette when it unloads the tape prior to ejection. This problem usually is caused by a dirty or worn idler (in machines that use an idler), and normally it should have shown up in the machine failing to pass the previous tests of fast-forwarding and rewinding, or failing to move the tape correctly in the "play" mode.   Try running the procedures in Chapters 13 and 14, as if the machine had failed these earlier tests.

If the machine pulls tape out at the ends, this indicates that there is some trouble with the sensors that are supposed to sense the end of the tape. This is an electronic problem beyond the scope of this book, but in Hitachi-category VCR's, this problem can often be corrected by replacing the tape end-sensor phototransistors, as explained in Chapter 11.

"Mommy finished fixing the VCR with the book.   She wants to know how you're doing with the toaster? . . . "

Chapter 16

TV Signal Does Not Pass Through VCR to TV Set in the "TV" Mode

When the VCR is plugged in and the TV/VCR switch is in the "TV" position, whatever is arriving on the antenna or cable that is connected to the "RF In" (or "In from Ant.") F-type terminal on the VCR should go straight through the VCR and right on out to the TV set connected to the "RF Out" (or "Out to TV") terminal.

In this position of the switch, then, you should be able to tune in and see on the TV set, all the same channels that you would see if you connected the same antenna or cable line directly to the TV set.

If this signal does not pass through the VCR to the TV set when this switch is in the "TV" position, or if the picture comes through extremely grainy, snowy, or otherwise bad, then either you have not got the TV set correctly tuned to the incoming channels, or something is wrong.

The first test to make is disconnecting the antenna or cable from the VCR input and connecting it instead directly to the TV set. Begin by trying this. Note: If your signal source is a subscription cable service and VCR is "cable-ready" but the TV set is not cable-ready, then you will need to connect a cable-box between the cable and the TV set to make this test.

If you do not receive channels when you hook the antenna or cable directly to the TV set, then the problem is not in the VCR. It is outside the VCR, either in the TV set, or in the antenna or cable.

But if you do receive stations when you make this test, then you have a problem in the VCR whose most likely causes are:

1. The (second) piece of cable connected between the VCR and the TV set, or the balun (if one is used), is bad.

2. The "antenna switcher" in the VCR is bad.

3. Elsewhere in the VCR, the circuitry that is supposed to control the "antenna switcher" is bad.

These problems are covered in the following sections.

2. Verify that the F-to-F Cable Used is Good

The piece of cable used to connect the VCR to the TV set may be bad. The easiest way to check this possibility is to substitute for it another piece of cable that you know is good. You could buy a new length of cable to try in its place, or borrow a piece that you know is good from a friend .

If you get a good picture when you connect a different piece of cable from the VCR to the TV set, then you know that the piece of cable you were originally using is bad and needs to be replaced.

But if the channels still do not come through, then there may be trouble in a part of your VCR called the "antenna switcher."

3.   Trouble With the Antenna Switcher?

The "antenna switcher" inside your VCR is a switch connected inside to the "RF Out" or ("Out to TV") F-terminal.   This switch has two positions.

When this switch is in one position (called the "VCR" position), it connects the output terminal to whatever the circuits deep in your VCR are trying to send out. If the machine is playing a cassette, it sends out, on channel 3 or 4, the picture and sound on the tape that is playing.   Or if the VCR is recording a program, or just tuned to an incoming channel, it sends out on channel 3 or 4 the program that it is receiving. This is what should happen when the switch is in the "VCR" position.

But when this switch is in the "TV" position, it should connect the VCR's output F-terminal directly to its input F-terminal.   In this position, whatever is coming into the VCR's input F-terminal should be sent straight to its output F-terminal with no change. So, a TV set connected to this output terminal should see all the incoming channels just as if the antenna or cable were directly connected to the TV set.

This switch on the output is called the "antenna switcher."   In the first early VCR's, the antenna switcher was a heavy mechanical switch that was physically moved back and forth between these two positions by a lever that moved when the operator pushed the TV/VCR switch from its one position to the other position.

But in later VCR's, a remote control "electronic" switch is used, which is switched back and forth between its two positions by electricity sent out to it on wires from another part of the VCR. This change in design simplifies the mechanism inside VCR's, but it has one disadvantage.

Inside the electronic switch, tiny fragile crystalline devices (usually what are called "small diodes") are used to do the remote-controlled switching back and forth.   These fragile parts normally never receive more than a very small electrical voltage, which is all they can withstand.

But if lightning hits near the wires of a community subscription cable system, for example, a momentary surge or spike of voltage can be sent for miles over the cable lines. If it comes down the cable feed to your house, it comes into the VCR's input terminal and hits the antenna switcher for a fraction of a second with more voltage than its fragile parts can stand.   There may be no other sign of damage, but inside the antenna switcher in the VCR some parts have been affected in a way that damages its function as a switch.

The result is that when things are back to normal on the cable, the antenna switcher will no longer bypass the channels directly to the TV set when you put the switch in the "TV" position.   Either nothing comes through the antenna switcher, or so much is lost from what comes through that the picture is grainy or snowy.   The other position of the switch (the "VCR" position) may, or may not, also be affected.

About 97% of the time, a damaged "antenna switcher" is the reason why the signal will not pass through the VCR to the TV set when the switch is in the "TV" position.   In these cases, the solution is to replace or repair the antenna switcher.   Most of the time this job must be done by a shop, but on a few models, you can order up a new antenna switcher and install it in place of the old one yourself.   Some of these are listed below.

The other 3% of the time the problem is not in the antenna switcher, but elsewhere in the VCR circuits that are supposed to control the antenna switcher by sending it electricity on certain wires to switch it back and forth.   If this is the cause of the problem, fixing it requires advanced electronic troubleshooting techniques beyond the scope of this book.

Where is the Antenna Switcher Located?

You know where the two F-type terminals come out of your VCR, right? The antenna switcher is located in the part on the inside of the VCR to which these terminals are directly attached.

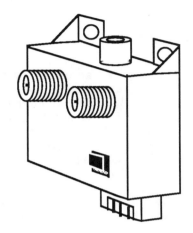

Typical Antenna Switcher Assembly

It is usually in a rectangular metal enclosure about one-third the size of a sardine can.

On some models, this part is soldered to circuit boards or attached to other large parts. On these models, replacing or repairing the antenna switcher requires advanced skills and is beyond the scope of this book.

On other models, this part is only attached by wires to the rest of the VCR. But sometimes, it is located in an inaccessible place requiring elaborate disassembly to reach. On these models, replacing the antenna switcher is a repair of intermediate difficulty. In this case too, repairing the problem is a little beyond the scope of this book, although a few adventurous readers might like the challenge of disassembling and reassembling all the parts without breaking anything.

But on a few happy models, like many Fisher-category machines, for example, the antenna switcher is easily removed by opening the cabinet and unscrewing four or five screws. On these happy models, you can easily replace the antenna switcher yourself. You need only remove the old one, get the part number, order up a new one, and install it.

If your VCR is one of those on which replacing the antenna switcher is too difficult, there is one more possibility that you might want to check out before

spending the money to take it to a shop to have the job done.

You can return to the Chapter 4 Universal Check-Out Procedure and run the rest of the tests described there.

If you find that the antenna switcher is the only part that is bad, and that it will still do everything else you want it to do (play? record programs?), and that the only problem is that you cannot watch one program while recording another, then you can make your own "antenna switch" from an A/B switch and another length of F-to-F cable, as described later in this chapter in the section entitled "Substituting an A/B Switch for the Antenna Switcher."

Antenna Switcher Combined with Tiny Transmitter

In some VCR's, the antenna switcher is in the same assembly as the VCR's little TV station mentioned in Chapter 2. This little transmitter is called the "RF modulator," so you will actually be ordering up a part called "an Antenna switcher/RF modulator," but don't let these big words scare you. It makes them more expensive to buy, but you still only need the replacement part number to order a new one, and a cross-slot screwdriver to install it.

Finding the Antenna Switcher

Look inside your VCR to determine whether the antenna switcher in your machine is positioned in a location that makes it accessible. If it is, get to it and see whether you can find any of the manufacturer's part numbers in the following list printed on it. If so, you can order a replacement from the original company, or a replacement at a discount price from Electrodynamics, Inc., whose address is given in Appendix II.

Akai

A replacement for the Akai E60-11-097-USA modulator is available from Electrodynamics as part number 06-1450.

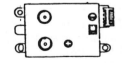

Emerson

You can order from Electrodynamics, an antenna switcher/RF modulator to replace the one in the Emerson model VCR-800 and in model VCR-801. The Electrodynamics part number is 06-1470 (Edc).

For the Emerson VCR 754, VCR 843, and VCR 952, which use the Emerson # MDF33VA3409 antenna switcher/RF modulator, you can use a #06-1480 (Edc) from Electrodynamics.

FISHER-CATEGORY

On many Fisher models, the antenna switcher/RF modulator is attached to a piece of plastic that is fastened to the inside back of the VCR with two screws.

Remove these screws first

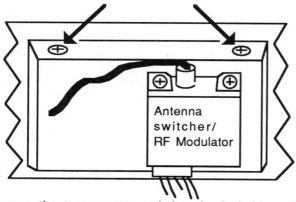

Antenna switcher/ RF Modulator

Remove the two screws and the plastic holder. The part number of the assembly should be printed on a label pasted to the outside of the metal enclosure.

The following are the most common ones used:

```
4-1164-011600 (SFC)
4-1164-051620 (SFC)  = CA7004 (Radio Shack)
4-1164-031610 (SFC)  = 06-1410 (Edc)
4-1164-031620 (SFC)  = 06-1420 (Edc)
4-1164-031600 (SFC)  = 06-1460 (Edc)
```

The number before the '=' symbol is the Sanyo-Fisher part number. The second number following the '=' symbol is the part number for ordering a low-price replacement from an alternative source. The '031600 model is soldered to the circuit board, and requires unsoldering and resoldering to change out.

HITACHI-CATEGORY

Hitachi antenna switchers are generally part of larger circuit boards, and must be sent to the shop to have new diodes installed by a technician.

J V C -CATEGORY

Many JVC and Zenith VCR's used JVC's #PU-52135 for the RF modulator. A replacement is available from Electrodynamics as part number 06-1440. Soldering may be required for this one.

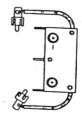

PANASONIC-CATEGORY

Panasonic-category antenna switchers are frequent lightning casualties. Many come in detachable subassemblies, but unfortunately they are located in a place that requires extensive risky disassembly to reach. Their replacement is probably best left to a shop, unless you have enough skill and patience to build a model sailing ship inside a bottle without breaking anything.

Lots of different assemblies were used in the various makes and models, so you would also need to find someone reliable to look up the correct part number for your particular unit.

SAMSUNG

From Electrodynamics you can order an antenna switcher/RF modulator to replace the one in Samsung models VT-210, VT-222, and VT-225. The Electrodynamics part number is 06-1470.

SANYO

A replacement for the Sanyo 143-9-4300-57500 antenna switcher/RF modulator is available at a discount from Electrodynamics as part number 06-1460. It is the same as the Fisher #4-1164-031600.

SHARP CATEGORY

Sharp antenna switchers/RF modulators come soldered to the bottom circuit board in a position difficult to reach. The level of difficulty in replacing them is intermediate to advanced.

TOSHIBA

A replacement for the Toshiba #70123083, for the Toshiba MSU-911, is available from Electrodynamics as part #06-1430.

Before Replacing the Antenna Switcher

One more thing to consider before ordering up a replacement antenna switcher is the possibility that your machine is in the 3% minority in which the problem is actually in the circuitry elsewhere in the VCR that is supposed to supply the electricity to switch the antenna switcher back and forth between "TV" and "VCR." It would be frustrating to order up a new antenna switcher and install it only to find that the problem of not letting the signal through in the TV position still remained.

Before ordering up this replacement part, a careful technician would eliminate this possibility by attaching his voltmeter to the appropriate wire coming into the antenna switcher, plugging in the VCR, powering up, and pushing the TV/VCR switch back and forth between its two positions while watching to see whether the voltage switched between 0 volts and some higher number, like 5 or 10 or 12 volts.

If it did, he would conclude that the circuits that are supposed to switch the antenna switcher are OK, and the problem is in the antenna switcher, which needs to be repaired or replaced.

Checking the Switching Control

But you probably do not have a voltmeter to run this test. So, what should you do? There are two possibilities.

If you don't mind taking a little bit of a gamble, you could play the odds, figure that the trouble is probably in the antenna switcher, take the risk of ordering up a new one, install it, and find out then whether your gamble paid off.

But if you hate to run the risk of buying an antenna switcher that you might possibly not need, and end up having to take the machine to a shop on top of that, for about $2 you can rig up a way to test whether the trouble is inside the antenna switcher or not.

Homemade Test Equipment

Instead of a voltmeter, we can use a small 12 volt lamp that comes with two wires attached to it, such as the #272-332B lamp assembly from Radio Shack, and some clip leads (see Section 19 of Appendix I). (Do NOT use a large 12 volt lamp, like an automobile lamp. It draws too much current and might damage the VCR.)

Unplug the VCR, and open the cabinet, if you have not already done so. Unplug the multi-wire connector from the antenna switcher module. Do you see the row of little holes where the male prongs go into the plug? Start at one end of the row, stick a straight pin into the first hole, and connect a clip lead to it as shown in the illustration. Attach the other end of the

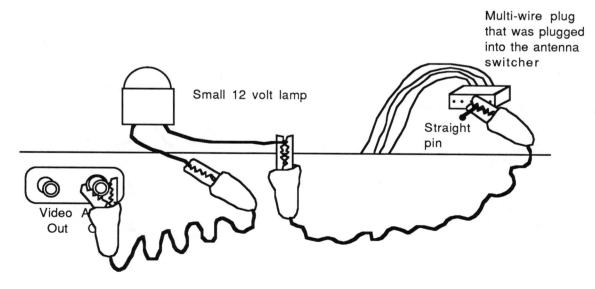

Small 12 volt lamp

Multi-wire plug that was plugged into the antenna switcher

Straight pin

Video A
Out

Simple Homemade Equipment to Verify that the Trouble is in the Antenna Switcher

clip lead to one of the wires from the lamp. Attach a second clip lead between the outside of one of the RCA phono plugs (what is called a "ground point") and the other wire from the lamp.

Set everything down carefully so that none of the bare ends of the wires or connections touches anything metalic. Plug in the VCR, push the power switch on the front cabinet piece, and then push the TV/VCR switch a few times, while observing what the lamp does. Then unplug the VCR, move the pin to the next hole, reattach the clip lead to the pin, plug in the VCR, power up, and again push the TV/VCR switch a few times. Repeat this test for every hole in the multi-wire plug.

If there is one hole in the plug where the lamp switches on and off as you push the TV/VCR switch, then the correct control signals are coming to the antenna switcher/RF modulator, and the problem is in it.

If the light does not come on and off for any of the holes, then you will need to take the VCR to a shop.

FISHER

On the widely used Fisher 4-1164-031610 antenna switcher/RF modulator, to save time, the place to check is the second pin from the left, pin 2, as the plug is normally positioned. Unplug it, insert a straight pin, attach the little lamp,

plug in the VCR, power up, and push the TV/VCR switch a few times.

If the light goes on and off as you push the switch, then the control power to the assembly is correct, and the problem is in the antenna switcher/RF modulator assembly.

On the 4-1164-231610 version which is soldered to the circuit board, the pin to check is the one marked "CONVCTL" in the printing on the circuit board.

Substituting an A/B Switch for the Antenna Switcher

If the only thing that your VCR will not do is let you watch one program while recording another, then you can make a substitute antenna switcher from an A/B switch and splitter connected as shown in the illustration.

With the A/B switch in one position, you can view whatever is coming out of the VCR, or what the VCR is recording, while in the other position you can view something else coming in on the antenna at the same time.

The drawback to this solution is that the strength of the signal coming from the antenna is cut in half by the splitter, so it will only work if you live in an area where the TV signals are very strong. However, you could use a spitter that has a built-in amplifier, so that there will be no loss of signal strength. You can get one from Electrodynamics as #5424 (Edc).

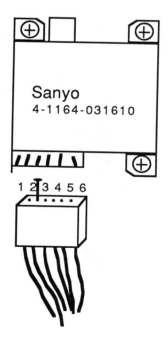

Testing the TV/VCR Control Switch

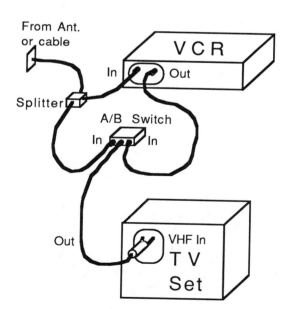

Substituting a Splitter and an A/B Switch for a Bad Antenna Switcher in the VCR

Chapter 17

VCR Will Not Pick Up TV Stations in "VCR" Mode

When the antenna or cable is connected to the "RF In" (or "In from Ant.") F-terminal, the TV/VCR switch is in the "VCR" position, and a TV set is connected to the "RF Out" (or "Out to TV") F-terminal, you should be able to select and tune active local channels with the channel selector on the VCR, and view them on the TV set when it is tuned to the output channel from the VCR (either channel 3 or 4).

If sound and picture are not coming through, then either there is a problem in the way the controls are set, or there is a malfunction in the VCR. This chapter deals with this situation.

This chapter assumes that the equipment passed the previous test in the check-out procedure, so that we know that signals are coming in on the antenna or cable, and we know that the piece of F-to-F cable hooked between the VCR and the TV set is good.

It is also assumed that you have correctly tuned the TV set to the VCR output channel (3 or 4) and tried adjusting the fine tuning of the TV set, as explained in Section 1 of Chapter 1, and moreover that you have tried adjusting the channel selector and fine tuning on the VCR to tune in a station, as explained in Section 2 of Chapter 1.

If there is some doubt whether the tuning process for the TV and VCR have been done correctly, you can try inserting a known good prerecorded tape, like a movie, and running the next test in the Chapter 4 check-out procedure. If a picture comes up when you play a cassette, this will make it easy to adjust the TV set's channel selector and fine-tune it to the output from the VCR's little transmitter without having to worry about adjusting the tuner on the VCR. Whether or not this attempt is successful, return to this chapter to deal with the problem of the failure of the VCR to tune in stations.

Try the VCR's Other Output Channel

The next test is to switch the output channel selector switch on the VCR to its other position (3 or 4), change the channel selector switch on the TV set to the same new number, and try again.

If a picture now comes up, switch both back to the previous position and see what happens. If it works on one output channel but not the other, then probably one of the VCR output channels is bad, and you can either go ahead and replace the little TV station in the VCR (the "RF modulator," it is called), or you can simply always use the other output position, the one that is good, and save the money.

Recheck Controls and Connections

Look to see whether the VCR has a "tuner/camera/aux." switch, and if it does, make certain that it is in the "tuner" position. Verify that nothing is plugged into the phono plugs on the VCR, and that no camera is attached to the machine.

Try selecting several different channels and tuning them in.

Check the UHF channels too. If there is an active UHF station in your area and the VCR has separate terminals on the back for UHF and VHF antennas, attach a "bow tie" antenna or other appropriate UHF antenna to the UHF input terminals and try to tune in a UHF channel.

If you find that the VCR will tune in a UHF station clearly but not a VHF station, then probably a part

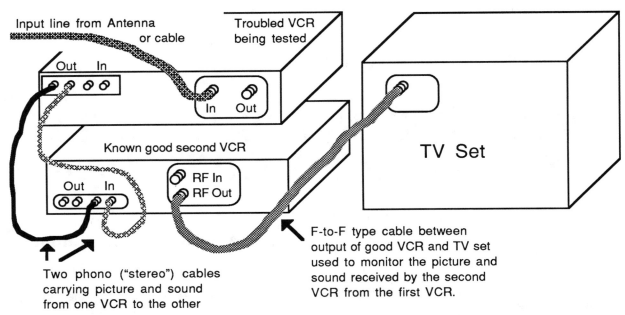

Input line from Antenna or cable

Troubled VCR being tested

Out   In

In   Out

Known good second VCR

Out   In

RF In
RF Out

TV Set

Two phono ("stereo") cables carrying picture and sound from one VCR to the other

F-to-F type cable between output of good VCR and TV set used to monitor the picture and sound received by the second VCR from the first VCR.

Using a Second VCR to Diagnose the Trouble

called the "antenna switcher" is bad. Turn to Chapter 16 to read about the antenna switcher, what it looks like, and when you can replace this part yourself, and when you cannot, depending on the model of the VCR.

Possible Causes of the Problem

There are several possible causes of the problem of nothing coming out when you put the TV/VCR switch in the "VCR" position, and try to tune in a channel. The trouble could be in:
1. VCR's antenna switcher (not the "antenna," but the "antenna switcher," which is something different from the antenna)
2. VCR's tiny transmitter ("RF modulator")
3. VCR's tuner-demodulator

If the problem is either of the first two possibilities, on some models you can fix the problem yourself by replacing the bad parts, but on other models, only a shop can make these replacements.

The machines on which you might be able to fix the problem yourself are the same as the models in which it is easy to replace a part called the "antenna switcher/RF modulator," listed in Chapter 16. If your VCR is not one of these, and the problem turns out to be a bad "RF modulator," you may also be able to avoid the expense of taking it to a shop by buying a separate RF modulator from Radio Shack and hooking your VCR to it, rather than getting the bad RF modulator in your VCR replaced.

Using Another VCR to Diagnose the Problem

If you can borrow or rent a second VCR that is working, you can use it instead of test equipment to localize the trouble in your VCR.

Unplug both VCR's.

Connect the two VCR's together as if you were going to copy a tape from your VCR to the second VCR.   For this test, you must use phono cords ("stereo cables") connected from the audio and video out phono plugs on your VCR to the audio and video in phono plugs on the second VCR, as shown in the illustration.

Disconnect the F-to-F cable to the TV set from your VCR, and reconnect it to the "RF Out" (or "Out to TV") F-terminal on the second VCR.   The set-up will now be as shown in the illustration, with your VCR in the top position, and the second VCR in the lower position. Note the channel on which the second VCR is set to output (3 or 4?) and reset the second VCR to output on the same channel that your VCR was set to output on. Leave the antenna, or subscription cable, connected to the input of your VCR.

Look and see whether the second VCR has a switch that says "tuner/aux." If so, set it to the "aux" position. If it has no such switch, then probably the switching is done automatically when you plug the phono cables into the phono plugs.

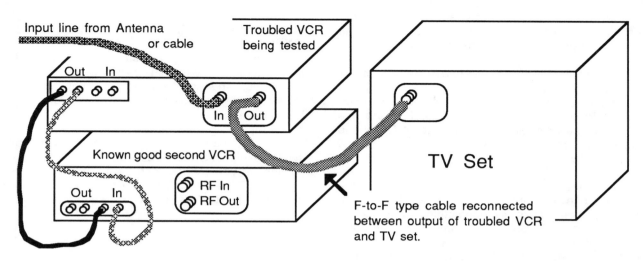

Input line from Antenna or cable

Out   In

Troubled VCR being tested

In   Out

Known good second VCR

Out   In

RF In
RF Out

TV Set

F-to-F type cable reconnected between output of troubled VCR and TV set.

TV Set Reconnected Directly to VCR Being Tested

Plug everything in, turn on all power switches, and turn the channel selector on the TV set to whichever channel (3 or 4) the second VCR is set to output on.

Now try to tune in a station with the channel selector and fine-tuning on your VCR.

If you still cannot tune in stations, go down to the next section below, "If Still No Picture." But if you can tune in stations now, then your VCR has a bad part in it. Your tuner (the "traffic cop") is OK, but there is trouble in the antenna switcher or little TV station ("RF modulator") in your VCR. This outcome establishes that the tuner is good but the antenna switcher/RF modulator in your VCR is bad and needs to be repaired or replaced.

The trouble in this case is not in the "antenna," but in a part in the VCR called the "antenna switcher/RF modulator," that comes in a little can, as shown in an illustration in Chapter 16. Go to the last section entitled, "Replacing the Bad Parts."

If Still No Picture

If you still cannot tune in stations, try putting a known good prerecorded cassette in your VCR, set the machine to the correct speed, hit "Play," and observe the results.

If a picture comes up — that is, if the TV set shows a picture — then while the cassette is playing, disconnect the F-to-F cable from the output F-terminal of the second machine and connect it to the output F-terminal of your VCR.

If you still see no picture when the TV set is connected directly to your VCR when it is playing a cassette, but when you used the second VCR, you saw it playing on the TV set, this also indicates that something is wrong with the antenna switcher or little TV station ("RF Modulator") in your VCR. Go to the last section entitled "Replacing the Bad Parts."

However, if a tape playing on your VCR comes through on the TV set when the TV set is connected to your VCR, but your VCR will not receive TV stations, then either the antenna switcher is not letting the stations come into your VCR, or else something is wrong with its tuner-traffic-cop. We do not know yet which is the problem, but the test in the next section may give us an answer.

Is the VCR Tuner Bad?

The next test only can be run on models that have the antenna switcher separated from the rest of the circuitry, and connected to it by a little cable about

Unplug this cable

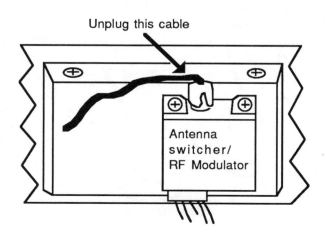

Antenna switcher/ RF Modulator

the diameter of a stereo cable with a phono plug at the end like on a stereo cable. For example, many Fisher models have a cable like this plugged into the antenna switcher/RF modulator.

We are going to unplug this cable and attach the antenna or cable line directly to it, bypassing the antenna switcher completely. To do this, we will need an adapter to connect the F-type terminal on the line from the antenna or cable company to the RCA phono plug on the end of the cable we disconnected from the antenna switcher.

In the case of the Fisher pictured above, the unplugged cable will present a male R C A phono connector, so you would need something to connect a male RCA plug to a male F-type connector — in other words, a third part that was female R C A on one end and female F-type on the other end would do the job. MCM sells such a connector as part #27-445. Or you could do the job some adapter and connector parts from Radio Shack: Join together a female-F-type-to-male-R C A-phono-plug connector (Radio Shack #278-252) and a female-R C A-phono-to-female-R C A-phono connector (Radio Shack #274-1553) You can use this pair to connect the line coming in from the antenna to the unplugged black wire that normally plugs into the suspected assembly.

On the other hand, on a different design where the cable to the antenna switcher unplugs at its other end, you would need something to connect a female RCA connector to a male F-type connector — a third part that was male RCA on one end and female F-type on the other end would work.   MCM sells such a connector as part #33-490. Or you can do the job with just the #278-252 adaptor from Radio Shack.

Unplug the VCR, unplug the cable to the antenna switcher, and use the adapter(s) to connect the antenna or cable line not to the antenna switcher, but

to the part to which it used to be attached via the cable you just unplugged. Connect the TV set to the RF output terminal from the second VCR.

Set everything in a position where no bare wires are touching any metal parts, plug in everything, power up, and see whether you now can bring in stations with the channel selector and tuning on your VCR.

If you can, then the problem is in the antenna switcher/RF modulator, and you should go to the next section. If you cannot, then the problem appears to be in your VCR's traffic cop tuner/demodulator, and you will probably need to take it to a shop.

Replacing the Bad Parts

If your VCR is not one of the models on which the antenna switcher/RF modulator is easily replaced listed in Chapter 16, you must either take it to a shop for repair, or else buy a little separate external RF modulator, like the Radio Shack #15-1273, and hook it between your audio and video out phono plugs and the TV set, as shown in the illustration. If the external modulator has a switch on it to choose between 75 ohm and 1000 ohm output, put it in the 75 ohm position.

You can connect the antenna or cable box to the "RF In" F-terminal on the VCR, or if the signal does not come through there due to the damaged antenna switcher/RF modulator in the VCR, you can run it to the A input of an A/B switch, connect the substitute external modulator to the B input (instead of the TV set), and connect the output of the switch to the TV set. Then you can use the A/B switch to select either the VCR or the antenna/cable as what you want to watch.

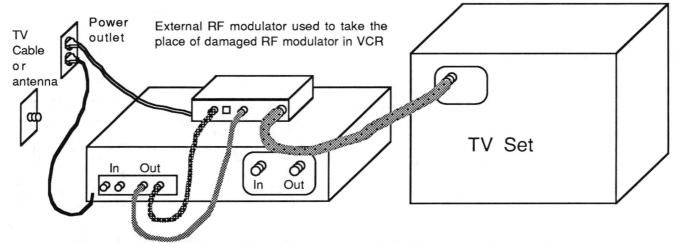

Connecting an External Modulator to Substitute for the Damaged One in VCR

Important: Use the check-out procedure
in Chapter 4 before using this chapter.

Chapter 18

Bad Picture or No Picture in Play Mode

This chapter covers situations where the VCR will run a cassette in the play mode without damaging the tape, but either there is no picture, or the picture contains white streaks or stationary horizontal bars of "video noise," or is bad in some other way. This chapter is organized around the different kinds of problems that may appear in the picture.

In addition to describing these problems in words, I will also try to portray them roughly with line drawings that represent the television screen. For clarity, these drawings will show only the lines or other distortion in the picture, and not the picture on which they are superimposed. For example, suppose the picture was supposed to be a face:

are the problem, by drawing just the distortion and leaving out the face and the rest of the picture, like this:

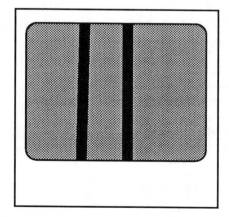

But suppose that there are two black vertical bars running down the screen as shown in the next illustration. We could make a picture that showed just the distortion, or the vertical lines in the picture that

When a drawing looks like this, it should be understood that there may be a picture, or part of a picture, in the background, not shown in the drawing. The bars are to be understood as appearing over the top of a picture in the area shown as gray.

Similarly, if the distortion appeared as white lines over the picture, like this,

Then the problem will be represented by omitting the face and putting gray where the picture appears:

This picture is to be understood as showing only the distorting white bars that are superimposed on whatever picture is behind them.

1. Diagnosing the Problem from the Appearance of the Picture

NOTE: For any of the problems listed in this section, try pushing down on the top of the cassette with a wooden stick while the cassette is playing. If the picture goes good when you press down on the cassette, but goes bad again when you release the pressure on it, go to Chapter 12, "Cassette Does Not Sit in VCR Correctly," and check to make certain that nothing is preventing the cassette box from going down all the way to rest on the pegs.

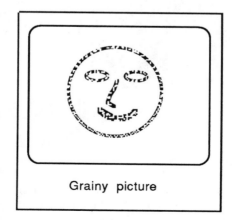

Grainy picture

Symptom: Entire picture is very grainy. Big objects in picture are blurry, but can be recognized; fine detail is missing.

Cause: Dirty heads, a damaged RF modulator, or other electronic problems.

Cure: Try cleaning the heads as explained in Chapter 3.

   If the problem persists, go to Chapter 17 and use the procedure there to test the RF modulator by connecting in a second VCR.

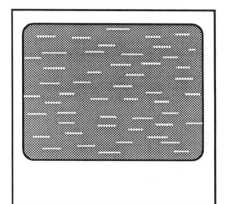

Symptom: Sound OK but lacey white lines superimposed on the picture during playback. The lines may be especially concentrated around light-colored areas and highlights in the underlying picture.

Cause: Tracking misadjustment, or bad tape, or dirty or worn video heads, or trouble in electrical connections or VCR amplifier circuits.

Cure: Try adjusting customer tracking control, replacing tape, and cleaning the video heads following the procedures in Chapter 3. If the problem persists, unplug and replug the multi-pin connector at end of wires running to the cylinder with video heads. As a last resort, take VCR to a shop to have amplifier circuits serviced or heads replaced.

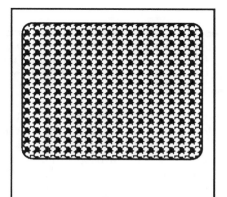

Symptom: Sound OK but screen full of black-and-white, salt-and-pepper snow during playback.

Cause and Cure: Same as above (cleaning heads, reseating multi-pin plug connecting wires from cylinder with video heads).

Symptom: The picture is visible, barely, but it looks as if it were being viewed through a grimy automobile windshield in the rain with bad wipers.

Cause: The picture is only being put on the screen half of the time, alternating with a salt-and-pepper snow during the other half of the time. (Video technicians call it "missing one of the two fields.") This is caused by one of the two heads being extremely dirty or broken.

Cure: A thorough head-cleaning, including possibly spraying out the heads, as explained in Chapter 3.

If repeated cleanings do not solve the problem, it may be necessary to take it to a shop and possibly to have the heads replaced.

Note: if the VCR has four or more heads, the picture may still be good in some speeds that use a different pair of heads.

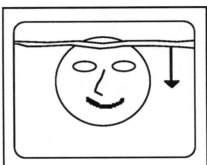

One or more horizontal lines float slowly down over picture

Symptom: One or more horizontal lines float slowly down the screen over the picture.

Cause: Scratches or other damage to the tape in the cassette.

Cure: Replace with a new cassette. There is no way to repair the scratched tape in the bad cassette.

One or more horizontal lines remain(s) stationary across picture on part of the screen

Symptom: One or more horizontal lines appears across the picture and remains more or less stationary. In the drawing, the lines are shown across the bottom half of the picture, but they can also appear across the top or middle part instead, or even across both top and bottom. Between the lines, the picture is more or less clear

Cause: The tape is not passing correctly around the cylinder with the video heads, due to a misalignment of the guides, or the back tension on the tape is too tight or too loose.

Cure: Try adjusting customer tracking control. If lines remain, clean tracks that P-guides run in and make certain that toe of P-guide is not catching on V-stopper, as explained in Section 2. If that fails, realign the P-guides or adjust the back-tension. See Section 2 of this chapter.

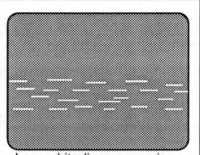

Lacy white lines appearing across part of picture move when tracking is adjusted

Symptom: Lines appear in a band across part (or all) of the picture. Lines change position or disappear when customer tracking control is adjusted.

Cause: VCR cannot track material on the tape correctly. If the tape was recorded on a different VCR, the problem may be in the other VCR, and there may be nothing that can be done now to make it play correctly on your VCR. However, if the problem appears on most tapes, including rental tapes and tapes that play OK on everyone else's machines, the tracking may need to be adjusted on your VCR.

Cure: Try adjusting customer tracking control.   If the problem is definitely in the VCR, see Section 4.

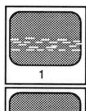

Broad band of lines appears, then disappears, then appears again, alternating off and on slowly over time.

Symptom: Horizontal lines appear in picture as if tracking were misadjusted, only they come and go in a slowly alternating pattern, good, then bad, then good again, then bad again, and so on. The sound may come off-and-on too.

Cause: Customer tracking control misadjustment, frills or damage to the edge of the tape in the cassette, or failure of the capstan to move the tape at the correct speed.

Cure: Check tape for damage. Completely clean the machine with special attention to the tape path, capstan, pinch roller, and capstan belt if the machine has one. If that does not fix the problem, and adjusting the customer tracking control does not make it go away, proceed to Section 5.

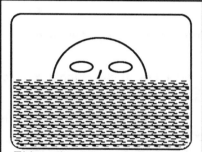

Either the top or bottom half of the screen is full of scratchy lines

Symptom: One half of the screen, either the top half or the bottom half, is filled with scratchy lines and snow, but the picture is fairly clear on the other half. Usually the top half is bad.

Cause: Dirty or corroded head-switching relay.

Cure: Clean or replace head-switching relay. See Section 6.

Vertical colored lines or rainbow appears on picture

Symptom: Vertical colored lines or a rainbow band appears over part of the picture. Two cases: (A) Rainbow is put into tapes when they are recorded on the VCR; (B) Rainbow appears when playing back known good tapes recorded on another machine.

Cause: (A) When a rainbow gets recorded into tapes during recording, this is due to incomplete erasure of the tape by erase head during recording. (B) When a rainbow appears when playing back known good tapes recorded on another VCR, it is caused by bad connections in the wires somewhere between the video heads and the circuit board the wires go to.

Cure: If the problem appears only at the beginning of a new recording, for a few seconds after the recording was begun, and then disappears, this is normal for many VCR's. To avoid this effect, start the tape going in the record mode a few seconds before the program material that you desire to record begins.

(A) If the problem persists during entire recordings, try unplugging and replugging the multi-wire connector to the erase head. (B) If there is a head-switching relay, clean it (see Section 6), and eliminate any bad connections in wires coming from the lower cylinder assembly. If problem still remains, the VCR will need professional service.

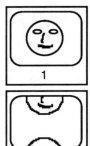

Picture
rolls
vertically

Symptom: Picture continuously rolls vertically on screen.

Cause: TV set is unable to deal with the signals from VCR that tell it when to start showing the next image in the sequence of pictures sent to it from the VCR. One by one, substitute a different video tape, a different TV set, and a different VCR to localize the trouble further. (This effect can also be the result of the copyguard system blocking the unauthorized rerecording of copyrighted material, like movies, as discussed in Section 3 of Chapter 1.)

Cure: Replace the weak tape, if that's the problem. If the problem is in the TV set, try adjusting the vertical hold, if the TV set offers this adjustment. If this does not fix the problem, you will have to take the TV set to a shop. If the trouble is in the VCR and cleaning the heads does not fix the problem, you will have to take the machine to a shop. (They may need to adjust the "head-switching point.")

Picture is
good, then
suddenly
jumps and
goes bad for
a second,
then is good
again, in a
regularly
repeating
pattern.

Symptom: Picture is good, then suddenly jumps and goes bad, then clears up, in a regularly repeating cycle.

Cause: Probably something hitting on the edge of the supply reel, or affecting the back-tension on the tape unevenly.

Cure: Change parts around supply reel, or their position, so that the back-tension is always even. See Section 7.

Picture is unstable and
tears at the top of screen

Sympton: Picture bends and tears at the top of the screen.

Cause: Usually due to a difference in back-tension between the VCR on which the tape was recorded and the VCR on which the tape is being played. One (or both) has the wrong back-tension.

Cure: Adjust back-tension on whichever machine is not correct. See Section 3

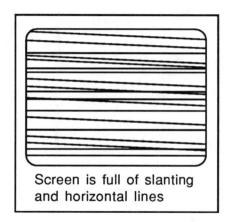

Screen is full of slanting and horizontal lines

Symptom: Sound is OK, but picture is just horizontal slanting lines. No clear picture between lines. (This is called "horizontal tearing.")

Cause: If problem appears during "Play" only, probably the cause is that the video heads are not spinning at the correct speed. If problem appears also when in "VCR" mode and tuned to any incoming station, the cause probably is that the video signal coming out of the VCR is too weak for the TV set to be able to display.

Cure: With either of the above causes, fixing the problem requires electronic troubleshooting.

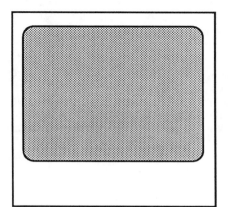

Symptom: Screen is uniformly gray. Not salt-and-pepper, as with dirty heads or no signal, but uniformly gray.

Cause: The picture is being muted electronically, or a video amplifier circuit has a trouble in it.

Cure: Requires electronic troubleshooting.

2. Aligning the Tape Path

If the VCR is a Beta, adjusting the height of the tape guides is difficult even for the experienced technician, so you should turn the problem over to a shop. But in many cases, it is possible for a do-it-yourselfer to correct tape path alignment problems in VHS machines.

One important secret of success is to determine exactly what is out of alignment first, and make only one adjustment. It is highly unlikely that more than one guide has gotten out of alignment, and if you change more than one adjustment, you probably are throwing the machine even farther out of adjustment, by introducing a second misadjustment.

Two misadjustments are not merely twice as hard to fix as one. It is more like ten times as hard,

because without the adjustment jigs they use at the factory, the only way a technician can adjust the tape path is by making a change in one adjustment, then checking whether that fixes the picture. If two adjustments are off, fixing neither one of them will, by itself, fix the picture, so it is extremely difficult to know when the first misadjustment has been corrected, since the picture still looks bad when it is correct, just as it looks when it is incorrect.

Secondly, it is vitally important to understand that there are some things in a VCR that must NEVER be touched or adjusted even though it looks like there is an adjustment screw for them. For example, technicians know that even a technician must never loosen the screws holding down the V-stoppers (explained in Chapter 2), or change the position of the angle brackets, because these can never be gotten back into their proper place again. They are installed with an extremely expensive special jig at the factory in Asia, and there is no way to get them back into their proper place if ever they are moved.

As a general rule, never touch any adjustment in a VCR unless you know exactly what it is, and you have specific reasons or instructions to change it. And even in that case, it is a good policy to keep track of exactly how much you turned it, and in what direction, so that you will have the best possible chance of getting it back to its original position in case changing it turns out not to fix the problem.

Another good rule is to return any adjustment exactly to its original position, if making it did not fix the problem. If you change an adjustment and it does not fix the problem, put it back exactly where it was before you made the adjustment. Do not leave it in the new position, assuming that a technician at a service center will be able to find it and restore it to its original position. This may be an impossible task.

If you make several wrong changes, even the best VCR technician may be unable to correct all the problems you added to the original problem. Remember that technicians are not at the factory and do not have all the elaborate special equipment or service information required to align everything in the first place.

Is the Toe of the Base of the P-guide hitting on the Edge of its V-stopper?

On many machines, depending on design, the two sliding base assemblies to which the P-guides are attached often each have a "toe" that is supposed to slide under a ledge on the forward side of the V-stopper.

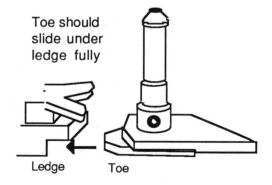

Toe should slide under ledge fully

Ledge   Toe

Sometimes the soft metal from which the base assembly is made wears away with time so that the shape of the edge of the toe changes so that instead of

sliding under the ledge as it is supposed to do, it

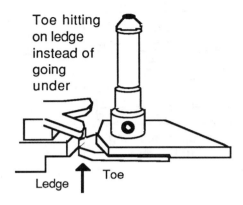

Toe hitting on ledge instead of going under

Ledge   Toe

catches on the edge, preventing the P-guide from going all the way up solid against the V-stopper. When this happens, the tape does not wrap like it should around the cylinder with the video heads, and lines and other distortion appear in the picture.

It takes only a moment to check whether this is the problem. With the top of the machine opened up so that you have access to the mechanism while it is playing a tape with the picture up on a TV screen, take a small wooden dowel or plastic object with no metal in it and watch each P-guide in relation to the V-stopper as you press down on the top of one P-guide and then the other. If you see or feel one P-guide move slightly down and up tighter against the V-stopper, and the picture immediately clears up, push "Stop" and repeat the experiment again several times.

If the same thing happens several times, look carefully to see whether the toe of the P-guide is catching against the ledge that it is supposed to go under on the V-stopper, and examine the toe of the P-guide to see if a little rough spot or spur has developed in the metal.

If this is not happening, the machine has passed this test, and you can skip to the next section.

But if the toe on the sliding base of the P-guide is definitely catching on the ledge of the V-stopper rather than going under it as it should, eject the tape, power down, unplug, and use cotton swab sticks moistened with alcohol to completely clean the tracks in which the P-guides slide along, paying special attention to removing any debris that may be caught under the ledge of the V-stopper at the end, or around the toe at the base of the P-guide. (Never unscrew or move the V-stoppers!) Relubricate the tracks with lithium grease if you do not have the special grease they use for this. Power up and see if the P-guides will now go all the way to up against the V-stoppers.

If one of the P-guides still stops short of going all

the way into the V-stopper, and you know that it is not due to dirt in the track, then the next step is to

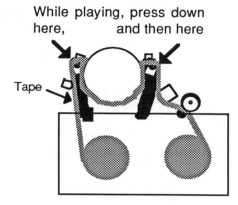

While playing, press down here, and then here

Tape

round off the toe of the P-guide. On some VCR's, like the Fisher, it is easiest to remove the P-guide along with the base to reshape the toe; on other machines, like RCA's and Hitachi's, you can only work from the top with the cassette loading basket removed to give you space. If the latter is your situation, skip the next paragraph.

If it is going to be easier to remove the P-guide assembly on your machine, open up the bottom of the VCR (see Appendix III for help getting the bottom open, if necessary). From the bottom, study how the bottom of the base assembly with the P-guide is attached to the mechanism that slides it along. Usually it will be held on by a C-ring. Remove the fastener (see Appendix I for help), and carefully count and lay out whatever washers and spacers are under the C-ring. Remove these and lay them out in order so that you will be able to replace them in the same order. When the base assembly is free, lift it out from the top side.

Take a small file with fine teeth and round off the top edge of the toe, just a little bit, not too much. Make it nice and smooth. Reinstall everything and try again. If it still catches, remove it again and file it some more. Do NOT touch or do anything to the V-stopper.

In my experience, it has always been possible to fix this problem by cleaning the track or reshaping the toe of the base assembly, but if this failed, the next step would be to call a parts source and order up a replacement P-guide base assembly. The one on the left side (viewed from the top) is called the "entrance P-guide base assembly," and the one on the right is the "exit P-guide base assembly." If you install a new one, you will have to align the P-guide height, as explained in the next section.

Aligning VHS P-guides

MITSUBISHI and VIDEO CONCEPTS:

If the VCR is a Mitsubishi or Video Concepts machine, and the problem is the appearance of wide stationary horizontal bars of what look like white and black scratches across some, or all, of the picture, bars that perhaps can be moved up or down by turning the tracking control, but that cannot be eliminated completely by adjusting the tracking control, then turn directly to the procedure for aligning Mitsubishi and Video Concepts tape guides under the special Mitsubishi repair procedure later in this section.

Tape Guide Alignment Procedure

Tools required: If the VCR is a Mitsubishi, Fisher, or Video Concepts machine, you will need a 1.5 mm allen head hex key, described in Section 11 of Appendix I. For most other brands, you will need a 0.89 mm allen head hex key, plus a special screwdriver with a notch cut into it, as described in Section 13 of Appendix I. You will also need some colored fingernail polish, or if it is an N E C or Vector Research machine, you will need lock-paint from the hardware store. And you will need a known good one or two-hour prerecorded tape.

A common problem is that as a VCR is used over the years, the action of the tape moving past the P guides tends to rotate them, causing them to become screwed in slightly farther than they should be, and changing their vertical position enough to produce distortion on the screen. I have often encountered this problem with Mitsubishi and Video Concepts VCR's, but it also happens with other brands.

To correct this problem, we will screw the offending P-guide back into correct position, and on Mitsubishi and Video Concepts machines, add a small dab of fingernail polish or lacquer paint to hold the P-guide in this position more permanently, which hopefully will forever prevent recurrence of the problem. All of this will be explained step-by-step, with diagrams, so don't worry. You should, however, quickly review the explanation of P-guides in Chapter 2, and also obtain the needed tools from one of the sources in Appendix II before starting.

Begin by cleaning the heads and other parts of the VCR following the procedures in Chapter 3.. Then, assuming that the same problem remains (as it probably will), return to this point in Chapter l8, and proceed with the realignment procedure as follows.

We are going to adjust the P-guides with the VCR connected to a TV set and running with a cassette in it, using changes that we can see in the picture on the TV set to guide us as we make the adjustments.

This will take some time, so we will need a known good prerecorded tape that is, say, at least an hour long, like a prerecorded movie.

It is extremely important to use a very long tape for this procedure, especially if you are working on a Mitsubishi, Fisher, or Video Concepts VCR, because one great danger is that the tape will come to an end while you still have the hex key inserted into the P-guide. If that happens, the VCR will try to unload the tape, pulling the P-guides (with your hex key stuck in one of them) back toward the cassette basket, jamming the key against the metal, and possibly bending or breaking the P-guide.

The chances of this accident can be minimized by using a very long prerecorded tape, and starting the procedure before half way through the tape. That should give you enough time. This accident is much less of a problem with other models of VCR's in which a notched screw-driver, rather than a hex key, is used to adjust the height of the P-guides, because the screwdriver will not stay caught in the slot if the VCR suddenly starts to unload. So, before you start the job, obtain a long prerecorded tape.

Initial Rough Adjustment

We will adjust the P-guides in two steps, first a rough, approximate adjustment guided by looking at the tape as it runs through the tape path and around the heads, secondly a fine adjustment guided by the way the picture looks on the TV set when the tape is playing.

Start by studying the lower cylinder (the round stationary part right below the turning cylinder that contains the video heads). Look carefully and you will see a little edge or ridge machined into the metal in a pattern that is very slightly spiral. The next illustration is designed to indicate what is meant by this ridge.

When the P-guides are correctly positioned, the

bottom edge of the tape is supposed to run along the edge of this ridge, down to, but not down over, the raised part.

Plug the VCR in, power up, put a full-length, long cassette into it, and being careful not to touch any of the electronic circuitry while the VCR is running, look down carefully at how the tape is running along the edge of this ridge.

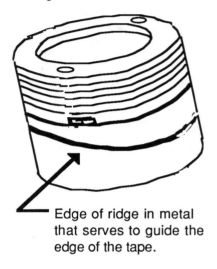

Edge of ridge in metal
that serves to guide the
edge of the tape.

*A Tape Guide Ridge is Machined
into the Lower Cylinder*

Is the edge of the tape bunching slightly against the edge of the ridge? If so, is this happening on both sides of the lower cylinder, or mostly on one side?

Usually, the problem is on the right side (as viewed from the front of the VCR, which is called the "exit" side). But sometimes it bunches on the left ("entrance") side. Watch carefully for a few minutes to form a well-based judgment about this.

If you have difficulty telling whether or not the tape is riding up on the edge of the ridge, here is a trick that may help. While looking down onto this area from a point above the VCR, shine a flashlight down on this area from the same point. The bottom cylinder is normally made of a bright, shiny metal, and you should be able to see a thin line of bright metal reflecting light all along the ridge.

While the tape is playing, if you see this bright edge along part of the spiral ridge, but the other part appears dark this is because the tape is bunching, or riding up onto the ridge. A P-guide needs to be adjusted so that a thin bright line of metal appears uniformly all along the lower edge of the tape while the tape is playing. This flashlight method works well, of course, only to detect tape that is running too low and covering the edge of the ridge, but this is the most common misalignment in tape position.

ment is gross, the
beside the edge of
pre-adjustment is
erial to learn how to
ut do not actually do
dure explained under

ρe is observed bunching on the right
, decide whether the tape is running too
.ne cylinder, or too far down. Usually the
is that it is running too low, too far down,
.eing bunched or bent slightly by the edge of the
.e, as it covers that edge.

If the tape is running too far down, it will be necessary to raise it very slightly by first loosening the locking screw in the exit P-guide, and then using the appropriate tool (either an allen hex key or a screwdriver with a notch cut into it), to rotate the part holding the P-guide in a counter-clockwise direction to raise it slightly; if the tape is running too high, you'll need to loosen the lock screw and rotate the P-guide in a clockwise direction to lower it.

First, unloosen the lock screw that is supposed to hold the P-guide in place and keep it from unscrewing. On some VCR's, like the Mitsubishi-category machines, this lock screw is a small allen head screw that is visible from the top.

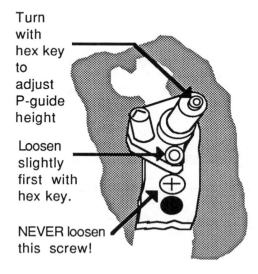

Turn
with
hex key
to
adjust
P-guide
height

Loosen
slightly
first with
hex key.

NEVER loosen
this screw!

Mitsubishi-type P-guide

In this situation, just insert the 1.5 mm allen key into this hex head and unscrew it about a half a turn. NEVER, NEVER, NEVER turn the cross-slot screw next to it, because this will disturb the position of the angle

bracket, and even a service center would have difficulty getting it back into correct position! But read the next few paragraphs first, before you loosen anything

In other VCR's, the lock screw is on the side of the P-guide that faces toward the back of the VCR. To gain access to this lock screw, push "Stop," and then as the VCR is unloading the tape, pull out the power plug from the wall when the P-guides are about half-way back on their curved tracks to the cassette. When the P-guides are half-way between their fully-loaded postion and their fully-unloaded position, they look like this:

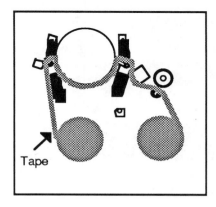

Tape Half-Loaded

If you have a dummy cassette box handy (see Section 17 of Appendix I), or if you use the procedures in Appendix V to trick the machine into playing with no cassette inside, you can pull the cord as it is loading (or unloading) with no tape to leave the mechanism in the same position with no tape in it,

P-Guides in Half-Loaded Position

which makes it even easier to reach the P-guide lock-screws.

With the P-guides stopped here, you can get the

appropriate tool (usually a 0.89 mm hex allen key) into the locking hex screw on the side. Turn it just a half turn in the counterclockwise direction to loosen it, enabling the threaded P-guide to be turned. Be certain that you have the hex key all the way in, and turn it slowly and gently, because a hex key this small is very easily stripped and damaged.

Here is a picture of a tiny allen wrench inserted into the adjusting screw on one P-Guide:

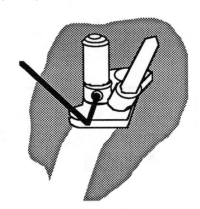

Allen Wrench Inserted into Lock-screw on P-Guide

After you get the wrench inserted into the tiny allen head screw, stop and think carefully which direction to turn it to unscrew it. If you turn it the wrong direction even a little, you will destroy your tool, and have to buy another set of allen wrenches. The tiny 0.89 mm allen key is very small and delicate. It is easy to damage it by forcing it too hard.

To unscrew, just push it gently with one finger in a counterclockwise direction as viewed from the head of the screw. The direction will appear opposite if you are facing the front of the VCR. Turn it about one-half turn. It might be a good idea at this point to turn the VCR around so as to face it from the back when loosening these lock screws on the side of the P-guides; that way you will be certain to turn it in the correct counter-clockwise direction to unloosen it.

Do this only on the one P-guide on the side where visual inspection of the tape as it was running showed that a adjustment is needed. Usually this will be the right-hand ("exit") P-guide. After slightly loosening the screw, plug the VCR back in, and let it unload the (either "real" or "illusory" tape) and push "Play" again.

Now insert the appropriate tool (either a hex allen key or a notched screwdriver) into the slot or allen head socket at the top of the P-guide. Make a careful note on scratch paper of the starting position of the tool before you turn it.

If the tape has been bunching on the ridge on the

lower cylinder, that is, going down too far, then screw the P-guide upwards (counter-clockwise) until the tape just stops bunching. Usually, you should not need to turn it more than half a turn. Screw it back and forth, in and out, up and down a quarter of a turn, noting the point where the tape just starts bunching against the ridge on the lower cylinder, and then raise it up above that point a tiny bit to the place where it first stops bunching.

It is unlikely that the tape is also bunching on the other side of the cylinder, but if it is, repeat the same process with the other P-guide. Loosen the lock screw, and screw or unscrew the P-guide a tiny bit until the tape is just about to start bunching. Remove the screwdriver or hex key.

Fine Adjustments

Now we must make whatever fine adjustments are needed to get a good picture. We will turn the P-guides up or down just enough to eliminate any lines from the picture on the TV set.

With the VCR connected to a TV set and the picture up on the screen as the tape is playing, first turn the tracking control on the front of the VCR to the position that seems to minimize the lines.

Next, note the position of most lines or scratch bands on the picture.

If they are in the top half of the picture on the TV, this indicates need to put your adjusting tool back into the P-guide on the left ("entrance") and screw it back and forth, up and down a tiny bit (one-quarter turn) while watching the picture on the TV. There should come a point where the scratches are minimized or disappear. If they disappear over a little range of turning, position the P-guide in the middle of this range.

On the other hand, if the worst lines are on the bottom half of the picture, start with the exit P-guide and adjust it for the point at which the lines on the bottom of the picture are minimized. Set it also in the middle of the range that gives the clearest picture.

It is unlikely, but possible, that you may need to work back and forth a few times between both P-guides to get a perfect picture. Remember this is fine tuning in which all adjustments are less than one half a turn.

If you cannot get rid of the lines in this way, try readjusting the tracking control on the front of the VCR and repeating the process.

If you still cannot get rid of the lines or bands of

scratches, then you will need to give up, restore every adjustment to its original position, and take the machine to a professional service technician. In most cases, you will be successful, however.

Locking Down the P-Guides

After you have finished adjusting the height of the P-guides, tighten any lock screws that you loosened earlier, using the same procedure as earlier to get access to them.

MITSUBISHI and VIDEO CONCEPTS

If the VCR is a Mitsubishi or Video Concepts machine, you can take measures at this point to minimize the likelihood of a recurrence of this problem. After you have made the adjustments that result in a clear picture, pull out the allen key, hit Stop, unload, and eject the tape. Put in a dummy cassette, or trick the machine into starting to load with no cassette in it, and pull the power cord when the empty P-guides are half-way between the loaded and the unloaded position.

Then leaving one P-guide in the correct position in which you placed it, completely unscrew the other P-guide, counting the turns, until it comes completely unscrewed and loose.

Paint the little threads at the bottom of the part that has come loose with fingernail paint, and screw it back in, again counting the turns as you go.

When you get it into approximately the same position as measured by the number of turns, remove the allen key, plug the VCR in, and wait for it to complete the unloading process. Insert a good cassette, push Play again, and while the tape is playing, readjust this P-guide that you just removed for a clear picture again. Then turn the locking hex screw to lock it down.

Next, repeat the same process with the other P-guide, removing it, painting it, and correctly repositioning it. NOTE: NEVER REMOVE BOTH P-GUIDES AT THE SAME TIME, AS ONE OF THEM NEEDS TO BE IN CORRECT ALIGNMENT IN ORDER TO BE ABLE TO ALIGN THE OTHER. IF YOU REMOVE BOTH AT THE SAME TIME, IT WILL BE EXTREMELY DIFFICULT OR IMPOSSIBLE EVER TO GET THE P-GUIDES CORRECTLY ALIGNED AGAIN!

N E C and Vector Research

On N E C and Vector Research machines, the brass piece at the base of the P-guide sometimes works loose in the gray metal base into which it is pushed. This looseness allows it to work up or down as tapes are loaded and unloaded.

You can adjust the P-guides as described above to get all the lines out of the picture, but two weeks later, the brass piece has moved in the base throwing it out of adjustment again.

The solution is to pull the loose P-guide (the one that keeps needing adjustment) completely out of the metal base into which it is pushed, clean the hole in the base and the outside of the brass piece with a good degreaser, apply lock-paint from the hardware store (and not just fingernail polish), and push it back in again part way. Plug in, power up, let the machine unload, insert a cassette, and push "Play."

Now adjust the height of the P-guide by watching the picture as you tap gently on the top of the reinstalled guide with the plastic handle of a screwdriver. As the guide slides little-by-little down into its base, there should come a point where the picture clears up. Stop there.

Make certain that there is still a little space left below the tiny hex screw on the side of the brass piece, between it and the base below into which you are pushing it, so that you can still make fine adjustments by screwing the guide up or down as usual.

Push "Stop," unload, power down, and let the machine sit and dry for twenty-four hours. Then power up and test again. You may need to loosen the hex screw and screw the guide up or down a little to make the picture completely clear. But now, when you lock the guide down again with the hex screw, the lock-paint will keep the whole thing from moving up or down in the sliding base that supports it.

All Other Brands

Other brands of VCR's do not require the locking steps in the previous paragraphs. If you have some other brand, it should be enough simply to stop the VCR in the half-unloaded position by pulling the power plug and gently screw the tiny allen-head set-screw down with the fragile little allen key. Be very gentle. Put a drop of fingernail paint on the little allen head locking screws to keep them from unscrewing.

This will probably solve your problem, but if not, there are two more things you can try before turning the VCR over to a professional service technician: adjusting the back-tension band and cleaning the head-switching relay in the few VCR's that have these components. These procedures are explained in Sections 3 and 6.

3.   Adjusting the Back-tension

To play properly, the supply reel must pull back on the tape slightly as the tape plays. This pull is called "back-tension." If there is not enough back-tension, or if there is too much back-tension, the machine will not play with a good picture, and also the heads will get dirty too soon.

On newer VCR's that use electronic means to control the back-tension, you cannot adjust the back-tension yourself without a service manual and test equipment. But many machines use a mechanical tension band around the supply reel to maintain back-tension. It looks as shown in the next illustration. How it works is explained in Section 18 of Chapter 2

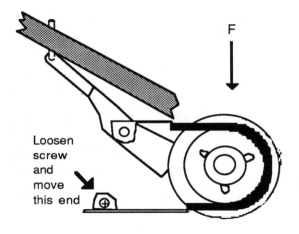

Loosen screw and move this end

If your VCR has this type of back-tension system, you can adjust the back-tension yourself by loosening the screw that holds down the stationary end of the back-tension band, moving the band slightly, and retightening the screw.   On some Fisher-category machines, this screw must be reached from below the reel table, instead of on top.

The problem is: How are you going to know when the back-tension is right? If you are having problems with the heads getting dirty too quickly, this will not show up immediately at the time you make the adjustment.

And if a bad picture gets better after you adjust the back-tension, this does not necessarily mean that the back-tension is now right. For a back-tension misadjustment can sometimes (appear to) compensate in the picture for other troubles elsewhere. And if you misadjust the back-tension in this situation, you may not discover that you have misadjusted it until later, when the heads get dirty. And if you make this

mistake, you have just added a second problem to the first problem, which is still uncorrected.

They sell special gauges that can be used to measure tape back-tension, but they cost as much as a new VCR, so that is not a feasible solution for the do-it-yourselfer.

There is a possible solution to this problem, but it requires that you have access to a TV set that has adjustments on the back that permit you either to shrink the size of the picture, or roll the picture up or down with the vertical hold adjustment, as shown in the next illustration, so that you can see the very bottom edge of the picture in the screen. On a TV set as normally adjusted, there is a little bit of the top and bottom edges of the picture that is off-screen, hidden above and below the part that you can see.

This part that is normally hidden is the part that you need to be able to see to adjust the back-tension. If you cannot find a TV set that will enable you to do this, then you probably will need to take the VCR to a shop. But if you can find a TV set that will enable you to do this, here is how you can use it to adjust the back tension.

If you look carefully and closely at the bottom edge of the picture while a cassette is playing, you should see a tiny bright point, or sparkle (like a single spark from a Fourth-of July sparkler) in the picture a short distance above the place where the picture ceases. It may be stationary, or it may move back and forth horizontally. (This is called the "head-

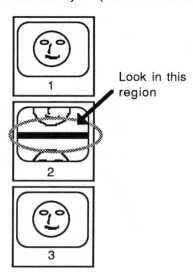

Look in this region

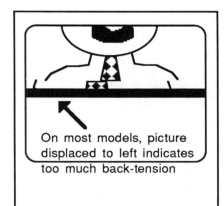

On most models, picture displaced to left indicates too much back-tension

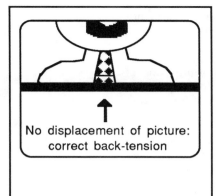

No displacement of picture: correct back-tension

On most models, picture displaced to right indicates too little back-tension

Using the Appearance of the Bottom of the Picture to Adjust the Back Tension to the Correct Amount

switching point.")

Now compare the picture above this point with the picture below this point. If the back-tension is correct, the two parts of the picture will line up, as shown in the center illustration.

But if the back-tension is wrong, the part of the picture below this point will be displaced some distance to the right or left of the corresponding part of the picture above this point. It will look as if someone took scissors, cut across the picture in a horizontal direction, and then slid the bottom piece slightly to the right or left, so that the two parts of the picture no longer line up.

The picture almost never lines up perfectly. If the parts above and below this point are within 1 inch (25 mm) of aligning, then it is close enough. You are probably not going to be able to get it any closer than this, and if the displacement is this small, your problem is something other than the back-tension. But if it is greater than 2 inches or so, then the back-tension needs adjustment.

Experiment by pushing on the arm that is touching the tape, moving it in the direction that tightens the band around the reel, and then moving it in the direction that loosens the band around the reel, to see which direction it needs to go to make the parts of the picture above and below the head switching point line up. If it needs to be looser, then when you loosen and move the other end of the tension band, you will need to move it in the direction that loosens it. And move it the other way if it needs to be tightened. Usually it needs to be loosened, as the felt on the band develops more friction with age and use.

To adjust the back-tension, first take a scribe and scratch a line in the metal or plastic along the side of the other end that you are going to move, so you have a reference point in case you move it too much. Then loosen the screw and move the back tension band a TINY bit, then tighten the screw down and take another look at the picture. If it is still not right, try again. Move the end of the band in the direction that tightens its grip around the supply reel to increase back tension. Move it in the direction that makes the loop of the back-tension band larger to loosen it and reduce the back-tension. Put a drop of fingernail paint on the screw when you are done.

If you cannot get the back-tension to adjust correctly, the felt on the band may be worn out and the whole band need to be replaced. The friction and back-tension usually seem to increase with age as the band wears.

4.   Tracking Problems

When the tracking is off in a VCR, the problem can look like dirty video heads, with white lines across the picture, or a whole screen full of snow. One difference, however, is that if it is a tracking problem, the lines or snow disappear (or at least move or change in some other way), when you turn the operator's tracking control on the top or front panel, while if the problem is dirty heads, it does not change when you turn the tracking control.

What if you have what appears to be a tracking problem, and the picture is better at some positions of the tracking control knob than at other positions, but it seems like you cannot turn the knob quite far enough to completely clear up the picture?

The first thing to do is check whether the problem is in the tape cassette or the VCR. It could be that the cassette tape you are trying to play was recorded on someone's machine on which the tracking was off. Then it might play OK on the VCR on which it was recorded, but not on other VCR's like yours. In such a case, it would be a big mistake to go into yours and change its internal tracking adjustments, because then normal tapes would not play correctly on your machine. So, try a number of other cassettes in your VCR before you conclude that it really has a problem.

If there is really a general problem with the tracking on your machine, it is possible that the physical position of the audio/control head (see Chapter 2 for an explanation of it) has changed slightly, throwing the tracking off.

Or some of the electronic components may have aged or corroded slightly, making the old position of an internal adjusting control no longer correct. On older machines, there is another tracking control adjustment inside the VCR, usually on a bottom circuit board, that does the same sort of thing as the external operator's tracking control, except more so. The easiest and most conservative strategy is to find this internal adjustment, and try turning it to see if that will correct the general tracking problem.

There is no one position in which this second internal tracking control is always located. It is located in different positions on different makes and models, and some machines do not have it at all.

You could spend $20 to $50 and wait three weeks to order up a service book for your particular model, and look in there to try to find out where it is. But chances are that you would still need to spend time looking over the circuit boards for a little hole labeled "tracking," or "SP tracking," "LP tracking," or some such thing. When you look through the little hole, what you see recessed in it is a slot in a little piece of metal that rotates through a total of about three-fourths of a turn to adjust tracking, just like the operator's tracking control on the outside does.

Look all over the bottom circuit board(s) to see whether you can find an adjustment like this. If you can find a sub-area of a circuit board that has "Servo" written on it, what you are looking for is probably in this area. If you do not find any such adjustment, it may be because your VCR does not have this internal adjustment, in which case, you will need to try adjusting the position of the audio/control head, or turn the problem over to a shop.

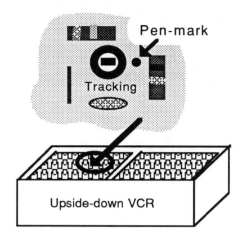

Typical Internal Tracking Control

Before turning this adjustment, with the VCR unplugged, spray some head cleaner into the hole where the adjusting tool goes. Let it dry, then power up and try playing a cassette. The whole problem may have been dirt and corrosion in the rotating control. If it still mistracks, the next step is to try adjusting the tracking control.

Here is a simple way to adjust the internal tracking control without test instruments. With a felt-tip pen, make a mark on the bottom circuit board exactly aligned with one end of the rectangular slot, as shown in the illustration. With a knife, whittle a chisel point on a cotton swab stick or the stick of a burned wooden match, so that it will fit into it. Take care not to cut yourself. Position the VCR so that you can reach the tracking control while it is playing, either by turning it upside down, or by balancing it securely on one end.

With the VCR connected to a TV set, plug in, power up, insert a cassette, and hit "Play." Turn the external, operator's tracking control to its midway position, even if the picture is better when it is set elsewhere. Be careful not to touch anything on the circuit board or bottom of the VCR with your hands, as you insert the chisel point on the little wooden stick into the rectangular hole in the internal tracking control. Pay attention to which direction you turn it, as you try turning it, say, clockwise, while watching to see whether the picture clears up. If so, leave the adjustment at its clearest point.

If it does not clear up, return the adjustment to its original position and then try turning it counterclockwise. Again, if the picture clears, leave the adjustment at the point that gives the clearest picture. If the picture does not clear up, return the control to its original position — tracking is not the problem. If the picture clears partially but not completely, you can leave the adjustment at the best position for a moment, while going back and turning the external tracking control to see whether that will finish clearing the picture. If so, leave it there. If not, return the internal control to its original position.

Changing the Horizontal Position of the Audio/Control Head

The tracking control(s) compensate for variations in the critical distance between the spinning video heads and the stationary audio/control head, which gets crucial timing information from the tape, as explained in Chapter 2.

As a last resort, we can try moving the audio/control head a tiny distance back and forth along the tape path. Since this head also picks up sound, and the procedure for adjusting its horizontal position is explained in Chapter 20 on problems with the sound, please turn to Chapter 20. It explains there how to adjust the horizontal position of the audio/control head. Section 8 of Chapter 20 deals specifically with this adjustment, but you probably will find it helpful to read first the preceding sections of Chapter 20. If that does not solve the problem, you will need to take it to a shop.

5. Capstan Speed Problems

If bands of scratchy lines appear in the picture, and then the picture clears, and then the lines appear again, off-again, on-again, slowly as the machine plays, this indicates that the capstan is not moving the tape at the correct speed.

Begin by determining whether the VCR has a belt-driven capstan or a direct-drive capstan. If you can find your reel table pictured in Appendix VI, look and see whether a capstan belt is shown. If there is no belt, then it is a direct-drive capstan.

If the VCR has a direct-drive capstan, there is little you can do except make certain that the tape path is clean, and that there is no little particle of something stuck somewhere in the tape path slowing down the tape as it moves across it. If that doesn't fix the problem, you will need to take it to a shop.

If the VCR has a belt-driven capstan, then you can try first lubricating the capstan oil bearing, as explained in Section 9 of the Chapter 3 cleaning procedure. If that does not solve the problem, try disassembling and relubricating the capstan, as explained in the section of Chapter 20 entitled, "Fixing Capstan Drag." You can also order up a new capstan belt and try replacing it. Additional information on Panasonic-category capstan belts and the meaning of the markings on them is found in the section of Chapter 14 on belts.

If your machine is an early vintage Panasonic, Sony, or Fisher-category model, there is one more trick you can try. Look around on the bottom circuit board for a little hole that is marked "capstan speed," or "capstan free run," or "C. free run," and through which you can see a little slotted adjuster that can be turned with a tiny screwdriver. This makes fine adjustments in the "free run" speed of the capstan.

Unplug the machine, remove the bottom cover and stand the machine on end with a cassette inside.

Use a flashlight to look through the hole and make a mark with a felt-tip pen at the edge of the hole in line with the end of the slot in the adjusting screw in its present position, so that you can return everything to its original position, if necessary.

Turn the adjustment about one-twelfth of a rotation in one direction, say clockwise (CW), plug back in, power up, and see whether the rapidity of the rate at which the picture changes back and forth between good and bad has increased or decreased. Unplug. If the rapidity decreased, turn the adjustment a tiny bit more in the same direction, power up, and .

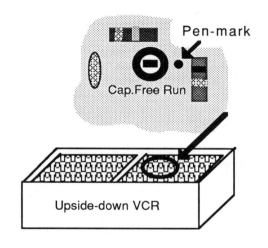

test again.   If the rapidity increased, turn the adjustment in the other direction and retest.

If you can find a point where the picture stays clear, your problem is solved.   But if you can only reduce the rate at which it alternates between good and bad, but cannot make it stay good, then you will need to take it to a shop.

6. Cleaning Head-switching Relays

A few VCR's, like the JVC HR 7650U, the Hitachi model VT9700, and some early RCA's, used a mechanical relay for switching in the picture circuits. When the contacts in this relay become dirty or corroded, they can cause either the top half, or the bottom half, of the picture to go bad, while the other half remains more-or-less OK.

If your machine is one of the rare models that uses a head-switching relay, it will be mounted on a little circuit board to the rear of the cylinder.   On the JVC, it is located inside a little square metal can.   On the Hitachi-category machines, you will see it in a little plastic enclosure about 3/4 inch (20 mm) long mounted on a circuit board in back of the cylinder. Carefully pop off the clear plastic cover with the tip of a scribe, and either spray out with head cleaner — or better, clean with a thin piece of paper moistened with head cleaner or alcohol, as described in Section 13 of Chapter 3.

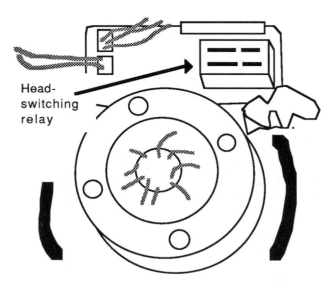

Head-switching relay

Position of Hitachi-category Head-switching Relay

7. Something Making Picture Jump

If the picture periodically suddenly jumps, or goes bad, for a second, then runs clear for a few seconds, then suddenly jumps and goes bad again in a repeating pattern, watch the supply reel carefully as the tape is playing.   If you see the supply reel jump, or fail to run smoothly, in time with the disturbance on the screen, then chances are that whatever is disturbing the smooth rotation of the supply reel is causing your problem.

Many things can cause the supply reel to jump, and you will need to look around to see what is doing it in your machine.   There is no set solution.   Something must be hitting against it, or it must be hitting against

something momentarily.   This is a situation in which the technique explained in Appendix V of tricking the VCR into going through the motions of playing with no cassette inside can be helpful to enable you to see what is going on underneath where the cassette normally rests.

There are two frequent causes of this problem. On some Hitachi-category machines, the back-tension band sometimes slips up too high, out of place, and disturbs the smooth rotation of the reel.   Check for this problem, and push it down if it has come out of place.

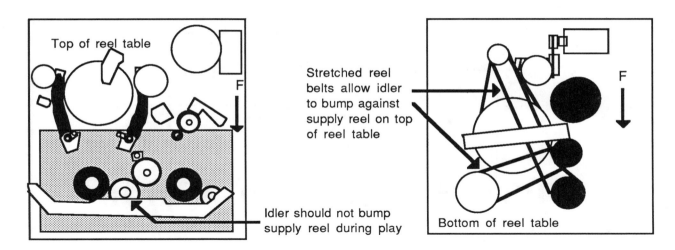

Top of reel table

F

Idler should not bump
supply reel during play

Stretched reel
belts allow idler
to bump against
supply reel on top
of reel table

F

Bottom of reel table

And on early Panasonic-category "tanks," the tension from one of the reel belts underneath the reel table is used to pull an idler back from touching the side of the supply reel on the topside of the reel table. When this belt begins to get stretched and tired, it allows the rotating idler wheel to come over and bump the side of the supply reel from time to time, disturbing the picture.

The solution is to replace the tired reel belt. You might as well replace both of them while you are at it. See Appendix VI for the part numbers.

*"I thought that maybe it was low on oil,
so I added a quart. . . ."*

Chapter 19

Picture with No Color

If a picture comes up when you try to play a tape, but there is no color, just black-and-white, there are several possibilities that you should check out before taking the VCR to a shop.

Adjust Fine Tuning on the TV Set

Try adjusting the fine tuning on the TV set, just as if you were having trouble tuning in a station. It may be that the fine tuning has drifted a little from where it should be.

If this does not restore the color, try changing the output channel selector switch on the VCR to its other position (3 or 4), and changing the TV set channel selector and fine tuning to this new channel number. (See Section 1 of Chapter 1 for how to do this.) The VCR may be weak in one output other channel.

Clean the Video Heads

Sometimes loss of color is caused by dirty video heads.

Follow the procedures in Chapter 3 to clean the video heads, allow to dry, and then test again.

Is VCR's Little TV Station Weak?

As explained in Chapter 1, every VCR has a tiny miniature TV station inside it transmitting picture and sound out on channel 3 or 4, as you choose by setting the output channel selector switch on the machine. Technicians call this little station "the RF modulator."

Sometimes this transmitter gets weak or slightly damaged, resulting in a picture with no color.

You can check the list in Chapter 16 to see whether your machine is one of the models on which a do-it-yourselfer can change out the little station ("RF modulator"). If it is, then turn to Chapter 17 to read how to connect a second borrowed or rented VCR to your VCR in such a way as to check this part to see whether it is bad.

As explained in Chapter 17, you connect the two machines together with stereo cables as if you were going to copy a tape from your machine onto a tape in the borrowed machine. You hook the TV set to the RF output on the second machine.

If you now get a colored picture when you play the same cassette in your machine, then the problem is in this part, the "RF modulator."

On some makes and models, you can replace it fairly easily, as explained in Chapter 16. For other models, you can purchase a separate external substitute RF modulator from Radio Shack, or you can take the machine to a shop to have the part replaced in it.

If you still do not get color, then advanced troubleshooting techniques will be required to fix the problem. The machine will have to go to the shop.

Couldn't fix it?

Professional Guaranteed VCR Repair Available —

What to do:

1. Check whether you already have, or can find, a strong, sturdy, corrugated cardboard box that is at least 8 inches longer, 8 inches wider, and 8 inches deeper than your VCR. (If you have no such box, the U-Haul "extra large" (24" x 24" x 18") box will hold even the largest VCR.) Pack tightly crushed newspaper evenly on all sides for padding (Do **NOT** ship it to the publisher of this book! VCRs are **not serviced** by the publisher.)
2. Dial toll-free 1-800-635-3001. When the operator answers, say **"Guaranteed VCR Service, please."** These words function as an operator-code to assure that your call and questions are correctly routed. You will be told current repair rates, methods of payment, and what to do next if you decide to ship the unit to us for repair.

Rules:

A. The party shipping the VCR pays shipping and insurance. You pay the cost of shipping it to us, and Guaranteed VCR Service pays the cost of shipping the machine back to you.
B. A 12% discount is given to shops or individuals who ship us more than *two separate and distinct* machines to be worked on within six months time. This discount begins on the *third* machine submitted for repair.

VCR Service is provided by Paragon Video Service, Inc. of Tampa, Florida

". . . Yeah, I know it says that it'd be fixed by last week, but our repair guy's out catching some of those killer waves we've been having this season, y'know? . . . " 243

Chapter 20

No Sound or Other Problems with the Sound

Problems with the sound, or "audio problems" as they are called, may require professional service, but often you can easily fix them yourself by doing some of the same simple things that an experienced VCR service technician would try first himself, before getting into the hard stuff.

1. Diagnosis

Let us first identify and categorize the exact nature of the audio problem accurately. The most common problems with sound in a VCR are:

1. Wavering, or slow dragging, sound, where it sounds like a record or tape that is not playing at quite the right speed, or where the speed fluctuates somewhat. In audio work, this is sometimes called "wow" or "flutter." See Section 2.
2. No sound, or almost no sound. The sound is totally absent, or at a very low level, barely audible when playing a tape. Or, the sound may go off and on again and again. See Sections 3 and 4.
3. Too fast, "Donald-duck" sound. See Section 6.
4. Sound that is out of synchronization with the picture. See Section 7.
5. Hum or other distortion accompanying the sound, or on top of the sound. Requires electronic troubleshooting.

With any of the above problems, begin by checking the possibility that the cassette itself, and not the VCR, is at fault. A worn or damaged tape can cause any of the above problems, and it would be silly to put time and effort into trying to fix the VCR if the problem is in the tape and not the VCR.

Do not make the mistake of thinking that because you have used a certain cassette many times before with no problem, therefore the cassette is not at fault. A worn-out, or damaged, tape could be the cause of the problem, and the audio trouble you are hearing could be the first sign of this. So get some other known, good tapes and try playing and recording on them.

If the problem shows up on rented movies played in normal or "SP" speed, rent or borrow some other tapes and see if the sound is also bad on them. If the problem only shows up when you make your own recordings in slow "SLP" or "EP" speed, buy a brand-new medium-priced Fuji or Panasonic tape cassette, and see if the problem happens with it too. If it does not, then you need to discard the cassettes with which you have trouble.

If the problem remains the same on other good test tapes, the next question to ask is whether you are certain that you definitely have a problem. If you know that the sound on your VCR used to be considerably better, then proceed to the section below under the heading that describes your problem. But if the difference is only a tiny little difference that you did not really notice until you listened to sound on your neightbor's brand-new $1000 super hi-fi VCR and you want to make your older model VCR sound like his, then you are probably attempting the impossible. There is a limit to how good you can make the audio sound on older machines with a lot of hours on them.

Still, though, it will not hurt to run the procedures below, and they may still significantly improve the sound on even a worn older VCR. In some cases, the results will be an impressive improvement.

After you have categorized your problem under one of the above descriptions, go directly to the corresponding section under the same heading in this chapter.

2. Wavering or Slow Sound

Although wavering or slow sound can be caused by trouble with a motor that drives the capstan, or in the circuits that are supposed to control the speed of the capstan motor, more often, this problem is caused by some trouble along the tape path that you can fix yourself. One extremely common cause of this problem is a tiny, almost invisible, particle of some material from a tape that has gotten stuck on the surface of the cylinder somewhere the tape runs over, or possibly sticking to one of the rollers, guides, or other heads (erase or audio head) that the tape runs across.

Before cleaning the VCR, however, let's just take a moment to eliminate one other possible cause of the problem. Open the VCR cabinet so that you can see the top of the reel table, referring to Appendix III, if you need help getting the cabinet open.

Find the capstan, and look carefully at the base of it. On some VCR's there is a metal clip ring at the base of the capstan, that is locked into a groove in the capstan shaft so that it cannot slide, but on many VCR's there is a plastic oil-retainer washer at the base of the capstan, which sometimes slides up the capstan shaft enough to touch the base of the tape, putting waves in the edge of the tape and causing trouble with playback performance.

The capstan should look as shown on the left in the drawing on this page. But, if it has ridden up, as shown in the drawing on the right, put two drops of light oil (like sewing machine oil or 3-in-1 oil) on the absorbant pad underneath and push it back down all the way. If you have no oil, skip the step of lubricating it, and just push it back down. If you found this problem, test to see if this fixes the audio trouble.

The next step should be to do a careful, thorough cleaning of the tape path with cotton swabs moistened with alcohol (see Appendix I), or chamois-tipped sticks moistened with freon, in the case of the video heads in the cylinder, following the procedures explained in Chapter 3.

Shine a very good light or strong flashlight on the video head cylinder assembly (see Chapter 2 for location), and inspect its surface very carefully, looking for a tiny, almost invisible black speck or streak stuck somewhere to the surface.

On VHS machines, you can rotate the top half of the cylinder with your finger to inspect all sides of it, but remember also to check the whole portion of the stationary lower cylinder that is above the fine lip or little horizontal cliff that you see machined into the metal. Look also at the shallow, fine parallel lines machined into the sides of the upper cylinder to make certain that no tiny deposit of black stuff is stuck down in one of these fine grooves, where it can still cause a wavering in the sound.

To clean the little video heads that barely stick out below the bottom edge of the top half of the cylinder, you must use a chamois-tipped special VCR cleaning stick, because anything else can cause expensive damage, but if you are very careful, you can use cotton swabs as a last resort to clear deposits from the rest of the cylinder surface, so long as you keep a safe distance from the video heads.

Push oil retainer washer
back down as far as it
will go.

Putting Plastic Oil Retainer Washer Back in Place.

Lubricating the Capstan shaft and Bearing

If some deposits stuck near the heads will not come off with freon, use alcohol on the chamois stick instead. If they use alcohol on the chamois stick to clean the video heads, many VCR technicians will go over it a final time with freon to remove any alcohol deposits, but it is no big problem if you don't. All other parts of the tape path should be cleaned with alcohol and cotton swabs, including the other heads. Inspect each part closely as you clean it. For example, I once found a tiny piece of scotch tape stuck to the audio/control head on a VCR causing sound problems. I have no idea how it got there, but when I removed it, the problem went away.

As you clean the VCR, look to see whether it has a friction-type back-tension band around the supply reel. This looks like a long strip of white felt attached to a thin, springy long metal strip that wraps around the supply reel, like this:

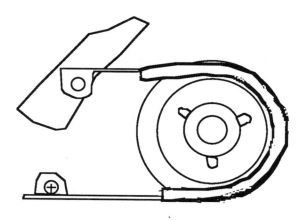

Back Tension Band

The function of this band is to maintain the right amount of tightness, or tension, in the tape as it runs through the mechanism. Some VCR's control the tension on the tape electronically, but if your VCR uses a band like this, be certain to clean the part of the reel that the felt presses against.

Sometimes a worn-out back-tension band, or worn or dirty felt, can also cause the back-tension to fluctuate, making the speed of the tape fluctuate, and cause flutter or waver in the sound, but in order to install a new back tension band, you would need not only the part number for the band on your model VCR, but a means to set the adjustments correctly to yield the right back-tension. Do not remove the back tension band, or loosen the screw(s) holding it in place, unless you have the equipment to set the back tension to the correct value when reinstalling the back-tension band. Section 3 of Chapter 18 explains how to use an older-type television set to adjust the back-tension to the correct amount.

At the base of the capstan on the top side of the reel table, you will see either a plastic oil-retaining washer, as described earlier, or a metal C-ring around the capstan shaft. If it is a metal C-ring, put two drops of light machine oil (like sewing machine or 3-in-1 oil) in the little reservoir right below the fixed C-ring; just let the two drops flow down. Do not add more oil than this, or you will create a new problem. If there is a plastic washer at the base of the capstan, catch the point of your scribe under the edge of the washer, slide it up the capstan shaft a short distance, push down the little piece of sponge around the capstan under the washer, if it lifted up too, put two drops of light machine oil on the sponge, and push the oil-retainer ring back down all the way.

Let the VCR dry from the cleaning for about thirty minutes, and then check the machine with your new tape. If the sound still wavers, clean and check it again, let dry again, and check again.

If the problem still remains, there is one more trick you can try, but it is a little more trouble, and it only solves the problem in a percentage of the cases. First, you must find out whether the capstan in your VCR is direct-drive or belt-driven.

Remove screw

Remove screw

Do not remove this screw. Leave it as is.

Removal of Belt-driven Capstan and Flywheel

If you can find your reel table in Appendix VI, just look and see whether a capstan belt is shown in a picture of the bottom of the reel table. If you cannot find your reel table pictured there, you will need to open up the bottom of the VCR to get at the underside of the reel table (using the procedures in Appendix III) to look. When you have the bottom open, look and see what kind of assembly is on the lower end of the capstan shaft.

If the capstan shaft is attached to a large-diameter flywheel that is driven by a belt coming from elsewhere, go on to the next paragraph. If the capstan is not turned by a belt, put the VCR back together again — you've done all that can be done by a nontechnician for this problem.

If the capstan is turned by a belt around it, slip the belt loose enough to be able to rotate the flywheel with your finger.

If it is stiff, or you feel any resistance at all, the capstan shaft on the other side of the flywheel may be too tight in the bearing that it goes through. This is especially a problem on some VCR's in the Hitachi category, where the shaft sometimes gets so tight in the bearing that it partly wears out the capstan motor from the extra labor required to turn the stiff bearing.

If the shaft is stiff in the bearing, and the oil you put on earlier did not fix the problem, you can remove the flywheel and attached capstan, clean the shaft with alcohol, and reassemble it.

To remove the shaft, first slide up the plastic oil retainer washer and the little piece of sponge (if any) from the top side of the reel table, as described earlier.

Then remove the bracket that extends over the capstan flywheel, being careful to make a drawing or marks on the bracket part with a felt-tip pen to indicate which end goes where, and which screw went in which hole (sometimes they are different lengths or thread sizes).

Clean off the capstan shaft that emerges out from the center of the flywheel piece that you just removed. Use alcohol and a clean cloth.

While you have the belt off the capstan, try wiggling from side to side the pulley and shaft that the other end of the belt goes to (at the capstan motor). If it is fairly loose from side to side, this could be your problem: a worn out bearing in the capstan motor. Unfortunately, there is no cheap solution to this problem: capstan motors are fairly expensive parts, but if the waver in the sound is bad enough to make the VCR unusable, you may decide, as a last resort, to order up one and install it.

When you reassemble the capstan and bearing , there should be a small amount of "end play": that is, the capstan should be able to slide up and down in the bearing a distance equal to about the thickness of a piece of paper or playing card. The little center adjusting screw, which probably has a dab of paint on it, controls this. Do not change it unless it is so tight, or so loose, that it is clearly wrong.

Remember to put the oil retainer rings back on the topside end of the shaft, wipe the oil off the exposed end of the shaft, put the capstan belt back on, and maybe add a drop of oil to the retainer sponge on the topside before you push it down all the way.

If the wavering sound problem still persists, you have a problem beyond the scope of this basic book.

3. No Sound or Faint Sound

Although electronic troubles in the circuitry can cause the problem of weak sound or no sound, this symptom is also often caused by conditions you can find and repair yourself.

One common cause of this problem is some switch on the front panel operating controls having gotten accidentally moved to a wrong position, by a child, a house guest, or a gremlin. Begin by systematically going over each switch, one by one, and checking its position. Do not assume that because you never touch a certain switch, that therefore it is still in the same position it was in when the VCR was working with good sound.

If there are some controls or switches on the front panel that you never touch because you do not know what they are for, and which you have always left in the same position they were in when you bought the VCR, get out your operator's manual (which you kept, right?) and verify that they are still in the correct position.

The next most common cause of no audio, or weak audio, believe it or not, is trouble in the contacts in the "audio out" RCA plug or jack that most people never use. This is the female plug that takes a stereo cable plug (called an "RCA phono plug") for people who want to hook their VCR to their hi-fi or stereo to get better sound quality, or for the use of people who have a "video monitor" that will take in sound directly from this plug.

The audio output jack is usually located right next to a "video out" jack that looks similar, but has a different label, and there may also be "audio in" and "video in" phono plugs in the same neighborhood.

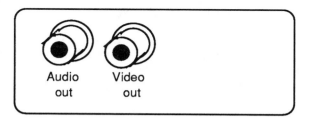

Audio and Video Out Phono Jacks on a VCR

The common problem is that there are mechanical contacts, like a little switch, inside these plugs that carry the sound and video information when nothing is plugged into these jacks, and with the passage of time, these switch contacts corrode a little bit, even (especially) if you never use them, creating a bad electrical connection.

Even when nothing is plugged into the jacks, sound (and video) are passing through the switch contacts on the inside, or if they are corroded, no sound is passing through.

The solution to this problem is easier than understanding it. Get a male R C A patch cord from your stereo, or elsewhere, and with the VCR unplugged, first spray some head cleaner or alcohol into the center hole, then plug and unplug the stereo cable in and out of the hole about twenty times. This will usually clean the switch contacts that this action opens and closes, with the result that the sound will be restored, if this was the problem.

Trouble with an Audio Relay?

A small number of fairly early machines in the Panasonic and Hitachi categories used mechanical relays in the audio circuits. With time, the contacts in these relays become dirty or corroded, resulting in sound that is bad. Often it is intermittent: bad, then good the next time you use the machine, then bad again, and so on. A technician encountering one of these relays should simply replace it, but this involves soldering and ordering up a part, so if your VCR is in the small group having an audio relay, you may just want to clean it to get by (at least until you can get a replacement installed).

On the small family of Hitachi-category machines that used an audio relay, it is located in a metal "can" or shield on the bottom circuit board. When you remove the bottom cover of the cabinet, it will be located in the position shown in the illustration.

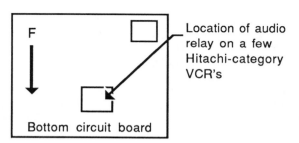

Pry off the cover of the can, remove the plastic cover of the relay, and use the relay-cleaning procedure in Chapter 3. The contacts may be on the side, instead of the top as shown in Chapter 3, but the procedure is the same.

Sound Comes Off and On Again and Again

Open the cassette flap and check for damage on edge of tape. Replace bad tape. If the machine damages another tape in the same way, check to make sure that the oil seal on the capstan is all the way down. (See Chapter 15 for other causes of damage to the tape.)

Cleaning the Mechanical Audio Switches on Early Panasonic-category VCR's

If the VCR is one of the early top-loading Matsushita models that has the metal mechanical switches that you push down along the front edge,

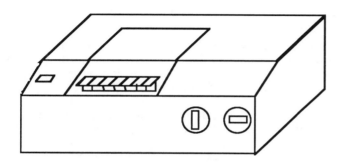

Early Matsushita "Tank" Model

there are some other mechanical switches inside the VCR that also can cause loss of audio.

One of these is the "camera-tuner" switch that you can spray and work a number of times without opening the cabinet.

The other problematic switches require removing the bottom of the cabinet (see Appendix III) and opening up the circuit boards on the bottom of the VCR to get at the switches.

Unplug the machine. When you turn one of these heavy "tanks" upside down, be sure to put a small pillow or pad by the "Off-On" power switch at the front left corner of the VCR to prevent the weight of the VCR from breaking the switch.

After removing the bottom cover, you will see two circuit boards hinged like doors that open along the center line of the VCR after you remove about six or seven screws, which are usually screws whose heads have been given a very slight redish-colored tint.

Depending on the particular model, there are one, two, or three sliding mechanical switches that get dirty or corroded as time passes, and cause loss of audio. Mounted on the circuit board, these switches look as shown in the next illustration. On one side of the switch or switches, stamped into the metal, will be a Panasonic part number like VSS0029, VSS0036, or VSS0037, depending on your model.

These switches are very inexpensive, and really should be unsoldered and replaced if you are having audio trouble, but since having to use a soldering iron would put this procedure beyond the elementary level of this basic book, let's just clean the switches by spraying them out (unless you know how to solder or want to learn).

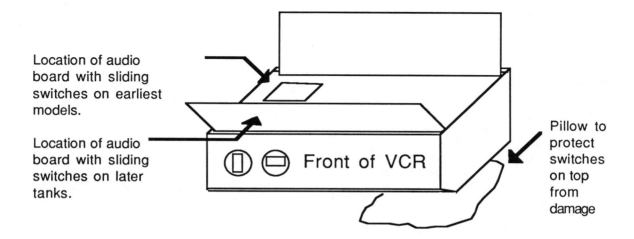

Location of audio board with sliding switches on earliest models.

Location of audio board with sliding switches on later tanks.

Front of VCR

Pillow to protect switches on top from damage

Opening up the Bottom of an Early Matsushita "Tank"

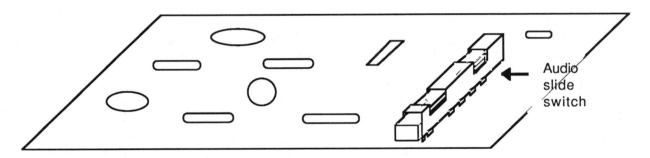

Audio Slide Switch on Early Matsushita "Tank"

As you prepare to spray them, copy down the numbers off the side of the switches on your model, so that if cleaning does not improve the level of the sound on your VCR, it will be easy to order correct new replacement parts if you later learn how to solder and want to replace them, and will not require opening up the machine again to get the part numbers.

Make certain that the machine is unplugged. Clean the switch or switches by tipping up the circuit board on which they are mounted, and spraying lots of freon, or putting some alcohol down inside them using an eyedropper or soda straw. Put the liquid into every place that you can see the plastic inner slider appear through openings cut in the metal, and at the ends, and then push on the plastic part that protrudes from one end.

Experiment a little and you will see that the inner plastic part is attached to a spring inside so that when you push on one of its ends, it springs back when you release it. Put the liquid into the first switch, then work it by pushing on the spring-loaded end about a fifty times. If anything will clean it off, this should do the job. If there is more than one such switch, do this for each.

Reassemble the VCR, noting that when you fold the doors with the circuit boards back down, the switch comes to rest in a position where it gets pushed by an adjoining part of the mechanism.

Power up, put in a cassette, and test to see whether the audio is back. If the audio has been restored, this was the problem. If the audio stays good for a month or two, then goes bad again, you know that cleaning is not enough, and that the switch(es) must be replaced.

4. Checking out the Audio Head on NonHi-Fi VCRs

On most VCR's, except for hi-fi models, the sound is recorded on a narrow strip along the top edge of the tape by a separate, stationary "audio/control" head, as explained in Chapter 2. For short, this is called the "A/C" head. The audio head on a VCR sometimes goes bad due to what is called "random component failure" — which really means that they don't know why they go bad. Fortunately, there is a simple, fairly accurate way to test for this.

Unplug the VCR. Remove the top of the cabinet and lift up, or open, whatever circuit boards and metal pieces prevent you from getting access to the audio head (See Chapter 2 for its location). Get some pieces of wood, plastic, or other nonmetalic material and prop open any circuit board that you needed to open out on its hinges, so that you can get access to the A/C head while you run a tape.

Get a small or narrow screwdriver with an insulated handle. Plug in and power up.

Load a known good prerecorded tape, put the VCR in Play, and verify that a picture from the tape is showing on your TV set, and that the volume on your TV set is set more-or-less to its normal level.

Now you hear no sound, or weak sound, because this is the trouble you're having, right? OK, you will see several little wires that come up to the back of the A/C head. Some of these wires carry sound to or from the head, and others carry timing pulses that are used to control the speed of the tape. Now take your screwdriver in hand, holding only the insulated end, and one after another, go through and touch the blade to the exposed metal terminal where each wire is connected to the A/C head. For one of these terminal points at least, you should hear a loud buzz or hum in the speaker at the TV whenever you touch it.

If you do not hear this noise from the speaker, it means that the VCR evidently has trouble somewhere in the audio circuitry, which, if not fixed by the measures described earlier in this chapter, probably will require professional service from a shop.

If you do hear a loud buzz when you touch one of the exposed wires, however, it means that the rest of your audio circuits are probably OK, and that you have a bad A/C head.

Replacement of the audio/control head is a little more difficult that most of the other, simpler procedures in this book, so you should read through the entire following general procedure, and also look over the layout of connected parts in your VCR carefully, before you make a final decision to call and get the part number and price, and order a replacement audio head. This is usually a quite expensive part. Do not remove the old head until you have an exactly similar replacement head in hand.

5. Procedure for Changing Out an Audio-Control Head

Once you have a replacement A/C head, before you remove the old head, power up, put a known good, but expendable, cassette in the VCR, hit "Play," and adjust the operator's tracking control on the front panel for best picture. Do something to mark or record or note this position of this control, so that you know that you can return it to precisely this position later, if it gets bumped out of position as you are handling the machine.

Eject the tape, unplug, and study carefully again how the A/C head is held in the VCR. Trace the wiring harness attached to the head back down to where it plugs into a circuit board and unplug it, carefully noting exactly how it plugged into the board, and double-checking to make certain that the replacement head you got is exactly similar.

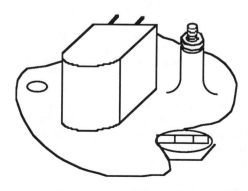

One Type of Audio Head Assembly

There will be a number of adjusting screws in the vicinity of the head that you should not touch or remove, and you need to find the right screw or nut to unloosen to remove the A/C head assembly. On many models, the A/C head is held in place by a single nut that is screwed down on a threaded shaft holding down a cast metal base that the head is attached to, as shown in the illustration.

The position of the nut — that is, exactly how far it is screwed down — determines the vertical distance, or height, of the A/C head above the reel table, which is a very critical adjustment.

If this is wrong when the new head is installed, not only will there still be no sound, but there will be lines and distortion in the picture too — a new problem that did not exist before.

So, as you remove this nut, you need to count exactly how many turns it was threaded down the shaft. Make a tiny mark on the nut with a felt tip pen, or scratch it with the scribe, and if you use a nut-driver to remove it, find a mark on it too, so that you can count carefully how many times you turned the nut before it came off. As you know, it should be turned counter-clockwise (CCW) to unscrew it.

First, though, put some soft tissue paper down all around the assembly to catch the nut in case you fail to grab it when it comes off, and to prevent it from falling down into the mechanism below (which would create a major problem, if you could not get the nut to fall back out of the mechanism by holding the VCR upside down over a clear floor and shaking it until the nut fell out).

Remove the nut (and any other mechanical part that prevents sliding the A/C head up and free). Write down the exact number of turns of the nut, so that you won't forget it. If a spring is attached or pressing on it, get it before it comes all the way loose, and make a sketch of how it was positioned, so that you can reinstall it in the exact same position.

On many models, the A/C head is mounted on a little platform that can swing back and forth horizontally a short distance, with a strong spring that always pushes it CCW until it comes to a stop against a funny looking nut

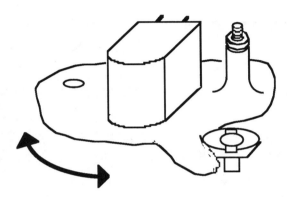

Spring-loaded Swinging Type of A/C Head Assembly

This nut, which looks like an upside-down, cut-off cone, controls the exact lateral, or horizontal position of the head, and you should take care not to turn this nut at the present time. (Because if you do, the picture tracking adjustments will be thrown off.) The A/C head may need to be swung a tiny distance clockwise in order for its base platform to clear the top of the big head of the funny conical nut, so that you can slide it up and free.

To install the new head, put the spring back in the same position it was in with the old head, and slip the new head assembly down over the threaded shaft that had the hex nut on it, making sure that the base platform is to the left of the funny adjusting nut, as it was on the old head. Screw the hex nut down the threaded shaft exactly the same number of turns that you counted when you removed it. Run the wires from the new head along the same path, and plug them into the same plugs that the wires from the old head were plugged into.

Now let's see whether this is your lucky day — will the new head work with no further adjustment?

Prop open any top circuit boards, plug in the VCR, power up, insert the cassette you removed, hit

"Play," and carefully watch as the VCR loads the tape and starts to play to make certain that neither edge of the tape gets caught or bent as it passes through the tape path, especially in the vicinity of the new head.

If there is no picture at all — only a blank, gray screen — put your nut driver or tool back on the same hex nut you were turning before, and carefully count as you gently turn it down clockwise one, then two, then three more turns, stopping after each turn to see if the picture has come up. After each turn, look carefully at the tape as it moves across the A/C head to make certain that neither the top nor the bottom edge is pinching, catching, or bending.

If as you turn the nut, you come to the point where it will turn no more, stop — do not force it! If there still is no picture, unscrew the nut (CCW) back the same number of additional turns that you just made, and then count and watch as you continue to unscrew it back up, one, then two, then three turns until the picture comes up. If there is still no picture, double-check to make certain that you have plugged the wires from the new head in correctly, and that no wire is broken or pinched. You can also try turning the operator's tracking control.

If a picture comes up, but it has lines, bars, or horizontal scratches across it, stop for a minute and watch them to determine whether (1) the picture goes through cycles of clearing up for a moment, then getting scratchy lines, then clearing, then going bad again, etc., repeating alternately in a cycle, or (2) the scratchy lines remain in the picture more or less constantly. If (2) is what you see, then skip the next subsection and go directly to the subsection entitled, "Adjusting the Horizontal Position of the Audio-Control Head."

Adjusting the Vertical Position of the Audio-Control Head

If (1), scratchy lines come and go repeatedly, off and on, then the trouble is probably that the height of the A/C head needs to be changed. (Because the part of the A/C that is supposed to pick up signals used to control the speed of the tape is failing to do so, causing the picture to go in and out of proper synchronization.)

If the tape is catching or bending at the top edge of the A/C head as it runs across the head, then you know that you need to turn the nut CCW to raise the head. If you see it catching at the bottom edge of the head, turn the nut CW to lower the head.

If you see the tape running smoothly everywhere, then we do not yet know whether the head needs to be

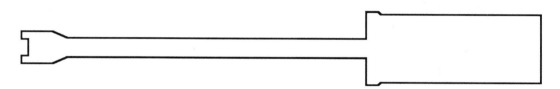

Home-made Tool for Adjusting Position of A/C Head

raised or lowered, so we will have to try one possibility, then the other, until we find the correct vertical position. Try unscrewing the nut you were just screwing, watching the picture and also listening for sound. If raising the nut a few turns does not clear up the problem in the picture, try turning the nut in the other direction for twice the number of turns you just finished making.

If the picture suddenly goes clear all the time, continue slowly turning the nut until you hear sound. Determine by trial and error the range of turns of the nut that gives the best picture and sound, and then put it in the middle of this range.

Adjusting the Horizontal Position of the Audio-Control Head

If (2) the picture has scratchy lines continuously, note again the position of the tracking knob on the front panel, and then try turning it in one direction, then the other, until the picture has the largest possible clear area.

If the scratchy lines now come and go, alternating with a clear picture, then go back and make the adjustments described in the preceding subsection.

But if the picture stays more or less the same, with scratches over part or all of it, but the sound is OK, then you probably need to move the A/C head a very tiny bit in the horizontal direction. This can be confirmed on VCR's that have the spring-driven swinging head set-up by using your finger to swing the head a very, very tiny distance in the clockwise direction, away from the funny conical adjusting nut described earlier.

If as you rotate or swing the head, the picture gets much clearer at one point, then this indicates that its side-to-side, horizontal position needs to be adjusted by turning the funny conical adjusting nut in one direction or the other.

If your VCR has the funny conical nut type of system for adjusting the horizontal position of the A/C head, and your flat-bladed screwdriver does not

fit into the grooves on the top of the funny nut because it hits on the high part of the threaded shaft coming up through the center, then turn to Section 13 of Appendix I to learn how to make a special tool for adjusting this nut by taking a file to a cheap screwdriver to make the tip look like the illustration.

Experienced VCR technicians adjust this nut while the VCR is playing a tape, but your first time, you should stop the VCR, unplug, turn the funny nut a half turn, then power up, hit "Play," observe the results, and continue to repeat this process until the picture clears. Before turning the funny nut, be sure to put the tracking knob back in the exact same position in which it played the same known good cassette when you tested it much earlier before removing the original A/C head.

In theory, you should be able to correct the tracking problem by moving the A/C head in either direction horizontally a small distance, the only difference being that if you move it in the same direction that the tape moves, the sound will be heard a tiny bit later in time than the picture (the lips move, and then a fraction of a second later, the words are heard), while if you move it in the opposite direction, the sound will start a little bit earlier.

However, you probably will be moving the A/C head such a short distance that the difference will be so small that you cannot even detect it. Therefore, in practice, you can begin by moving the head in whichever direction it moves most easily. In VCR's with the funny conical adjusting nut, usually it is easiest to begin by turning this nut CCW.

Double-check to make certain that you have put the tracking knob back into the position in which you noted that it gave a good picture with the old head. Move the A/C head a little bit by turning the adjusting screw a half-turn, or pushing it with a rotary prying tool, check the picture, move it another tiny bit, check again, etc.

If your VCR has the conical-nut type of adjustment system, you might need to know later how many rotations you turned it, so keep count and write it down.

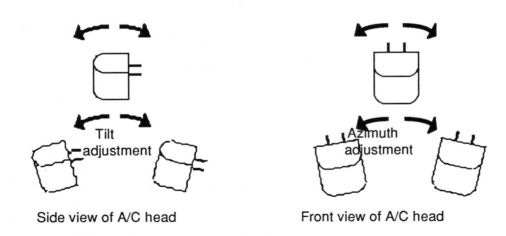

Side view of A/C head   Front view of A/C head

The Meaning of "Tilt" and "Azimuth" Adjustments

If the picture has not mostly cleared up after a few tiny moves of the screw in one direction, put it back into the original horizontal position from which you started, and then try moving a little at a time in the other direction. When the picture gets as clear of scratchy bars as possible, try readjusting the front-panel tracking control knob.

If a small adjustment of the tracking control knob finishes clearing up the picture, or if the position of the knob that clears the picture is close to the center "detente" position, then you can figure that the adjustment is "close enough" and quit.

But if only a big adjustment of the front-panel tracking knob makes the picture clear, and it is clear only when the knob is close to one extreme, then you will need to go back to adjusting the horizontal position some more until you get it to a position where the picture is good when the tracking knob is close to its original position.

Adjusting the "Tilt" and "Azimuth" of the Audio/Control Head

If the picture is now OK but the sound is a little muffled or weak, you can try improving it by adjusting now the "tilt" and "azimuth" of the A/C head. Changing the angle of the head from front to back is adjusting the "tilt," while changing the angle of the head from side-to-side (as viewed from the front of the head) changes its "azimuth."

The amount that the tilt or azimuth of the head is

changed during alignment is shown greatly exaggerated in the drawing for purposes of illustration. Actually, the amount of adjustment is so small that it is almost invisible to the eye.

There will be one adjusting screw for the tilt adjustment, and another adjusting screw for the azimuth adjustment.

It is best to make these adjustments very slowly, counting half-turns as you do so. This way, you can always return the adjusting screw to its original position, if necessary. Listen carefully to the sound over the speaker for a little while after each half-turn.

If it is not improving, return the screw to its original position, and then try making and counting half-turns in the opposite direction. If it does not improve in that direction either, return the screw to its original position, and try the other kind of adjustment.

Be careful not to turn any screw to the point where it is in danger of coming out, because putting it back could be a LOT of trouble (assuming that you can even manage to find the screw if it falls out).

If the picture starts to go bad when you make any adjustment, turn the screw the other way to put it back where it was.

Leave the adjustment screws in the position that gives the best sound and picture. Put a drop of nail polish or lacquer on each screw-head after you have finished.

6. Too Fast "Donald-Duck" Sound

This symptom indicates that the tape is moving too fast because of some trouble in the circuitry that is supposed to control tape speed. It is almost always due to some problem in the electronic circuits beyond the scope of this book.

But before taking the VCR to the shop, you can try checking the top-most part of the audio/control head to see whether there might be something sticking to it, so as to interfere with the ability of the head to pick up the control pulses from their track along the top edge of the tape.

TEAC AND FUNAI: Technicians see notes on Funai 1 for the solution to a common cause of this problem on these models.

7. Sound that is Out of Synchronization with the Picture

If there is only one tape on which you have this problem, and all other tapes have properly synchronized sound, probably the trouble is with the tape and not your VCR: the tape may have been made originally on a misadjusted machine. But if you hear the words before you see the speaker's lips move on most tapes, including rental tapes, then the audio/control (A/C) head needs to be moved slightly further along the tape in the direction of tape movement toward the take-up reel — that is, moving it a distance in the same direction that the tape moves.

On the other hand, if the sound comes too soon before the action, you can correct this problem by moving the A/C head a tiny distance in the opposite direction than the tape moves.

There are two things that you should know before changing the horizontal position of the A/C head to correct this problem.

First, if your present problem is that the sound is OK on tapes that you have recorded yourself on the same VCR, but the sound is out-of-synchronization on tapes recorded on other machines and rental tapes, then after you move the A/C head so that others' tapes play back in correct sync on your VCR, all your collection of old tapes that you recorded on this VCR will now play back out-of-sync.

This is no problem if you only use your VCR to watch rented movies, and you always erase and reuse your self-recorded tapes, but if you have built up a large, priceless collection of tapes of old TV and film classics that you have recorded yourself on this VCR

as it is now, then you may wish to leave the machine as it is now for playing your old tapes, and dub copies of them using a second correctly adjusted VCR to record the copy, before changing your machine — or you may even want to get a second VCR and leave your present VCR as it is. See Chapter 1 for more on dubbing.

Secondly, you should know that the same physical head that picks up the sound also controls important aspects of the picture. You know how a picture goes bad when you turn the front-panel tracking knob through its full range of motion, while viewing a tape as it plays, right? Usually, the picture is bad at some positions, or amounts of rotation, of the knob, and good at others. (If you have never tried this experiment, you really should do so at your first opportunity.)

Well, the same thing will happen to the picture as you change the horizontal position of the A/C head: at some positions, the picture will look as though the tracking is misadjusted, and then at other positions, as you continue to move the head horizontally, it will go good again, then bad again, then good again, and so on. The picture will alternate good and bad as you move the head along horizontally. It may take several attempts and several trials before you succeed in finding a new position where the picture is good and the sound is in sync with the picture.

If, after reading the preceding two paragraphs, you feel that you still want to attempt to correct the problem of sound being out-of-sync on your VCR, begin by getting a tape of the sort that you normally watch on which the problem is especially easy to see.

While playing this tape, adjust the front-panel tracking control for best picture, and note its precise position. Watch the screen while listening to the sound for a while to determine for sure whether the sound is coming too early or too late.

Note and keep track of where the A/C head was before you loosen or move it. If the head is moved by turning an adjusting nut, keep a careful count of how many turns and in what direction you move it each time, so that you always can put it back where it started from, if necessary. If the A/C head is moved by being pried with a special tool after loosening the hold-down screw(s), with the machine unplugged, first use the sharp tip of a scribe or ice-pick to scratch an outline of the position of the movable base in the metal of the reel table under the base.

After observing the problem, figure out which direction you will need to move the A/C head before you start to move it. The rule is that if the sound is coming early, move the head a little in the same direction that the tape moves when playing. And if the sound is late, coming after the action or lip motion, move the head up the tape, opposite to the direction the tape moves in while playing.

Turn to the earlier section of this chapter on adjusting the horizontal position of the A/C head and follow the instructions there. Use trial and error, moving the head, then putting the VCR in Play and observing picture and sound, then repeating this process until the desired result is obtained. Expect to have the picture go bad and then good again, as you move the head along.

"... Hey, Dad!. . The manual says that VHS
and Beta aren't the same after all!. . ."

Chapter 21

VCR Does Not Record

This chapter covers situations where the VCR will not record WHEN IT HAS PASSED ALL THE PRECEDING TESTS IN THE CHAPTER 4 UNIVERSAL CHECK-OUT PROCEDURE.

If you have not run all the Chapter 4 tests yet, you must do so before turning to this chapter, because some of the possible causes of failure to record (such as a damaged "antenna switcher," to mention only one example) are detected in earlier tests and covered not here, but in earlier chapters. The procedures in this chapter assume that the failure-to-record problem is not due to any of the possible causes that would have been detected in earlier tests and covered in earlier chapters.

It also is assumed that on the cassette on which you cannot record, the "record safety tab" is not broken off. If you are uncertain what this tab is and how to check it, see the subsection entitled "Recording a Program" in Section 2 of Chapter 1.

On most VCR's, to put the machine into record, it is necessary to push two buttons, "Play" and "Record," at the same time. If there is any possiblity that you are not pushing the buttons in the correct way to put it into the recording mode, get an owner's manual and double-check how to put it into record.

The next question is this: When you try to make a recording, does the VCR record neither the sound nor the picture, or does it record one without the other (like sound, but no picture)?

If the VCR records neither sound nor picture, go directly to the section entitled "Record Safety Switch Problems."

If the machine records sound but no picture, the most likely cause of the failure is a problem either with the heads, or in the electronics. Cleaning the heads several times, using the procedure in Chapter 3, and unplugging and replugging the multi-wire plug connecting the heads to the rest of the circuitry is about all you can do. If that does not fix the problem, the machine will need to go to the shop.

1. Record Safety Switch Problems

If neither sound nor picture is being recorded, the indications are that the VCR is never really going into the record mode at all (even if it says "Record" on the front panel display).

This could be due to (1) the little computer never receiving the instruction to go into the record mode when you press the buttons, or (2) some trouble in the little computer or its support circuitry, or (3) a problem with the record safety switch. The third possible cause is the only one that you can repair yourself without advanced electronic trouble-shooting techniques.

Inside the VCR attached to a switch, there is a little "feeler" that presses against the part of the cassette box that has the record safety tab.

Inside the switch there are contacts that are usually not making contact when no cassette is in the VCR, but when a cassette box with an intact record safety tab is loaded into the VCR, the part of the cassette box where the tab is located pushes against this feeler and causes the switch to "close," making an electrical connection.   A signal is then sent via some wires to the VCR's little brain telling it that it is OK to record on this cassette if you tell it to do so.

If the record safety tab is broken off, this feeler does not get pushed in such a way as to close the contacts when you insert the cassette.   The contacts remain open, and the VCR's little computer interprets this as a signal that it should not record on the cassette, no matter what the operator may tell it to do.

Question: What is going to happen if the feeler gets bent, or the contacts get corroded, so that electrical contact is not made inside the switch when a cassette with intact safety tab is inserted? Answer: The little computer does not get a signal that it is OK to record on the tape, so it will not record on it, no matter what you tell it to do.   Possibly this is your problem.

This would be a simple problem which would have a simple solution (fix or replace the record safety switch) except for one thing: On VHS machines, this switch is located at the right front edge of the reel table,

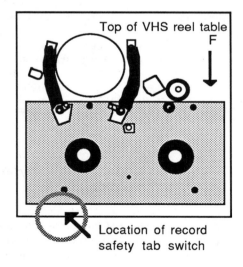

Top of VHS reel table

F

Location of record safety tab switch

in a position where it is difficult to get at.   On some models, you can simply unscrew the switch from the cassette front-loader or from the edge of the reel table, and replace it, but on other models more disassembly is required, and sometimes even the whole reel table must be removed from the VCR to gain access to the record safety switch.

We would not want to put a lot of time and trouble into disassembling the VCR to get at the record safety switch and then find out that the problem was deeper in the electronics and not in the switch at all.   So, let's try a couple of short-cut tests first.

Warped Feeler

There usually is a plastic piece that is pushed upon by the plastic part of the cassette box where the record safety tab is located.   Sometimes this piece gets worn down with use, or it warps in time and bends, with the result that it no longer pushes the switch enough to make contact when a cassette with an intact record safety tab is inserted.   Consequently, the VCR "thinks" that it has a protected cassette inside, and will not record on it.

If the machine is a VHS, try this.   Get some thin cardboard, like the cover of a matchbook, and with scissors cut out a little rectangle about the size of the record safety tab.   Put this on top of the intact record safety tab on a cassette box that has not had the tab broken off, and put some clear tape over it to hold it on.   Now when the cassette box is inserted, it will push the feeler a little bit more than before.   Insert the cassette and test to see whether the machine will now record on it.

If it still will not record, eject, and attach a second thin piece of cardboard over the first, raising that part of the cassette even more, and test again.

If the VCR still will not record, go to the section below entitled, "Is there a Problem with the Switch Contacts?"

If it now does record, then we know that the problem definitely is in the switch.   You have two choices.   You can just attach cardboard squares to all the cassettes that you want to use to make recordings and live with the warped switch, or you can order up a new part and replace it.   Look at the switch in your VCR to see whether soldering will be required to replace it.   (How to solder is explained in Section 20 of Appendix I.)

If the part number is not given in Appendix VI, you will need to call the number listed in Appendix II to get it.   You can call either the company whose name is on your machine, or the company under whose name your reel table is pictured.

Is there a Problem with the Switch Contacts?

If the feeler is closing the switch contacts, but the contacts are dirty or corroded so that they do not make good electrical connections, the message that the little computer receives will be the same as if the contacts were never closed at all. It will think that the cassette has a broken safety tab and that it should not record on it.

One way to test for this problem, obviously, is to install a new switch and see if that solves the problem. The trouble with this way of testing is that if bad switch contacts turn out not to be the problem, your money and time would be wasted.

Another way to test this possibility is to use a clip-lead to connect together directly the two wires attached to the record safety switch. Now the little computer gets the message that it is OK to record on the cassette. See the next section for how to do this.

If bad switch contacts are the problem, the machine should now record OK when the switch wires or terminals are connected together with a clip lead.

If the machine still does not record, then the problem is deeper in the electronics, and can only be fixed by a shop.

Attaching a Clip-lead to the Switch

If you can reach the place where the wires attach to the switch, either from above or below the reel table, then you may be able to attach the two alligator clips directly to metal terminals where the wires are attached to the switch. Just take a clip-lead and connect one end of it to one terminal, and the other end of it to the other terminal on the switch, in effect connecting the two terminals together.

If this cannot be done because the terminals are too difficult to reach without extensive disassembly, or for any other reason, then if you can find the wires running from the switch to the circuit board that it attaches to, you can use the following trick.

Unplug the machine. Get two straight pins of the type used in sewing. Through each of the two wires, push a straight pin all the way, so that its little head comes right up against the insulation on the wire. If you pushed the pin through the middle of the side of the wire, it should be making contact with the metal wire inside the insulation. Now connect the alligator clips of a clip-lead to the two pins, connecting them together electrically.

Make certain that the metal of the clip leads, and the pins, cannot touch any metal or circuit parts before you power up again and test to see whether the machine now will record. You can run the machine standing on one end, with the bottom cover removed, if this is more convenient.

If the machine still will not record, then the trouble is in the electronics, and it will need to go to a shop.

But if the machine now records, then you know that the trouble is in the switch contacts. You have three options:

(1) You can just leave the clip lead in place so that you can now make recordings. This is fine, but remember that the record safety switch has now been disabled, so that if you have any treasured, priceless recordings with the safety tab broken off, you had better hide them from the kids and the gremlins, because they are no longer protected by the safety tab. If someone inserts one in the machine and pushes "Record" (or "Record" and "Play" together), the machine will record on them, and erase the previous recording.

(2) you can disassemble the VCR to the switch, clean the switch contacts with the same procedure used in Chapter 3 to clean relay switch contacts, and test to see whether this is enough.

(3) You can order up a new switch and replace the bad one. But first, look at the switch in your VCR to see whether soldering will be required to replace it. (How to solder is explained in Section 20 of Appendix I.)

If the part number is not given in Appendix VI, you will need to call the telephone number listed in Appendix II to get it. You can call either the company whose name is on your machine, or the company under whose name your reel table is pictured.

Appendix I

The Use of Tools and Materials for Working on VCR's

1. Preliminaries

This appendix describes all of the tools used in the various procedures throughout this book, and tells where you can obtain them. Some of the tools in this list, such as Phillips screwdrivers, are required for almost every procedure in this book. Other tools, such as "hex keys" or a soldering iron, are seldom needed, but are described here for the sake of completeness.

Most repairs in this book require only a few of the tools and materials described in the following list, so don't worry, you don't need to buy them all (unless you are going into the VCR repair business). You will only need a few of the following, but a complete list is provided here as a reference:

> Containers (like paper cups), pencil & paper
> Number 2 Cross-slot or Phillips screwdriver
> Number 1 Cross-slot or Phillips screwdriver
> Number 0 Cross-slot or Phillips screwdriver
> Cotton swabs (NOT to be used on video heads!)
> Rubbing alcohol (90% or stronger)
> Special chamois-tipped head-cleaning sticks
> Freon video head-cleaner
> Magnetic Phillips screwdriver
> Parts "picker-upper" tool
> Scribe
> Set of small flat-bladed ("jewelers") screw drivers
> Small needle-nosed pliers
> Hex keys
> Set of nut-drivers or small wrenches
> Notched screwdrivers (homemade special positioning tools)
> Spout can of light ("sewing machine") oil
> Light (molytone) grease
> Heavy grease

> Epoxy (steel)
> Lock paint or small bottle of fingernail paint
> Dummy cassette (homemade)
> Test tapes (can use discarded advertisement preview trailers sent to video rental stores for promotional purposes)
> Nine-volt battery and connectors
> Soldering iron, stand, sponge, solder, and solder wick

The following pages describe each of these tools and materials, and tell where you can order them, if you cannot find them where you are locally. If you live near a city, you probably can purchase most of these tools locally, if you need them, but names of sources from whom you can have any of these tools shipped to you, if that is more convenient, are also given in the following sections.

In a few cases, the use of some of these tools for special techniques, like C-ring removal and replacement, and desoldering and resoldering, is also explained for readers not familiar with these techniques.

In all disassemblies, many different kinds of screws and other small parts will be removed one after another. It is vital that these go back into the exact same places in the reverse order of disassembly when you reassemble. At the time you remove them, it seems easy to just set the screws, etc., down on the table in a pile. This is easy to do, but a big mistake, because when you go to reassemble, you will find that you cannot remember which of the different kinds of screws went where. And if you put the wrong screw in the wrong place, you can do very expensive damage to the VCR.

To avoid this problem, get some sort of containers that you can number and put in order. I like to use small paper cups, because I can put each type of screw into a cup as I disassemble, and then nest another paper cup in on top of them, and put the next screws from the next part of the disassembly into the nested cup. This preserves the order and prevents spilling them. Other technicians like to use a box with a lot of compartments, like a fishing tackle box, or just an old egg carton, numbering the compartments. Whichever system you use, draw a sketch and make a note of where the screws in each container compartment came from at the time you remove them. Later you will be glad that you did this.

2. Cross-slot screwdrivers

All VCR screws have normal threading — that is, you turn them counterclockwise to remove them,

Turn screws counterclockwise
(CCW) to remove them

Turn screws clockwise (CW)
to screw them back in

and clockwise to screw them back in again.

With few exceptions, most screws in VCR's, including cabinet screws, have Japanese-style Phillips head slots, called "cross-slot screws," and require

appropriate screwdrivers for removal and replacement. These resemble the familiar Phillips screws with a cross or "+" shaped slot rather than just a single, straight-line slot, in the head.

Superficially, the Japanese cross-slot screws look just like regular Phillips-head screws, and usually you can use regular Phillips-head screwdrivers from your hardware store to deal adequately with most of these screws. But there is a tiny, microscopic difference in the shape of the cross in the head of the cross-slot screw, and screwdrivers made especially for these VCR screws grip better than do Phillips screwdrivers from the local hardware store.

Most of the time it does not matter whether you use an ordinary Phillips screwdriver, or a special cross-slot screwdriver, but every once in a while in a VCR, you encounter a screw that is fastened down extra-tight, and if you are not paying careful attention to what you are doing, you may tear some of the metal around the cross-slot in an unsuccessful effort to remove it.

If this happens, it can mean big trouble, because now you have got a screw that was difficult to remove to start with (because it was fastened in so tightly), which has been made even more difficult to remove because now the screwdriver slips rather than catches in the slot, and the more you try to turn it, the worse it gets. Less than 1% of VCR screws are in this category, but when you encounter one of them, having the special cross-slot screwdriver rather than the local Phillips makes all the difference!

If you do not want to invest in special cross-slot screwdrivers, you probably can always avoid ever being caught in a bad situation by concentrating very carefully on what is happening each time you remove, for the first time, a screw that is screwed down into metal in a VCR. The primary villains, for some reason, seem to be screws fastening cassette basket assemblies to the metal reel table.

These drawings are about half-size. Go by the number printed into the plastic handle of the screwdriver, and NOT by the overall size of the screwdriver.

Size #0

Size #1

Size #2

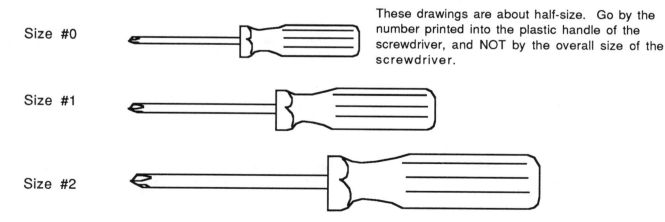

Three sizes of VCR Screwdrivers

When removing such screws, press the screwdriver down into the slot of the screw as hard as you can — I mean, really lean down hard on it! — and watch carefully what is happening to the head of the screw as you SLOWLY start to turn the screwdriver in a counter-clockwise direction. If you see or feel the screwdriver start to rise up out of the slot, stop and don't turn it any more, because if you turn the screwdriver when the screw is not coming with it, you will damage the slot in the head, making it extremely difficult to get out.

You can try again leaning on it even harder, but again, do not continue to turn the screwdriver if the screw is not moving with it. If you run into this situation, you've encountered one of those rare screws that really requires a special cross-slot screwdriver.

The special screwdrivers are available as tools with RCA parts numbers. The RCA part number for the special #1 size is 144398, and for the #2 size is 144399. You can order them from a supplier selling RCA parts and tools, such as Pan Son (find address in Appendix II). They currently retail for around $6 each (but they are worth it).

The only repair procedure in this book that requires a #0 Phillips screwdriver is replacing the cassette basket assembly in the Panasonic-category two-piece portable VCR's described in Section 8 - 10 of Chapter 7. A #0 Phillips screwdriver comes in a set of six precision miniature screwdrivers sold as Number 22-140 by MCM (address in Appendix I) for about $2. The Xcelite XST-100 miniature Phillips works too, and it has a magnetic tip; it can be ordered from Pan Son.

3. Cotton Swabs

Wooden sticks with cotton balls on the end are good for cleaning everything except video heads. They must never be used on video heads, because they can catch and damage them, but for cleaning the other heads, idlers, belts, and other parts of the mechanism, they're great. You can use

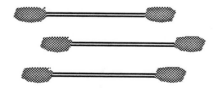

Cotton Swab Sticks

Q-tips from your local store, or order cotton swabs with longer sticks in packages of 100 from MCM as #21-060.

4. Alcohol

The best kind of alcohol to use for cleaning off the tape path, rubber parts, and general degreasing is rubbing alcohol in a solution stronger than 90% (that is, more than 180 proof but don't drink it) from your drug store, if you can get it.

Rubbing Alcohol Used for Cleaning

The more common 70% solution is a lot harder to clean with, because it does not cut grease nearly as well. Methol alcohol would also cut the grease, but it is poisonous and probably should be avoided for this close work because of its fumes.

5. Video Head Cleaning Sticks

As explained in Chapter 2, "How a VCR Works," the video heads perform the crucial function of recording the TV picture onto the tape and then reading it back off the tape in playback. These parts are extremely tiny, delicate, fragile, and unrepairable if bent, broken, or knocked out of alignment. And they are quite expensive to replace.

Only one tool ever should be used to clean video heads: the special little plastic sticks tipped with doeskin or chamois made especially for that purpose. NEVER use Q-tips or cotton swabs to attempt to clean video heads, because a fiber or strand of cotton may catch the edge of the tiny head and bend it out of alignment far enough to render it functionally useless forever after.

Swabs tipped with foam are also sold as video head cleaners, but their use is controversial and risky. The foam is so soft that it does not apply enough friction often to clean a dirty head, and the temptation to press down harder on them is dangerous because they usually have a cotton swab under the outside layer of foam, which again could catch and damage a head.

Dry-type head cleaning cassettes only work in some cases, usually only when the heads are not very dirty, and the wet-type head-cleaning cassettes are highly dangerous to use, because they also can break a video head or pull it out of alignment.

So, to repeat myself, the only safe and generally effective tools to use to clean video heads are the special chamois-tipped plastic sticks designed

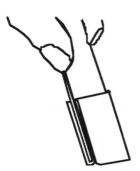

Special Chamois-tipped Plastic Stick for Cleaning Video Heads

for that purpose. They are available as Panasonic part #VFK27, or from RCA in packages of three as part #144589 (RCA). You can get them from Electrodynamics in packages of five as #07-1980. A version with a 45-degree bend in the plastic stick is available from MCM as #21-710. The best brand to buy is Chemtronics.

6. Head Cleaning Liquid

In their service literature, Sony and Hitachi both list alcohol as acceptable for cleaning video heads. But most technicians prefer to use Freon to clean heads, because it dries with less residue.

Freon is even more useful when purchased in spray cans with a long, thin spray tube included for optional use, because then you can also use it

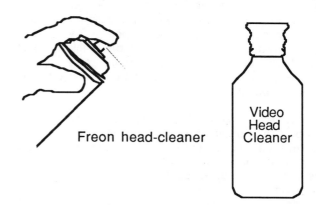

Freon head-cleaner   Video
                             Head
                             Cleaner

Freon-type Head Cleaning Liquids

to spray out heads, switch contacts, and other areas that you cannot get into without disassembly in any other way. It is available as a 20 oz. spray can from Electrodynamics as part #03-6020, and from Pan Son in a 16 oz. can as #1230009. It also can be bought in nonpressurized cans from Pan Son as 1230007.

7. Magnetic Screwdriver or a Parts "Picker-upper"

A few crucial screws in VCR's are located in places where it is quite difficult to reach them with your fingers because of the surrounding mechanism, but you can reach them with a screwdriver or other long, thin tool. In such cases, a special tool is needed to lift the screw out as it is unscrewed, and to start the screw turning in its threads in the hole when you reinstall it. There are two different tools that can be used for this purpose, a magnetic screwdriver, and a little mechanical parts "picker-upper."

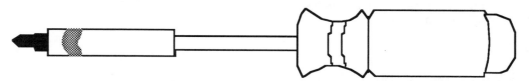

Magnetic Screwdriver

Parts "Picker-Upper"

As its name implies, a magnetic screwdriver has a tip that is made from a magnet, so that screws made out of any type of metal attracted by a magnet will stick to the tip of the screwdriver, allowing you to pull out the screw along with the screwdriver when working in close quarters. And when you wish to put the screw back in, you can simply stick it to the magnetized tip of the screwdriver, and use the screwdriver to reach in and start the screw turning in the proper place.

Magnetic Phillips screwdrivers often do not have as good a tip as regular unmagnetized Phillips screwdrivers, so it is a good idea to first start unscrewing the screw with a regular screwdriver, especially if it is at all tight, and then as it is turning freely and about to come out, switch to the magnetic screwdriver to finish unscrewing and removing it. And when reinstalling the screw, once you have used the magnetic screwdriver to get it started, you can switch to a regular cross-slot or Phillips screwdriver with a better-shaped nonmagnetic tip to finish running the screw down tightly.

A magnetic screwdriver with several interchangeable tips is available from MCM (address in Appendix II) as part #22-645, or if you want the fancier ratcheting magnetic screwdriver, as part #22-640.

There are two situations in which a magnetic screwdriver fails to work. It does not work if the screw does not have enough of the right kind of metal in it to be attracted by a magnet (and some screws in VCR's are like this) and it does not work if the space through which you need to reach is so narrow that the wider shaft of a magnetic screwdriver like the MCM models cannot fit through the space.

In either of these dilemmas, a little mechanical "parts picker-upper" will do the job. This tool has three thin little fingers, or grippers, that come out of the end of a thin shaft when you press on the other end, gripping the screw (or any other little part) and holding it fast until you press on the end again. The larger versions of this tool used by mechanics are too big for VCR work, but MCM has a miniature version that sells for about $2 as part #22-125.

8. Scribe

An astonishingly useful tool in working on VCR's is a scribe of the type used by automotive mechanics to scratch marks on metal pieces. It is a long, thin tool made from extremely hard metal, with a sharp straight point at one end, and usually a hook or L-bend at the other end.

In VCR work, it is also used to scratch lines in metal and plastic to mark the position of critically aligned parts before removing them, and also for getting under and prying loose tiny parts like oil seals. Technicians also use them for pulling wires through narrow places. You should be able to buy a scribe at your local outlet for auto parts and tools.

9. Miniature Flat Blade Screwdrivers

A set of miniature flat blade screwdrivers will be useful for removing the two types of little metal C-rings or locking rings often used in VCR's to attach idlers and gears to the axles or shafts on which they

Two Common Types of Locking Metalic Rings

turn. There are special tools (part #VFK0335 (Pan)) that look somewhat like pliers made especially for removing and reinstalling "grip rings" that look like the one on the right, but a small flat-bladed screwdriver is a suitable substitute for the grip ring type on the right, and needed anyway for the C-ring type on the left.

These rings are removed with a small flat-bladed screwdriver as follows. With the C-ring, the screwdriver is inserted into either of the little spaces

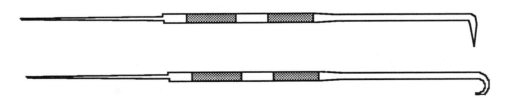

Scribes

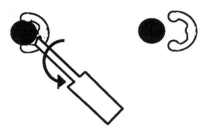

To remove C-ring, insert blade of small screwdriver in space between C-ring and shaft, cover with cloth (not shown) and twist.

Using a Small Flat Blade Screwdriver to Remove a Metal C-Ring

between the C-ring and the metal axle.

When removing the other type of fastener, grip rings or "compression rings", it is a wise precaution to first stuff tissue paper into any nearby holes or opeings in the mechanism, because these fasteners have a tendency to escape, once they are freed and it takes a lot of time to find a replacement once lost. You cannot expect to find them at a hardware store.

To remove compression-type fasteners, like the one on the right in the earlier illustration, choose the blade size that just fits into the widest part of the gap between its two prongs or extending arms. insert it, and put a cloth

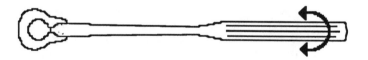

Using a Small Flat Blade Screwdriver to Remove a Metal Compression Fastener

down close over the fastener and screwdriver (to prevent the fastener from flying off and getting lost when it springs loose). With precautions in place,

twist the screwdriver slightly and slide the fastener off the axle.

The C-ring type on the left usually clips into a notch cut in the metal of the axle, but the grip-ring type on the right are usually held on by friction alone, with no notch in the shaft onto which they are fastened.

To replace a grip or compression ring, again use a small screwdriver to spread the ring slightly and slide it down over the shaft onto which it goes. To replace the C-ring type on the left, use needle-nosed pliers, as described in the next section.

Various kinds of sets of jeweler's screwdrivers are widely available. You can get a set of five miniature straightblade screwdrivers from MCM as #22-130,

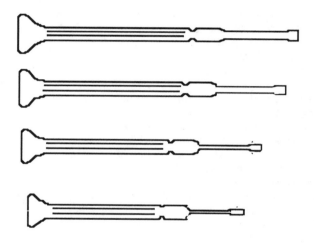

Set of Jeweler's Screwdrivers

or a six-piece set with four flat blades and a #0 and #1 Phillips as #22-140. These sets cost around $2.

10.   Small Needle-Nosed Pliers

For reinstalling metal C-rings, and for both removing and reinstalling the flexible plastic or nylon

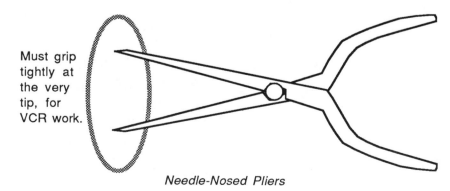

Must grip tightly at the very tip, for VCR work.

Needle-Nosed Pliers

lock-washers used by Fisher and other companies to hold

Flexible Type Locking Fastener

moving parts on axles, you will need some good small needle-nosed pliers. Flexible lock-washers are removed by gripping them near the cut, twisting slightly, pulling out a tiny bit, and sort of unwrapping them from the shaft they are on.

These washers are quite thin and difficult to grip with a tool because they are small, and they need to be gripped quite firmly. So when you buy a set of needle-nosed pliers for VCR work, examine them very carefully first to make certain that they have a good bite, a firm grip, right at their extreme tip. Many needle-nosed pliers make contact somewhere in the middle of the needle-nose, and the two parts never quite make contact at the extreme tip. Hold a few up to the light so that you can see through the space between the jaws when they are closed, and you probably will see what I mean.

A good test of the adequacy of a pair of these pliers for VCR work is that you should be able to grip firmly a piece of thin paper with just the tip end alone, gripping it firmly enough to be able to rip out a tiny piece about the size of the head of a straight pin. In order to get a pair that passed this test, I had to file away some of the metal along the middle of the jaws on a pair that flunked. You may need to do the

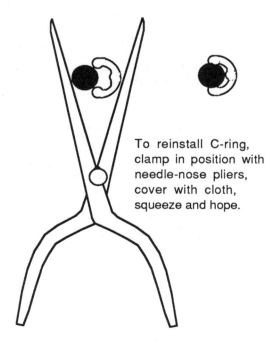

To reinstall C-ring, clamp in position with needle-nose pliers, cover with cloth, squeeze and hope.

Using Needle-Nosed Pliers to Reinstall a Metal C-Ring

same. It does not take long. Or you may get lucky and find a pair that already passes the test.

To reinstall metal C-rings, line the C-ring up with the groove in the shaft that it goes onto, hold it in place with needle-nosed pliers with one hand while covering with a cloth (to catch the C-ring in case it decides to fly away), and then squeeze gently to force the C-ring to spread and go into the groove.

11. Hex Keys

In VCR's, a few parts and adjustments require Allen, or hex, keys to change. The female part has a recessed six-sided hole in it, into which fits a corresponding six-sided male tool, called a "hex key" or "Allen wrench."

Hex (Allen) Nut and Key

Hex keys are available in hardware and auto parts stores, but UNFORTUNATELY, NOT IN THE SIZES NEEDED FOR VCR WORK.

The two sizes generally needed for this work are 0.9 mm (.87 mm, to be exact), and 1.5 mm. They are needed to adjust the P-guides in some VHS models, to remove the idler in an early Fisher model, to adjust Mitsubishi P-guides, and in a larger size, to remove the loading motor in an early Panasonic-category "tank" model. Otherwise, you do not need them.

The .9 mm and the 1.5 mm hex keys are available from Panasonic as part #VFK0146 and #VFK76 respectively, but probably the best buy is a set of eight in a pouch ranging in size from 0.71 to 4.0 mm (and including the crucial .9 mm and 1.5 mm) available as RCA part #144401. You can order these from Pan Son or other RCA parts suppliers.

12. Set of Nut Drivers or Small Wrenches

For removing some audio heads and a few other applications, a set of small nut drivers or normal hex wrenches in small metric sizes is useful.

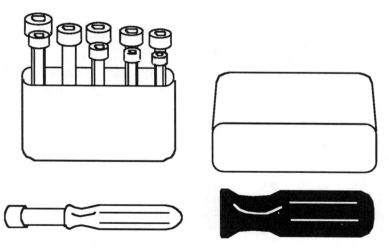

Nut Drivers

13. Notched Screwdrivers or Special Positioning Tools

Special tools are required to adjust (1) the height of the P-guides on many models, (2) to adjust the position of the audio head on Panasonic-category VCR's, and (3) to adjust the back-tension in some Fishers.

These tools are expensive to buy ready-made, but you can make them yourself by using a file to reshape the tip of inexpensive flat-blade screwdrivers from a discount store.

Three shapes that may be needed are:

(1) P-Guide
Adjustment

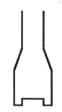

(2) Audio & Control
Head Adjustment

(3) Audio & Control
Head Adjustment

Tips of Special VCR Adjusting Tools

If you need one of these shapes, take a file and cut the appropriate shape into the end of the blade of a regular flat-blade screwdriver. The tools that I made myself in this way actually seem to be stronger and easier to use than the expensive ready-made tools sold as (1) RCA #144389, (2) RCA#148936, and (3) RCA#144387, respectively.

14. Light Oil and Grease

NEVER make the mistake of oiling parts in a VCR that service literature does not specifically tell you to oil. If you go in and start squirting or spraying oil around everywhere, you will just have to pay a professional service technician for the time it takes him to remove it and replace the rubber parts that got oil onto them. About the only part that ever needs oil is the capstan bearing, which can use one or two drops of light oil, like sewing machine oil, every couple of years.

There is a special light "Molytone" grease made for plastic and nylon moving parts, but it is expensive, and the only procedure in this book that requires something like this is the one on the plastic or nylon gears in JVC-category machines that need to be removed and have a special light grease put on them. This grease is specially made for plastics, and is available as RCA #144590 or Panasonic #MOR265.

15. Epoxy

Usually when a plastic or nylon part breaks in a VCR, it does not work to try to glue it back together, even with superglue. But there are situations in which epoxy can be used to reinforce existing plastic parts against future strain. The type of epoxy that has worked best for me is the gray "Epoxy Steel" made by Duro, widely available in hardware stores.

16. Lock Paint or Fingernail Polish

You will notice small areas of what look like red or green fingernail polish applied to the heads of some screws and nuts. This is "lock paint." It is used to prevent it from changing position due to vibration.

When you must remove one of these screws or nuts to change a part or make an adjustment, naturally the lock paint bond breaks when you move the fastener. It is supposed to do this. But when you are finished, you should apply lock paint again to hold the fastener in its new position.

Also, on some Mitsubishi-made and NEC-made VCR's, there are places on the P-guides where the factory should have applied lock paint (because they come loose in time), but did not. Here the repair procedure involves readjusting and locking the parts down with lock paint for the first time.

Although special lock paint is sold for VCR work, you can use a plain bottle of fingernail paint (any color, of course) to do the job.

17. Dummy Cassette

From an old, worn-out cassette, you can easily make an extremely useful tool for troubleshooting certain kinds of problems. It is called a "dummy cassette," and it is used to trick the VCR's little computer into thinking that there is a regular cassette in the machine, so that it will try to go through all its normal operations — rewind, fast-forward, play, record, etc. — as though there were a tape in the machine, when actually there is no tape in the machine — only an empty cassette box.

Being able to "trick" the machine in this way can be useful in diagnosing problems in these various

functions, so that you can run the machine without having to worry about tape getting in your way.

In order to do this, it is necessary to rig up an empty cassette box in such a way that the VCR will load it like a normal cassette and its little computer will think that there is a tape that it is rewinding, playing, etc. The cassette box itself serves to fool any "cassette-in" sensor, because when the empty box is loaded, this sensor sees that a cassette box has come in, and it does not detect that there is no tape inside the cassette.

But the tape end-sensors on VHS machines are more difficult to fool. (If you are interested in making a dummy cassette only for Beta machines, skip the next six paragraphs.)

As explained in Chapter 2, in VHS machines, there is a little light source on a tiny pedestal that goes into a hole provided for it in the VHS cassette box when it goes down onto the reel table. In early VCR's, the light was of the visible type, but on later models, they went to an infra-red light source (a "light emitting diode," in case you're interested).

There are little passageways for this light to pass from the center of the cassette out toward both ends. There are two more little windows in the ends of the cassette box, under the edge of the cassette door flap, and when light shines out these windows on each end, it strikes a sensor in the VCR or in the front-loading assembly that sends a signal to the machine's brain telling it whether light is reaching that sensor.

The windows are positioned so that when a normal cassette is in the machine, the tape in the cassette blocks the light from getting out one or both of the two end windows. This enables the little computer to tell that there is a tape in the cassette that is in the VCR.

To both ends of every VHS tape, there is spliced a short length of clear plastic tape or leader that the light can shine through. When the tape reaches its end in either direction, as the opaque tape moves out of the way and the clear leader moves in front of one of the end windows, it allows light suddenly to pass through, striking the photosensor, and telling the VCR's brain that the tape has come to an end in one direction or the other. Then the little computer will initiate appropriate behavior, like ceasing to rewind, fast-forward, play, or record. Stopping automatically in this way prevents one end of the tape from getting pulled out of the cassette by the mechanism.

Now if we just put an empty cassette box into the VCR, light would shine through the little windows on both ends (because there would be no tape to block the light), the two sensors would signal that both ends of the tape have been reached at the same time, and the little computer would know that something is not right, and would shut down and refuse to try to operate the mechanism. And, of course, in a sense, this is the correct thing for it to do.

But if we want to observe how the mechanism operates during play, rewind, etc., this defeats our purpose. So, we will put something opaque in front of the windows on our dummy cassette, so that the light will not shine through in either direction, and this will make the little computer think that there is a tape in the cassette box, and that it is somewhere in the middle of the tape. Here's how we can do this.

How to Make a Dummy Cassette

The first step is to turn the cassette box upside-down, and remove all the screws that hold the top and the bottom half together.

Separate the two halves, and remove the spools, tape, and miscellaneous small plastic and metal parts.

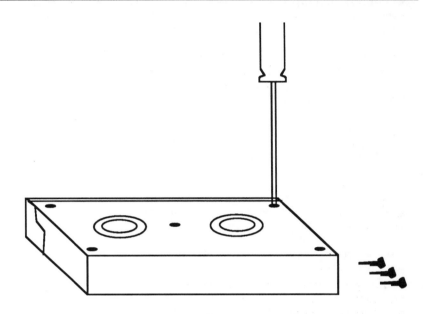

Removing the Screws Holding the Cassette Box Together

For a VHS dummy, take out everything that comes loose; for a Beta dummy cassette, notice how the front door flap hinges are held in, because a Beta dummy must have the front door reattached in order to work.

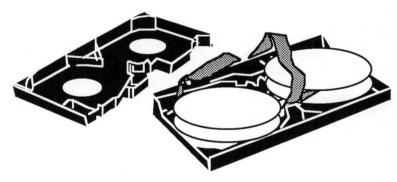

Open the Cassette and Remove Contents

At this point, you must decide whether or not you want to break open the top cover of the cassette box. It is often useful to have holes in the top side of the box through which you can reach in and feel (or measure) how much how forcefully the reels turn in play, rewind, fast-forward, etc, if you need to. That is a reason for making holes in the top cover, as described below.

Cassette Box Emptied of Contents

On the other hand, in a certain minority of VCR's, especially those with cassette-in detector switches located along the middle of the top of the cassette basket, making holes in the top of the cassette box can cause the dummy cassette to jam in the front-loader, due to the plastic fingers, or feelers, on these switches catching in the holes as the cassette passes by. For this situation, it would be better to have a dummy cassette with the top lid still intact.

Another consideration is that once you make holes, they are permanent, whereas if you do not make holes in the top of the box today, you still have the option of doing so in the future if you need to. What some people really need are two dummy cassettes, one with holes in the top, and another without.

You will need to make your decision based on your requirements. If you want to make holes in the top, the procedure is in the next paragraph; if you do not, skip the next paragraph, and go directly to the next page in this section.

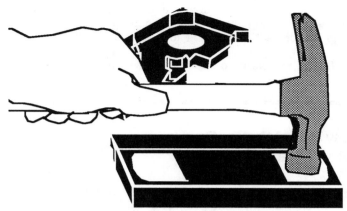

Breaking Out the Plastic Windows to Make a VHS Dummy Cassette

If you want to make holes to monitor the turning force on the reels, it only takes a few minutes to turn over the top half and smash out the plastic window or windows with a hammer, taking care to protect your eyes from flying parts. Don't hit it hard enough to crack the black plastic case around the windows. On a Beta, break out the one plastic window.

Now use pliers to enlarge the openings, so that you will be able to reach in and feel how much "torque" (turning power) the reels have on them in the different modes of operation.

For a Beta dummy cassette, skip the next paragraph.

For VHS, go to the next page.

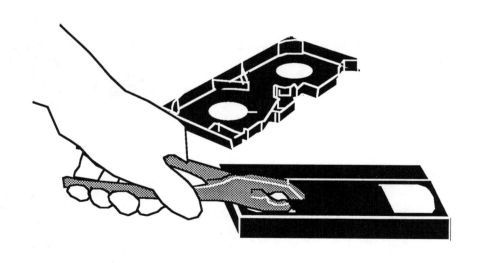

Enlarging the Opening with Pliers

For a VHS dummy cassette, now turn over the top half, and cover the little windows or passageways for light with black plastic tape, or cardboard and tape, so that the light cannot shine through. (Remember, we are preparing to trick the VCR into "playing" a nonexistent tape.)

Reassemble the two halves minus the insides using the same screws removed earlier. On Beta's, remember to put the door back into the cassette box, or else it might not work, and skip the next paragraph.

Cover Passageways for Light with Opaque Material

When you load the VHS dummy cassette into a VHS machine, it should go through all its paces, just as if there were a tape in the machine. NOTE: On some SHARP-category VCR's, to complete the deception it is necessary to turn the take-up reel counterclockwise by hand during the time that you see the supply reel turning CCW (counter-clockwise) for a moment right after you load the dummy cassette. (Some Sharp microcomputers are so "sharp" that they turn the supply reel as if rewinding tape for a moment, and check through another sensor to make sure that the take-up reel is turning in the same direction, as it would if there were a tape between them. By turning the take-up reel, you complete the deception.)

When you load the Beta dummy cassette into a Beta machine, you may also need to "trick" the machine's "tape slack sensor" in order to make it go through the motions as if loading and playing a tape. How to do this is explained in Appendix V, "How to Trick a VCR into Running With No Tape Inside for Diagnostic Purposes."

18. Test Tapes

When you are trying to diagnose, repair, and check the repair of a trouble, you often need to have a real tape with correctly recorded material on it to do the job. At the same time, you do not want to use an expensive prerecorded movie, or a priceless tape from your own collection, because there is a good chance that the tape may get damaged in the process, before you get the problem solved.

It would be foolhardy to buy a special test tape for this purpose, because video test tapes are even more expensive, and more easily damaged, than prerecorded movies. If you have a recorded movie that is ruined because it is damaged in one place, but good everywhere else, you could use it for test purposes, so long as you did not run the damaged portion of the tape through the VCR again.

Another good source of test tapes is a video tape rental store with which you are on good terms. The wholesalers of rental movie tapes often send to these stores short promotional "trailer" preview tapes, with short excerpts from upcoming rental movies beautifully recorded on them. These trailers have no real value after they have been viewed by the people at the store, and are often simply thrown away. If you can get them to save you some of these trailer cassettes instead of throwing them away, they make excellent test tapes, and you may be able to get them for a song.

19. Nine-volt Battery and Connectors

For testing motors and incandescent VHS sensor lamps, a small nine-volt battery is useful. Get one of the little rectangular type, about 1 X 1 3/4 inches (25 X 48 mm) in size, with one male, and one female, round snap on one end, available at most grocery or hardware stores. You also can use a Radio Shack battery #23-464.

You also will need something to connect the battery to whatever you want to test by applying

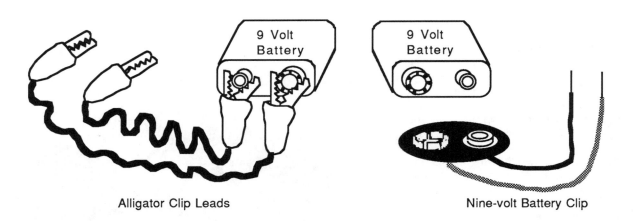

Alligator Clip Leads   Nine-volt Battery Clip

Nine-volt Battery with Different Kinds of Connectors

power. You can use standard technician"s "clip leads," which are just wires with alligator clips attached to each end, available from Radio Shack as #278-1157 for the medium size in packs of eight, attaching these directly to the snaps on the end of the battery.

If you get tired of these connectors coming unclipped from the battery all the time, you can get fancy and buy special 9-volt battery clips, which are designed to connect firmly to the snaps on the end of the battery, and then connect clip leads to the wires coming from the battery clips. These are available from Radio Shack as #270-325, but you have to buy a package of five., and you will still probably need the alligator clip leads too.

To use this apparatus to apply power to something, you can clip one of the alligator clips to a scribe, and another to a small screwdriver, to reach in and apply power to a motor or lamp, after you have double-checked to make certain that the VCR is unplugged.

20. Soldering Tools and Materials

Although almost none of the repair procedures in this basic book requires soldering, possession of a soldering iron and mastery of basic desoldering and resoldering skills marks the beginning of the transition from elementary to intermediate levels of VCR repair. There are some common repairs involving soldering that are expensive if you have a shop do them for you, but so easy to do yourself, that I have included some of them in the book, and added this brief section on soldering for those adventurous folks who want, or need, to learn to solder.

For VCR work, a very small "pencil-type" soldering iron, a stand to hold it when it is hot, a small sponge, some solder, and some good "solder wick" are all you need.

The heating power of the iron should be between 25 and 45 watts. One suitable soldering iron for these purposes is the Ungar Princess iron, with a 45 watt

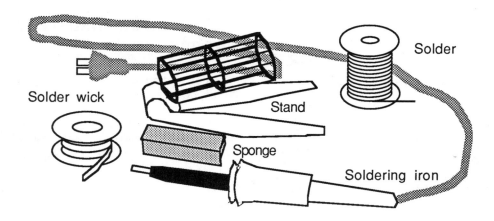

Soldering Iron, Materials, and Accessories

screw-in heater element, and an Ungar #6951 small chisel bit, or a pointed bit, if you are also going to use the iron to melt small holes through plastic.

There is no need to buy an expensive iron for home repair use, and stay away from all the gadgets sold as "desoldering aides" — solder wick is the best, as well as the cheapest, desoldering system you can buy. MCM sells a 25 watt Ungar iron for around $12, and tips (pencil tip: #21-506; chisel tip: #21-507; long taper chisel tip: 21-508) for about $3. They also have Tenma brand irons, a 30 watt iron #21-170 for about $3, and a 25 watt iron #21-160 with screw-in tip design for about $5, plus $2 for each replacement tip (#21-435).

A stand to hold the soldering iron when hot is available from MCM as #21-180 for about $5.

Choice of solder is important: if you do not use the right kind of solder, the acid it contains can cause trouble later. I recommend only Ersin multicore or Sav-a-bit solder. MCM sells 100 feet of Ersin solder for about $6 as #21-330. And I would never use any brand of solder wick but Chemtronics Chem-Wik — other brands may look the same, but I have not found that they work as well. MCM sells a good size of Chem-Wik for about $1.50 as #21-320.

Soldering Procedure

How do you solder? First, set up a fan to blow the solder smoke out a window (this smoke is very bad for your eyes).

Next, to start with, you need to "tin" the tip on your new soldering iron following the manufacturer's directions. Basically you touch the iron as soon as it heats to the end of a string of solder that you unwind from the roll. The solder melts, and you make sure that you get the whole end of the tip covered with melted solder. Then you just let it sit for a couple of minutes, cooking. Now put on a second coat of solder, let it sit for another minute, wipe on a damp sponge, and apply some more solder. Let it cook for another couple of minutes, and it's ready to go.

In most repair situations, you must first unsolder the old part before you can solder in the new one, so let's begin with desoldering. It is necessary not only to melt the old solder, but to remove almost all of it while it is hot, because it hardens so fast when you remove the soldering iron that you never could set the iron down and pull the part loose before the solder hardened into solid metal again.

There are a lot of "desoldering systems" around, but the best, for my money, is still "solder wick," which is used as follows.

First, while the iron is heating, work the end of the strand from the cool roll of solder wick with your fingers, so that its "weave" loosens up a little, so as to absorb more solder. Do not cut it off the roll until after it is full of solder that it has absorbed.

Put the end of the strand on the soldered junction or connection that you wish to desolder, wipe the hot iron on the sponge, and put it down on top of the wick pressing down firmly and watching for the solder suddenly to be melted and absorbed up into the wick.

When this happens, lift up the iron and wick at the same time (so that the wick does not end up soldered to what you are working on). Examine the joint carefully to determine where solder remains that still needs to be removed. Repeat the process until only a thin bright coat of solder remains where the old solder lump used to be, and the part is free. If the part is soldered down in more than one place, repeat the process for each point of attachment. Now you can remove the part.

Soldering in the new part is much easier. Put the part in place, and bend any wires or little metal legs so that they will stay in place until you are able to solder them.

Wipe the tip of the soldering iron on a wet sponge. Now put the soldering iron down on the junction or point of contact between the parts you are connecting, and after about two seconds, touch the end of a strand of solder directly to the tip of the iron where it contacts the parts.

The solder should instantly melt, and serve to finish carrying heat to the two parts. As soon as this happens, lift the iron and solder and allow to cool.

A good solder joint should not look like a bead of water on a waxed surface. Instead, it should make flush contact all along the surface underneath. The surface of the solder when it hardens should look smooth. If it has tiny lines in it like miniature crystals or frost, then you need to reheat the solder — it did not get hot enough or cool correctly.

Unplug the iron as soon as you are through. It ruins the tip to leave an iron standing hot for a long time with no fresh solder being regularly applied to it.

Appendix II

Where to Get Parts and Service Literature for Fixing VCR's

LIST OF MANUFACTURERS, DISTRIBUTORS, AND OTHER PARTS SOURCES

How to Use this Appendix

This appendix provides a list of manufacturers, distributors, and parts sources for VCR's. If you have gotten the number of the part you need from elsewhere in this book, the part number will be followed by an abbreviation in parentheses. Look this abbreviation up in the Introduction, find out what it abbreviates, and then look under that name. For example, if the part number is VDVS0122 (Mat), the "Mat" abbreviates "Matsushita," and means that it is a Panasonic part number, the same as (Pan). Or, for example, if you were looking for an idler for a Midland, the Introduction might tell you to look under "Samsung," and when you found the reel table, it might give a part number followed by "(Shp)," the abbreviation for Sharp. In that case, order the part from a Sharp parts supplier.

If you are trying to replace a cabinet piece (like a broken cabinet front), ordering an operator's manual or a service manual for your particular machine, then you should look under the brand name of your machine and contact the company whose name is on your VCR. Service manuals for VCR's are the size of a book, and cost between $18 and $55 each.

Key to Symbols Used

Forms of payment accepted by the supplier at the time this list was compiled are abbreviated as

follows:

M C	Mastercard
Visa	Visa
A E	American Express
Dis	Discover card
P C	Personal Check
M O	Money Order
C C	Certified check
C.O.D	You pay the person who delivers

When a supplier requires a certain minimum purchase in order to ship, this is indicated by the word "Min." followed by an amount, written over the telephone numbers. Usually they will ship less than the minimum order, if you are willing to pay an additional charge. When "No min." appears above the telephone numbers, it means that there is no minimum limit on orders from this supplier.

Each different parts source has its own policy regarding whether they will look up parts numbers for you, and if so, whether they will give these numbers out over the telephone or send them to you by mail. Each also has its own policy about the forms of payment they will accept. Some will take a check, some require money orders. Some will ship C.O.D., and some will not. Each of the parts sources listed in this appendix says that it will sell parts directly to consumers. If you have trouble with any of them, a letter to the author, c/o Worthington Publishing Company would be most appreciated.

The *procedure* that each parts supplier was following at the time of the writing of this appendix is indicated by abbreviations that have the following meanings:

Proc. 1: The supplier gives you price(s) and other necessary information over the telephone, you use this information to mail them an order with payment, and they send you the part(s).

Proc. 2: The supplier takes the order over the telephone and ships directly to you based on a credit card or other form of credit.

Proc. 3: The supplier takes information from you over the telephone, sends you a form with prices and other necessary information regarding the parts, you mail the form back to them with an acceptable form of payment, and they send you the part(s).

If the parts you need come to less than the supplier's minimum order, consider ordering other rubber parts that you might need for your machine in the future, or a duplicate of the part you're presently ordering, or cleaning supplies, if it is a general supplier. You could also go in with a friend who has a machine for which the supplier also carries parts, ordering spare parts for his machine too, and splitting the cost.

If the parts supplier tells you that the part number you have given "crosses to another part number," go with that. It is all that is available, and it may actually be an improved version of the part (like a rubber belt made from better rubber).

If you are in a hurry to get the part, and you have the part number, and you live in a city that is large enough to have local electronics parts suppliers, you might save time to take a minute to look up their telephone number in the yellow pages under "Electronic Equipment and Supplies" or "Electronics Parts Suppliers" and give them a call. Tell them you have the part number.

Parts lines sometimes tend to overflow on days after days=off, so consider the calling time.

Aiwa	Union Electronic Distributors 16012 Cottage Grove South Hollend, IL 60473	MC, Visa, AE, PC, MO, Proc.2 or Proc.1, no min. (800) 648-6657 In Illinois: (708) 333-4100
	Tandy Consumer Parts 7439 Airport Freeway Ft. Worth, TX 76118	PC, MO, Proc. 1 or MC, Visa, AE, Dis, Proc. 2, $5 min. Will look up part numbers (800) 243 1311
Akai	Mitsubishi Electric Sales, Inc 5757 Plaza Dr Cypress, CA 90630-0007	MC, Visa, PC, MO, Proc.1 or Proc.2, $6.50 min Parts: (800) 553-7278 Tech. Assist.for Servicers: (800) 552-8324 Tech. Assist.for Consumers: (714) 220-1464
	Tandy Consumer Parts 7439 Airport Freeway Ft. Worth, TX 76118	PC, MO, Proc. 1 or MC, Visa, AE, Dis, Proc. 2, $5 min. Will look up part numbers (800) 243 1311
Broksonic	Hatzlachh Supply Inc. 935 Broadway, 6th Fl. New York, NY 10010 See Shintom also	C.O.D., cash or CC, Proc. 2, No min (212) 254-9012

Canon

Canon USA, Inc.
100 Jamesburg Road
Jamesburg, NJ 08831
and

MC, Visa, AE, $10 min; Proc. 2
sometimes C.O.D. for less than $10
Cabinet parts & service lit.: (800) 828-4040
Internal Parts for servicers: (908) 521-7000

Tandy Consumer Parts
7439 Airport Freeway
Ft. Worth, TX 76118

PC, MO, Proc. 1 or
MC, Visa, AE, Dis, Proc. 2, $5 min.
Will look up part numbers
(800) 243 1311

Circuit City

See Shintom

Colt

See Shintom

Curtis Mathes

Curtis Mathes Corporation
P.O. Box 2160
Athens, TX 75751-2160

C.O.D., Proc 2, $10 min.
(903) 675-6886

Daytron

Daewoo Intl. America Corp.
100 Daewoo Pl.
Carlstadt, NJ 07072

(201) 935-8700
Parts unavailable at time of writing;
Daewoo says to contact retailer
from whom VCR was purchased for
exchange unit or latest information

Denon

Denon America Inc.
222 New Road
Parsippany, NJ 07054

M O or Certified check; Proc 1; No min.
(201) 882-7490

Dual

Dual Parts and Service
122 Dupont Street
Plainview, NY 11803

CC or MO; Proc.1
(516) 349-7757

South Street Service Co.
202 South Street
Oyster Bay, NY 11771

Visa, MC, C.O.D., Proc 2
(516) 922-0337

Adcom

Adcom Service Corporation
11 Elkins Road
East Brunswick, NJ 08816

(201) 390-1130

Dynatech

No information available

**Electrodynamics
(Edc)**

Electrodynamics, Inc.
135 Eileen Way
P.O. Box 9022
Syosset, NY 11791-9022

MC, Visa, C.O.D., Proc.2, $20 min
(800) 426-6423

Elmo

Elmo Manufacturing Corp.
70 New Hyde Park
New Hyde Park, NY 11040

MC, Visa, C.O.D.; Proc. 2; $10 min.
(800) 654-7628

Emerson

Fox International Ltd., Inc.
23600 Aurora Road
Bedford Heights, Ohio 44146

MC, Visa, MO, PC, Proc.2 or 3, No min.
Parts orders: (800) 321-6993

Fisher

Parts Department
Sanyo-Fisher Company
1200 W. Artesia Blvd.
Compton,CA 90224

MC, Visa, Dis, Proc.2, min $10 or CC, MO, PC, Proc.3
Parts (310) 537-5830 or (310 605-6741
Servicers only: (310) 605-6742
Fax: (310) 605-6744

Tritronics
1306 Continental Drive
Abingdon, MD 21009

M C, Visa, Proc. 2 or P C or M O, Proc.1; Min. $5
Information (410) 676-7300
To order (800) 638-3328
FAX: (800) 888-3293
In the South: (800) 365-8030
Information: (305) 938-8030

Tandy Consumer Parts
7439 Airport Freeway
Ft. Worth, TX 76118

PC, MO, Proc. 1 or
MC, Visa, AE, Dis, Proc. 2, $5 min.
Will look up part numbers
(800) 243 1311

G M B Electronics Supply
140 Terminal Road
Setauket, NY 11733

MC, Visa, MO; Proc.2 or 3, No min.
Parts orders: (800) 874-1765
Parts information: (516) 689-3400

Funai

Attn Parts
Symphonic Parts Department
100 North Street
Teterboro, NJ 07608

MC, Visa, Proc. 1 or C C or M O, Proc.1, No min
For VCR parts: (800) 242-7158

Tandy Consumer Parts
7439 Airport Freeway
Ft. Worth, TX 76118

PC, MO, Proc. 1 or
MC, Visa, AE, Dis, Proc. 2, $5 min.
Will look up part numbers
(800) 243 1311

G E

For Service Literature only:
Thomson Consumer Electronics
Technical Publications
10003 Bunsen Way
Louisville, KY 40299

MC, Visa, Proc.2; PC,MO, Proc.1; no min.
(502) 491-8110
G E Tech. Services:(913) 541-0402

Tritronics
1306 Continental Drive
Abingdon, MD 21009

M C, Visa, Proc. 2 or P C or M O, Proc.1; Min. $5
Information (410) 676-7300
To order (800) 638-3328
FAX: (800) 888-3293
In the South: (800) 365-8030
Information: (305) 938-8030

Andrews Electronics
25158 Avenue Stanford
Valencia, CA 91355

Visa, MC; Proc. 2; $15 min.
PC, MO; Proc. 3; no min.
Will research part numbers
(800) 274-4666

Union Electronic Distributors
16012 Cottage Grove
South Hollend, IL 60473

MC, Visa, AE, PC, MO,
Proc.2 or Proc.1, no min.
(800) 648-6657
In Illinois: (708) 333-4100

Tandy Consumer Parts
7439 Airport Freeway
Ft. Worth, TX 76118

PC, MO, Proc. 1 or
MC, Visa, AE, Dis, Proc. 2, $5 min.
Will look up part numbers (800) 243 1311

Fox International Ltd., Inc.
23600 Aurora Road
Bedford Heights, Ohio 44146

MC, Visa, MO, PC, Proc.2 or 3, No min.
Parts orders: (800) 321-6993

G M B Electronics Supply
140 Terminal Road
Setauket, NY 11733

MC, Visa, MO; Proc.2 or 3, No min.
Parts orders: (800) 874-1765
Parts information: (516) 689-3400

If you already have the part number:
Vance Baldwin Inc.
2207 S. Andrews Ave.
Ft. Lauderdale, FL 33316

M C, Visa, Proc. 2, No Min.
(800) 432-8542
(800) 232-1007
(800) 367-1641

Goldstar

Tandy Consumer Parts
7439 Airport Freeway
Ft. Worth, TX 76118

PC, MO, Proc. 1 or
MC, Visa, AE, Dis, Proc. 2, $5 min.
Will look up part numbers
(800) 243 1311

Fox International Ltd., Inc.
140 Terminal Road
Setauket, NY 11733

MC, Visa, MO; Proc.2 or 3, No min.
Parts orders: (516) 689-3400

Goldstar Parts
P.O. Box 6166
Huntsville, AL 35824

Check or M O, Proc. 1 No min
(800) 562-0244

Astro Technology
1510 Lunt Avenue
Elk Grove Village, IL 60007

Check or M O, Proc. 1 No min
(800) 252-7788
Parts research (800 222 6457

G M B Electronics Supply
140 Terminal Road
Setauket, NY 11733

MC, Visa, MO; Proc.2 or 3, No min.
Parts orders: (800) 874-1765
Parts information: (516) 689-3400

As a last resort only:
Shine Electronics
11-22 45th Road
Long Island City, NY 11101

C.O.D cash only Proc. 2 or MO Proc 1; No min.
In Eastern U.S.: (800) 221-0404
In NY: (718) 786-2229 or 786-6698

Harmon Kardon

Harmon Kardon, Inc.
240 Crossways Park West
Woodbury, NY 11797

MO; Proc.1; No min.
(516) 496-3400, ext. 242, 212, or 270

Hitachi

Andrews Electronics
25158 Avenue Stanford
Valencia, CA 91355

Visa, MC; Proc. 2; $15 min.
PC, MO; Proc. 3; no min.
Will research part numbers
(800) 274-4666

Panson Electronics
268 Norman Avenue
Greenpoint, NY 11222

MC, Visa, PC, or MO, Proc.1 or Proc. 2, Min. $5
Parts orders or to look up part nos.(800) 255-5229
(718) 383-3400

Tandy Consumer Parts
7439 Airport Freeway
Ft. Worth, TX 76118

PC, MO, Proc. 1 or
MC, Visa, AE, Dis, Proc. 2, $5 min.
Will look up part numbers
(800) 243 1311

G M B Electronics Supply
140 Terminal Road
Setauket, NY 11733

MC, Visa, MO; Proc.2 or 3, No min.
Parts orders: (800) 874-1765
Parts information: (516) 689-3400

Fox International Ltd., Inc.
23600 Aurora Road
Bedford Heights, Ohio 44146

MC, Visa, MO, PC, Proc.2 or 3, No min.
Parts orders: (800) 321-6993

Union Electronic Distributors
16012 Cottage Grove
South Hollend, IL 60473

MC, Visa, AE, PC, MO,
Proc.2 or Proc.1, no min.
(800) 648-6657
In Illinois: (708) 333-4100

P.I. Burks Co.
P.O. Box 1015
Louisville, KY 40201

M C, Visa, Dis, Proc. 2 or P C, M O, Proc.1; no min.
(800) 274-2875

If you already have the part number:
Vance Baldwin Inc.
2207 S. Andrews Ave.
Ft. Lauderdale, FL 33316

M C, Visa, Proc. 2, No Min.
(800) 432-8542
(800) 232-1007
(800) 367-1641

Ikko   Last address: National Sound Inc.
330 W. 38th St.
New York, NY 10018

No current number available

Instant Replay   Instant Replay Co.
2601 S. Bayshore Dr. Suite 1050
Miami, FL 33133

MC, Visa, COD, MO, PC; Proc. 1 or 2
(305) 854-8777, Parts Dept.

J C Penney   J C Penney Co
Parts Distribution Center 9496-1
6840 Morrow, GA 30260-3006

MC, Visa, AE: Proc. 2; No Min.
Parts: (800) 222-6161
Price & Inquiries on parts: (404) 961-8400

Jensen   Attn: Parts Department
25 Tri-State
International Office Center, Suite 400
Lincolnshire, IL 60069

PC or MO, Proc 1, No min.
(800) 323-4815
In Illinois: (708) 317 3800

Tandy Consumer Parts
7439 Airport Freeway
Ft. Worth, TX 76118

PC, MO, Proc. 1 or
MC, Visa, AE, Dis, Proc. 2, $5 min.
Will look up part numbers
(800) 243 1311

J V C   J V C Company of America
Southwest Office
407 Garden Oaks Blvd.
Houston, TX 77018

Tech. Assistance for servicers (713) 694-3331
Tech. Assistance for consumers (201) 808-2100
or (800) 882-2345, engineering dept.

Panson Electronics
268 Norman Avenue
Greenpoint, NY 11222

MC, Visa, PC, or MO, Proc.1 or Proc. 2, Min. $5
Parts orders or to look up part nos.(800) 255-5229
(718) 383-3400

Tandy Consumer Parts
7439 Airport Freeway
Ft. Worth, TX 76118

PC, MO, Proc. 1 or
MC, Visa, AE, Dis, Proc. 2, $5 min.
Will look up part numbers
(800) 243 1311

	P.I. Burks Co. P.O. Box 1015 Louisville, KY 40201	M C, Visa, Dis, Proc. 2 or P C, M O, Proc.1; no min. (800) 274-2875
K L H	See Shintom	
Kenwood	Kenwood Parts Department East Coast Transistor, Inc. P.O. 238 2 Marlborough Rd. West Hempstead, NY 11552	Visa, MC, PC, MO; Proc.1 or 2; $6 min. (800) 328-0388
	Tandy Consumer Parts 7439 Airport Freeway Ft. Worth, TX 76118	PC, MO, Proc. 1 or MC, Visa, AE, Dis, Proc. 2, $5 min. Will look up part numbers (800) 243 1311
Kodak	Eastman Kodak Co. Parts Service Dept. Bldg 601 800 Lee Road Rochester, NY 14650	C.O.D., P C, M O, Proc.2; sometimes will bill net 30 days; No Min. Parts: (800) 431-7278 or (716) 724-7278 24 hrs Tech. Serv. Info. : (800) 558-5028 Information Center (800) 242-2424 Product Status/Equip. Repair Ctr.: (800) 237-5398
Lloyds	Lloyds 200 Clearview Ave. Edison, NJ 08818	PC, MO, Proc. 1 or Pan3; No min. (908) 225-2030
Logik	See Shintom	
M C M	MCM 650 E. Congress Park Dr. Centerville, OH 45459-4072	Visa, MC, or cash COD Proc. 2 or Proc.1; $20 min on COD; $25 min on credit card National Sales: (800) 543-4330 Canada: (800) 824-9491 (513) 434-0031
Magnavox	Attn: Parts Department Philips Consumer Electronics P.O. Box 967 Greeneville, TN 37744-0967	Inquiry for part number only, parts identification service: (615) 636-5859 When you have the part #, MC, Visa, AE, Dis, Proc.1 or 2. PC or MO. no min. Parts orders: (800) 851-8885 Information Center (615) 475-8869 Tech. Assistance (615) 475-6141 or (312) 827-9408 or (201) 935-2700 Main number and Data (615) 475-3801
	Tritronics 1306 Continental Drive Abingdon, MD 21009	M C, Visa, Proc. 2 or P C or M O, Proc.1; Min. $5 Information (410) 676-7300 To order (800) 638-3328 FAX: (800) 888-3293 In the South: (800) 365-8030 Information: (305) 938-8030

Tandy Consumer Parts
7439 Airport Freeway
Ft. Worth, TX 76118

PC, MO, Proc. 1 or
MC, Visa, AE, Dis, Proc. 2, $5 min.
Will look up part numbers
(800) 243 1311

G M B Electronics Supply
140 Terminal Road
Setauket, NY 11733

MC, Visa, MO; Proc.2 or 3, No min.
Parts orders: (800) 874-1765
Parts information: (516) 689-3400

If you already have the part number:
Vance Baldwin Inc.
2207 S. Andrews Ave.
Ft. Lauderdale, FL 33316

M C, Visa, Proc. 2, No Min.
(800) 432-8542
(800) 232-1007
(800) 367-1641

Marantz

Marantz Co.
20525 Nordhoff St.
Chatsworth, CA 91311

(818) 998-9333

Tandy Consumer Parts
7439 Airport Freeway
Ft. Worth, TX 76118

PC, MO, Proc. 1 or
MC, Visa, AE, Dis, Proc. 2, $5 min.
Will look up part numbers
(800) 243 1311

Matsushita

Information Service   For locations of parts & service ctrs (800) 447-4700

Matsushita Services Co.
Service Literature Division
P.O. Box 1815
Elgin, IL 60121
For more, see Panasonic

PC or MO, Proc.1, No Min
For service literature only: (817) 640-2532

Memorex

Tandy Consumer Parts
7439 Airport Freeway
Ft. Worth, TX 76118

PC, MO, Proc. 1 or
MC, Visa, AE, Dis, Proc. 2, $5 min.
Will look up part numbers
(800) 243 1311

National Parts Dept.
900 E. North Side Dr.
Ft. Worth, TX 76102

M C, Visa, A E, Dis, or C.O.D.Cash, Proc.2, or
P C, M O, Proc. 1; Min. $3.35
Parts orders & parts information: (800) 442-2425
Main number: (817) 870-5600

Minolta

Minolta Corporation
Attn: Video Parts Department
101 Williams Dr.
Ramsay, NJ 07446

C.O.D., P C, M O, Proc.1 or 2; No min
Parts Orders (201) 825-4000, Parts extension,
3244 or 3245
Tech. Asst. extension 5376

Mitsubishi

Mitsubishi Electric Sales, Inc
5757 Plaza Dr
Cypress, CA 90630-0007

MC, Visa,PC, MO, Proc.1 or Proc.2,$6.50 min.
Parts: (800) 553-7278
Tech. Assist.for Servicers: (800) 552-8324
Tech. Assist.for Consumers: (714) 220-1464

Tandy Consumer Parts
7439 Airport Freeway
Ft. Worth, TX 76118

PC, MO, Proc. 1 or
MC, Visa, AE, Dis, Proc. 2, $5 min.
Will look up part numbers
(800) 243 1311

G M B Electronics Supply
140 Terminal Road
Setauket, NY 11733

MC, Visa, MO; Proc.2 or 3, No min.
Parts orders: (800) 874-1765
Parts information: (516) 689-3400

Fox International Ltd., Inc.
140 Terminal Road
Setauket, NY 11733

MC, Visa, MO, PC, Proc.2 or 3, No min.
Parts orders: (516) 689-3400

Montgomery Wards

Montgomery Wards Parts Dept.
5750 McDermott Dr.
Berkeley, IL 60163

Visa, M C, A E, Dis, M W, Proc. 2 or
Ck or M O, Proc. 1; No min.
(800) 323-1965

Multitech

Trans World Electronics
186 University Parkway
Pomona CA 91768

C C or M O, Proc. 1, or MC, VISA, COD, Proc. 2
No min.
(800) 822-1236
In Calif.: (818) 333-2011

Tandy Consumer Parts
7439 Airport Freeway
Ft. Worth, TX 76118

PC, MO, Proc. 1 or
MC, Visa, AE, Dis, Proc. 2, $5 min.
Will look up part numbers
(800) 243 1311

N E C

NEC Parts Department
1255 Michael Drive
Wood Dale, IL 60191

MC, Visa, AE, PC, Mo;Proc 2 or 1; No min.
(708) 860-0335 (expect a long wait)
FAX (800) 356-2415

Tandy Consumer Parts
7439 Airport Freeway
Ft. Worth, TX 76118

PC, MO, Proc. 1 or
MC, Visa, AE, Dis, Proc. 2, $5 min.
Will look up part numbers
(800) 243 1311

Olympus

Olympus Corp.
Consumer Products Group
145 Crossways Park West
Woodbury, NY 11797

Under $15, M O, P C, Proc. 1
Over $15, C.O.D. Proc. 2 or bill later
For ordering only: (800) 622-6372, ext. 350
or (800) 247-9674
(516) 364-3000, ext. 350

PTS Electronics

Rebuilds tuners and RF modulators
PTS Electronics
5233 Hwy 37 South
Bloomington, IN 47402

MC, Visa, or COD
(800) 333-7871

Prime Electronics

For custom rebuilt pinch rollers, idlers, & other parts
Prime Electronics
P.O. Box 28
Whitewater, WI 53190

MC, Visa, C.O.D., M O, PC;
Proc.2 or 1; Min $10
(800) 558-9572
In Wisconsin: (800) 242-9553

Panasonic

Matsushita Services Co.
Parts Division
50 Meadowland Parkway
Secaucus, NJ 07094

Matsushita Services Co.
Service Literature Division
P.O. Box 1815
Elgin, IL 60121

PC or MO, Proc.1, No Min
For service literature only: (817) 640-2532
To replace Owner's operating instructions manuals, look
under "Matsushita" in this appendix

Tritronics
1306 Continental Drive
Abingdon, MD 21009

M C, Visa, Proc. 2 or P C or M O, Proc.1; Min. $5
Information (410) 676-7300
To order (800) 638-3328
FAX: (800) 888-3293
In the South: (800) 365-8030
Information: (305) 938-8030

Panson Electronics
268 Norman Avenue
Greenpoint, NY 11222

MC, Visa, PC, or MO, Proc.1 or Proc. 2, Min. $5
Parts orders or to look up part nos.(800) 255-5229
(718) 383-3400

Andrews Electronics
25158 Avenue Stanford
Valencia, CA 91355

Visa, MC; Proc. 2; $15 min.
PC, MO; Proc. 3; no min.
Will research part numbers
(800) 274-4666

Tandy Consumer Parts
7439 Airport Freeway
Ft. Worth, TX 76118

PC, MO, Proc. 1 or
MC, Visa, AE, Dis, Proc. 2, $5 min.
Will look up part numbers
(800) 243 1311

G M B Electronics Supply
140 Terminal Road
Setauket, NY 11733

MC, Visa, MO; Proc.2 or 3, No min.
Parts orders: (800) 874-1765
Parts information: (516) 689-3400

Fox International Ltd., Inc.
23600 Aurora Road
Bedford Heights, Ohio 44146

MC, Visa, MO, PC, Proc.2 or 3, No min.
Parts orders: (800) 321-6993

Union Electronic Distributors
16012 Cottage Grove
South Hollend, IL 60473

MC, Visa, AE, PC, MO,
Proc.2 or Proc.1, no min.
(800) 648-6657
In Illinois: (708) 333-4100

If you already have the part number:
Vance Baldwin Inc.
2207 S. Andrews Ave.
Ft. Lauderdale, FL 33316

M C, Visa, Proc. 2, No Min.
(800) 432-8542
(800) 232-1007
(800) 367-1641

Pentax

Pentax Corp.
35 Inverness Drive East
Englewood, CO 80112

C.O.D. (P C, M O), Proc. 2 or
P C, M O, Proc. 1; $10 min.
(800) 877-0155, ext. 272, 273, or 274
(303) 799-8000

Pfanstiehl

Pfanstiehl Corp.
3300 Washington Street
Waukegan, IL 60085

MC, Visa, Proc. 2
(800) 323-9446

Philco

Attn: Parts Department
Philips Consumer Electronics
P.O. Box 967
Greeneville, TN 37744-0967

Inquiry for part number only, parts
identification service: (615) 636-5859
When you have the part #, MC, Visa, AE, Dis,
Proc.1 or 2. PC or MO. $10 min.
Parts & manuals orders: (800) 851-8885
Information Center (615) 475-8869
Tech. Assistance (615) 475-6141 or
(312) 827-9408 or (201) 935-2700
Main number and Data (615) 475-3801

Tritronics
1306 Continental Drive
Abingdon, MD 21009

M C, Visa, Proc. 2 or P C or M O, Proc.1; Min. $5
Information (410) 676-7300
To order (800) 638-3328
FAX: (800) 888-3293
In the South: (800) 365-8030
Information: (305) 938-8030

Tandy Consumer Parts
7439 Airport Freeway
Ft. Worth, TX 76118

PC, MO, Proc. 1 or
MC, Visa, AE, Dis, Proc. 2, $5 min.
Will look up part numbers
(800) 243 1311

G M B Electronics Supply
140 Terminal Road
Setauket, NY 11733

MC, Visa, MO; Proc.2 or 3, No min.
Parts orders: (800) 874-1765
Parts information: (516) 689-3400

Pioneer

Pioneer Electronics Inc.
Parts Department
Dept. 49592
Los Angeles, CA 90088

VISA, MC, PC, MO; Proc. 2 or 1; no min
will look up part numbers
(800) 228-7221
(310) 835-6177

Tandy Consumer Parts
7439 Airport Freeway
Ft. Worth, TX 76118

PC, MO, Proc. 1 or
MC, Visa, AE, Dis, Proc. 2, $5 min.
Will look up part numbers
(800) 243 1311

Portland

See Daytron

Quasar

Information Services

Matsushita Services Co.
Parts Division
50 Meadowland Parkway
Secaucus, NJ 07094

(800) 447-4700
PC or MO, Proc.1, No Min
For service literature only: (817) 640-2532

Matsushita Services Co.
Service Literature Division
P.O. Box 1815
Elgin, IL 60121

To replace Owner's operating instructions manuals, look
under "Matsushita" in this appendix

Panson Electronics
268 Norman Avenue
Greenpoint, NY 11222

MC, Visa, PC, or MO, Proc.1 or Proc. 2, Min. $5
Parts orders or to look up part nos.(800) 255-5229
(718) 383-3400

Tandy Consumer Parts
7439 Airport Freeway
Ft. Worth, TX 76118

PC, MO, Proc. 1 or
MC, Visa, AE, Dis, Proc. 2, $5 min.
Will look up part numbers
(800) 243 1311

Andrews Electronics
25158 Avenue Stanford
Valencia, CA 91355

Visa, MC; Proc. 2; $15 min.
PC, MO; Proc. 3; no min.
Will research part numbers
(800) 274-4666

Fox International Ltd., Inc.
23600 Aurora Road
Bedford Heights, Ohio 44146

MC, Visa, MO, PC, Proc.2 or 3, No min.
Parts orders: (800) 321-6993

© 1990 Stephen N. Thomas

G M B Electronics Supply
140 Terminal Road
Setauket, NY 11733

MC, Visa, MO; Proc.2 or 3, No min..
Parts orders: (800) 874-1765
Parts information: (516) 689-3400

Union Electronic Distributors
16012 Cottage Grove
South Hollend, IL 60473

MC, Visa, AE, PC, MO,
Proc.2 or Proc.1, no min.
(800) 648-6657
In Illinois: (708) 333-4100

If you already have the part number:
Vance Baldwin Inc.
2207 S. Andrews Ave.
Ft. Lauderdale, FL 33316

M C, Visa, Proc. 2, No Min.
(800) 432-8542
(800) 232-1007
(800) 367-1641

Radio Shack See Realistic

Randix Last known address:
Randix Industries, Ltd.
3 Tech Circle
Natick, MA 01670

No number available

R C A Thomson RCA Consumer Electronics
600 N. Sherman Dr.
Indianapolis, IN 46206

Technical Assistance Only: (913) 541-0402

Thomson Consumer Electronics
Technical Publications
10003 Bunsen Way
Louisville, KY 40299

MC, Visa, Proc.2; PC,MO, Proc.1; no min.
(502) 491-8110
Tech. Services:(913) 541-0402

Tritronics
1306 Continental Drive
Abingdon, MD 21009

M C, Visa, Proc. 2 or P C or M O, Proc.1; Min. $5
Information (410) 676-7300
To order (800) 638-3328
FAX: (800) 888-3293
In the South: (800) 365-8030
Information: (305) 938-8030

Andrews Electronics
25158 Avenue Stanford
Valencia, CA 91355

Visa, MC; Proc. 2; $15 min.
PC, MO; Proc. 3; no min.
Will research part numbers
(800) 274-4666

Panson Electronics
268 Norman Avenue
Greenpoint, NY 11222

MC, Visa, PC, or MO, Proc.1 or Proc. 2, Min. $5
Parts orders or to look up part nos.(800) 255-5229
(718) 383-3400

Tandy Consumer Parts
7439 Airport Freeway
Ft. Worth, TX 76118

PC, MO, Proc. 1 or
MC, Visa, AE, Dis, Proc. 2, $5 min.
Will look up part numbers
(800) 243 1311

G M B Electronics Supply
140 Terminal Road
Setauket, NY 11733

MC, Visa, MO; Proc.2 or 3, No min.
Parts orders: (800) 874-1765
Parts information: (516) 689-3400

Fox International Ltd., Inc.
23600 Aurora Road
Bedford Heights, Ohio 44146

MC, Visa, MO, PC, Proc.2 or 3, No min.
Parts orders: (800) 321-6993

Union Electronic Distributors
16012 Cottage Grove
South Hollend, IL 60473

MC, Visa, AE, PC, MO,
Proc.2 or Proc.1, no min.
(800) 648-6657
In Illinois: (708) 333-4100

If you already have the part number:

Vance Baldwin Inc.
2207 S. Andrews Ave.
Ft. Lauderdale, FL 33316

M C, Visa, Proc. 2, No Min.
(800) 432-8542
(800) 232-1007
(800) 367-1641

Realistic

Attn: Customer Service Dept.
National Parts Dept.
900 E. North Side Dr.
Ft. Worth, TX 76102

M C, Visa, A E, Dis, or C.O.D.
(P C or M O), Proc.2, or
P C, M O, Proc. 1; Min. $5
Parts orders & parts information: (800) 442-2425
Main number: (817) 870-5600

Sampo

Sampo Corporation of America
Parts and Service
5550 Peachtree Industrial Blvd.
Norcross, GA 30071

P C or M O, Proc. 1, No min.
Will look up part numbers
(404) 449-6220

G M B.
2700 Middle Country Road
Centereach, NY 11720

C.O.D. (cash), Proc. 2, Min. $10
Parts information (516) 585-8111
Parts orders national (800) 874-1765
Parts orders NY only (800) 874-1764

Samsung

Samsung Electronics America, Inc.
1860 Broadwick St.
Rancho Dominguez, CA 90220

Tech. Assistance: (800) 833-6616

Parts and Service Literature:

Fox International Ltd., Inc.
23600 Aurora Road
Bedford Heights, Ohio 44146

MC, Visa, MO, PC, Proc.2 or 3, No min.
Parts orders: (800) 321-6993

Tandy Consumer Parts
7439 Airport Freeway
Ft. Worth, TX 76118

PC, MO, Proc. 1 or
MC, Visa, AE, Dis, Proc. 2, $5 min.
Will look up part numbers
(800) 243 1311

G M B Electronics Supply
140 Terminal Road
Setauket, NY 11733

MC, Visa, MO; Proc.2 or 3, No min.
Parts orders: (800) 874-1765
Parts information: (516) 689-3400

Sansui

Sansui Electronics Corp.
Parts Department
17150 South Margay Ave.
P.O. 4687
Carson, CA 90746

C C or M O, Proc. 1; No min.
Servicers only (800) 421-1500
In Calif.: (213) 604-7300

Tandy Consumer Parts
7439 Airport Freeway
Ft. Worth, TX 76118

PC, MO, Proc. 1 or
MC, Visa, AE, Dis, Proc. 2, $5 min.
Will look up part numbers
(800) 243 1311

Fox International Ltd., Inc.
23600 Aurora Road
Bedford Heights, Ohio 44146

MC, Visa, MO, PC, Proc.2 or 3, No min.
Parts orders: (800) 321-6993

Sanyo

Parts Department
Sanyo-Fisher Company
1200 W. Artesia Blvd.
Compton,CA 90224
Tandy Consumer Parts
7439 Airport Freeway
Ft. Worth, TX 76118

MC, Visa, Dis, Proc.2, min $10 or CC, MO, PC, Proc.3
Parts (310) 537-5830 or (310 605-6741
Servicers only: (310) 605-6742
Fax: (310) 605-6744
PC, MO, Proc. 1 or
MC, Visa, AE, Dis, Proc. 2, $5 min.
Will look up part numbers
(800) 243 1311

Tritronics
1306 Continental Drive
Abingdon, MD 21009

M C, Visa, Proc. 2 or P C or M O, Proc.1; Min. $5
Information (410) 676-7300
To order (800) 638-3328
FAX: (800) 888-3293
In the South: (800) 365-8030
Information: (305) 938-8030

G M B Electronics Supply
140 Terminal Road
Setauket, NY 11733

MC, Visa, MO; Proc.2 or 3, No min.
Parts orders: (800) 874-1765
Parts information: (516) 689-3400

Union Electronic Distributors
16012 Cottage Grove
South Hollend, IL 60473

MC, Visa, AE, PC, MO,
Proc.2 or Proc.1, no min.
(800) 648-6657
In Illinois: (708) 333-4100

Scott

H.H. Scott and Emerson Radio
Highway 41 South & Country Rd 100 West
Princeton, Indiana 47670
North Bergen, NJ 07047

Parts: (812) 386-3200, ext. 1
Tech. Assist. (800) 388-8333
General (201) 662-2000

Tandy Consumer Parts
7439 Airport Freeway
Ft. Worth, TX 76118

PC, MO, Proc. 1 or
MC, Visa, AE, Dis, Proc. 2, $5 min.
Will look up part numbers
(800) 243 1311

Fox International Ltd., Inc.
23600 Aurora Road
Bedford Heights, Ohio 44146

MC, Visa, MO, PC, Proc.2 or 3, No min.
Parts orders: (800) 321-6993

Sears See Local Telephone Book to
try locally first, or contact:

Sears, Roebuck & Co.
National Parts Department
Sears Tower
Chicago, IL 60684

MC, Visa, Dis, Sears Chg.Card, Proc.2
Parts Orders Nat'l: (800) 366-7278
Sears General: (312) 875-2500

Sharp

Tritronics
1306 Continental Drive
Abingdon, MD 21009

M C, Visa, Proc. 2 or P C or M O, Proc.1; Min. $5
Information (410) 676-7300
To order (800) 638-3328
FAX: (800) 888-3293
In the South: (800) 365-8030
Information: (305) 938-8030

	Andrews Electronics 25158 Avenue Stanford Valencia, CA 91355	Visa, MC; Proc. 2; $15 min. PC, MO; Proc. 3; no min. Will research part numbers (800) 274-4666
	Tandy Consumer Parts 7439 Airport Freeway Ft. Worth, TX 76118	PC, MO, Proc. 1 or MC, Visa, AE, Dis, Proc. 2, $5 min. Will look up part numbers (800) 243 1311
	Fox International Ltd., Inc. 23600 Aurora Road Bedford Heights, Ohio 44146	MC, Visa, MO, PC, Proc.2 or 3, No min. Parts orders: (800) 321-6993
Shintom	Panson Electronics 268 Norman Avenue Greenpoint, NY 11222 Shintom West Corporation 20435 South Western Ave. Torrance, CA 90501	MC, Visa, PC, or MO, Proc.1 or Proc. 2, Min. $5 Parts orders or to look up part nos.(800) 255-5229 (718) 383-3400 (310) 328-7200
	High Quality International USA Inc. 683 New York Drive Pomona, CA 91768-3313	PC or C.O.D cash.; Proc.1 Parts: (800) 333-1098 (714) 629-6508, ext. 115 or 116
	AmKotrom 11612 E. Washington Blvd. Ste. H Whittier, CA 90606	MC, Visa, C.O.D. cash; Proc.2; no min. (800) 344-3882
	Tandy Consumer Parts 7439 Airport Freeway Ft. Worth, TX 76118	PC, MO, Proc. 1 or MC, Visa, AE, Dis, Proc. 2, $5 min. Will look up part numbers (800) 243 1311
Signature	See Montgomery Ward	
Singer	See Shintom	
Sony	Andrews Electronics 25158 Avenue Stanford Valencia, CA 91355	Visa, MC; Proc. 2; $15 min. PC, MO; Proc. 3; no min. Will research part numbers (800) 274-4666
	P.I. Burks Co. P.O. Box 1015 Louisville, KY 40201	M C, Visa, Dis, Proc. 2 or P C, M O, Proc.1; no min. (800) 274-2875
	Joseph Electronics 8830 N. Milwaukee Ave. Niles, IL 60648	Proc. 1, P C or M O; Proc. 2, M C, Visa, Dis. (800) 323-5925 In Illinois call: (708) 297-4200
	Tandy Consumer Parts 7439 Airport Freeway Ft. Worth, TX 76118	PC, MO, Proc. 1 or MC, Visa, AE, Dis, Proc. 2, $5 min. Will look up part numbers (800) 243 1311

Fox International Ltd., Inc.
752 South Sherman
Richardson, TX 75081

MC, Visa, MO, PC, Proc.2 or 3, $8 min.
Parts orders: (800) 331-2501

Union Electronic Distributors
16012 Cottage Grove
South Hollend, IL 60473

MC, Visa, AE, PC, MO,
Proc.2 or Proc.1, no min.
(800) 648-6657
In Illinois: (708) 333-4100

Sound Design

Tandy Consumer Parts
7439 Airport Freeway
Ft. Worth, TX 76118

PC, MO, Proc. 1 or
MC, Visa, AE, Dis, Proc. 2, $5 min.
Will look up part numbers
(800) 243 1311

Supra

Central Electronics
153-11 Northern Boulevard
Flushing, NY 11354

C.O.D. cash, Proc. 2; No min.
Very limited Parts and Service (718) 539-4327

Sylvania

Attn: Parts Department
Philips Consumer Electronics
P.O. Box 967
Greeneville, TN 37744-0967

Inquiry for part number only, parts
identification service: (615) 636-5859
When you have the part #, MC, Visa, AE, Dis, C.O.D.
Proc.1 or 2. PC or MO. no min.
Parts & manuals orders: (800) 851-8885
Information Center (615) 475-8869
Tech. Assistance (615) 475-6141 or
(312) 827-9408 or (201) 935-2700
Main number and Data (615) 475-3801

Tritronics
1306 Continental Drive
Abingdon, MD 21009

M C, Visa, Proc. 2 or P C or M O, Proc.1; Min. $5
Information (410) 676-7300
To order (800) 638-3328
FAX: (800) 888-3293
In the South: (800) 365-8030
Information: (305) 938-8030

Tandy Consumer Parts
7439 Airport Freeway
Ft. Worth, TX 76118

PC, MO, Proc. 1 or
MC, Visa, AE, Dis, Proc. 2, $5 min.
Will look up part numbers
(800) 243 1311

Fox International Ltd., Inc.
23600 Aurora Road
Bedford Heights, Ohio 44146

MC, Visa, MO, PC, Proc.2 or 3, No min.
Parts orders: (800) 321-6993

G M B Electronics Supply
140 Terminal Road
Setauket, NY 11733

MC, Visa, MO; Proc.2 or 3, No min.
Parts orders: (800) 874-1765
Parts information: (516) 689-3400

If you already have the part number:
Vance Baldwin Inc.
2207 S. Andrews Ave.
Ft. Lauderdale, FL 33316

M C, Visa, Proc. 2, No Min.
(800) 432-8542
(800) 232-1007
(800) 367-1641

Symphonic

Attn Parts
Symphonic Parts Department
100 North Street
Teterboro, NJ 07608

MC, Visa, Proc. 1 or C C or M O, Proc.1, No min
For VCR parts: (800) 242-7158

	Tandy Consumer Parts 7439 Airport Freeway Ft. Worth, TX 76118	PC, MO, Proc. 1 or MC, Visa, AE, Dis, Proc. 2, $5 min. Will look up part numbers (800) 243 1311
	Trans World Electronics 186 University Parkway Pomona CA 91768	C C or M O, Proc. 1, or MC, VISA, COD, Proc. 2 No min. (800) 822-1236 In Calif.: (818) 333-2011
Tatung	Tatung Company of America 2850 El Presidio St. Long Beach, CA 90810	M O, P C, Proc. 1, No min. (310) 637-2105
Teac	Teac Corporation of America 7733 Telegraph Road Montebello, CA 90640	M C, Visa, Min $10, Proc.2; P C, M O, Proc. 1, No min. (213) 726-0303, ext. 840
	Tandy Consumer Parts 7439 Airport Freeway Ft. Worth, TX 76118	PC, MO, Proc. 1 or MC, Visa, AE, Dis, Proc. 2, $5 min. Will look up part numbers (800) 243 1311
Technics	Andrews Electronics 25158 Avenue Stanford Valencia, CA 91355	Visa, MC; Proc. 2; $15 min. PC, MO; Proc. 3; no min. Will research part numbers (800) 274-4666
	Union Electronic Distributors 16012 Cottage Grove South Hollend, IL 60473	MC, Visa, AE, PC, MO, Proc.2 or Proc.1, no min. (800) 648-6657 In Illinois: (708) 333-4100
If you already have the part number:	Vance Baldwin Inc. 2207 S. Andrews Ave. Ft. Lauderdale, FL 33316	M C, Visa, Proc. 2, No Min. (800) 432-8542 (800) 232-1007 (800) 367-1641
Teknika	Teknika Electronics Corporation 353 Route 46W Fairfield, New Jersey 07006	C C or M O, Proc. 1, No Min. Parts: (201) 575-0380 Customer Hot Line: (800) 835-6452 In N J: (800) 962-1272
	Tandy Consumer Parts 7439 Airport Freeway Ft. Worth, TX 76118	PC, MO, Proc. 1 or MC, Visa, AE, Dis, Proc. 2, $5 min. Will look up part numbers (800) 243 1311
	Joseph Electronics 8830 N. Milwaukee Ave. Niles, IL 60648	Under $20, Proc. 3, P C or M O; Over $20, Proc. 2, M C, Visa, Dis. (800) 323-5925 In Illinois call: (312) 297-4200
T M K	Toyomenka America Inc. Light Equipment Division 357 County Avenue Secaucus, NJ 07094	M O or C C; Proc.1; no min (201) 348-1100

Tandy Consumer Parts
7439 Airport Freeway
Ft. Worth, TX 76118

PC, MO, Proc. 1 or
MC, Visa, AE, Dis, Proc. 2, $5 min.
Will look up part numbers
(800) 243 1311

Toshiba

Andrews Electronics
25158 Avenue Stanford
Valencia, CA 91355

Visa, MC; Proc. 2; $15 min.
PC, MO; Proc. 3; no min.
Will research part numbers
(800) 274-4666

Tandy Consumer Parts
7439 Airport Freeway
Ft. Worth, TX 76118

PC, MO, Proc. 1 or
MC, Visa, AE, Dis, Proc. 2, $5 min.
Will look up part numbers
(800) 243 1311

Fox International Ltd., Inc.
23600 Aurora Road
Bedford Heights, Ohio 44146

MC, Visa, MO, PC, Proc.2 or 3, No min.
Parts orders: (800) 321-6993

Totevision

Totevision
969 Thomas Street
Seattle, WA 98109

M C, Visa, AE, C.O.D.; Proc 2 or
PC, Proc. 1; No min
(206) 682-4343
FAX (206) 623-6609

Toyomenka

See T M K

Unitech

Last known address:
Unitech
13327 Paxton
Pacoima, CA 91331

No telephone number could be found
for this company

Vector Research

Vector Research
1230 Calle Suerte
Camarillo, CA 93012

PC, Proc. 1, No min.Call with part
numbers only, or to order a service manual from
which you can get part numbers: (805) 987-1312
FAX (805) 987-1956

Video Concepts

American Home Video Corporation
Englewood, CO 80111

No telephone number could be found
for this company

National Parts Dept.
900 E. North Side Dr.
Ft. Worth, TX 76102

M C, Visa, A E, Dis, or C.O.D.Cash, Proc.2, or
P C, M O, Proc. 1; Min. $3.35
Parts orders & parts information: (800) 442-2425
Main number: (817) 870-5600

Yamaha

Attn: Audio (Video) Parts
Yamaha Electronics Corp., USA
6660 Orangethorpe Ave.
Buena Park, CA 90620

M C or Visa, Proc. 2, or
C C, M O, Proc.1; Min.$5
Parts: (714) 994-3312
Other: (800) 492-6242

Tandy Consumer Parts
7439 Airport Freeway
Ft. Worth, TX 76118

PC, MO, Proc. 1 or
MC, Visa, AE, Dis, Proc. 2, $5 min.
Will look up part numbers
(800) 243 1311

Zenith

Tandy Consumer Parts
7439 Airport Freeway
Ft. Worth, TX 76118

PC, MO, Proc. 1 or
MC, Visa, AE, Dis, Proc. 2, $5 min.
Will look up part numbers
(800) 243 1311

Andrews Electronics
25158 Avenue Stanford
Valencia, CA 91355

Visa, MC; Proc. 2; $15 min.
PC, MO; Proc. 3; no min.
Will research part numbers
(800) 274-4666

G M B Electronics Supply
140 Terminal Road
Setauket, NY 11733

MC, Visa, MO; Proc.2 or 3, No min.
Parts orders: (800) 874-1765
Parts information: (516) 689-3400

Fox International Ltd., Inc.
23600 Aurora Road
Bedford Heights, Ohio 44146

MC, Visa, MO, PC, Proc.2 or 3, No min.
Parts orders: (800) 321-6993

Union Electronic Distributors
16012 Cottage Grove
South Hollend, IL 60473

MC, Visa, AE, PC, MO,
Proc.2 or Proc.1, no min.
(800) 648-6657
In Illinois: (708) 333-4100

If you already have the part number:
Vance Baldwin Inc.
2207 S. Andrews Ave.
Ft. Lauderdale, FL 33316

M C, Visa, Proc. 2, No Min.
(800) 432-8542
(800) 232-1007
(800) 367-1641

". . . Don't give me any of that 'bad microprocessor in system control' jazz,
Sonny. It's just a worn loading belt, and you know it."

Appendix III

Getting In: How to Open VCR Cabinets

It's funny, but one of the most difficult parts of the job of repairing a VCR for a beginner may be getting the cabinet open. All brands use screws to fasten the top, back, and bottom covers of their cabinets, but sometimes something in addition to removing these screws is required to get the darned thing open.

1. Preliminaries

Look at the screws visible around the top, sides, back, and bottom of the VCR. Most cabinet screws require a "+" tipped cross-slot or Phillips screwdriver of the #1 or #2 size, but a few brands used ordinary flat slotted screws on the cassette basket lid and cabinet top. Even if your VCR has straight slotted cabinet screws, you may find that once you get the cabinet open, you still must have a cross-slotted screwdriver to progress any further. For more information on the kind of screwdrivers needed to deal with these screws, see Section 2 of Appendix I.

> Whenever you remove a screw from a VCR, be extremely careful not to let the screw drop down into the machine. If you drop a screw into it, and you cannot get the screw back out, it can damage the circuitry or the mechanism when the VCR is operated in the future. Whenever you are removing a screw and there is a possible dangerous place into which it could fall nearby, it is wise to cover the whole area with a soft paper, tissue, or cloth to catch the screw in case it does fall.

If you ever drop a screw or other small part into a VCR, first take a lot of time to look for it carefully everyplace you can see. Mentally divide the area into small squares, and systematically search them one at a time. If you cannot find the lost screw or part, as a

last resort, hold the VCR upside-down over a large clean smooth floor (like a big linoleum floor) and shake the VCR until (hopefully) the screw or part finally falls out.

If it never comes out, you will just have to proceed without it and pray that it does not damage anything. It is not feasible to dismantle the entire mechanism to look for it, and even if you did, this would be even more dangerous, because it would be too difficult to get the deep mechanism back together correctly, so that it would work.

Procedure:

Get a good stack of paper cups, or a box with lots of dividers (like a couple of egg cartons, or a tackle box) and a supply of note paper. Number the screws as you remove them. When you get ready to remove the next screw from the back, top, ends, front, or bottom, make a rough sketch of that part of the cabinet, write its number on the sketch where that screw came from, and put it in a paper cup on which you write the same number.

Anyone can remove screws; the trick is being able to put each screw back in the right place, in an sequence that is the reverse of the order in which they were removed. Many different kinds of screws are used in a VCR, and it is vitally important that each screw be put back exactly where it was. On some models, there are places where if you put the wrong screw in a hole and screw it down, it can go through and puncture a circuit board inside (because the correct screw is shorter), virtually destroying the VCR.

It is, of course, possible, that the problem in your VCR will turn out to be something beyond the scope of this basic manual, and that you will need to reassemble the unit and take it to a professional service technician to do the job. Some repair shops do not like to work on VCR's that the customer has worked on first, and in some cases, will even refuse a job if they can tell that the customer opened up the machine. Partly this is because they do not want customers doing the easy jobs themselves, and leaving only the difficult jobs to the technician, because that takes "easy money" away from the shops. And partly this is because they are afraid that the customer may have gone into the VCR with no guiding knowledge or service literature, and may have turned some adjustments, or broken something additional to the original problem, making a second trouble in the machine, which makes it not just twice, but about ten-times as difficult to repair.

With this book, you should not make this mistake, so this should not be a problem, but the shops, of course, do not know that you are different from the average do-it-yourselfer who tries to fix a VCR with no knowledge, so they may still be unhappy about your opening it up first. If you take careful notes, and get all the screws back into their right places, it will not be apparent that you ever opened up the machine yourself, in case you do need to take it to a shop later.

Do not count on your memory to remember where the screws went. Make clear notes on paper as you go along. It may seem like this way will take longer, but actually it will save a lot of time on the reassembly end, because you won't have to puzzle out where the screws are supposed to go when you are reassembling, nor will you need to take it apart again to try to get it back together right. Remember the children's story about how the turtle passes the rabbit on the back lap. Now with tools, paper, and pencil in hand, let's proceed.

Make certain that the unit is unplugged.

Put old newspapers, a towel, or other protection down on the working surface to avoid scratching the cabinet later when you turn it over upside down.

If the VCR is a front-loader, go to Section 4. If it is a top-loader, go to section 3, unless it is an early Panasonic-category machine with the mechanical push-down switches, known affectionately to their owners as "tanks" because of their heavy construction and durability, that look like this,

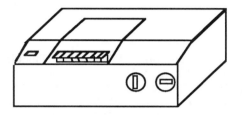

Exterior of Earliest Panasonic Category VCR's

In this case, you will need to start by going to Section 2 to remove the back cover.

2. Remove Back Cover

On second-generation and later VCR's, the back cover is part of the same piece of metal or plastic as the top cover, but on the earliest Panasonic-category VCR's that look like the previous illustration, there is a separate back cover that needs to be removed first, before the top cover can be removed.

If your VCR is one of this type, look at the back of the VCR. Do you see a number of cross-slot screws

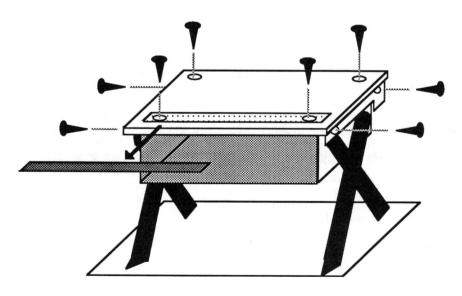

Typical Locations of Screws Fastening On the Cassette Basket Lid

down inside recessed holes? Usually there are five of them. You will need to remove them all and put them in the first container. Then you can pull the back straight off. As you remove the back, the power cord and the plug can be pulled through the big hole they go through in the back cover, so that the back cover can be completely detached and set aside.

On these models, unlike later machines, the top of the cabinet is held on by about six screws that go down through the top. It is unnecessary to remove the cassette lid before removing the top cover, and it is probably better to leave the lid on for now, until it must be removed for some specific reason, so go now to Section 4.

3. Remove Lid to Cassette Basket on Top Loaders

On many top-loaders, the lid to the cassette basket is slightly larger than the hole through which the rest of the basket pops up, so it is usually necessary to remove the cassette lid in order to get the cabinet top off.

Cassette lids are always fastened on by screws that either go down from the top, or go in from the sides. They may be located in any of the places shown in the following composite illustration:

If the machine is a top-loader, and the plastic lid to the cassette compartment is held on by screws visible from the top, remove the screws holding on this lid, leaving the top down while you unscrew the screws, as further protection against their falling down into the machine. On some VCR's, the screws that attach the lid are under a rubber or flexible

plastic strip that is pressed down into a recessed area on the lid; simply pry this strip up, lift it out, and remove the screws underneath. If the cassette lid is held on by screws at its sides, it is probably best to leave it attached for now.

If you need to remove a plastic lid held on by screws along the sides of the lid, lift the lid by pressing "Eject" in order to remove the screws. BE SUPER-CAREFUL NOT TO LET ANY OF THESE SCREWS DROP INSIDE THE MACHINE! If this happens, you will have to spend a lot of time later removing the dropped screw, which cannot be left inside the machine because it may jam the delicate mechanism or short out and damage critical circuitry. Hold your hand under the screw as you remove it, to catch it if it falls. If the VCR is an early Panasonic-category "tank" type, you do not need to remove the cassette lid to remove the top cover.

If you have the problem that the VCR will not eject a tape, and you cannot remove the screws to get the lid off with the cassette down, go to Chapter 7 and use the procedure later in this appendix to open up the bottom of the reel table to manually eject the cassette basket.

Put the screws you remove into a separate numbered paper cup.

4. Removing the Top of the Cabinet

The top of VCR cabinets are held on by screws in any of the following locations: screwed into the back of the VCR, into the sides, up from the bottom, and down through the top itself. Screws holding on the top

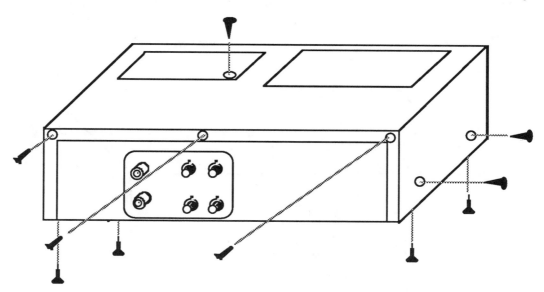

Locations of Screws Attaching the Cabinet Top on Various Models

of the cabinet can be found in any of the places shown in the composite illustration. No VCR has cabinet screws in all of those locations, of course, but each model will have them in some of these locations. On machines built by Shintom, often used for rentals, you will need to order the special tool # MB50 (Shi) from a Shintom supplier, or use needle nosed pliers to remove the special fasteners.

If the VCR is NOT a later-generation Panasonic, Quarsar, Magnavox, Sylvania, Philco, G.E., Curtis Mathes, J.C. Penney, Canon, or other Panasonic-category unit, skip directly to the next paragraph below. If it is a later-generation Panasonic-category machine, it may be necessary to remove some bottom screws in order to get the top cabinet part off, because screws from the bottom extend up into a flange of the cover. If the VCR is one of these brands, turn it over and look at the screws along the edges of the bottom. If there are arrows stamped into the metal covering the bottom that point to some of the screws along the side edges, remove these screws and place them into another compartment or cup nestled in the preceding cup, and note in your record the number of this container and the "bottom-sides" location from which they came. Turn the VCR rightside-up again.

Look at the back of the VCR to see if there are some small screws around the antenna and TV output connectors, in addition to other screws at strategic points on the back surface. Do NOT remove any small flat-headed screws grouped around the connectors for the antenna input and the connectors for the output to the TV. They are not holding the cabinet on. Remove ONLY large screws that go through holes in the edge of the cabinet top and into the back of the VCR, and put them into a separate compartment or paper cup nested inside the previous cup to perserve the order in which you removed the screws.

Make a note on your record of the contents of the compartments or containers, like this: "(1) cassette lid screws — two; (2) back cabinet screws — how many, etc. Look at the screws as you remove them. If any is shorter or smaller than the other(s), make a note of which hole it came from. Turn the VCR rightside-up again.

Remove any screws from the sides of the VCR (unless it is a Panasonic-category "tank," on most models of which you can remove the top without taking screws out of the sides). Put in separate container.

Remove any screws visible on the top of the cabinet. If the VCR has a little plastic hinged door over the channel frequency adjusters on the top, open this door and remove any screw visible here. Put it into a separate container.

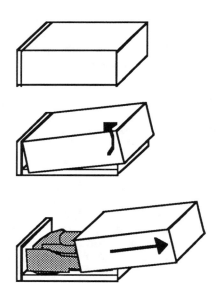

Direction of Movement for Removal of Most Cabinets

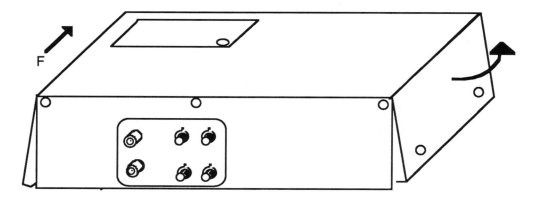

Spreading the Sides Slightly to Remove a Fisher-Sanyo Cabinet Top

Now comes the challenging part. If you are lucky, the top piece (including sides and back as all one piece, on most VCR's), can be rotated up from the back, pulled back slightly, and lifted up and away. On a few top-loaders, you may need to push the cassette basket back down (either part of the way or all of the way) to slide the top off. But if you are not so lucky and the cabinet part resists removal in this fashion, you may be looking at a unit on which the top must be released in a different way. The procedure for machines made by different manufacturers differs, as follows:

FISHER AND SANYO

On Fisher and Sanyo machines, it may be necessary to use your fingernail to get in at the edges of the cabinet along the bottom edge on the side, and pry the cabinet outward, spreading it slightly, while pulling back and up.

Some early Panasonic-category machines (like the Panasonic PV1600, etc.) have little hooks molded in the plastic ends of the cabinet, which engage in slots cut in the edges of the top cabinet. They cannot be seen until the cabinet top is removed, so you would not know that they are there unless you had read about this, or had previous experience with them.

On these VCR's, in order to remove the top, it is necessary first to remove the two sides by taking out the two screws in the middle of each side. When the

screws have been removed from the top too, the top then can be lifted straight off.

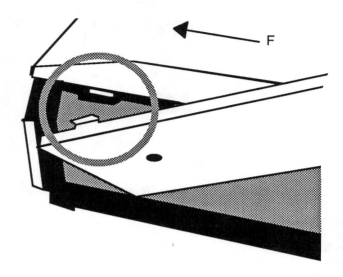

Tabs on Side Pieces Holding on the Cabinet Top

On some models, there are feet or catches along the bottom edge of the sides that grip into notches molded in the chassis below, while on other models, there are little hooks molded in the plastic edges of the chassis that engage in slots cut into the folded-in edges of the top cabinet metal. On these models, the top is released by sliding it straight back about 1/4

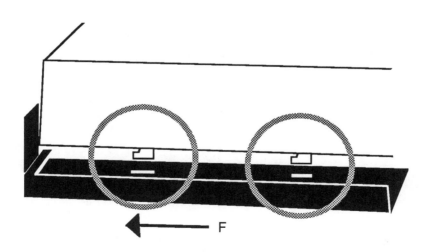

Locking Tabs that Prevent the Cabinet Top from Being Rotated Off

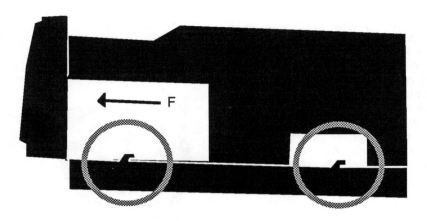

Locking Tabs Coming Up from the Chassis that Prevent the Cabinet Top from Being Rotated Off

inch, after

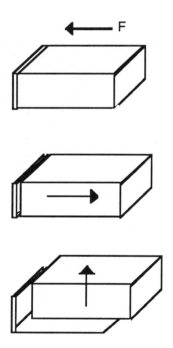

Back-and-Up Motion to Release Cabinet Top Held by Side Tabs

which it can be lifted up and away.

HITACHI

Be careful when reassembling some early Hitachi-category machines, like the VT 76A, because the screws that go into the sides of the cabinet are of slightly different length on the two sides, and if you put the longer screw back into the wrong hole, it will

stick through on the inside and short out the power supply for the whole machine. Note which screw goes where when disassembling, so that you can get it back into the right hole.

MITSUBISHI

Some Mitsubishi-category VCR's have hidden little metal tabs or teeth along the front edge of the cabinet top piece that lock into hidden slots along the top back edge of the front cabinet piece. The picture in the next illustration shows how the teeth on the cabinet top look, once it is off the VCR — but of course, you do not see this when the cabinet is still together.

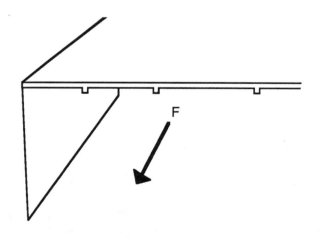

Locking Teeth on Mitsubishi-Made Cabinet Top

You can only see them once the cabinet top is removed. A slightly different procedure is required to lift this type of cabinet top free. After removing any screws along the sides of the cabinet, the cabinet top must be lifted almost straight up to free these

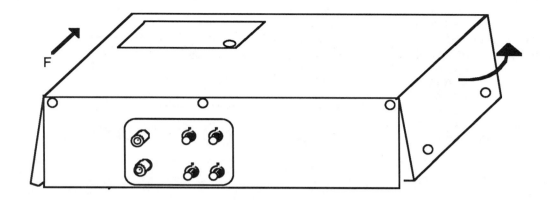

Spreading the Sides Slightly and Lifting the Top Cabinet Straight Up

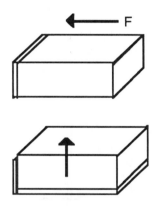

Straight-up Motion to Release Cabinet Top Held by Locking Front Tabs

locking tabs. If this is the first time the cabinet has been opened since it was manufactured, it may take a some wiggling and struggling with the cabinet top to get it free.

On some MITSUBISHI-category VCR's, it is also necessary to spread the sides a little, especially up near the front, and lift the top cabinet straight up.

SHARP and MONTGOMERY WARD: On these VCR's, especially when they have never been opened before, a metal lip at the front of the top side of the cabinet extends in under the plastic front panel. This lips needs to be worked free, in order to get the top off.

Sometimes you can do this by placing the VCR on the table facing you, gripping the top cabinet firmly at the sides and wiggling it from side to side while pulling it back from the front and away from you. Sometimes it is necessary to insert some tool like your fingernail or a wood popsicle stick in the groove between the top cabinet and the edge of the front part and prying a gap

between the two. A metal screwdriver should not be used except as a last resort, since it may scratch the paint or plastic.

5. Gaining Access to the Heads and Top of the Reel Table

On many models, you need only remove the top of the cabinet, and you can see the top of the reel table with the various heads, etc. But on some models, it is necessary to open up a small board with components soldered to it (a "circuit board"), and remove a piece of formed sheet metal that serves to minimize electrical interference with other electronic circuits (a metal "shield"), in order to get at the top of the reel table.

If there is a circuit board in the way, take a felt-tip pen and trace the outline of each screw head you see on top of it. If a little springy piece of brass or some other metal is being held down to the circuit board by one of these screws, trace its outline on the circuit board with the pen, and draw an arrow on the piece of metal pointing toward the front of the VCR (to help you get it back in place correctly on reassembly). If there is a wire coming to a terminal with an eye, held down to the circuit board by a screw, trace the outline of this terminal on the circuit board.

Remove all the marked screws, and put them in a separate numbered container. The circuit board should now fold open on little plastic hinges. Prop it open with a little wooden stick, catching the stick under one of the larger, stronger components soldered to the circuit board. If the board is not on hinges, lift it up and move it out of the way as far as the wires attached to it will allow. It should be unnecessary to detach any of the wires.

EMERSON AND MITSUBISHI-CATEGORY VCR'S: On a few older Mitsubishi-category machines, there is an enormous circuit board that covers the whole top of the machine, and which is attached to metal parts that are also attached to front circuit boards. It is a real nuisance, but on these models, it is necessary to remove the front cabinet piece in order to fold up this circuit board and get to the top of the reel table. If this is your situation, follow the procedure in the section later in this appendix for removing front cabinet panels, then return to this point.

After you remove the front cabinet piece, look to see whether the top circuit board is attached to the front circuit board. If so, you will need to remove any screws holding down the front circuit board too. Then you can fold up the whole assembly with both circuit boards together.

If you see a piece of sheet metal over part of the back half of the reel table, draw circles around any screws holding it down, and remove these, putting them in a separate container.

You should now have access to the various heads and other parts on the top of the reel table.

6. Removing Bottom Cover

The bottom cover must be removed to clean or work on the bottom side of the reel table, and on most models, to replace the loading belts. See Appendix VI for the location of loading belts on your model. And sometimes the bottom cover needs to be removed in order manually to eject a stuck tape.

Some older model VCR's cannot be safely turned over once the top of the cabinet is removed, because a crucial part sticks up to far, and if you set the machine upside down with the top of the cabinet removed, some of the weight will go on this part and damage it. For example, some of the early Panasonic-category "tanks" had a power switch that stuck out far enough that it could get broken if the machine were set upside down on it. And on some SONY-category Beta's (like the Zenith VR8500PT), the black plastic fan on the top of the upper assembly projects up far enough from the chassis, that the force placed on it by the weight of the machine when upside down could throw the cylinder and tape path out of alignment.

If there is any question or doubt, place a yardstick or straightedge across the top of the machine with the top removed to see whether there are any dangerously projecting crucial parts before turning it over. If not, go to the next paragraph. If there are, you have two choices. You can place a pillow or wooden block in a position where it can hold

up a safe part of the chassis, keeping the weight off of the delicate part, as follows:

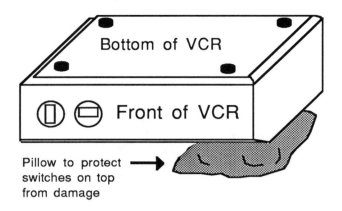

Pillow Protecting the Power Switch from the Weight of Upside-down Early Panasonic-type VCR

Or you can instead place the VCR in what technicians and service literature call "the servicing position," which means balancing the machine on one end, usually

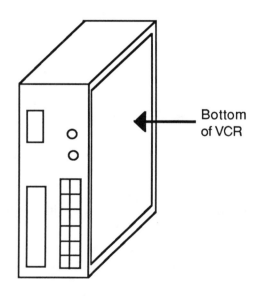

Official "Servicing Position" for Many Models

the heavier end, and working on it in a vertical position. Some well-regarded authorities take the position that VCR's should always be serviced in this position, but personally I have found that many models can fall over easily in this position, so I only use it when the upside-down horizontal position will not work at all.

Double-check to make certain that the VCR is unplugged. Turn the VCR over. With a felt-tip pen, draw an arrow pointing toward the front of the VCR on

Tabs Holding Down Circuit Board

the bottom cover, and put an "F" (for "Front") beside it. Now look and see whether there are any screws whose heads have been tinted red. With the felt-tip pen, put a small dot or other mark beside each screw you see that is NOT tinted red. The nonred screws hold on the bottom cover.

Remove each of these screws, and as you do so, look closely at each screw. Some may be longer or shorter than the rest. If so, put a "L" beside the hole that a long screw comes out of (usually there is only one), and put an "S" beside the holes that any screws shorter than the rest come out of. These marks will be used later to get these special screws back into the correct holes. Put bottom cover screws into their own numbered container.

JVC-category VCR's: On some later-generation JVC-category machines, there is one screw that goes from beside the big thing on top (a "transformer") down to screw into the bottom cover. And four of the screws that hold on the bottom cover go through rubber feet on the bottom of the cabinet. Remove these.

Now you can lift off the bottom cover.

7. Opening Up Bottom Circuit Board(s)

On some models, once you remove the bottom cover, you find yourself looking at the bottom of the reel table. On other models, a board with components soldered to it (a "circuit board"), will stand between you and the bottom of the reel table. On models with bottom circuit boards, the circuit boards are attached to hinges along one edge. You do not really remove these circuit boards — rather, you remove some screws or plastic locking tabs, and then open the boards out on their hinges.

The difficult aspect is that on some models, it is also necessary to remove the cabinet front (with the controls on it) before you can open up the bottom circuit board, while on other models, you can open up the bottom circuit board without removing the front

cabinet piece. I know of no simple formula for saying beforehand whether a new machine will, or will not, require removing the cabinet front in order to open out the bottom circuit board. One good strategy, though, is to start the process of removing the bottom circuit board, unfastening any locking tabs and removing any hold-down screws, which would need to be done anyway, and then observing whether it will be necessary to remove the front cabinet piece in order to get the bottom circuit board open.

NOTE: If the VCR is an early Panasonic-category "tank," open the bottom circuit boards in the manner shown in the picture in Section 4 of Chapter 20.

Look over the bottom circuit board carefully, and notice whether there are any screws with their heads tinted red. If so, put a circle around each of them. Some manufacturers use the red color coding to enable us to tell which of the screws is actually attaching the bottom circuit board. After marking them remove any red-headed screws.

If none of the screws is red-headed, then your model does not use this color coding. Some, or all, of the black or brass screws you see going into the bottom circuit board are serving to hold it on. Put circles around the ones that seem most likely to be serving this function, and remove them.

Now try wiggling the bottom circuit board. If it still seems to be held down fast, look for plastic locking tabs around the edge of the circuit board and also coming through holes molded for them in the bottom circuit board. Free these by pushing them back and lifting up the edge of the circuit board enough to catch on the edge of the tab and hold it back.

Continue unlocking tabs and removing screws until the bottom circuit board comes free enough that you can move it slightly from side to side in a horizontal plane. You possibly may now be able to open it up on its hinges, or you may find that it will not open because the front edge is caught in under the bottom edge of the front cabinet panel. Before removing the front cabinet panel, find the hinges for the circuit board and examine them to see whether

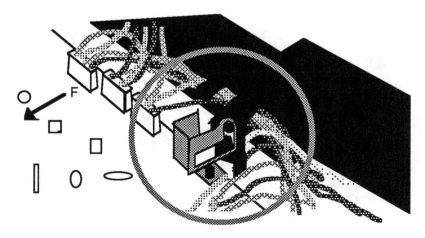

L-Slotted Bottom Circuit Board Hinge

perhaps they are of a design that permits you to slide the bottom circuit board back toward the rear of the machine enough to enable it to clear the edge of the front panel.

If you can see that the front panel obviously must be removed, you can go straight to the next section. But if the hinges are of a design like those shown in the circles in the two illustrations, then they may permit you to slide the bottom circuit board back

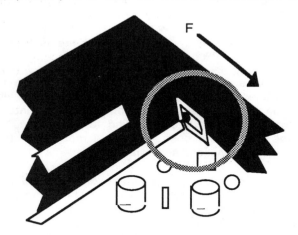

Hinge that Permits Sliding the Circuit Board Back Before Opening

a little toward the rear of the machine, and then open the circuit board on the hinge without needing to remove the front cabinet. If the hinge is L-slotted, as shown in the illustration, then you lift up the edge of the circuit board enough for the plastic peg to reach the corner of the L-shaped slot before sliding the board back and then opening it on the hinge.

Hinges that look as shown in the next illustration allow the whole circuit board edge to come free when

you slide the board toward the rear of the machine.

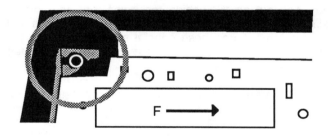

Unfastening-type Hinge

If the circuit board does not slide to the rear, or if it slides but not far enough to clear the edge of the front cabinet panel, then it will be necessary to go to the next section to remove the front of the VCR cabinet.

8. Removing the Front of the VCR Cabinet

Sometimes it is necessary to remove the front cabinet panel in order to open up a bottom circuit board to gain access to the bottom of the reel table. It is also usually necessary to remove the front cabinet in order to disassemble and realign the cassette-loading assembly in a front-loading VCR.

Here is the procedure.

Make certain that the VCR is unplugged.

With the VCR upside down, remove any screws that go through plastic tabs from the bottom of the front cover to attach it to the bottom of the VCR. Also, detach any plastic latching tabs by pulling the larger part

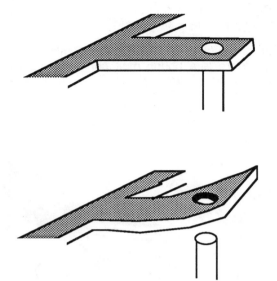

Releasing Plastic Front-Cabinet Locking Tabs

up over the plastic pins that go through the hole in the middle. Some VCR's have plastic tabs sitting loose in slots along the bottom of the front; these work as hinges, not locking tabs. Do not try to unfasten them yet. It will rotate down and away from the VCR on these "hinges," when you have unfastened the top of the front cabinet.

Turn VCR rightside-up.

Set the tracking control to the center position. IF THE HEAD OF THE KNOB IS BIGGER THAN THE HOLE IN THE CABINET THAT IT COMES THROUGH, remove any plastic knobs, such as tracking or level-control knobs, that come out through the front cabinet . Open any little doors or flaps on the front cabinet panel and check inside for knobs. Pull the knobs straight off, going from left to right, and putting each in its own container. Knobs that come out through holes larger than the knob obviously do not need to be removed.

MITSUBISHI-category VCR's: Some Mitsubishi-category machines (for example, the HS 318UR) have a screw close to the middle of the front cabinet that must be removed before the front cover will come off. Take this screw out, and put it in its own container.

If the VCR has a metal cross-brace running above the top of the front of the front-loader assembly (as do many Hitachi-category VCR's), use a felt-tip pen to draw an arrow pointing forward on it to help get it back on again correctly when you reassemble.

Look along the rear edge of the top of the front cabinet to see what holds it on. If it is held by locking tabs, release them as shown in the previous illustration. If it is held by screws, a small piece of metal may be held down under a middle screw. Write "F" (short for "front") on it with an arrow pointing to the front of the machine before removing it, so that you can reassemble it in the right orientation. Remove any screw(s) holding the front cabinet part on. If there are screws going down through plastic tabs, remove them from left to right, one by one. Put them in a container, number, and note where they came from. If some are different from the others, put them in a separate container.

Check to see whether the front cover has plastic locking tabs on the

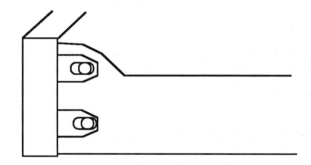

Tabs Locking the Ends of the Front Cabinet to the Sides of the Chassis

ends (or sides) that latch onto the sides of the VCR main chassis. Also, some HITACHI-category VCR's had a plastic trim piece that ran along the bottom of

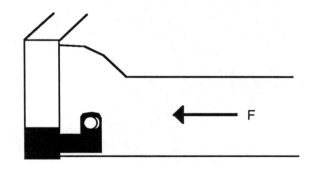

Locking Tab for Lower Part of Front Cabinet on some Hitachi-category VCR's

the front cover and was latched to the side with plastic tabs. Unfasten any plastic tabs holding on the ends of the front cabinet piece.

On some PANASONIC-category VCR's (like the G.E. 4010X, for example), and on older J V C-category machines, you must remove a panel with plastic interchangeable channel numerals under a little door at the edge of the front cabinet on the right before removing the front cabinet piece.

It should now be possible to pull off the front cover gently. On some models, the front of the cabinet pulls straight off. On other models, the cabinet front rotates down and off around some plastic tabs that act as hinges on the bottom. If so, lift the plastic lock tabs on the top and rotate the front cover down toward the bottom of the machine. Then pull the cover free from the plastic hinges on the bottom.

As you remove the front cover, look behind it as you start to pull it loose from the VCR. Watch for any little piece(s) of formed sheet metal that might be clipped onto the VCR under the panel and pull them loose, observing where they came from. Note where they came from with a felt tip pen on the machine, and in your notes.

If there are any loose buttons or slide switches, or pieces of felt or thin plastic with rectangular holes cut in them that were previously held in position by the front cabinet, remove them too, put them into a container, and note where they came from. Depending on the model, there may or may not be some of these, but if there are, catch them now, or you may lose them or overlook them on reassembly later.

If there are any wires with a push-on connector attached to thin sheet metal in back of the plastic cover, or to a multi-pin connector, disconnect it and note where it was connected as follows.

Unfastening Ribbon Connectors

On some JVC, Zenith, and other models, including some Panasonic-category machines, there may be a

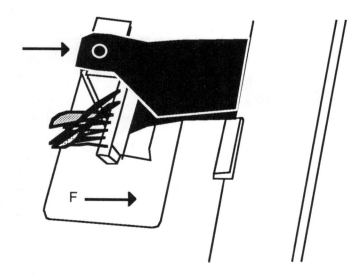

Unfastened Connector from Front Panel

thin, flexible piece of metal foil (a grounding connector) coming back from the front cover to a place where it is screwed down under a springy brass piece of metal with a single screw into a metal bracket on the bottom. If your model has this, write "F" with arrow pointing toward the front on the springy metal part, unscrew the screw, and put it and the springy part in a numbered container.

On the bottom of some JVC's, Zenith's, and other models, there is a ribbon multi-connector coming from the front cabinet part and running into a rectangular plastic connector. On some of these, you disconnect the ribbon by lifting up the end of the plastic connector closest to the right side of the VCR. It swings on a hinge on the end, like this

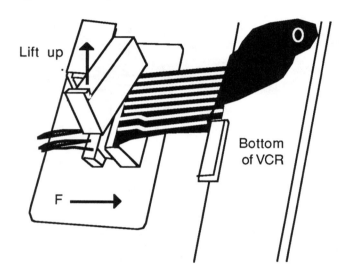

Disconnecting a Zenith-JVC Ribbon Connector

On some Panasonic-category machines and other models, *both ends* of the plastic connector pull up equally a short distance, and this frees the ribbon. Pull the ribbon out of the connector carefully.

Unplugging Multi-wire Plugs

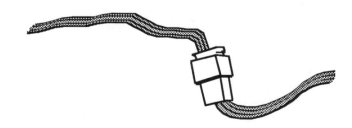

Multi-Wire Connector Used in Some Models

On some models, from the top of the cabinet part, a multi-wired connector may come out that needs to be unplugged before the cabinet part can be removed. There may be several wires running from the front cabinet piece into the top of the VCR where they plug into a connector as shown in the illustration. Disconnect this connector and make a note in your disassembly record that you did so. You may need to insert a scribe or other pointed tool into this plug to loosen it, as illustrated:

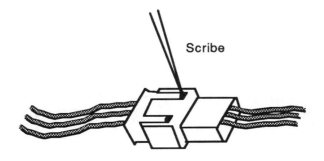

Using a Scribe to Disconnect a Multi-Wire Connector

Removing Front-Loading Assembly on Front-Loaders

On some Hitachi-category and other machines, there may be a horizontal metal brace that runs across the top of the front entrance to the front-loader that simply lifts free, once the front cabinet has been removed. Before lifting it free, though, it would be a good idea to draw a little arrow pointing forward on the top of it with a felt-tip pen, to help you get it back in correctly on reassembly later.

Sharp: On some Sharp-category models, there are two large screws at the bottom of the front-loader opening, fastening a long black plastic part to the VCR cabinet, which need to be removed.

On all models, there will be from zero to four screws, often with redish-tinted heads, holding the cassette loading assembly in, either screwed down below directly into the metal of the reel table, or else clamping the top of the front-loader assembly down to plastic pylons that are part of the plastic cabinet box, as in some machines in the JVC category and Fisher categories (see below). After these screws have been removed, on some models, the front-loader can be lifted straight up out of the VCR, while on Sharp and some other models, you must first slide it horizontally a short distance to free two little feet molded into the sides of the assembly that catch in places provided for them in the reel table. There is more on removing these screws in the next two paragraphs.

Remove the assembly as a whole. Never try just to remove only the top part of the front-loader.

To remove the cassette loading assembly, unplug the multi-wire connector, which is usually attached to the right rear corner and then locate the screws that are holding the assembly in the VCR. On most brands and models, there are two, three, or four screws that hold the bottom of the cassette loading assembly down to the reel table; if you look carefully down at the bottom along the sides of the assembly, you should be able to figure out which screws are holding down the assembly.

JVC AND FISHER: On some models made by JVC, and by FISHER, the screws holding in the cassette loading mechanism have their heads at the top, rather than at the bottom, of the assembly, and they screw into plastic pylons that are part of the cabinet rather than into the metal reel table. They are the outermost four screws.

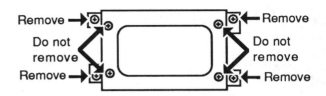

Position of Screws to be Removed on Front-loaders whose hold-down Screws are Up on Top Rather than Down at the Bottom

Don't loosen the inboard four screws that hold the front-loading assembly together, for we want to remove the assembly as a whole. Never try just to remove only the top part of the front-loader.

FISHER: If the machine is one of the few models with wires coming to the front loading assembly from circuit boards on both the right and left sides, (like, for example, model FVH 730), then removal of the front-loading assembly should be left to a service center. It is too difficult for this book.

Wherever they are, loosen them with a cross-slot screwdriver and lift them out taking care not to let them drop down into the mechanism. A magnetic screwdriver or parts "picker-upper" may be helpful here. (See Appendix I) There is more on these screws in the next paragraph.

For some unknown reason, the screws that fasten into the metal reel table sometimes seem to be in extremely tight. The robot who put them in must eat Wheaties for breakfast, or something. Here the special cross-slot screwdrives described in Appendix I are great because they grip these screws better than ordinary hardware-store Phillips' head screwdrivers.

The danger is that if the screwdriver does not grip well, and you try to unscrew it (CCW, of course), it may slip and round off the corners of the cross-shaped slot, so that you can never get the screw out.

You need the larger size screwdriver, #2, for these screws.

Whether you get the special screwdriver, or use the ordinary hardware store variety, be sure to press the screwdriver down into the slots very hard before you start to turn it the first time. Lean on it with about forty pounds (18 Kg.) of force before you start to turn it the first time. Turn it slowly and firmly, watching carefully to make certain that the screwdriver is not starting to slip and rise up out of the slot where it can slip and round the edges of the slot. After the screws have been broken free the first time, you should not need to use such force again on them.

Sharp: On some Sharp-category VCR's, there may also be two large screws, probably with black colored heads, along the front lower edge of the opening into which you insert the cassettes, which also need to be removed in order to remove the assembly.

Locate and remove the screws holding down the cassette basket assembly. On some JVC and Zenith models, they do not screw into the reel table. Instead, they go through a top piece of sheet metal into four plastic posts. One of these often has a springy piece of brass sheet metal with a wire attached to it. Use a felt-tip pen to draw the outline of the wire terminal on the brass piece under it, and draw an arrow pointing forward on the springy brass piece. Unscrew the screws, taking extreme care not to let any of them drop into the mechanism below. It is very important that you not let the screws drop into the mechanism.

With the screws out, remove the cassette basket assembly. On some models, you remove this by just wiggling it a little and pulling up on it. Some assemblies can then be lifted straight up and out of the VCR, while others you need to slide a short distance either forward (like the Sharp-category) or toward the rear (some Hitachi and other brands) in order to loosen little feet that catch through slots in the reel table before the assembly can be lifted free. You must slide some other front-loaders straight back, toward the rear of the machine, or forward, or tip them out after removing the screws

EMERSON

On the Emerson B, the trick is to unplug the multi-wire connector running to the front-loading assembly, tip up the rear of the front-loading assemply, and then lift it up and out.

FISHER

If the machine is one of the few models with wires coming to the front-loading assembly from circuit boards on both right and left sides (like, for example, the Fisher FVH 730), then removal of the front-loading assembly should be left to a service center. It is too difficult for this book.

HITACHI

You may need to slide the front-loader straight back, toward the rear of the machine, before you can lift it out.

SHARP

You must slide some Sharp-category and other cassette front-loading assemblies a short distance *toward the front* of the machine before little feet at the bottom come free of slots in the reel table beneath, so that they can be lifted out.

As the front-loading assembly starts to come loose, unfasten the multi-wired connector attached to it. If you have trouble, you can do this by holding the basket assembly in one hand and pulling on all the wires together while wiggling them, until the multi-wire plug starts to come unplugged. When a small space opens up between the two halves of the plug, you can insert the blade of a tiny flat screwdriver, and twist it to pry the plug the rest of the way open, working along the length of plug. When it comes free, lift the cassette assembly out.

Appendix IV

Generic Belts and Common Idler Part Numbers

If you cannot locate a part number, or find a source for, a belt or idler elsewhere in this book, the information in this appendix may help solve the problem.

Idlers

To use the information in this chapter to find the part number for an idler, look up the brand name of your VCR in the list in the Introduction, find the category in which it is placed, and then look in this appendix at the idlers listed under that category.

In some cases, the entire idler, including both the plastic wheel and the rubber belt, is pictured, and in other cases, only the rubber tire is listed. When only the rubber idler tire is pictured, the letter 'T' will appear in the center. In this case, the part number followed by the abbreviation of the name of the company (for example, "(Pan)," "(Sam)"), is the part number for the complete idler, while a number beginning with the prefix "04" and followed by "(Edc)" is the part number for just the tire alone, purchased from Electrodynamics.

Remove your idler and hold it up to the pictures to find its part number. If only the idler tire is shown, you may need to pull the rubber tire off your idler and compare it with the pictures. All idlers are shown actual size in the drawings.

In most categories, the idlers are sufficiently different in size that there should be no confusion. But if you run into a situation where you cannot figure out which of two possible choices is your idler, and you cannot determine this from the information in Chapter VI either, you might consider just ordering one of each of the two possibilities, and making the final determination when you have the parts in hand. Most idlers are so inexpensive that this strategy will still

be a whole lot cheaper than taking the machine to a shop. Most idlers are so inexpensive, in fact, that you may have a problem meeting the minimum purchase requirement of the supplier. Consider ordering some cleaning chemicals and tools at the same time, or go together with a friend and order parts for his machine too.

When you have found the part number, look in Appendix II for the name and telephone number of suppliers of parts in those categories.

Generic Belts

If you cannot get a replacement part number for a stretched or broken belt, or if you have a part number but the belt is no longer available, you probably can get by with a generic replacement belt of the same size, or approximately the same size.

To order a generic replacement belt, you will need to measure the inside circumference and thickness of the belt you are replacing. A belt measuring gauge is included at the end of this appendix for this purpose.

To measure the inside circumference of the old belt, stretch it between two pencils as shown, with one pencil centered on the black circle, and note the number of the other circle on which the end of the second pencil falls when the slack is stretched out of the belt.

If the old belt is broken, you will need to tape the broken ends together long enough to make the measurement, or get another person to hold the two ends together while you make the measurement.

Do not overstretch while measuring. The new belt should be three to five percent smaller than the old belt in circumference.

Next measure the thickness of the belt by holding it up against the belt width gauge. If the belt is not square, measure both dimensions.

Write down all your measurements and call one of the sources listed for generic belts. If they don't have exactly the same size, choose the next smaller circumference, and the next larger width

Common Idlers

Akai (see also Mitsubishi)

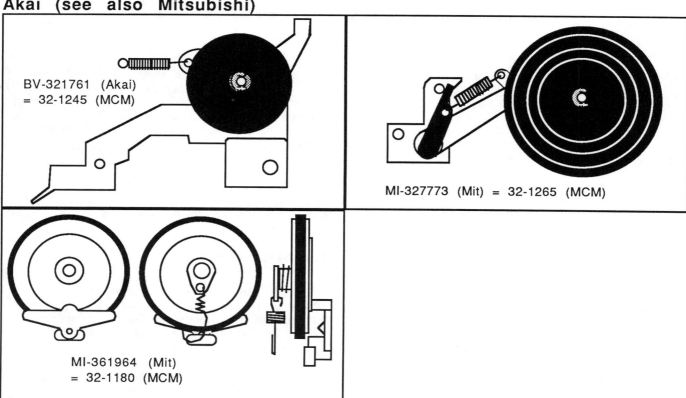

BV-321761 (Akai)
= 32-1245 (MCM)

MI-327773 (Mit) = 32-1265 (MCM)

MI-361964 (Mit)
= 32-1180 (MCM)

Emerson

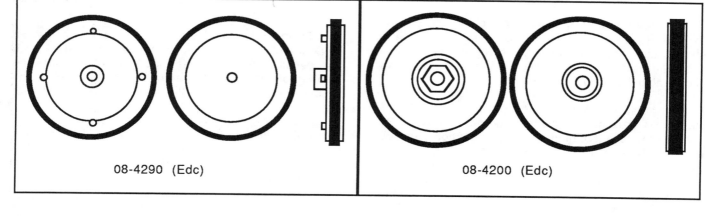

08-4290 (Edc)

08-4200 (Edc)

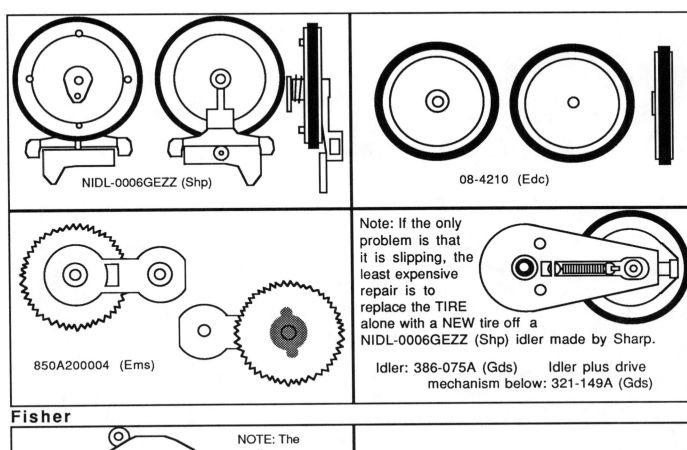

NIDL-0006GEZZ (Shp)

08-4210 (Edc)

850A200004 (Ems)

Note: If the only problem is that it is slipping, the least expensive repair is to replace the TIRE alone with a NEW tire off a NIDL-0006GEZZ (Shp) idler made by Sharp.

Idler: 386-075A (Gds)   Idler plus drive mechanism below: 321-149A (Gds)

Fisher

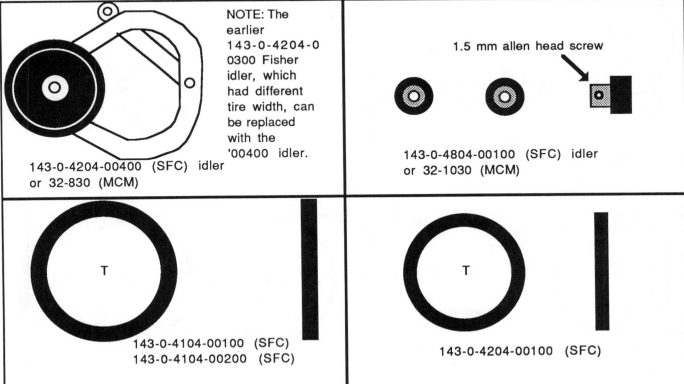

143-0-4204-00400 (SFC) idler
or 32-830 (MCM)

NOTE: The earlier 143-0-4204-0 0300 Fisher idler, which had different tire width, can be replaced with the '00400 idler.

1.5 mm allen head screw

143-0-4804-00100 (SFC) idler
or 32-1030 (MCM)

T

143-0-4104-00100 (SFC)
143-0-4104-00200 (SFC)

T

143-0-4204-00100 (SFC)

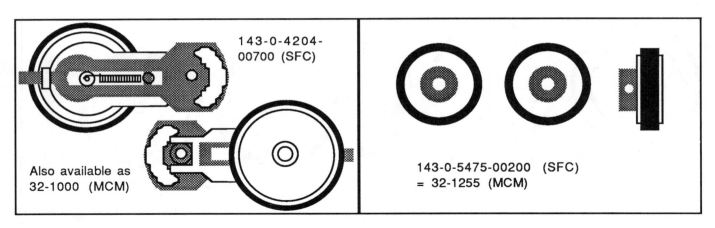

143-0-4204-00700 (SFC)

Also available as 32-1000 (MCM)

143-0-5475-00200 (SFC) = 32-1255 (MCM)

Goldstar

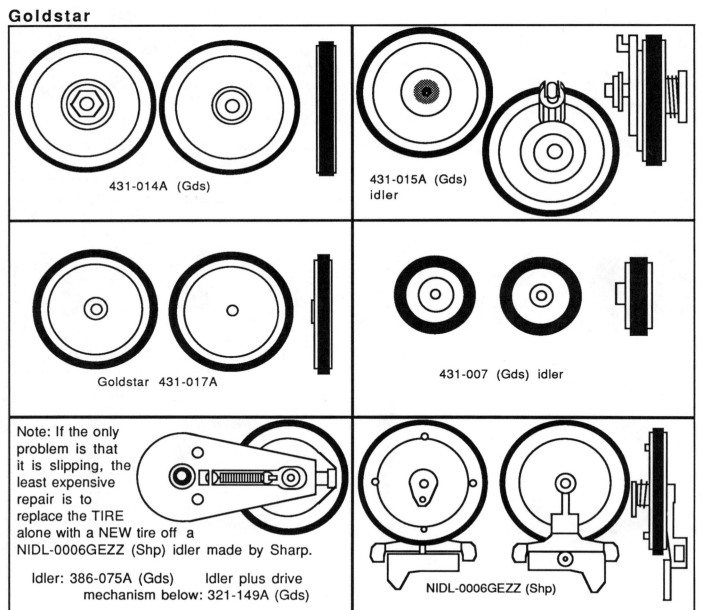

431-014A (Gds)

431-015A (Gds) idler

Goldstar 431-017A

431-007 (Gds) idler

Note: If the only problem is that it is slipping, the least expensive repair is to replace the TIRE alone with a NEW tire off a NIDL-0006GEZZ (Shp) idler made by Sharp.

Idler: 386-075A (Gds)   Idler plus drive mechanism below: 321-149A (Gds)

NIDL-0006GEZZ (Shp)

Hitachi-category

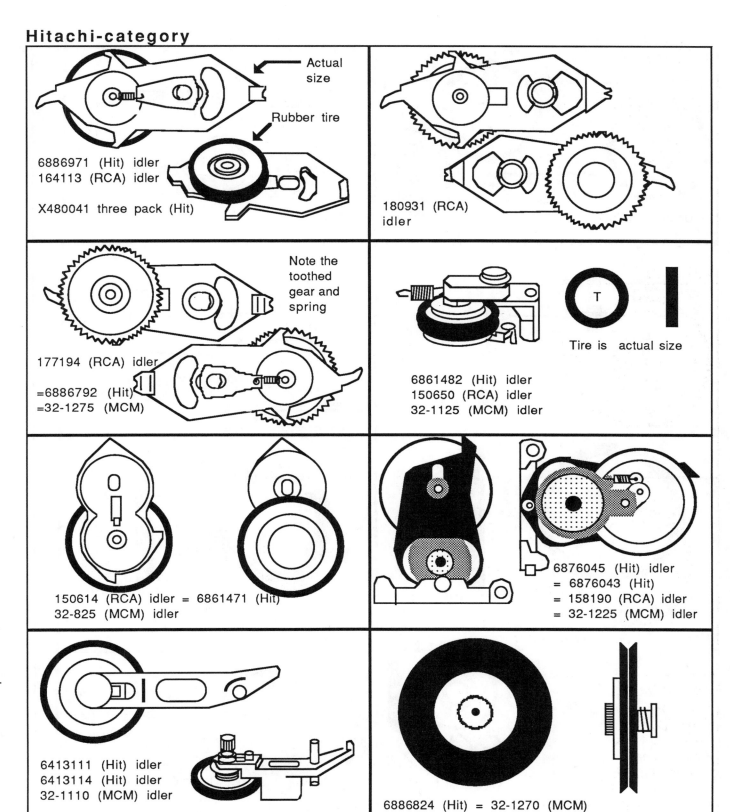

Actual size

Rubber tire

6886971 (Hit) idler
164113 (RCA) idler

X480041 three pack (Hit)

180931 (RCA)
idler

Note the toothed gear and spring

177194 (RCA) idler
=6886792 (Hit)
=32-1275 (MCM)

6861482 (Hit) idler
150650 (RCA) idler
32-1125 (MCM) idler

Tire is actual size

150614 (RCA) idler = 6861471 (Hit)
32-825 (MCM) idler

6876045 (Hit) idler
= 6876043 (Hit)
= 158190 (RCA) idler
= 32-1225 (MCM) idler

6413111 (Hit) idler
6413114 (Hit) idler
32-1110 (MCM) idler

6886824 (Hit) = 32-1270 (MCM)

J V C category

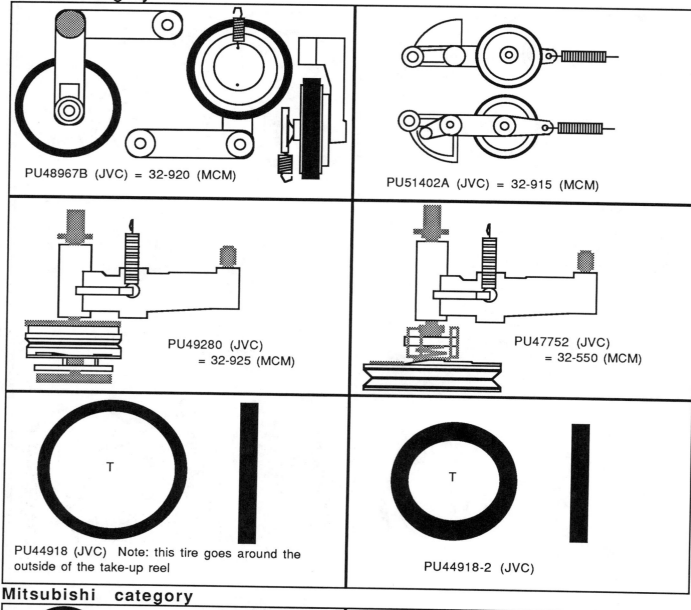

PU48967B (JVC) = 32-920 (MCM)

PU51402A (JVC) = 32-915 (MCM)

PU49280 (JVC)
= 32-925 (MCM)

PU47752 (JVC)
= 32-550 (MCM)

PU44918 (JVC) Note: this tire goes around the outside of the take-up reel

PU44918-2 (JVC)

Mitsubishi category

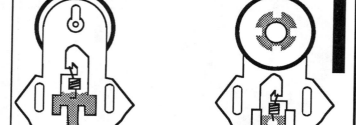

522B01701 (Mit) idler
04-0470 (Edc) tire

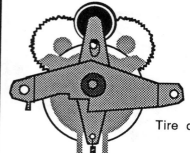

522P00201 (Mit) idler assembly shown at 75% of actual size

Tire only: 08-5475 (Edc)

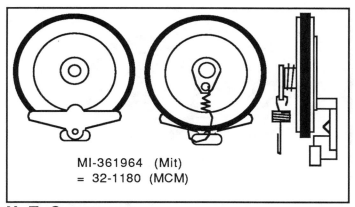

MI-361964 (Mit)
= 32-1180 (MCM)

N E C category

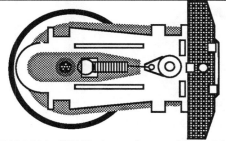

16180801 (NEC) (top view only) = 32-1220 (MCM)
(For idler plus entire clutch assembly underneath,
the part number is 016-17-8179 (NEC).)

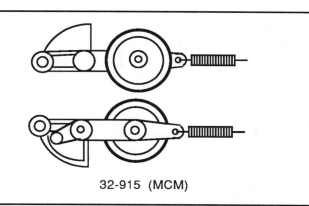

32-915 (MCM)

Panasonic category

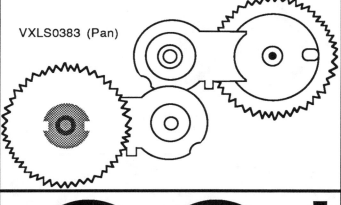

VXLS0383 (Pan)

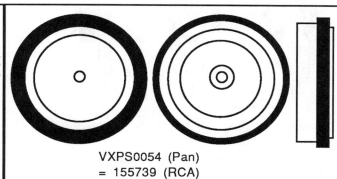

VXPS0054 (Pan)
= 155739 (RCA)

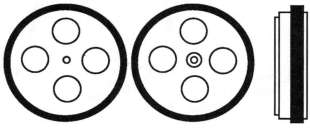

VXPS0069 (Pan) idler = VXP0401 (Pan)
= 32-810 (MCM)

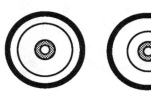

VXPS0075 (Pan) idler

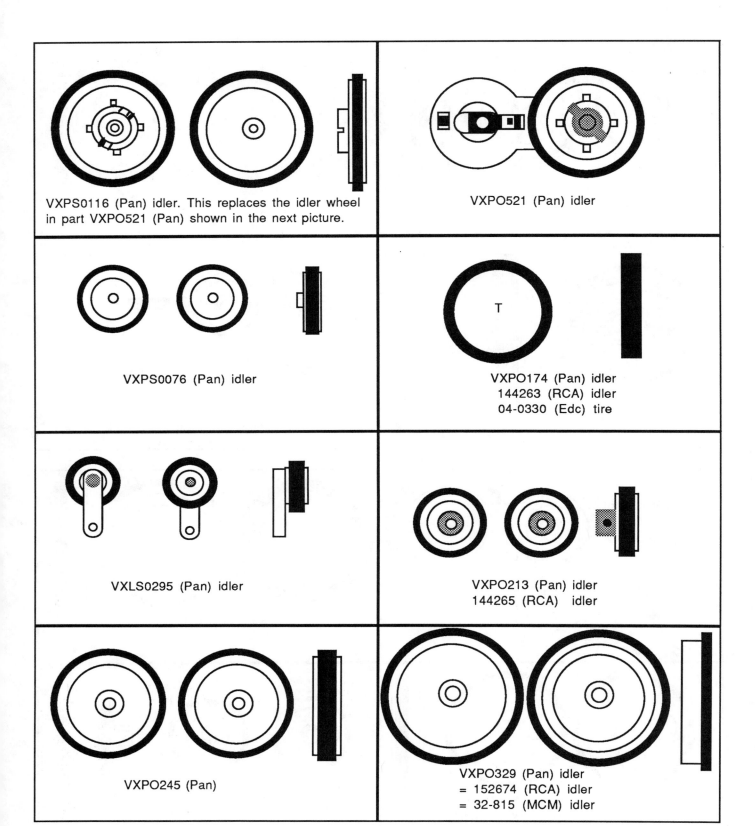

VXPS0116 (Pan) idler. This replaces the idler wheel in part VXPO521 (Pan) shown in the next picture.

VXPO521 (Pan) idler

VXPS0076 (Pan) idler

VXPO174 (Pan) idler
144263 (RCA) idler
04-0330 (Edc) tire

VXLS0295 (Pan) idler

VXPO213 (Pan) idler
144265 (RCA) idler

VXPO245 (Pan)

VXPO329 (Pan) idler
= 152674 (RCA) idler
= 32-815 (MCM) idler

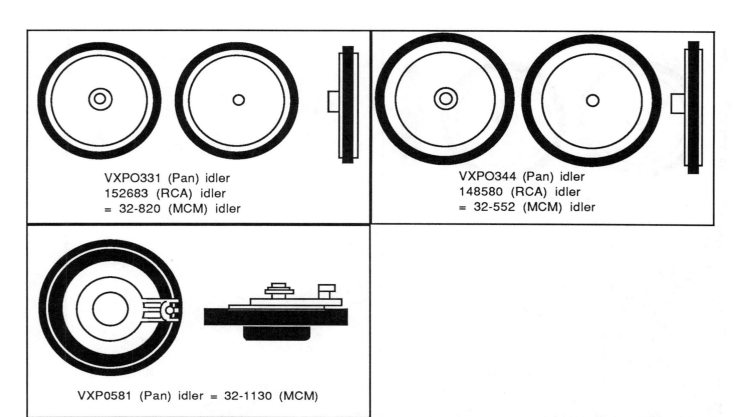

VXPO331 (Pan) idler
152683 (RCA) idler
= 32-820 (MCM) idler

VXPO344 (Pan) idler
148580 (RCA) idler
= 32-552 (MCM) idler

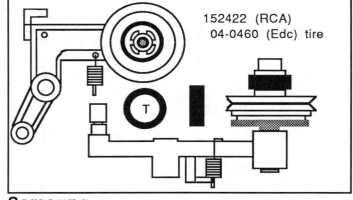

VXP0581 (Pan) idler = 32-1130 (MCM)

R C A (see also Hitachi, Panasonic, and Samsung categories)

152422 (RCA)
04-0460 (Edc) tire

Samsung

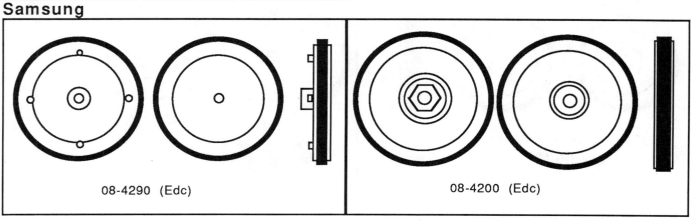

08-4290 (Edc)

08-4200 (Edc)

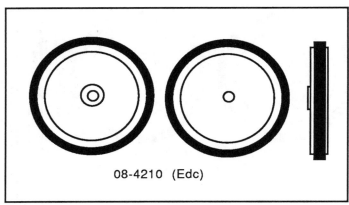

08-4210 (Edc)

Sanyo

143-0-662T-01201 (SFC)
or 32-1020 (MCM).
Can just replace idler wheel
& tire: 32-910 (MCM)

143-0-662T-01202   (SFC)
or 32-1020 (MCM).
Can just replace idler wheel
& tire: 32-910 (MCM)

143-0-662T-10350   (SFC)
or 32-1015 (MCM). Can
just replace idler wheel
& tire: 32-910 (MCM)

This subassembly is the part that usually goes bad in
the 143-0-662T-01201 (SFC) idler assy, and the
143-0-662T-01202 (SFC) idler assy, and the143-0
- 662T-10350 (SFC) idler assy. Its part number is
143-0-741T-20001(SFC)
or you can replace it
with 32-910 (MCM)
or tire 32-660 (MCM)

143-0-551T-01600 (SFC)
32-1135 (MCM)

143-0-662T-15620   (SFC)
= 32-1250 (MCM)

Sanyo (cont.)

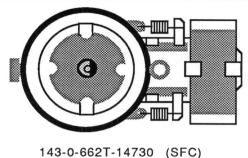

143-0-662T-14730 (SFC)
= 32-1240 (MCM)

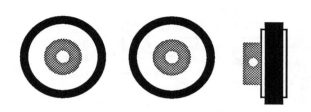

143-0-5475-00200 (SFC)
= 32-1255 (MCM)

Sansui

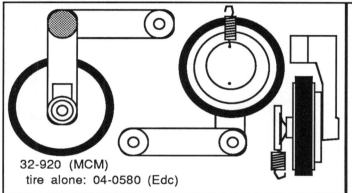

32-920 (MCM)
tire alone: 04-0580 (Edc)

32-915 (MCM)

Sharp

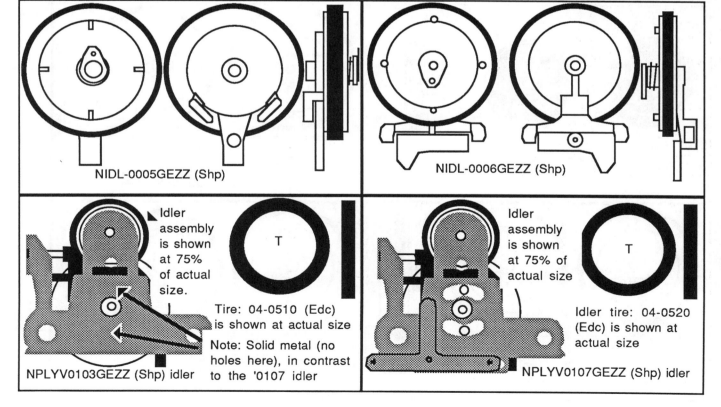

NIDL-0005GEZZ (Shp)

NIDL-0006GEZZ (Shp)

Idler assembly is shown at 75% of actual size.

Tire: 04-0510 (Edc) is shown at actual size

Note: Solid metal (no holes here), in contrast to the '0107 idler

NPLYV0103GEZZ (Shp) idler

Idler assembly is shown at 75% of actual size

Idler tire: 04-0520 (Edc) is shown at actual size

NPLYV0107GEZZ (Shp) idler

Sharp (cont.)

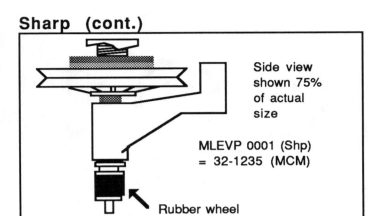

Side view shown 75% of actual size

MLEVP 0001 (Shp) = 32-1235 (MCM)

Rubber wheel

Sony category

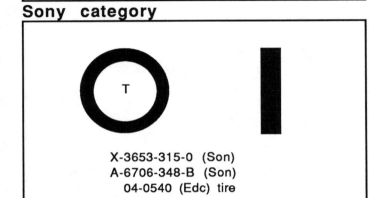

X-3653-315-0 (Son)
A-6706-348-B (Son)
04-0540 (Edc) tire

Toshiba

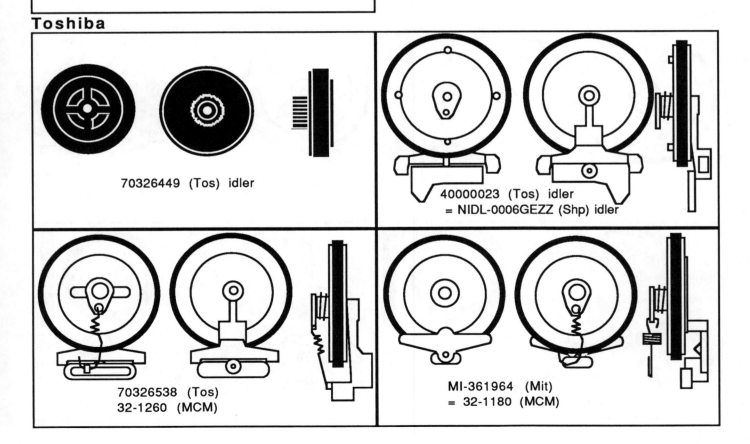

70326449 (Tos) idler

40000023 (Tos) idler = NIDL-0006GEZZ (Shp) idler

70326538 (Tos)
32-1260 (MCM)

MI-361964 (Mit) = 32-1180 (MCM)

Universal Generic Belt Measuring Gauges

Note: This measurement gauge is intended for use as a last resort, only if you cannot get an exact replacement from a supplier. Look first in Appendix VI for a picture of your reel table and a part number for the belt you need.

How to Measure the Inside Circumference of VCR Belts

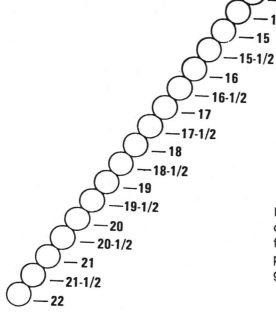

Position belt with black circle at the center. If belt is broken, tape back together temporarily to make the measurement, or else have someone hold the ends butted together while you measure. Put the eraser end of a pencil down on the black circle. Put the eraser end of a second pencil down also inside the belt. Pull second pencil in direction of numbered circles until all slack is eliminated from the belt. Do not stretch beyond this. Note number that second pencil is on. This is the inside circumference of the belt in inches. The new belt should be about 5 % smaller than the worn belt, but you will probably be forced to use a belt that is only approximately the same size anyway, since the number of generic belt sizes available is limited.

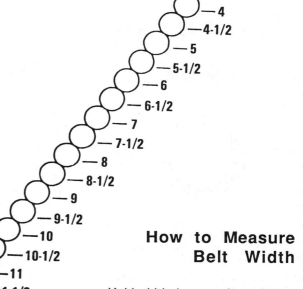

—3
—3-1/2
—4
—4-1/2
—5
—5-1/2
—6
—6-1/2
—7
—7-1/2
—8
—8-1/2
—9
—9-1/2
—10
—10-1/2
—11
—11-1/2
—12
—12-1/2
—13
—13-1/2
—14
—14-1/2
—15
—15-1/2
—16
—16-1/2
—17
—17-1/2
—18
—18-1/2
—19
—19-1/2
—20
—20-1/2
—21
—21-1/2
—22

How to Measure Belt Width

Hold old belt up to the stripes below, and write down the number of the stripe that is nearest in size. If it is in-between, go with the next larger size. If the belt is wider in one dimension than the other, measure both widths. If it is a round belt, the single measurement will give its diameter. The figures given are in fractions of inches.

Belt Width Gauge

.020
.025
.031
.040
.046 - ³/₆₄
.050
.062 - ¹/₁₆

.075
.093 - ³/₃₂
.100
.125 - ⅛
.140 - ⁹/₆₄

.156 - ⁵/₃₂
.187 - ³/₁₆
.203 - ¹³/₆₄

.218 - ⁷/₃₂
.250 - ¼

If you have the measured dimensions of the belt(s) you need, you can order replacement generic belt(s) of that size from PRB, from MCM, or from Vance-Baldwin/Workman. The addresses, phone numbers, and order procedures for these companies are given in Appendix II.

Appendix V

How to Trick a VCR into Running With No Tape Inside for Diagnostic Purposes

Some chapters of this book tell the reader to use the "trick of making the VCR run when there is no cassette inside it." This appendix is a supplement to the book explaining how to do this. This technique is sometimes extremely helpful when you are trying to troubleshoot a mechanical problem in a VCR's Play, Rewind, or other modes of operation.

For example, suppose the take-up reel is not turning, or the machine is not rewinding correctly, and you suspect that the problem is in an idler or motor that is located under where the cassette rests. You need to observe what is happening during the time it is running, but whenever it is running, a cassette is always covering the parts you need to see. In such a case, it is useful to be able to make the machinery run with *no cassette* in the machine, so that you can watch and figure out exactly where the trouble is.

However, it takes know-how to be able to make a VCR run with no cassette inside it. It certainly would not work just to hit the "Play" or "Rewind" button with no cassette in the machine. Even a perfectly working VCR will just sit there and do nothing if you do this, because it has a little computer inside that has sensors to monitor what is going on inside the machine, and it is permanently programmed to refuse to try to move a tape unless all its sensors report that a cassette is inside the machine.

You could insert an empty cassette box rigged up as a "dummy cassette," constructed as described in Section 17 of Appendix I, which would make it run, all right, and would enable you to see what was going on in some other parts of the machine, but still would not enable you to see what was going on underneath the cassette because there now would be an empty cassette box in the way. So, some other kind of trick is needed when the problem has this location.

To make it run as if there were a cassette inside, it is necessary to "fool" the little computer into thinking that the VCR has a cassette inside when actually it does not. This appendix tells you how to do this. — that is, how to trick it into thinking that a cassette is inside, and so, make it go through all its paces, when actually it is empty.

This trick will be explained in the following sections. But first, let me mention three things. For this trick, it is assumed that the cabinet top has already been removed, and that you have been referred to this appendix by some earlier chapter in the book in the process of hunting down and correcting a problem. If the machine will not move at all, then you need to be reading some other part of this book. This procedure assumes that although you may have trouble, if you insert a real cassette and push a button, the machine can at least move the mechanism. You cannot trick someone into going outdoors if he is paralyzed and cannot move at all. Similarly, the trick I am going to explain will not work if some trouble in the machine prevents it from doing anything at all even if you load a normal cassette and push "Rewind" or "Play." If the VCR is totally unable to move, then you need to use another procedure to fix the problem.

The procedures for carrying out this deception vary, depending on whether it is a top-loader or a front-loader, and on whether it is a VHS or Beta machine. Go to whichever section is appropriate to your situation: (1) VHS Top-loaders, (2) VHS Front-loaders, (3) Beta Top-Loaders, (4) Beta Front-Loaders.

1. VHS Top-loaders

In most VHS machines, there are three sensors that may need to be fooled in order to trick it into operating as if there were a tape inside, when actually it is empty: the tape-end sensors, the cassette-down latch and switch, and possibly a reel-motion detector. We will go through these one-by-one, and do something to make each sensor circuit think that it detects the presence of a tape (when actually there is no cassette in the machine). We do not disconnect the sensors (the little computer could detect that, and might shut everything down). Instead, we will activate each of the sensors as if a cassette were loaded into the VCR.

On a VHS *top loader*, when a cassette is inserted into the cassette basket and pushed down into the VCR for normal operation, the tape end-sensors, which have been receiving light from the sensor lamp, located between the two reels, are the first sensors to detect the tape. (See Chapter 2 for an explanation of the operation of the VHS sensor lamp and tape end-sensors.)

As the cassette descends, it blocks this light from reaching the end-sensors. For the end-sensors, everything goes black for a moment as the cassette descends, and they relay this fact as the first signal to the little computer indicating that a cassette has been loaded. So, we will outfox this system by leaving the empty cassette basket up, while we put some very thick opaque pieces of paper, or tape, over the end-sensors on each end. Something about the size of small business cards is good — that's big enough that we won't forget to remove them later (and then wonder why the little computer cannot tell when the end of the tape has been reached).

So, on a VHS top-loader, first plug it in and turn on the power switch. If the basket is not already up, raise it, and then put in the pieces of paper to block the light. The best place to put them is right in front of each end-sensor.

The cassette top-loader assembly has been omitted from this illustration for clarity, but it normally would still be in the machine (just in the "up" position), unless you needed to take it out for repair or something. Trying to cover the light itself does not work so well; it sometimes works, but other times there is enough light coming in from the lamp above your table that the end-sensors "see" it instead, and defeat your purpose. Each sensor is usually inside a small black plastic enclosure with a round port-hole window pointing toward the sensor light. On some models, the light is invisible infra-red light, so you cannot see it, but the sensors can. Other models use plain visible light from a little light bulb.

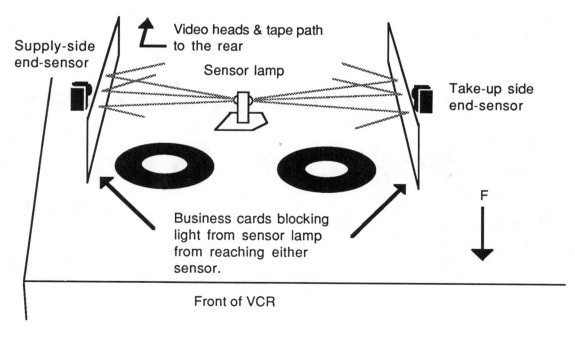

Blocking the End-Sensors from Receiving Light from the Sensor Lamp to Make the VCR Think a Cassette is Loaded

On early Panasonic-category "tanks," blocking the light to the end-sensors is all you need to do; after blocking the light, push "Rewind," "Fast-forward," "Play," etc., and the machine should go through its motions. But on most other VHS machines, we will need also to trick a mechanical switch that is normally moved and switched by the physical action of the descending cassette basket. In the simplest case, this switch is part of the mechanism that latches and holds the basket down.

Look on the bottom of the cassette basket and see whether you can see a little arm coming down with a small horizontal pin attached that goes into a waiting "jaw" in the mechanism below. If you can, take a scribe or similar tool and push it into the jaw until it snaps closed. On some models, the part that latches the basket down is not accessible from the top, and is not what trips the mechanism anyway. In this case, look for a small push-button switch mounted on the reel table underneath where the basket comes down, or for a prong sticking straight down like a bayonet from the bottom of the basket, a prong that goes through a hole in the reel table to activate a switch underneath. In either case, find a way to do the same thing to the switch yourself that the cassette basket would normally do as it descends.

Having done this, try pushing, say, "Rewind" and watch what happens. If the mechanism starts to move, you've succeeded in tricking it, and you can now go on to troubleshoot it, or whatever led you to this appendix in the first place. If the mechanism does not start to move, then another switch someplace on, underneath, or at the side of the reel table, must need to be activated to complete the deception. You will have to continue looking until you find it.

On some models, you may need to twirl one of the reels slowly with your fingers to keep the VCR operating with no tape. Just rotate the reel that is not being motor-driven, twirling it in the same direction that the tape would pull it if there were a tape in the machine — that is, the same direction (clockwise or counter-clockwise) that the other reel is moving.

2.   VHS Front-loaders

To make a front-loading VCR operate with no tape inside, it is necessary to lift the front-loading assembly up from the reel table, which means that the screws holding it down must be removed and the front-loader lifted out, which usually means that the front cabinet panel also must be removed. Naturally, this is done with the VCR unplugged. See Chapters 8 and 10, and Appendices III and VII if you need help in removing the front-loading assembly.

NOTE: If, when you remove the top of the cabinet, you can see that the wires coming to the front-loading assembly are not grouped together, but come to different points on it from both sides, or are very

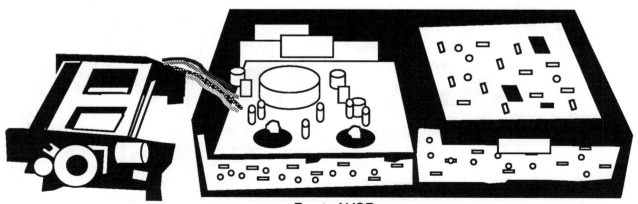

Front of VCR

Empty Front-Loader with Wires Attached Still Connected to VCR

Folded newspaper

Empty Front-Loader with Wires Attached Still Connected to VCR

short, with little or no slack, then it will be impossible to use the following procedure.

In particular, do not try to remove the front-loader on VCR's like the Fisher FVH 730 on which the wires connected to the front-loader are not long enough to reach the front-loader if it is set off to the side. However, you may still be able to trick the mechanism in such VCR's into playing with an empty "dummy cassette" inside, made as described in Section17 of Appendix I   And on Samsungs and later model R C A units, where the cassette front loader is run through gears by a motor located below the mechanical chassis, you will need to remove the screws but leave the front-loader still in the VCR until you have powered up and the unit has accepted a cassette, then push stop, then lift out the whole front-loader with the cassette still inside it.

Leave the wires connected to the front-loading assembly when you remove it. If the wires come into the reel-table side (seen as you face the front of the VCR), then probably they will be long enough to permit you to set the front-loading assembly by the closest end of the VCR, as shown in the illustration.

If the wires from the front-loader connect on the other side, then put a few layers of folder newspaper over the circuit boards, or whatever is on that side of the VCR, and set the front-loader on top of the paper with the wires still attached, as shown in the next illustration. Because the purpose of the paper is to insulate the loading assembly from the electronic circuits underneath, and a *metal* staple in a magazine, for example, might short out some components, and cause an additional problem, do not use paper with any metal in it Folded newspaper is best.

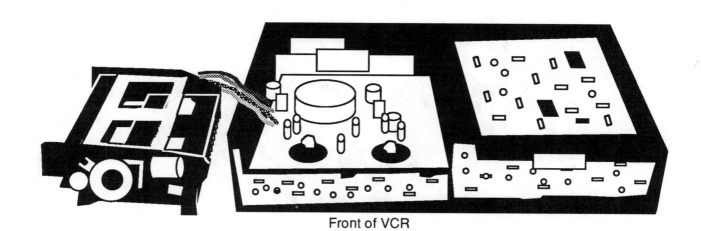

Front of VCR

Cassette Loaded Into Detached but Connected Front-Loader

Folded newspaper

Cassette Loaded into Front-Loader Resting on a Circuit Board

In the illustrations, the front-loader is drawn sitting rightside-up, but this is not necessary. Most front-loaders will work just as well upside-down (they do not work by gravity). If the wires to the front-loader are so short that you can only set the front-loader upside down with the wires still attached, or if the front-loader is too "tippy" when it is out of the machine in a rightside-up position, then set it upside down. The only important thing is to position it so that you can get at the front opening to put a cassette in.

Now plug in the VCR. The clock display should come on. Look at the front cabinet cover that you removed, and see where the power button is. Find the little push-button switch on the front circuit board that is underneath this button, and use the eraser of a wooden pencil, or a wooden stick, to push on it. The

power light, or whatever your VCR does when you power it up, now should come on.

Now steady the top of the removed front-loader with one hand, and with the other hand, take an ordinary normal cassette and push it gently into the opening, just as you do when you load a cassette. The mechanism should be activated and pull the cassette in as if it were loading it (which is what the little computer thinks it is doing)

Tricking the Sensors

Any "cassette-in" switches located in the front-loading assembly are getting switched by the presence of the cassette in the front-loader, so they send a

Press down on this "cassette in" lever

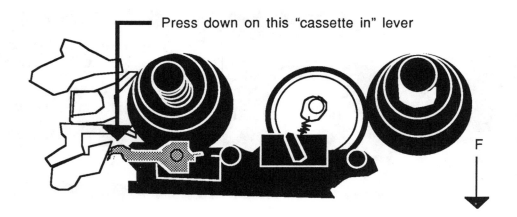

Cassette-Down Switch Mounted on some Sharp Mechanical Chasses

"cassette is in the VCR" signal through the wires, which are still attached to the front-loader, to the little computer (even though actually the whole assembly is sitting off to the side of the reel table.

And the light sensor mounted in the sides of the front-loading assembly see darkness, because they are covered by the loaded tape. So they also tell the little computer that there is a tape in the machine.

Observing the Hidden Operations

A cassette is inside the front-loader, which is still attached to the rest of the VCR by its wires, but the cassette is not covering the area of the machine that we wish to watch. It is now free and open to view, unlike previously when the loaded cassette sat down over it. The normal cassette is not blocking the view of this area, because the cassette is now sitting off to the side of the reel table, instead of sitting down on the reel table. Look at where the cassette used to sit. Now there is no cassette sitting there, right? So the machine thinks it has a cassette inside it, but the cassette is not covering the area we need to look at. That area now can be watched once we finish the deception to get the VCR running. Look at the area on the reel table underneath the place where the front loading assembly was sitting, and notice that it is now uncovered.

Completing the Deception

Look at the reel table under where the cassette would normally rest to see whether there is a little push-button switch that the cassette would normally depress when it is loaded. For example, in some SHARP-category VCR's, there is a little sensor that looks as shown in the next illustration.

If you see any such switch, hold it down with a stick, so the little computer will think there is a cassette in the VCR. On most models, this "cassette in" switch or sensor is part of the front-loading assembly, and is automatically activated when the cassette was loaded into it earlier — just loading the cassette is enough to complete the deception on many models, but if there is a switch on the reel table, press it like a cassette would do.

Now try pressing "Rewind" to see whether the little computer has been fooled and will start up the mechanism. On some models, you may need to twirl one of the reels slowly with your fingers to keep the VCR operating with no tape. Just rotate the reel that is not being motor-driven, twirling it in the same direction that the tape would be pulling it if a tape were in the machine — that is, the same direction (clockwise or counter-clockwise) that the other reel is moving.

If the mechanism does not come to life, turn off any lights above the work table and try again. (The light may be reaching the light-sensors in the front-loader, even though there is a cassette in it.)

If the mechanism still does not come to life, or if it comes to life but only ejects the cassette, unplug the VCR. While it is unplugged, manually load the cassette into the front-loader by turning the little

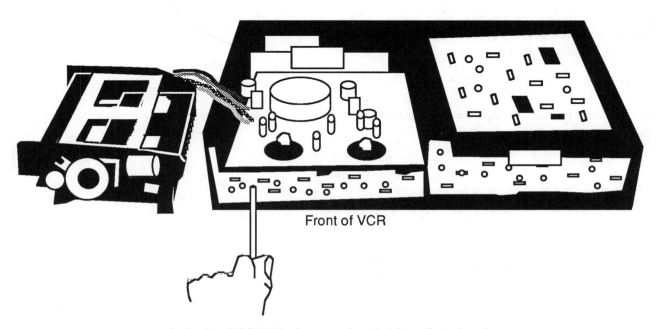

Front of VCR

Activating VCR With Cassette Loaded Into Front-Loader

Folded newspaper

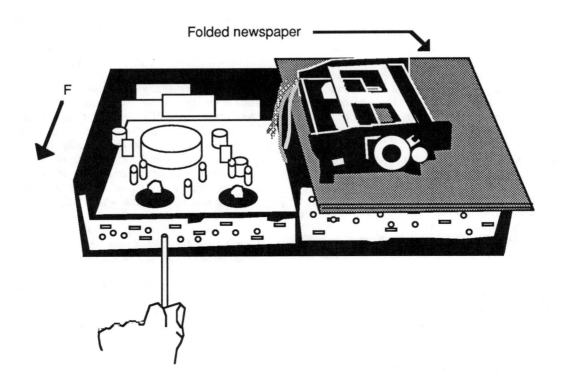

F

Activating VCR With Cassette Loaded Into Front-Loader

motor on the front-loader as you gently push the cassette in. Now get ready to turn the other stationary reel the moment the other reel starts to rotate when you plug it in. (The little computer sometimes tests everything when the VCR is first plugged in with a cassette in it by rotating one of the reels and observing whether the other reel also turns, as it would if there were a cassette in the VCR because the two would be connected by the tape. If the little computer detects the presence of a cassette in the front-loader, rotates one of the reels, and the other reel does not turn, then the computer knows that you're trying to fool it. We outsmart it by turning the other reel, just as a tape would do. Plug it

in, and if one of the reels turns, turn the other reel.

Power up again, and if the little computer rotates one of the reels, you rotate the other reel. Now try pushing "Rewind." If it ran on earlier tests when the front-loader was in place and you loaded a cassette, it should run now. If it does not, there must be a switch on the reel table under where the front-loader normally goes that still needs to be pushed and held down.

When you get it running, you can put it into "Play," or whatever mode you wish to see it run in.

3. Beta Top Loaders

In Beta machines, there are generally two sensor circuits that need to be circumvented: the cassette-down switch and the tape-slack detector. Basically we will go through these one-by-one, and do something to make each sensor circuit think that it detects the presence of a tape (when actually there is no tape in the machine). We do not disconnect the

sensors (the little computer could detect that, and shut everything down). Instead, we will activate each of the sensors in order, just as if a cassette were loaded into the VCR.

We will also need to trick a mechanical switch that gets moved and switched by the physical action of the descending cassette basket. In the simplest case, this switch is part of the mechanism that latches and holds the basket down.

On Beta top-loaders, look under the cassette basket on the reel table for a vertical lever that is attached to a switch (a "cassette detection switch").

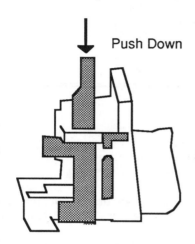

Push Down

Beta Cassette Detection Lever

Pushing this lever down a short distance should activate a switch. Hold it down with some tape or a clip lead,

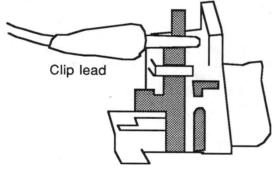

Clip lead

*Holding Beta Cassette Detection Lever
Down With a Clip Lead*

or with your finger, if you've got nothing better to do with your time.

On Beta machines, secondly, you will need to disarm the tape slack sensor, which shuts off the mechanism if it detects too much slack in the tape. This switch is located in different places on different Beta models, but one common location for it is near the left rear corner of the cylinder assembly. It looks like an arm that can swing out, with a long vertical pin attached to it that would rub against the side of the passing tape, if a tape were passing it.

Since we want to run the machine with no tape in it, the sensor switch is going to send back the signal

that the tape is too loose, shutting off the machine unless we take evasive action. The trick here is to do something that will hold the lever back from swinging out into the tape path. Some Beta models have a hole provided in the lever, which

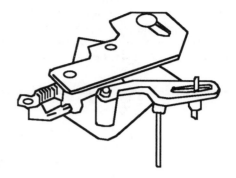

Beta Tape Slack Sensor Switch

lines up with a hole in the small stationary part above it, so that you can take something like a toothpick or matchstick, and stick it through the two

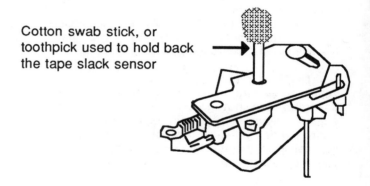

Cotton swab stick, or toothpick used to hold back the tape slack sensor

*Keeping the Tape Slack Sensor from Swinging in
and Shutting Off the Machine*

holes in a way that will hold the lever back. Or you could use a little bit of tape or thread to hold the lever back.

Look on the bottom of the cassette basket and see whether you can see a little arm coming down with a small horizontal pin attached that goes into a waiting "jaw" in the mechanism below. If so, take a scribe or similar tool and push it into the jaw until it snaps closed. On some models, the part that latches the basket down is not accessible from the top, and is not what sends the signal to the little computer, anyway. In this case, look for a small push-button switch mounted on the reel table underneath where the basket comes down, or for a prong sticking straight down like a bayonet from the bottom of the basket, that goes through a hole in the reel table to activate a switch underneath. In either case, find a way to do the same thing to the switch yourself that the cassette basket would normally do as it descends.

Having done this, try pushing, say, "Rewind" and watch what happens. If the mechanism starts to move, you've succeeded in tricking it, and you can now go on to troubleshoot it, or do whatever you were doing that led you to this appendix in the first place. If the mechanism does not start to move, then there must be another switch someplace on, underneath, or at the side of the reel table, that needs to be activated to complete the deception. You will have to continue looking until you find it.

On some models, you may need to twirl one of the reels slowly with your fingers to keep the VCR operating with no tape. Just rotate the reel that is not being motor-driven, twirling it in the same direction that the tape would be pulling it if there were a tape in the machine — that is, the same direction (clockwise or counter-clockwise) that the other reel is moving.

4. Beta Front-loaders

To make a front-loading VCR operate with no tape inside, it is necessary to lift the front-loading assembly up from the reel table, which means that the screws holding it down must be removed and the front-loader lifted out, which usually means that the front cabinet panel also must be removed. Naturally, this is done with the VCR unplugged. See Chapters 8 and 10, and Appendixes III and VII if you need help in removing the front-loading assembly.

NOTE: If, when you remove the top of the cabinet, you can see that the wires coming to the front-loading assembly are not grouped together, but come to different points on it from both sides, or are very short, with little or no slack, then it will be impossible to use the following procedure. However, you may still be able to trick the mechanism in such

VCR's into playing with an empty "dummy cassette" inside, as described in Section 17 of Appendix I.

Leave the wires connected to the front-loading assembly when you remove it, however. If the wires come into the reel-table side (seen as you face the front of the VCR), then probably they will be long enough to permit you to set the front-loading assembly by the closest end of the VCR, as shown in the illustration.

If the wires from the front-loader connect on the other side, then put a few layers of folded newspaper over the circuit boards, or whatever is on that side of the VCR, and set the front-loader on top of the paper with the wires still attached, as shown in the next illustration.

Front of VCR

Empty Front-Loader with Wires Attached Still Connected to Beta VCR

Folded newspaper

F

Empty Front-Loader with Wires Attached
Still Connected to Beta VCR

Do not use paper with any metal in it because this is supposed to insulate the loading assembly from the electronic circuits underneath, and a metal staple in a magazine, for example, might short out some components, and cause an additional problem beyond the scope of this book to fix. Folded newspaper is best.

In the illustrations, the front-loader is drawn sitting rightside-up, but this is not necessary. Most front-loaders will work just as well upside-down (they do not work by gravity). If the wires to the front-loader are so short that you can only set the front-loader upside-down with the wires still attached, or if the front-loader is too "tippy" when it is out of the machine in a rightside-up position, then set it upside down. The only important thing is to

position it so that you can have access to the front opening to put a cassette in.

Now plug in the VCR. The clock display should come on. Look at the front cabinet cover that you removed, and see where the power button is. Find the little push-button switch on the front circuit board that is underneath this button, and use the eraser of a wooden pencil, or a wooden stick, to push it. The power light, or whatever your VCR does when you power it up, now should come on.

Now steady the top of the removed front-loader with one hand, and with the other hand, take an ordinary cassette and push it gently into the opening, just as you do when you load a cassette. The mechanism should be activated and pull the cassette in

Front of VCR

Cassette Loaded Into Detached but Connected Beta Front-Loader

*Cassette Loaded into Beta Front-Loader
Resting on a Circuit Board*

as if it were loading it onto the reel table (which is what the little computer thinks it is doing). There is now a cassette inside the front-loader, which is still attached to the rest of the VCR by its wires, but the place where it normally would sit is now free and open to view.

Look at the reel table under where the cassette would normally rest to see whether there is a little push-button switch that the cassette would normally depress when it is loaded. Look under where the cassette basket would come

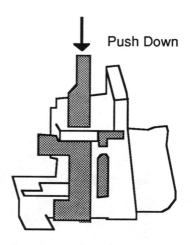

Beta Cassette Detection Lever

down for a vertical lever attached to a switch (a "cassette-in" switch"). On most models, the "cassette-in" switch or sensor is part of the front-loading assembly, and is automatically activated when the cassette is loaded into it earlier. Loading the cassette into the detached, but still-connected, front-loader is enough to complete the deception on many

models. But if there is a switch on the reel table, press it like a cassette would. Pushing this lever down a short distance should activate a switch. Hold it down with some tape or a clip lead,

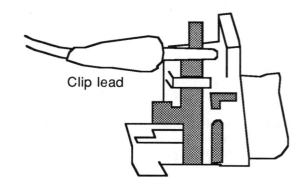

*Holding Beta Cassette Detection Lever Down With a
Clip Lead*

or with your finger, if you've got nothing better to do with your time.

On Beta machines, you may also need to disarm the tape-slack sensor, which shuts off the mechanism if it detects too much slack in the tape. This switch is located in different places on different Beta models, but one common location is near the left rear corner of the cylinder assembly. It looks like an arm that can swing out, with a long vertical pin attached to it that would rub against the side of the passing tape, if a tape were in the VCR. Since we want to operate the machine with no tape in it, the sensor switch is going to send back the signal that the tape is too loose, shutting off the machine unless we take evasive action. The trick here is to do something that will hold the lever back from swinging out into the tape path.

Some Beta models have a hole provided in the lever, which

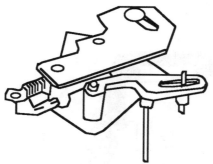

Beta Tape Slack Sensor Switch

lines up with a hole in the small stationary part above it, so that you can take something like a toothpick or matchstick, and stick it through the two holes in a

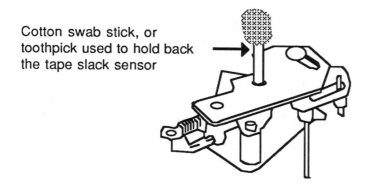

Cotton swab stick, or toothpick used to hold back the tape slack sensor

Keeping the Tape Slack Sensor from Swinging in and Shutting Off the Machine

way that will hold the lever back. Or you could use a little bit of tape or thread to hold the lever back.

Now try pressing "Rewind" to see whether the little computer has been fooled and will start up the mechanism. On some models, you may need to twirl one of the reels slowly with your fingers to keep the VCR operating with no tape. Just rotate the reel that is not being motor-driven, twirling it in the same direction that the tape would be pulling it if there were a tape in the machine — that is, the same direction (clockwise or counter-clockwise) that the other reel is moving.

If the mechanism does not come to life, or if it comes to life but only ejects the cassette, unplug the VCR. If it ejected the tape, while it is unplugged, manually load the cassette into the front-loader by turning the little motor on the front-loader as you gently push the cassette in. Now get ready to turn the stationary reel the moment the other reel starts to rotate when you plug it in. (The little computer sometimes tests everything when the VCR is first plugged in with a cassette in it by rotating one of the reels and observing whether the other reel also turns, as it would if there were a cassette in the VCR because the two would be connected by the tape. If the little computer detects the presence of a cassette in the front-loader, rotates one of the reels, and the other reel does not turn, then it knows you're trying to fool it. We outsmart it by turning the other reel, just as a tape would do.) Plug it in, and if one of the reels turns, turn the other reel.

Turn on the power again, and if the little computer rotates one of the reels, you rotate the

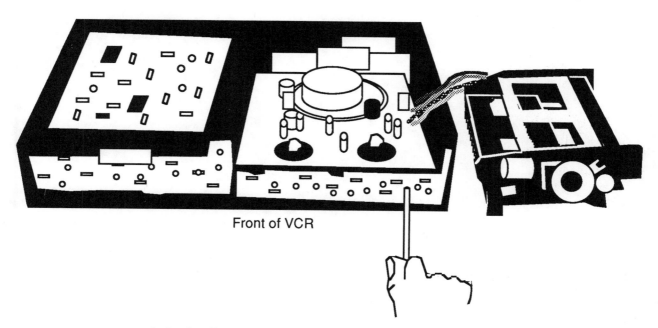

Front of VCR

Activating Beta VCR With Cassette Loaded Into Front-Loader

Folded newspaper

F

Activating Beta VCR With Cassette Loaded Into Front-Loader

other reel. Now try pushing "Rewind." If it ran on earlier tests when the front-loader was in place and you loaded a cassette, it should run now. If it does not, there must be a switch on the reel table under where the front-loader normally goes that still needs to be pushed and held down.

When you get it running, you can put it into "Play," or whatever mode you wish to see it operate in.

Appendix VI

Mechanisms in Common VCR Models

How To Use Appendix VI

The mechanisms pictured in this Apppendix usually are listed under a name that is different from the brand name on the VCR. The diagram for your VCR should be looked at and used in combination with the instructions on this and the following pages. To use these diagrams, turn first to the "Introduction" at the beginning of this book and find the name under which you should look for your brand. For example, if your VCR is a Quasar, look under Panasonic; if it is a Zenith, look under either JVC or Sony, depending on whether it is a VHS or Beta. And so on.

After you find the name(s) to look under, compare your opened-up VCR with the pictures listed under that name in the following pages until you find a diagram that corresponds to your VCR. If several names are given, look under, then look through the diagrams under all those various manufacturers until you find the diagram that resembles your mechanism.

Symbolism Used in Pictures

When a top view of the mechanism is shown, the part of the mechanism underneath the cassette basket (and under the cassette, if there is one in the machine) is indicated by a shaded rectangle, like this:

> This represents the area covered by the cassette basket (and cassette, if one is loaded inside the VCR).

You will be able to see the parts located in the shaded area only when the cassette basket is up in the ejected position; if a cassette is stuck in your VCR, you will not be able to see the hidden parts until you can raise the cassette basket and eject the cassette.

The straight arrow with a letter "F" points toward the front of the machine in the diagram. It is analogous to the arrow on a map that shows which direction

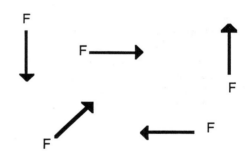

Symbols Used to Show Direction of the Front of the VCR on the Diagram

is North. The front of the VCR (the part with the controls, etc.) usually, but not always, is at the bottom of the picture.

The curved arrows show the parts and directions you rotate them to make the VCR unload a tape and and reel it in. You turn the parts in the directions the arrows point to UNLOAD the tape loading mechanism, and REEL IN loose tape. Rotating the part connect to the loading mechanism in the direction opposite to the arrow makes the VCR mechanism move in the direction to load a tape.

A small rectangle containing the letters "P R S" represents a "pinch-roller solenoid."

Position the Machine for Access

Set the machine securely on its end while you are doing this, so that you can look back and forth between the top and bottom of the mechanism while you unload the mechanism. Get someone else to help you hold it, or take great care to secure it against falling over and breaking one of the circuit boards that you have opened out, as these would be prohibitively expensive to replace!

Consult the diagram for your machine that appears later in this appendix as you continue below.

When the Machine is Jammed

Before we start moving parts, a word of caution: You should be able to turn the parts referred to below with the blade of a small screwdriver or the tip of your little pinky finger. If they seem stuck tight when the mechanism is definitely not in the fully unloaded position (as you can tell by looking at the position of the P-guides (VHS) or of the loading ring (Beta) on the topside), then look for a foreign object stuck in the mechanism. If you find no foreign object but the mechanism is locked up tight in a partly, or fully, loaded position, then you probably need to have a technician look at it.

But in most cases, the gear or pulley will rotate easily and you will see the loading ring or P-guides slowly move back toward the fully unloaded position as you rotate the pulley or gear in the direction indicated in the diagram.

Do this now: Start moving the part which it says in the diagram for your machine "unloads the tape." (Remember that "unloading the tape" refers to the action that takes place when the P-guides (or loading ring) are pulled back to their fully-unloaded position.) At first you probably will see no movement of the tape-loading mechanism, but as you continue to move the indicated part, finally you will see the loading mechanism start to move in the direction that unloads the tape.

Don't Go Too Far

Watch to tell when the loading mechanism first gets back to its fully unloaded position, because it is possible on some models to go too far and overshoot, in which case the mechanism on some models still will not eject, and on some Sharp and Montgomery Ward VCR's, it may go into a position in which, when you try to power it up later, the little computer inside will not be able to figure out what position the mechanism is in, or what to do next, and so, will shut everything down.

Reeling in Loose Tape

On a few models, any loose loop of tape will *automatically* be reeled back into the cassette when you unload the tape-loading mechanism, but on most VCR's, reeling in the loose tape requires a separate process. On some VCR's, *turning a second pulley* causes the tape to be pulled back into the cassette as you unload the mechanism. On some models, it is necessary to *move a mechanical arm* as well as turning the pulley to reel tape back into the cassette. And on many models, it is necessary to *use a battery to run a reel motor* in order to wind in the loose tape.

If the tape is all wrapped and tangled around the cylinder, you'll have to use your fingers to unwrap and free it. (The most common cause of this disaster is putting a cassette into the VCR too soon after use of a wet-type head-cleaning tape.)

Go Back and Forth

You can go back and forth between unloading the tape-loading mechanism a little, as explained first in the diagrams, then winding back in some loose tape, as explained secondly, then unloading a little more, and so on. What to turn to reel in the loose tape is also shown in the diagrams, and the direction to turn it is shown with an arrow. On some models, you can reel tape in by turning the pulley in either direction: one direction winds the tape onto the supply reel, and the other direction winds it onto the take-up reel. Go back and forth until the tape-loading mechanism is all the way back to the fully-unloaded position.

"Last Resort" if Idler Slips

Be aware, however, that if the "idler", a wheel with a rubber tire that communicates the rotation to the reels, is oily or worn down, then this procedure may fail to work, and you will need to use the last-resort trick of jamming the cassette door open with a stick, as explained in Chapters 7 and 8.

Battery for Reel Motor

On machines with a separate reel motor, it may be necessary to connect a little nine-volt battery to the reel motor in order to make it turn and reel in loose tape. When this is necessary, it is also shown on the diagrams. If you do not have a battery and clip leads as described in Section 19 of Appendix I, you can still jam the cassette door open, eject the cassette or remove it along with the loading assembly, and reel the loose tape in with your finger.

Unloading the Cassette

To eject the cassette from a *top-loading* VCR, go to Chapter 7. To eject the cassette from a *front-loading* VCR, go now to Section 5 of Chapter 8.

Meaning of Alphabetic Letters after Name

The alphabetic letters appearing after many brand names in this appendix — for example, "Fisher A," "Fisher B," "Fisher C," — are just arbitrary symbols used in this appendix as a means of separating different models appearing under the same brand name. They have no meaning outside this appendix. (If you called a parts supplier and said that you had a "Fisher B," they would not know what you were talking about.)

When calling parts suppliers, use either the make and model number printed on the back of your VCR, or one of the model numbers shown in fine print after the alphabetic letter.

Part Numbers in Diagrams

The numbers at the ends of arrows are the part numbers of the parts to which the arrows point. The manufacturer whose part number is given is indicated by the abbreviation shown in parentheses after the part number. For example, "VDVO122 (Mat)" means Matsushita part number VDV0122. The meanings of the abbreviations is listed in the Introduction to this book. For example, Matsushita is the same as Panasonic.

The part with this number will also fit on any other brand name VCR that uses the same mechanism. For instance, the Matsushita part in the example would also work on Quasar, Philco, G E, and other brands with the same mechanism. The same part usually is also available from these other companies too, but they will generally assign it a different part number.

Generic Belts

The symbol "N L A" beside an arrow means that this part is no longer available from the company under whose name the diagram is listed.

In the case of belts, it may still be possible to get a generic belt that will do the job. If you still have the old belt, go to Appendix IV and use the belt-measuring table to get a size for the belt. Then order up a generic belt of that size from one of the belt suppliers listed there. Their phone numbers and addresses are in Appendix II.

Diagrams of Common Mechanisms

Akai A   Includes Akai VS-515U/UM, VS-525U, VS-555U/UM, VS-565U, VS-5151U, VS 270U/UM, among others. See also Mitsubishi.

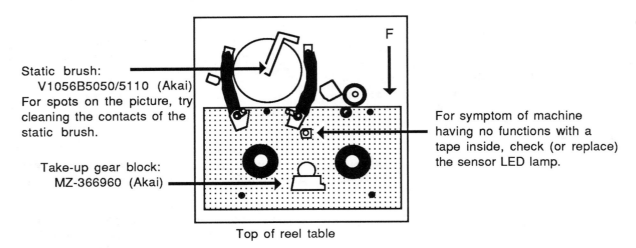

Static brush:
V1056B5050/5110 (Akai)
For spots on the picture, try cleaning the contacts of the static brush.

For symptom of machine having no functions with a tape inside, check (or replace) the sensor LED lamp.

Take-up gear block:
MZ-366960 (Akai)

Top of reel table

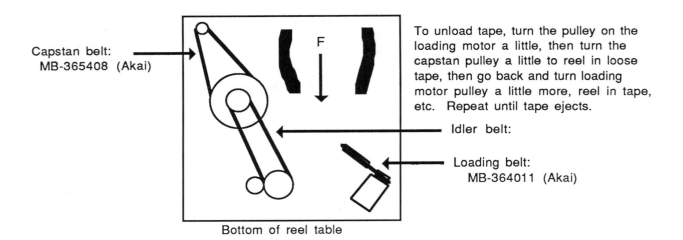

Capstan belt:
MB-365408 (Akai)

To unload tape, turn the pulley on the loading motor a little, then turn the capstan pulley a little to reel in loose tape, then go back and turn loading motor pulley a little more, reel in tape, etc. Repeat until tape ejects.

Idler belt:

Loading belt:
MB-364011 (Akai)

Bottom of reel table

Emerson   For Emerson look below & also under Mitsubishi and Goldstar
Emerson A   Includes Emerson VCP 661, VCR 754, VCR 755, VCR 872, VCR 873, VCR 874, VCR 951A, VCR 951HA, VCR 952, VCS 955A, VCS 966HA, VCS 972, and VCS 977A among others

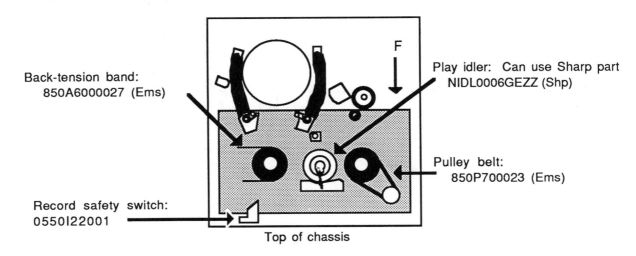

F

If unit keeps shutting off, or eating tapes, replace the Toothed idler assembly:850A200004 (Ems)

Common problem: When you hit Play, it loads the tape but won't run, and goes into Pause. Check or replace Limited Post Lever Assy: 850A600039 (Ems)

Back-tension band: 850A600045 (Ems)

Record safety switch: 0550111011 (Ems)

Another common problem is the clock running fast or slow. Try replacing the crystal on the timer or operation circuit board. It may be numbered X601. Part #10064R1903 (Ems)

Top of mechanical chassis

If the unit has no power, and will not load or unload the tape, most of the time the problem is that the power regulator integrated circuit, or loading driver IC, has gone bad. For most of the regulators on these models, the part number is on the integrated circuit itself. Look for a large black plastic part with leads attached on the power supply circuit board, and numbers on it beginning with "STK" (for example, STK5332).

F

Reel belt: 850P600124 (Ems) To unload tape, turn loading motor pulley CW (as shown here).

To wind tape back into cassette, turn pulley that reel belt goes around in either direction.

Loading belt: 850P00125 (Ems)

Loading motor as seen from rear

Bottom of mechanical chassis

Emerson B   Includes Emerson VCP 700, VCR 850, VCR 870, VCR 880X, VCR 951, VCR 951H, VCS 955, VCS 966H; and Scott SVR 330S among others

F

Back-tension band: 850A6000027 (Ems)

Play idler: Can use Sharp part NIDL0006GEZZ (Shp)

Pulley belt: 850P700023 (Ems)

Record safety switch: 0550122001

Top of chassis

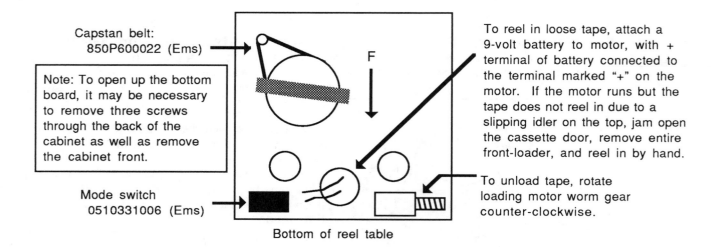

Capstan belt:
850P600022 (Ems)

Note: To open up the bottom board, it may be necessary to remove three screws through the back of the cabinet as well as remove the cabinet front.

Mode switch
0510331006 (Ems)

To reel in loose tape, attach a 9-volt battery to motor, with + terminal of battery connected to the terminal marked "+" on the motor. If the motor runs but the tape does not reel in due to a slipping idler on the top, jam open the cassette door, remove entire front-loader, and reel in by hand.

To unload tape, rotate loading motor worm gear counter-clockwise.

Bottom of reel table

Emerson C   Includes Emerson VCR 800

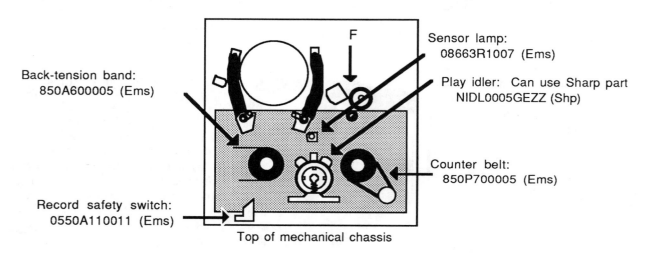

Back-tension band:
850A600005 (Ems)

Sensor lamp:
08663R1007 (Ems)

Play idler: Can use Sharp part
NIDL0005GEZZ (Shp)

Counter belt:
850P700005 (Ems)

Record safety switch:
0550A110011 (Ems)

Top of mechanical chassis

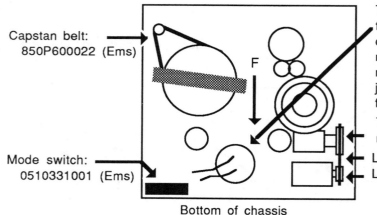

Capstan belt:
850P600022 (Ems)

Mode switch:
0510331001 (Ems)

To reel in loose tape, attach a 9-volt battery to motor, with + terminal of battery connected to the terminal marked "+" on the motor. If the motor runs but the tape does not reel in due to a slipping idler on the top, jam open the cassette door, remove entire front-loader, and reel in by hand.

To unload tape, rotate loading motor pulley counter-clockwise.

Loading belt: 850P300005 (Ems)
Loading motor: 1596358001 (Ems)

Bottom of chassis

Fisher   A

Includes Fisher FVH 515, FVH 530, among others

To unload tape, turn the loading motor pulley CW (that is, CW as viewed from the pulley side of the motor — that is, as viewed from the rear of the VCR).

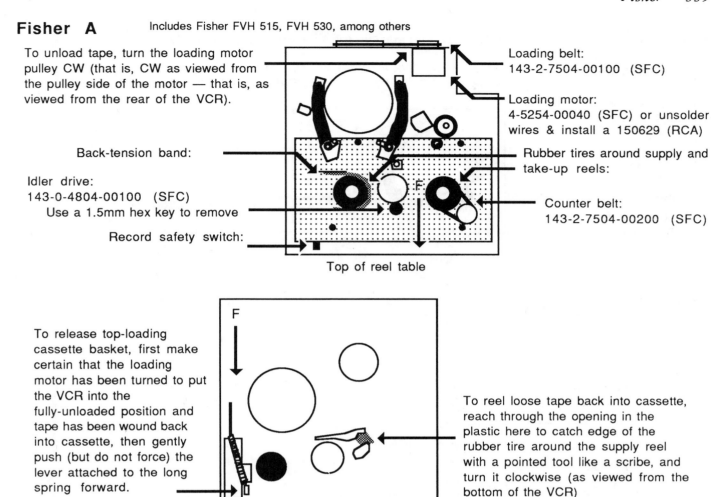

Loading belt: 143-2-7504-00100 (SFC)

Loading motor: 4-5254-00040 (SFC) or unsolder wires & install a 150629 (RCA)

Back-tension band:

Idler drive: 143-0-4804-00100 (SFC)
Use a 1.5mm hex key to remove

Rubber tires around supply and take-up reels:

Counter belt: 143-2-7504-00200 (SFC)

Record safety switch:

Top of reel table

To release top-loading cassette basket, first make certain that the loading motor has been turned to put the VCR into the fully-unloaded position and tape has been wound back into cassette, then gently push (but do not force) the lever attached to the long spring forward.

To reel loose tape back into cassette, reach through the opening in the plastic here to catch edge of the rubber tire around the supply reel with a pointed tool like a scribe, and turn it clockwise (as viewed from the bottom of the VCR)

Bottom view

Fisher   B

Includes Fisher FVH 615, FVH 715, FVH 720, FVH 721, FVH 722, FVH 725, FVH 805, FVH 805A, FVH 810, FVH 815, FVH 816, FVH 820, FVH 822, FVH 822, FVH 825, FVH 830, FVH 839, FVH 840, FVH 904, FVH 904A, FVH 905, FVH 906, FVH 907, FVH 919, FVH 920, FVH 922, FVH 930, FVH 940, FVH 960; Realistic Model 14; and Sears 564.53071450, & 53284550 among others

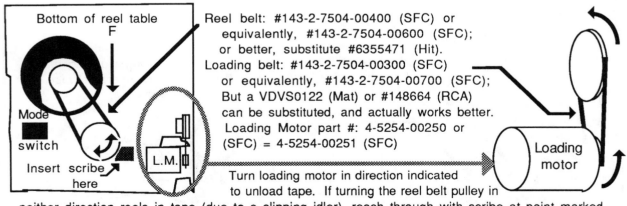

Bottom of reel table

Reel belt: #143-2-7504-00400 (SFC) or equivalently, #143-2-7504-00600 (SFC); or better, substitute #6355471 (Hit).
Loading belt: #143-2-7504-00300 (SFC) or equivalently, #143-2-7504-00700 (SFC); But a VDVS0122 (Mat) or #148664 (RCA) can be substituted, and actually works better.
Loading Motor part #: 4-5254-00250 or (SFC) = 4-5254-00251 (SFC)

Mode switch

Insert scribe here

L.M.

Loading motor

Turn loading motor in direction indicated to unload tape. If turning the reel belt pulley in neither direction reels in tape (due to a slipping idler), reach through with scribe at point marked ◼ , catch teeth on edge of supply reel, and with scribe turn reel CW, as seen from below.

The part number for the belt in the cassette front-loader depends on the model as follows
in FVH 715, FVH 720, FVH 721, FVH 722, FVH 725, FVH 815, FVH 820, FVH 830, FVH 839, FVH 840, FVH 960:
   143-2-7504-00500 (SFC)
in FVH 805, FVH 805A, FVH 810, FVH 816, FVH 822, FVH 825, FVH 922, FVH 930: 143-2-7504-00800 (SFC)
In FVH 904, FVH 904A, FVH 905, FVH 905A, FVH 906, FVH 907, FVH 919, FVH 920, FVH 940: 143-2-7504-01100 (SFC)

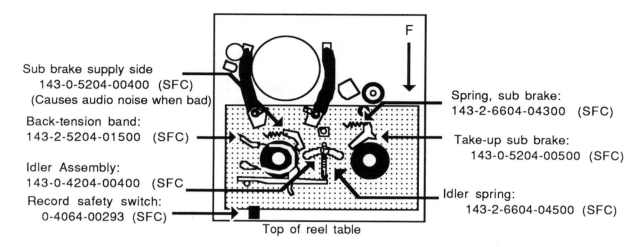

Sub brake supply side
   143-0-5204-00400 (SFC)
   (Causes audio noise when bad)

Back-tension band:
   143-2-5204-01500 (SFC)

Idler Assembly:
   143-0-4204-00400 (SFC

Record safety switch:
   0-4064-00293 (SFC)

Spring, sub brake:
   143-2-6604-04300 (SFC)

Take-up sub brake:
   143-0-5204-00500 (SFC)

Idler spring:
   143-2-6604-04500 (SFC)

Top of reel table

Fisher C

Includes Fisher FVH 950, FVH 980, FVH 990, FVH 4000, FVH 4000Z, FVH 4050, FVH 4100, FVH 4100Z, FVH 4200, FVH 5400, FVH-D5600

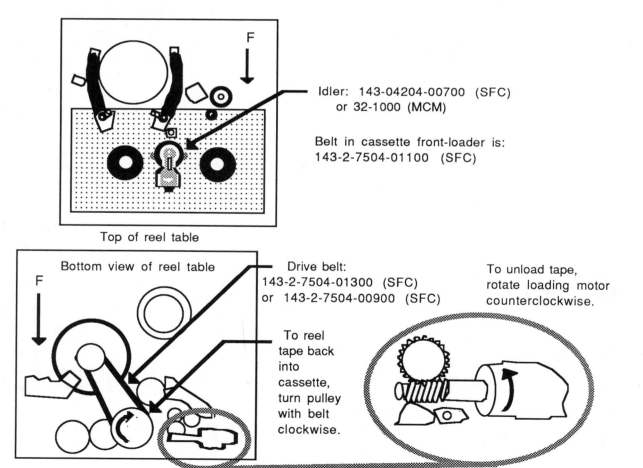

Idler: 143-04204-00700 (SFC)
   or 32-1000 (MCM)

Belt in cassette front-loader is:
143-2-7504-01100 (SFC)

Top of reel table

Bottom view of reel table

Drive belt:
143-2-7504-01300 (SFC)
or 143-2-7504-00900 (SFC)

To reel tape back into cassette, turn pulley with belt clockwise.

To unload tape, rotate loading motor counterclockwise.

Funai

Includes Dynatech VR-71; Multitech MV-050SM, MV-070, MV-075, MV-080, MV-089, MV-189, MV-289; Sound Design 8005; Symphonic 4500, 4900A, 5000, 5200, 5800; Teac MV350,   among others

If the machine shuts off, goes to different modes of operation by itself, eats tape, or ejects cassette with the tape hanging out, try replacing the mode switch (on the bottom).

If the capstan runs the tape too fast on any speed, try replacing IC 306, #BA 718.

Back-tension band:

Record safety switch:

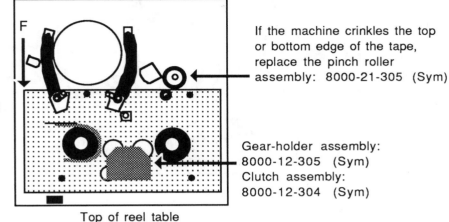

If the machine crinkles the top or bottom edge of the tape, replace the pinch roller assembly: 8000-21-305 (Sym)

Gear-holder assembly: 8000-12-305 (Sym)
Clutch assembly: 8000-12-304 (Sym)

Top of reel table

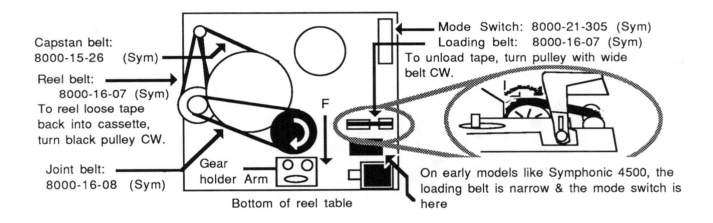

Capstan belt: 8000-15-26   (Sym)

Reel belt: 8000-16-07 (Sym)
To reel loose tape back into cassette, turn black pulley CW.

Joint belt: 8000-16-08 (Sym)

Gear holder Arm

Bottom of reel table

Mode Switch: 8000-21-305 (Sym)
Loading belt: 8000-16-07 (Sym)
To unload tape, turn pulley with wide belt CW.

On early models like Symphonic 4500, the loading belt is narrow & the mode switch is here

GE — Look under Hitachi and Panasonic

Goldstar A

Includes Goldstar GHV 41 FM, GHV 1000M, GHV 58FM, GHV 1200M, GHV 1210M, VCP 4000, among others.

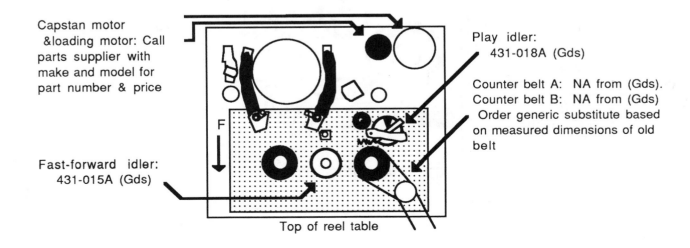

Capstan motor &loading motor: Call parts supplier with make and model for part number & price

Fast-forward idler: 431-015A (Gds)

Play idler: 431-018A (Gds)

Counter belt A: NA from (Gds).
Counter belt B: NA from (Gds)
Order generic substitute based on measured dimensions of old belt

Top of reel table

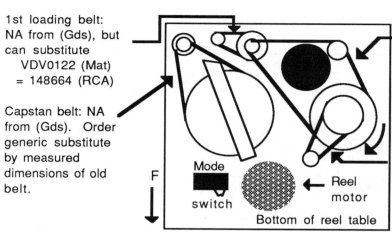

1st loading belt:
NA from (Gds), but
can substitute
   VDV0122 (Mat)
   = 148664 (RCA)

Capstan belt: NA
from (Gds). Order
generic substitute
by measured
dimensions of old
belt.

2nd loading belt: NA from (Gds), but
   can substitute   VDVS0020 (Mat)
   = 152412 (RCA)

To unload tape, turn pulley in direction
indicated. To reel in loose tape, attach 9
volt battery to reel motor in either
direction.

3rd "kick pulley" belt: NA from
(Gds), but can substitute
   VDVS0028 (Mat)
   = 154676 (RCA)

Mode switch

Reel motor

Bottom of reel table

Goldstar B

Includes Emerson VCP 662; Goldstar CV 5500, GHV-51, GHV-55, GHV 1233M, GHV 1240, GHV 1245M, GHV 1241, GHV1243M, GHV 1250, GHV 1265, GHV 1270, GHV1280, GHV 1285, GHV 1400M, GHV 1600, 5210, 8200, and VCP-4100, among others

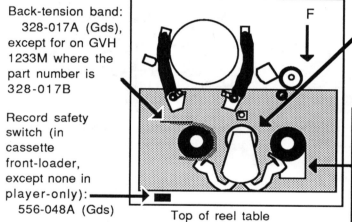

Back-tension band:
   328-017A (Gds),
except for on GVH
1233M where the
part number is
328-017B

Record safety
switch (in
cassette
front-loader,
except none in
player-only):
   556-048A (Gds)

Top of reel table

Idler assembly:321-149A (Gds) except on GHV
   1265 & GHV 1600 where it is 321-256A (Gds)
Idler: 386-075A (Gds) except on GHV 1265
   & GHV 1600 where it is 386-139A (Gds)
And on the GHV 1270 the idler is 456-007B (Gds)
   and the idler tire is 434-046A (Gds)

If VCR starts to play, then shuts down, and the
idler is moving the take-up reel as it should, first
try lifting the take-up reel and dusting the
reflective disk on the bottom with a cotton swab.
If this fails to cure the problem,   replace the
Reel sensor circuit board:
   513-194D (Gds) or 511-217A (Gds)

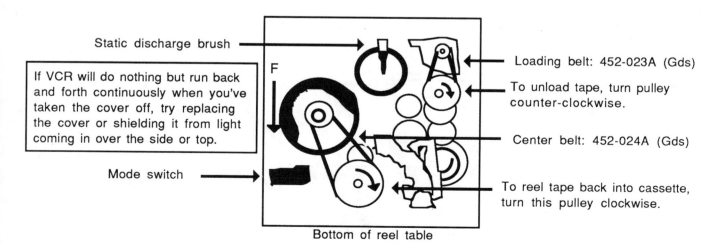

Static discharge brush

If VCR will do nothing but run back
and forth continuously when you've
taken the cover off, try replacing
the cover or shielding it from light
coming in over the side or top.

Mode switch

Loading belt: 452-023A (Gds)

To unload tape, turn pulley
counter-clockwise.

Center belt: 452-024A (Gds)

To reel tape back into cassette,
turn this pulley clockwise.

Bottom of reel table

Goldstar C
Includes J.C. Penny 686-6080; Goldstar GHV-1214M (but must use a schematic from JC Penny)

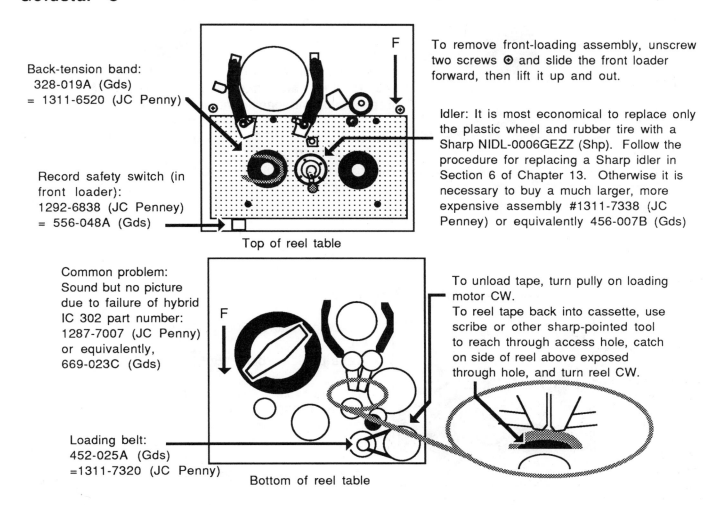

Back-tension band:
328-019A (Gds)
= 1311-6520 (JC Penny)

Record safety switch (in
front loader):
1292-6838 (JC Penney)
= 556-048A (Gds)

Top of reel table

To remove front-loading assembly, unscrew two screws ⊙ and slide the front loader forward, then lift it up and out.

Idler: It is most economical to replace only the plastic wheel and rubber tire with a Sharp NIDL-0006GEZZ (Shp). Follow the procedure for replacing a Sharp idler in Section 6 of Chapter 13. Otherwise it is necessary to buy a much larger, more expensive assembly #1311-7338 (JC Penney) or equivalently 456-007B (Gds)

Common problem:
Sound but no picture due to failure of hybrid IC 302 part number: 1287-7007 (JC Penny) or equivalently, 669-023C (Gds)

Loading belt:
452-025A (Gds)
=1311-7320 (JC Penny)

Bottom of reel table

To unload tape, turn pully on loading motor CW.
To reel tape back into cassette, use scribe or other sharp-pointed tool to reach through access hole, catch on side of reel above exposed through hole, and turn reel CW.

Harmon Kardon — For decks, look under Mitsubishi and N E C

Hitachi A
Includes Hitachi VT 56A, VT 56A(W); RCA VJT 500, VJT 700, VKT 500, VLT 700HF, among others

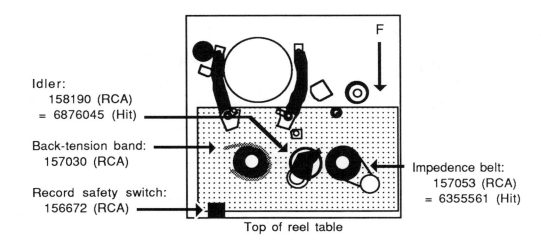

Idler:
158190 (RCA)
= 6876045 (Hit)

Back-tension band:
157030 (RCA)

Record safety switch:
156672 (RCA)

Impedence belt:
157053 (RCA)
= 6355561 (Hit)

Top of reel table

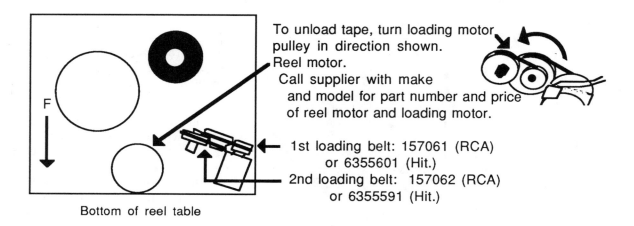

To unload tape, turn loading motor pulley in direction shown.
Reel motor.
  Call supplier with make and model for part number and price of reel motor and loading motor.

1st loading belt: 157061 (RCA) or 6355601 (Hit.)
2nd loading belt: 157062 (RCA) or 6355591 (Hit.)

Bottom of reel table

Hitachi B

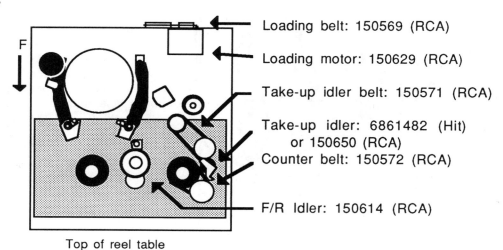

Loading belt: 150569 (RCA)

Loading motor: 150629 (RCA)

Take-up idler belt: 150571 (RCA)

Take-up idler: 6861482 (Hit) or 150650 (RCA)
Counter belt: 150572 (RCA)

F/R Idler: 150614 (RCA)

Top of reel table

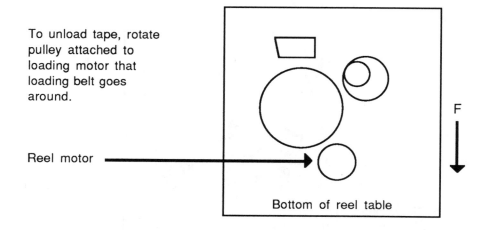

To unload tape, rotate pulley attached to loading motor that loading belt goes around.

Reel motor

Bottom of reel table

© 1990 Stephen N. Thomas

Hitachi C   Includes Hitachi VT5000, VT5600A, VT5800A, among others

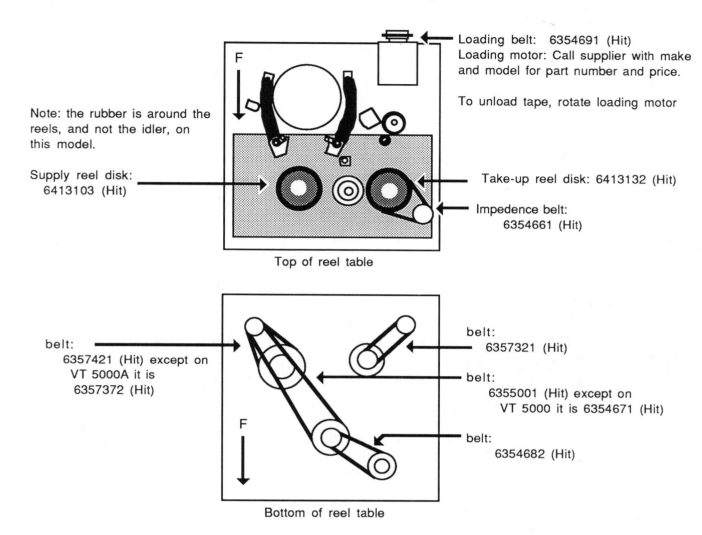

Loading belt: 6354691 (Hit)
Loading motor: Call supplier with make and model for part number and price.

To unload tape, rotate loading motor

Note: the rubber is around the reels, and not the idler, on this model.

Supply reel disk: 6413103 (Hit)

Take-up reel disk: 6413132 (Hit)

Impedence belt: 6354661 (Hit)

Top of reel table

belt: 6357421 (Hit) except on VT 5000A it is 6357372 (Hit)

belt: 6357321 (Hit)

belt: 6355001 (Hit) except on VT 5000 it is 6354671 (Hit)

belt: 6354682 (Hit)

Bottom of reel table

Hitachi D   Includes Hitachi VT 98A among others

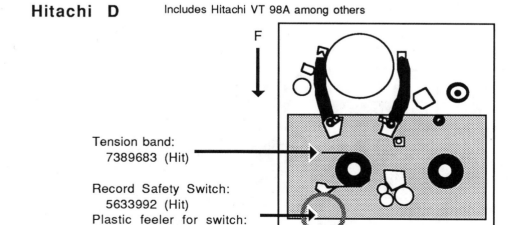

Belt in cassette loader: 6356401 (Hit)

Tension band: 7389683 (Hit)

Record Safety Switch: 5633992 (Hit)
Plastic feeler for switch: 6882951 (Hit)

Top of reel table

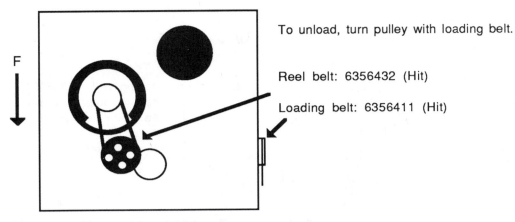

To unload, turn pulley with loading belt.

Reel belt: 6356432 (Hit)

Loading belt: 6356411 (Hit)

Bottom of reel table

Hitachi E

Includes Hitachi VT11A, VT11AR, VT12AX, VT12AY, VT15A, VT16A, VT16AY, VT18A, VT19A, VT33A, VT34A, VT35A, VT43AY, VT44AY, VT45AY, VT88A, VT89A, VT330A(W); RCA VJT 250, VJT255, VJT275, VJT400, VJT425, VKT 275, VKT300, VKT400, VKT425, VKT426, VKT430, VKT550, VKT650, VKT700; and Sears 934.53131350, among others

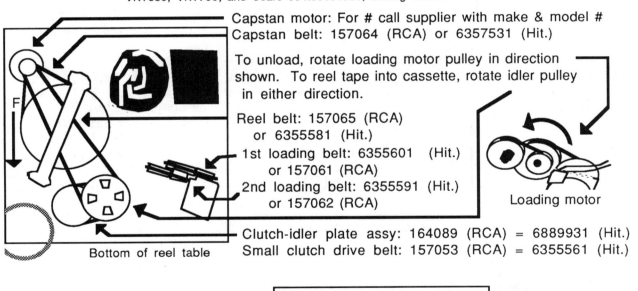

Capstan motor: For # call supplier with make & model #
Capstan belt: 157064 (RCA) or 6357531 (Hit.)

To unload, rotate loading motor pulley in direction shown. To reel tape into cassette, rotate idler pulley in either direction.

Reel belt: 157065 (RCA)
  or 6355581 (Hit.)
1st loading belt: 6355601   (Hit.)
  or 157061 (RCA)
2nd loading belt: 6355591 (Hit.)
  or 157062 (RCA)

Loading motor

Clutch-idler plate assy: 164089 (RCA) = 6889931 (Hit.)
Small clutch drive belt: 157053 (RCA) = 6355561 (Hit.)

Bottom of reel table

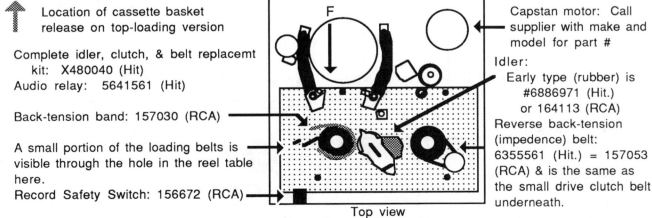

Location of cassette basket release on top-loading version

Complete idler, clutch, & belt replacemt kit: X480040 (Hit)
Audio relay: 5641561 (Hit)

Back-tension band: 157030 (RCA)

A small portion of the loading belts is visible through the hole in the reel table here.
Record Safety Switch: 156672 (RCA)

Capstan motor: Call supplier with make and model for part #
Idler:
  Early type (rubber) is #6886971 (Hit.)
    or 164113 (RCA)
Reverse back-tension (impedence) belt: 6355561 (Hit.) = 157053 (RCA) & is the same as the small drive clutch belt underneath.

Top view

Hitachi   F

Includes: Hitachi VT60A, VT61A, VT62A, VT63A, VT64A, VT65A, VT71A, VT74AY, VT78AY, VT86A, VT87A; J C Penney 686-5067; RCA VKT 385; VLT 250, VLT 260, VLT 270, VLT 375, VLT 385, VLT 395, VLT 450, VLT 460, VLT 470, VLT 600HF, VLT602HF, VLT 625HF, VLT 650HF

Complete idler, clutch, & belt replacement kit: X480049 (Hit)

Back-tension band: 157030 (RCA)

A small portion of the loading belts is visible through the hole in the reel table here.

Record Safety Switch: 156672 (RCA)

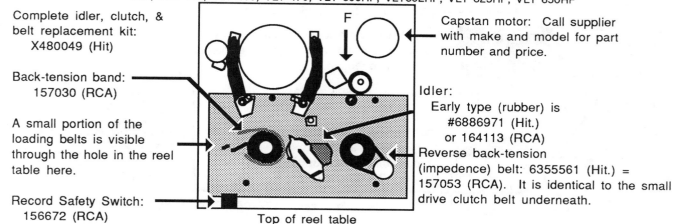

Top of reel table

Capstan motor: Call supplier with make and model for part number and price.

Idler:
   Early type (rubber) is
   #6886971 (Hit.)
   or 164113 (RCA)
Reverse back-tension (impedence) belt: 6355561 (Hit.) = 157053 (RCA). It is identical to the small drive clutch belt underneath.

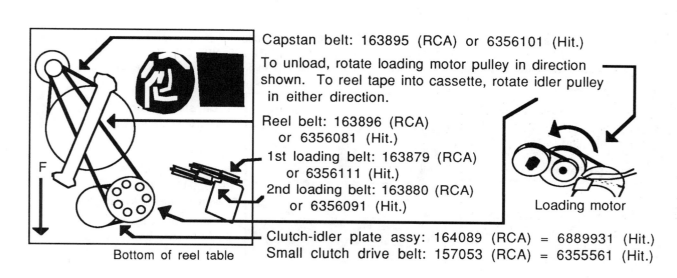

Bottom of reel table

Capstan belt: 163895 (RCA) or 6356101 (Hit.)

To unload, rotate loading motor pulley in direction shown. To reel tape into cassette, rotate idler pulley in either direction.

Reel belt: 163896 (RCA) or 6356081 (Hit.)
1st loading belt: 163879 (RCA) or 6356111 (Hit.)
2nd loading belt: 163880 (RCA) or 6356091 (Hit.)

Loading motor

Clutch-idler plate assy: 164089 (RCA) = 6889931 (Hit.)
Small clutch drive belt: 157053 (RCA) = 6355561 (Hit.)

Hitachi   G

Includes Hitachi VT 1100A, VT 1100AR, VT1110A, VT 1200A, VT 1300A, VT 1310A, VT 1320A, VT 1350A, VT 1370A, VT 1400A, VT 1410A, VT 1430A, VT 1450A, VT 1710A, VT 1800A; RCA VMT 295, VMT 385, VMT 385A, VMT 389, VMT 390, VMT 390A, VMT 400, VMT 590, VMT 595, VMT 630HF, VMT 670HF among others

Complete idler, clutch, & belt replacemt. kit: X480077 (Hit)

Back-tension band: 157030 (RCA)

A small portion of the loading belts is visible through the hole in the reel table here.

Record Safety Switch: 156672 (RCA)

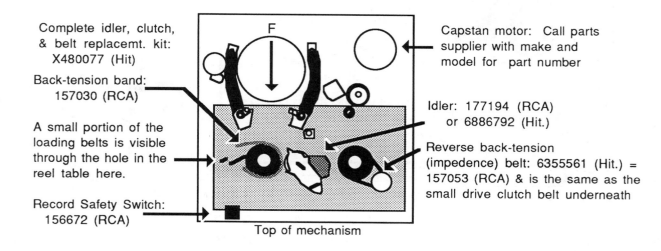

Top of mechanism

Capstan motor: Call parts supplier with make and model for part number

Idler: 177194 (RCA) or 6886792 (Hit.)

Reverse back-tension (impedence) belt: 6355561 (Hit.) = 157053 (RCA) & is the same as the small drive clutch belt underneath

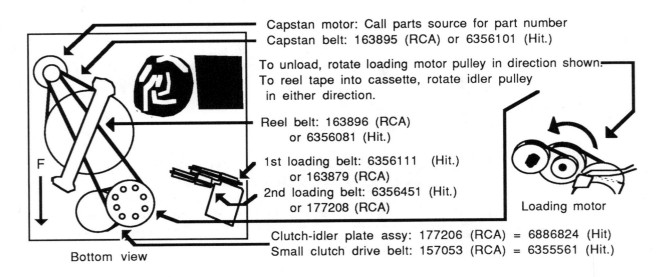

Capstan motor: Call parts source for part number
Capstan belt: 163895 (RCA) or 6356101 (Hit.)

To unload, rotate loading motor pulley in direction shown.
To reel tape into cassette, rotate idler pulley
in either direction.

Reel belt: 163896 (RCA)
  or 6356081 (Hit.)

1st loading belt: 6356111   (Hit.)
  or 163879 (RCA)
2nd loading belt: 6356451 (Hit.)
  or 177208 (RCA)

Loading motor

Clutch-idler plate assy: 177206 (RCA) = 6886824 (Hit)
Small clutch drive belt: 157053 (RCA) = 6355561 (Hit.)

Bottom view

Hitachi H

Includes Hitachi VT 2000A, VT 2010A, VT 2050A, VT 2060, VT 2100A, VT 2150A, VT 2200A, VT 2350A; RCA VPT 250, VPT 289, VPT 290, VPT 291, VPT 292, VPT 293, VPT 294, VPT 295, VPT 296, VPT 385, VPT 390, VPT 395, VPT 396, VPT 490, VPT 495, and VPT 595 among others.

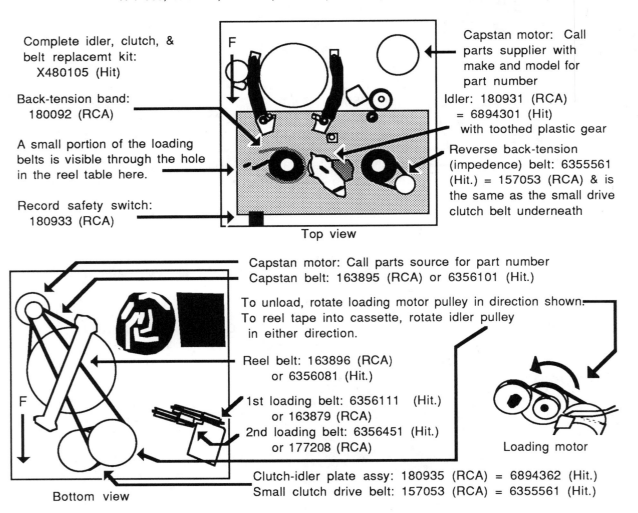

Complete idler, clutch, &
belt replacemt kit:
  X480105 (Hit)

Back-tension band:
  180092 (RCA)

A small portion of the loading
belts is visible through the hole
in the reel table here.

Record safety switch:
  180933 (RCA)

Capstan motor: Call
parts supplier with
make and model for
part number

Idler: 180931 (RCA)
  = 6894301 (Hit)
  with toothed plastic gear

Reverse back-tension
(impedence) belt: 6355561
(Hit.) = 157053 (RCA) & is
the same as the small drive
clutch belt underneath

Top view

Capstan motor: Call parts source for part number
Capstan belt: 163895 (RCA) or 6356101 (Hit.)

To unload, rotate loading motor pulley in direction shown.
To reel tape into cassette, rotate idler pulley
in either direction.

Reel belt: 163896 (RCA)
  or 6356081 (Hit.)

1st loading belt: 6356111   (Hit.)
  or 163879 (RCA)
2nd loading belt: 6356451 (Hit.)
  or 177208 (RCA)

Loading motor

Clutch-idler plate assy: 180935 (RCA) = 6894362 (Hit.)
Small clutch drive belt: 157053 (RCA) = 6355561 (Hit.)

Bottom view

Instant Replay — For decks, see Panasonic, Hitachi, and Funai

J C Penney — Look under Hitachi, Panasonic, and Goldstar

Note: Prime Electronics. (see Appendix II) has full generic belt kits for early J V C 's

JVC A   Includes J V C HR3300, HR3600, HR6700, among others

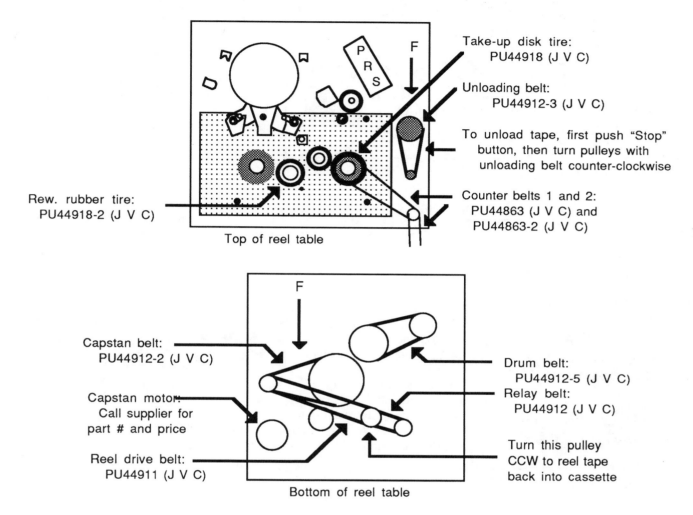

Take-up disk tire:
PU44918 (J V C)

Unloading belt:
PU44912-3 (J V C)

To unload tape, first push "Stop" button, then turn pulleys with unloading belt counter-clockwise

Counter belts 1 and 2:
PU44863 (J V C) and
PU44863-2 (J V C)

Rew. rubber tire:
PU44918-2 (J V C)

Top of reel table

Capstan belt:
PU44912-2 (J V C)

Capstan motor:
Call supplier for part # and price

Reel drive belt:
PU44911 (J V C)

Drum belt:
PU44912-5 (J V C)

Relay belt:
PU44912 (J V C)

Turn this pulley CCW to reel tape back into cassette

Bottom of reel table

JVC B   Includes JVC HR 7100U, HR 7300, and HR7650, among others

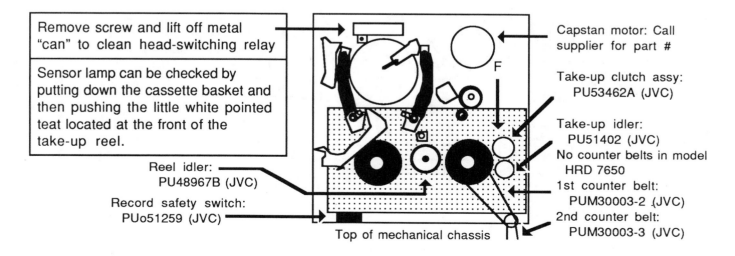

Remove screw and lift off metal "can" to clean head-switching relay

Sensor lamp can be checked by putting down the cassette basket and then pushing the little white pointed teat located at the front of the take-up reel.

Reel idler:
PU48967B (JVC)

Record safety switch:
PUo51259 (JVC)

Capstan motor: Call supplier for part #

Take-up clutch assy:
PU53462A (JVC)

Take-up idler:
PU51402 (JVC)
No counter belts in model HRD 7650

1st counter belt:
PUM30003-2 (JVC)

2nd counter belt:
PUM30003-3 (JVC)

Top of mechanical chassis

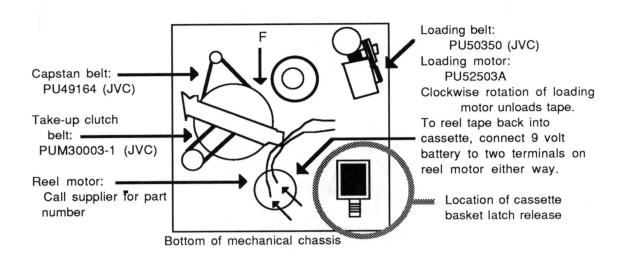

F

Capstan belt:
    PU49164 (JVC)

Take-up clutch
    belt:
    PUM30003-1 (JVC)

Reel motor:
    Call supplier for part
    number

Loading belt:
    PU50350 (JVC)
Loading motor:
    PU52503A
Clockwise rotation of loading
    motor unloads tape.
To reel tape back into
cassette, connect 9 volt
battery to two terminals on
reel motor either way.

Location of cassette
basket latch release

Bottom of mechanical chassis

JVC C

Includes JVC HRD 140U, HRD150U, HRD250U, HRD555U, HRD564, HRD565U, HRD566U, HRD725U, HRD756U; Toshiba AH-1950, M-5800, M-5820; Zenith VR 1800, VR 1805, VR 2000, VR 2100, VR 3000, VR 3010, VR 3100, VR 3200, VR 3220, VR 3250, VR 4000, VR 4100 among others

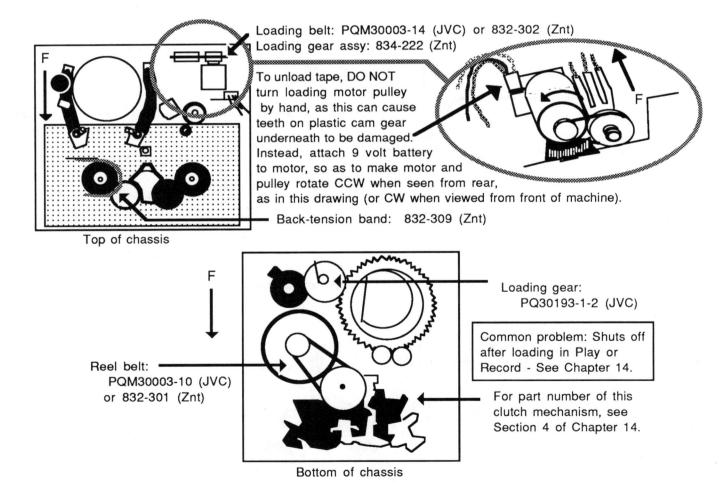

F

Loading belt: PQM30003-14 (JVC) or 832-302 (Znt)
Loading gear assy: 834-222 (Znt)

To unload tape, DO NOT
turn loading motor pulley
by hand, as this can cause
teeth on plastic cam gear
underneath to be damaged.
Instead, attach 9 volt battery
to motor, so as to make motor and
pulley rotate CCW when seen from rear,
as in this drawing (or CW when viewed from front of machine).

F

Back-tension band: 832-309 (Znt)

Top of chassis

F

Reel belt:
    PQM30003-10 (JVC)
    or 832-301 (Znt)

Loading gear:
    PQ30193-1-2 (JVC)

Common problem: Shuts off
after loading in Play or
Record - See Chapter 14.

For part number of this
clutch mechanism, see
Section 4 of Chapter 14.

Bottom of chassis

JVC D

Includes JVC HRD 120U, HRD 130U, HRD 131U, HRD 220U, HRD 225U, among others

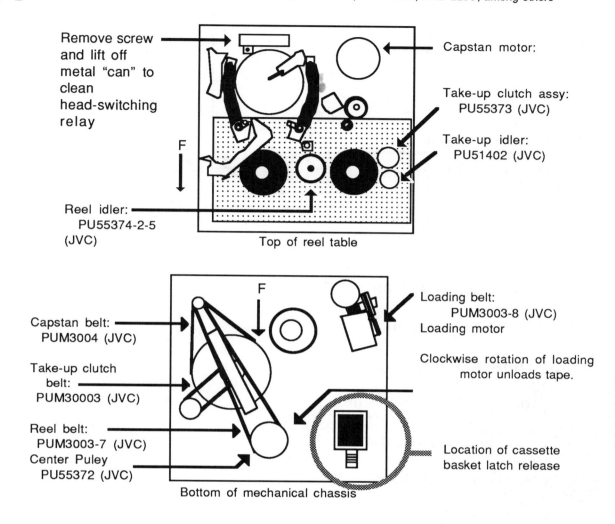

Remove screw and lift off metal "can" to clean head-switching relay

Capstan motor:

Take-up clutch assy: PU55373 (JVC)

Take-up idler: PU51402 (JVC)

Reel idler: PU55374-2-5 (JVC)

Top of reel table

Capstan belt: PUM3004 (JVC)

Take-up clutch belt: PUM30003 (JVC)

Reel belt: PUM3003-7 (JVC)
Center Puley PU55372 (JVC)

Loading belt: PUM3003-8 (JVC)
Loading motor

Clockwise rotation of loading motor unloads tape.

Location of cassette basket latch release

Bottom of mechanical chassis

JVC E

Includes JVC HR-D170U, HR-D180U, HR-D200U, HR-D230U, HR-D217U, HR-D310U, HR-D320U, HR-D360U, HR-D370U, HR-D400U, HR-D440U, HR-D470U, HR-D530U, HR-D570U, HR-D630U, HR-D750U, HR-S5500U; Zenith VR1810, VR1820, VR1820-1, VR1825, VR1825-1, VR1830, VR1860, VR1860-1, VR1862, VR1862-1, VR1870, VR1870-1, VR2220, VR2220-1, VR2225, VR2225-1, VR2230, VR2235, VR2300, and VR3300, among others

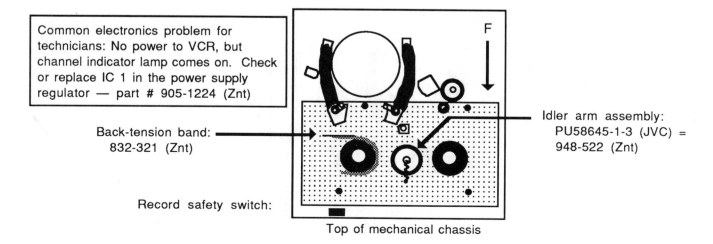

Common electronics problem for technicians: No power to VCR, but channel indicator lamp comes on. Check or replace IC 1 in the power supply regulator — part # 905-1224 (Znt)

Back-tension band: 832-321 (Znt)

Idler arm assembly: PU58645-1-3 (JVC) = 948-522 (Znt)

Record safety switch:

Top of mechanical chassis

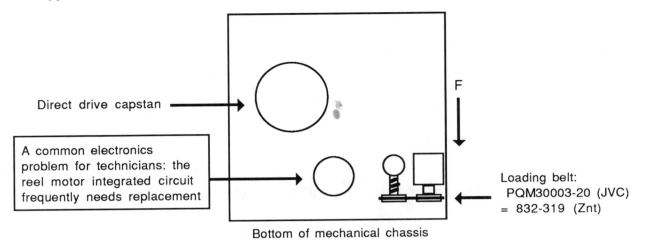

Direct drive capstan ➤

A common electronics problem for technicians: the reel motor integrated circuit frequently needs replacement ➤

F

Loading belt: PQM30003-20 (JVC) = 832-319 (Znt)

Bottom of mechanical chassis

K L H — Look under Shintom

Kenwood — Look under J V C and N E C

Kodak — Look under Panasonic

Lloyds — Look under N E C and Funai

Magnavox — Look under Panasonic

Memorex — Look under Goldstar

Midland — Look under Samsung

Minolta — Look under Hitachi

MGA — Look under Mitsubishi

Mitsubishi A Includes Mitsubishi HS300U and others

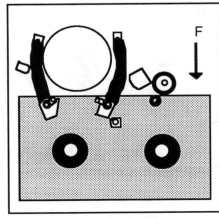

F

Top of reel table

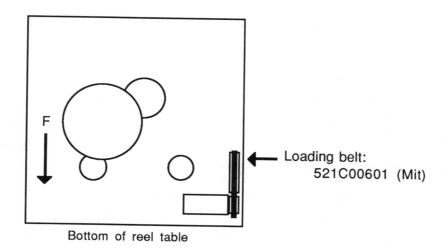

Bottom of reel table

Loading belt:
521C00601 (Mit)

Mitsubishi   B   Includes Mitsubishi HS304UR, HS305UR, HS320UR, HS330UR, among others

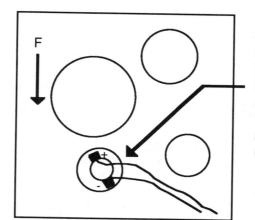

Reel motor:

Bottom of reel table

To reel tape back into cassette, attach 9-volt battery to reel motor terminals underneath VCR, attaching the terminal marked "+" to + terminal of battery, and "-" terminal of motor to minus terminal of battery.

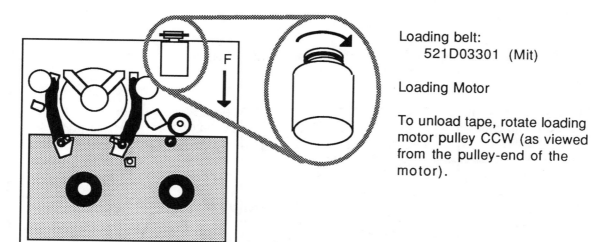

Top of reel table

Loading belt:
521D03301 (Mit)

Loading Motor

To unload tape, rotate loading motor pulley CCW (as viewed from the pulley-end of the motor).

Mitsubishi   C   Includes Mitsubishi HS-328UR, HS-337UR, HS-338UR, HS-339UR, HS-347UR, HS-348UR, HS-349UR, HS-359UR, HS-402UR, HS-411UR, HS-412UR, HS-413UR, HS-421UR, HS-422UR, HS-423UR, among others

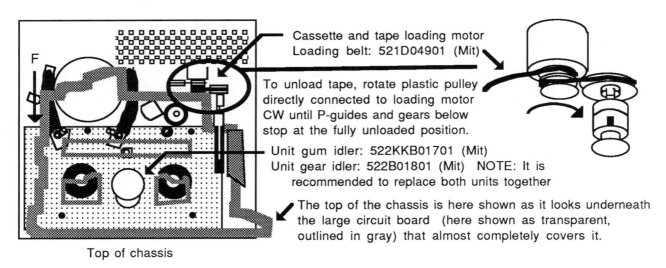

Cassette and tape loading motor
Loading belt: 521D04901 (Mit)

To unload tape, rotate plastic pulley directly connected to loading motor CW until P-guides and gears below stop at the fully unloaded position.

Unit gum idler: 522KKB01701 (Mit)
Unit gear idler: 522B01801 (Mit)   NOTE: It is recommended to replace both units together

The top of the chassis is here shown as it looks underneath the large circuit board (here shown as transparent, outlined in gray) that almost completely covers it.

Top of chassis

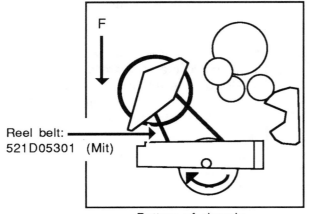

Reel belt:
521D05301 (Mit)

Bottom of chassis

Turn reel pulley CW to reel tape into cassette. To eject cassette from front loader, pull and twist plunger on cassette basket drive assembly rear-ward until it engages with the same plastic pulley that was turned in the firststep, and turnCW. After ejecting the tape, when the VCR is powered up again and you try to load a cassette, it will spit it out again. To get around this, manually load a cassette by pushing it in while rotating the same pulley CCW until it clicks. Plug in, power up, hit eject.

Mitsubishi   D   Includes Emerson VCR900, VCR 910, VCR 950; Mitsubishi HS-306UR, HS-315UR, HS-316UR, HS-317UR, HS-318UR, HS-400UR, HS 710UR among others

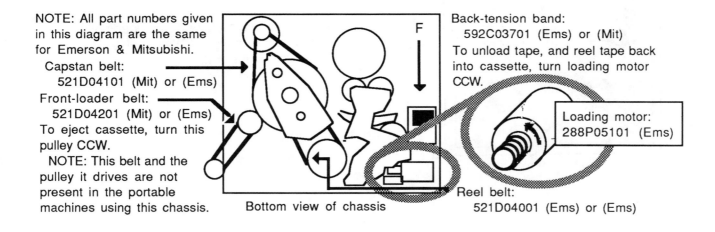

NOTE: All part numbers given in this diagram are the same for Emerson & Mitsubishi.
   Capstan belt:
      521D04101 (Mit) or (Ems)
   Front-loader belt:
      521D04201 (Mit) or (Ems)
   To eject cassette, turn this pulley CCW.
   NOTE: This belt and the pulley it drives are not present in the portable machines using this chassis.

Bottom view of chassis

Back-tension band:
   592C03701 (Ems) or (Mit)
To unload tape, and reel tape back into cassette, turn loading motor CCW.

Loading motor:
288P05101 (Ems)

Reel belt:
521D04001 (Ems) or (Ems)

Location of head-switching relay. Clean for bad picture, especially a picture with rainbow, as explained in Chapter 18, Section 6. For narrow lines at the top or bottom of the picture, adjust the tape guides, as explained in Section 2 of the same chapter.

Unit reel idler assy:
522P00201 (Mit)
or 522P00201 (Ems)
Record Safety Switch:
439C01501 (Mit)

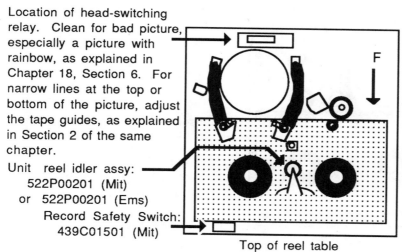

Top of reel table

Note: A common problem on this model is the capstan not running, or running at too slow a speed, or refusing to take a tape, or refusing to eject, or ejecting too slowly, due to a bad integrated circuit BA 6219, the part number for which is 266P13501 (Mit). Soldering is required for its replacement. It may be found attached to a metal part of the chassis, possibly at the edge or top rear.

Another common problem is the timer clock gaining time. If you have this problem, replace the timer IC 8AO. You can order the IC by description; just give your VCR model number.

Mitsubishi E Includes Mitsubishi 303UR among others

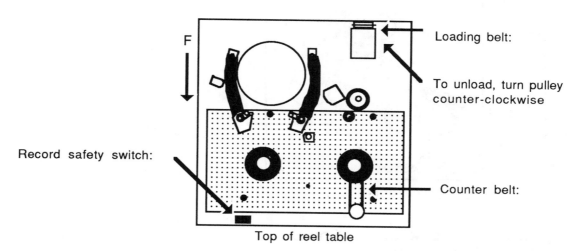

Record safety switch:

Loading belt:

To unload, turn pulley counter-clockwise

Counter belt:

Top of reel table

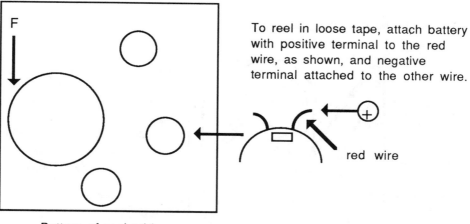

To reel in loose tape, attach battery with positive terminal to the red wire, as shown, and negative terminal attached to the other wire.

red wire

Bottom of reel table

Mitsubishi F   Includes Mitsubishi HS 310U among others

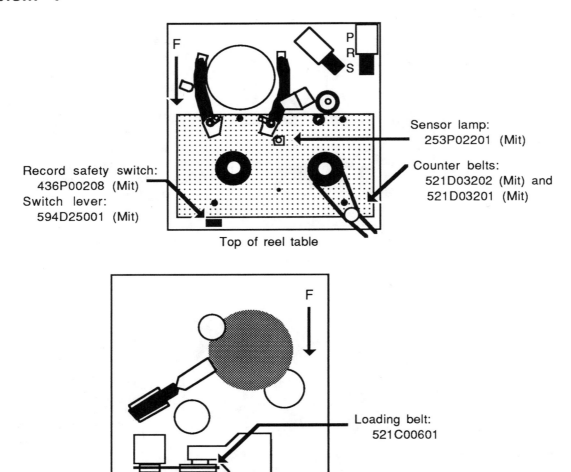

Sensor lamp:
253P02201 (Mit)

Record safety switch:
436P00208 (Mit)
Switch lever:
594D25001 (Mit)

Counter belts:
521D03202 (Mit) and
521D03201 (Mit)

Top of reel table

Loading belt:
521C00601

Bottom of reel table

Montgomery Ward — Look under Sharp
Multitech — Look under N E C and Funai

NEC A   Includes NEC N830, N831EC, N833E/C, N833EU, among others.

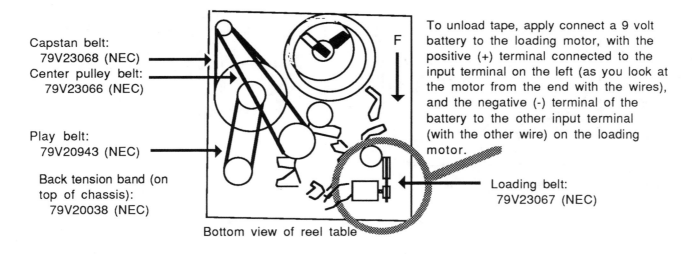

Capstan belt:
79V23068 (NEC)
Center pulley belt:
79V23066 (NEC)

Play belt:
79V20943 (NEC)

Back tension band (on
top of chassis):
79V20038 (NEC)

To unload tape, apply connect a 9 volt
battery to the loading motor, with the
positive (+) terminal connected to the
input terminal on the left (as you look at
the motor from the end with the wires),
and the negative (-) terminal of the
battery to the other input terminal
(with the other wire) on the loading
motor.

Loading belt:
79V23067 (NEC)

Bottom view of reel table

NEC B

Includes NEC DX-2000U/C, DX2500U/C, DX-3500U, DX 5000U, DS-8000U, N901, N902U, N906U, N911U, N912U, N915U/C, N916U, N925U/C, N926U, N930U, N945U, N946U, N951U, N955U/C, N956U, N965U, and N966U among others

Loading belt ("cam loading belt"): 16629251 (NEC)

Idler:
016-18-0801 (NEC)

Tension band:
16177851 (NEC)

Record Safety switch:
65907086 (NEC)

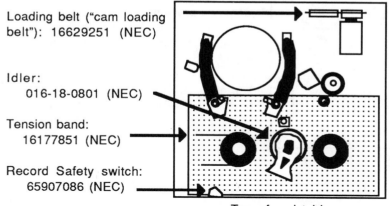

Top of reel table

Do not rotate loading motor by hand, because this may damage the gears underneath. Unload tape by attaching 9 volt battery to loading motor.

Front-loader belt:
16441731 (NEC)

Complete cassette front-loading assembly:
16177652 (NEC)

Common problem: Horizontal lines on picture caused by the base of the P-guides coming loose and moving slightly up or down in the pot metal part that slides along the two tracks to load the tape. Solution: Remove one at a time, paint with Lock-Tite, reinsert, adjust as explained in Section 2 of Chapter 18.

Drive belt:
16442161 (NEC)

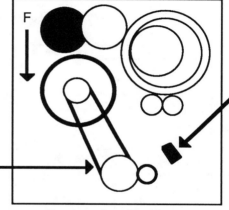

Bottom of reel table

To reel in loose tape, reach through rectangualr opening with scribe, catch in bottom of reel, and rotate reel clockwise as viewed from the bottom.

NEC C

Includes 895U/C, 958U/C, 968U/C, and 978U/C among others

Idler-clutch assembly:
16185482 (NEC)
= 16118548 (NEC)

Back-tension band:
16185812 (NEC)

Record safety switch:
82756PB1 (NEC)

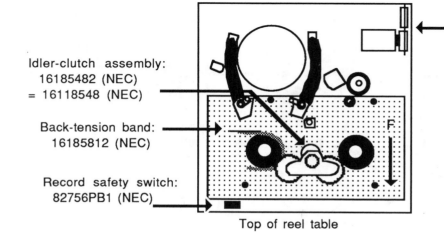

Top of reel table

Loading belt
16629251 (NEC)
To unload tape connect 9 volt battery to loading motor

Do not rotate loading motor by hand, because this may damage the gears underneath. Unload tape by attaching 9 volt battery to loading motor.

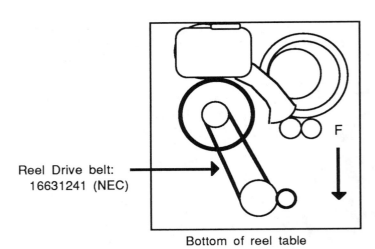

Bottom of reel table

Front loader belt:
16441731 (NEC)

Common problem: Horizontal lines on picture caused by the base of the P-guides coming loose and moving slightly up or down in the pot metal part that slides along the two tracks to load the tape. Solution: Remove one at a time, paint with Lock-Tite, reinsert, adjust as explained in Section 2 of Chapter 18.

Reel Drive belt:
16631241 (NEC)

Panasonic A

Includes Curtis Mathes C718; JC Penney 686-5001; GE 1VCR9000;Magnavox VH8200, VJ8220, VJ8225; Panasonic PV1000, PV1000A, PV1100; Philco V1000, V1100; Quasar VH 5000, VH 5010; Sylvania VC2400, VC2450, VC2500; RCA VBT200, VCT200, VCT201, VCT300, VCT310, VCT400, VCT400X, VCT555, VCT600, VDT201, VDT350, among others.

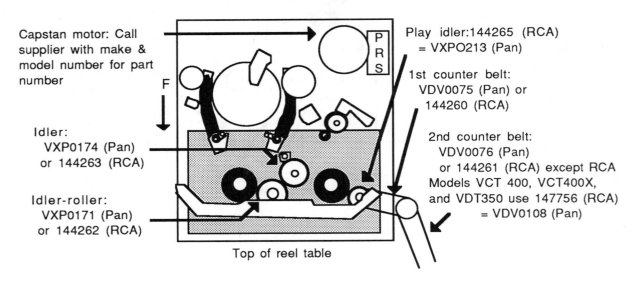

Capstan motor: Call supplier with make & model number for part number

Idler:
  VXP0174 (Pan)
  or 144263 (RCA)

Idler-roller:
  VXP0171 (Pan)
  or 144262 (RCA)

Play idler:144265 (RCA)
  = VXPO213 (Pan)

1st counter belt:
  VDV0075 (Pan) or
  144260 (RCA)

2nd counter belt:
  VDV0076 (Pan)
  or 144261 (RCA) except RCA
  Models VCT 400, VCT400X,
  and VDT350 use 147756 (RCA)
  = VDV0108 (Pan)

Top of reel table

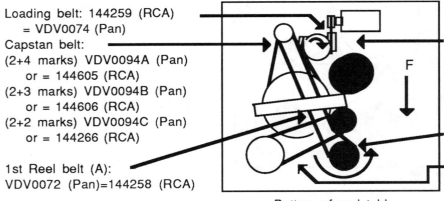

Loading belt: 144259 (RCA)
  = VDV0074 (Pan)
Capstan belt:
(2+4 marks) VDV0094A (Pan)
  or = 144605 (RCA)
(2+3 marks) VDV0094B (Pan)
  or = 144606 (RCA)
(2+2 marks) VDV0094C (Pan)
  or = 144266 (RCA)

1st Reel belt (A):
VDV0072 (Pan)=144258 (RCA)

To unload tape, turn pulley attached to loading motor clockwise so that part with arrow rotates CW, as shown. To reel tape back into cassette, turn this pulley CCW.

2nd reel belt (B):
  VDV0092 (Pan)
  or 144267 (RCA)

Bottom of reel table

Panasonic B

Includes Curtis Mathes D729, F740; JC Penney 686-5003, 686-5006, 686-5007, 686-5011; Magnavox VR8222, VK8222, VJ8225, VK8227, VR8310; Panasonic PV 1200, PV 1210, PV 1500, PV 1600; Philco V1010, V1300, V1500; Quasar VH5015, VH5020, VH5100, VH5150; RCA VCT400, VDT350, VDT501, VDT525, VDT555, VDT 600, VET180; Sylvania VC2200, VC2700, VC3000 among others.

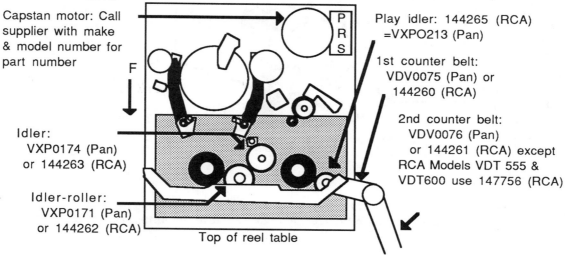

Capstan motor: Call supplier with make & model number for part number

Idler:
VXP0174 (Pan)
or 144263 (RCA)

Idler-roller:
VXP0171 (Pan)
or 144262 (RCA)

Top of reel table

Play idler: 144265 (RCA)
=VXPO213 (Pan)

1st counter belt:
VDV0075 (Pan) or
144260 (RCA)

2nd counter belt:
VDV0076 (Pan)
or 144261 (RCA) except
RCA Models VDT 555 &
VDT600 use 147756 (RCA)

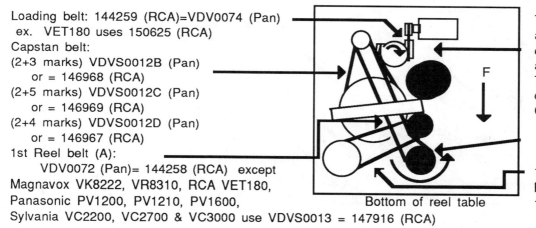

Loading belt: 144259 (RCA)=VDV0074 (Pan)
ex. VET180 uses 150625 (RCA)
Capstan belt:
(2+3 marks) VDVS0012B (Pan)
or = 146968 (RCA)
(2+5 marks) VDVS0012C (Pan)
or = 146969 (RCA)
(2+4 marks) VDVS0012D (Pan)
or = 146967 (RCA)
1st Reel belt (A):
VDV0072 (Pan)= 144258 (RCA) except
Magnavox VK8222, VR8310, RCA VET180,
Panasonic PV1200, PV1210, PV1600,
Sylvania VC2200, VC2700 & VC3000 use VDVS0013 = 147916 (RCA)

Bottom of reel table

To unload tape, turn pulley attached to loading motor clockwise so that part with arrow rotates CW, as shown. To reel tape back into cassette, turn this pulley CCW.

2nd reel belt (B):
VDV0092 (Pan) or
144267 (RCA) except on
RCA VDT625 where belt is
147883 (RCA)

Panasonic C

Includes GE 1VCR5014; Magnavox VK8222, VK8229; Panasonic PV 1650;Quasar VH5155; RCA VDT625

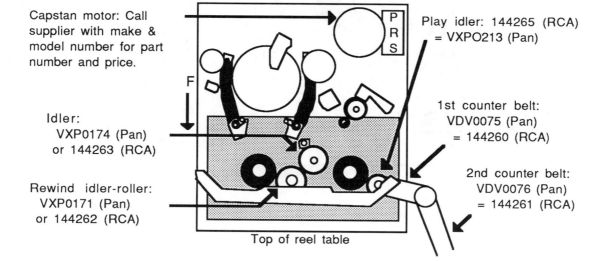

Capstan motor: Call supplier with make & model number for part number and price.

Idler:
VXP0174 (Pan)
or 144263 (RCA)

Rewind idler-roller:
VXP0171 (Pan)
or 144262 (RCA)

Top of reel table

Play idler: 144265 (RCA)
= VXPO213 (Pan)

1st counter belt:
VDV0075 (Pan)
= 144260 (RCA)

2nd counter belt:
VDV0076 (Pan)
= 144261 (RCA)

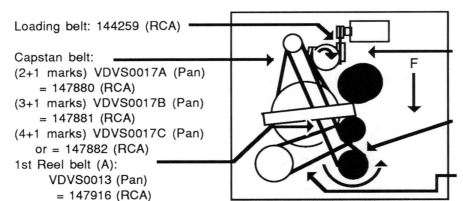

Loading belt: 144259 (RCA)

Capstan belt:
(2+1 marks) VDVS0017A (Pan)
   = 147880 (RCA)
(3+1 marks) VDVS0017B (Pan)
   = 147881 (RCA)
(4+1 marks) VDVS0017C (Pan)
   or = 147882 (RCA)
1st Reel belt (A):
   VDVS0013 (Pan)
   = 147916 (RCA)

To unload tape, turn pulley attached to loading motor clockwise so that part with arrow rotates CW, as shown. To reel tape back into cassette, turn this pulley CCW.

2nd reel belt (B):
   VDVS0016 (Pan) or
147883 (RCA)

Bottom of reel table

Panasonic D

Includes Curtis Mathes HV729, HV753; JC Penney 686-5021, 626-5016, 686-5020; Magnavox VR8306, VR8318; Panasonic PV 1265, PV 1280, PV 1310, PV 1510; Philco V1001, V1333; Quasar VH5022, VH5120, VH5125, VH5220; Sylvania VC2215, VC2950; RCA VGT205, VGT207, VGT225, and others

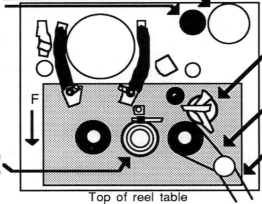

Capstan motor & loading motor: Call parts supplier with make and model for part number & price

Fast-forward idler:
VXPS0054 (Pan)
= 155739 (RCA)

Play idler: VXP0331 (Pan)
   = 152683 (RCA)

Counter belt A:
   155813 (RCA) = VDVS0032 (Pan)

Counter belt B:
   VDVS0033 (Pan) = 155826 (RCA)
except RCA VGT225, Panasonic PV1310, Philco V1333, Magnavox VR8318, Curtis Math HV753, Quasar 5125, and Sylvania VC2950 use VDVS0034 (Pan)=156172 (RCA)

Top of reel table

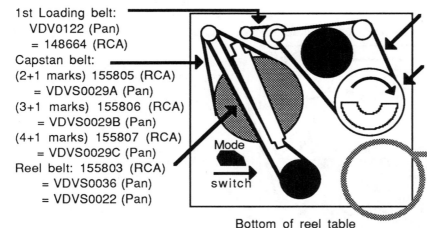

1st Loading belt:
   VDV0122 (Pan)
   = 148664 (RCA)
Capstan belt:
(2+1 marks) 155805 (RCA)
   = VDVS0029A (Pan)
(3+1 marks) 155806 (RCA)
   = VDVS0029B (Pan)
(4+1 marks) 155807 (RCA)
   = VDVS0029C (Pan)
Reel belt: 155803 (RCA)
   = VDVS0036 (Pan)
   = VDVS0022 (Pan)

Mode switch

2nd loading belt: VDVS0020 (Pan)
   = 152412 (RCA)
To unload, turn this pulley CW. On front loader, stop when mode switch is centered at notch. It is not feasible to reel tape back into cassette. Jam cassette door open, remove assembly, and reel in by hand.

Location of cassette basket latch release on top-loading version. To release catch, turn pulley all the way in the direction shown above.

Bottom of reel table

Panasonic E Includes Curtis Mathes HV753, F737, G750; JC Penney 686-5010, 686-5013; Magnavox VR8320, VR8325, VR8330, VR8335, VR8340, VR8345, VR8345; Montgomery Ward GEN 10523; Panasonic PV 1300, PV 1370, PV 1400, PV 1470, PV 1750, PV 1770; Philco V1440, V1441, V1550, V1551; Quasar VH5030, VH5040, VH5160, VH5210, VH5310, VH5610; Sylvania VC2800, VC2900, VC3100, VC3110, VC3610; RCA VET200, VET250, VET253, VET450, VET650, VFT250, VFT450, VFT650, VFT654 among others

Idler: VXP0344 (Pan)
= 148580 (RCA)

Back-tension band:
148809 (RCA)

Reel Motor Assembly
(peeking out through the top,
but mostly underneath):
VXPS0064 (Pan)
= 152426 (RCA)

Play belt: VDVS0026 (Pan)
= 152413 (RCA)

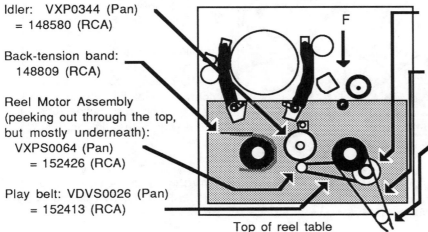

Top of reel table

Play arm unit:
VXLS0081 (Pan)
= 152422 (RCA)

1st Counter belt (A)
VDV0217 (Pan)
= 148668 (RCA)

2nd Counter belt (B)
VDV0218 (Pan)
= 149432 (RCA), except
Panasonic PV1750,PV 1770, RCA
VET650, VFT 650, VFT654, and
Sylvania VC3610 use 149639(RCA)

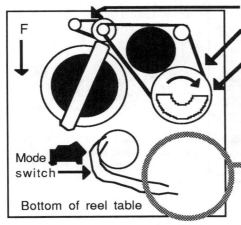

F

Mode switch

Bottom of reel table

1st loading belt: VDV0122 (Pan) = 148664 (RCA)
2nd loading belt: VDVS0020 (Pan) = 152412 (RCA)

To unload tape, gently turn pulley CW. For front-loader, stop when mode switch lines up with little notch; for top-loader, turn pulley until mode switch goes all the way in the direction shown. To reel tape back into cassette, follow yellow and blue wires from reel motor to connector, or penetrate insulation with sharp pins or scribe and connect 9 volt battery to two wires either way.

Location of cassete basket latch on top loading versions. To release catch, turn pulley above all the way in the direction shown above.

Panasonic F Includes Curtis Mathes G748, H749; JC Penney 686-5012; Magnavox VR8315, VR8316; Panasonic PV 1270, PV 1275; Philco V1011, V1012; Quasar VH5011, VH5021, VH5110; Sylvania VC2210, VC2212; RCA VFT 190, VFT193, VGT 180, VGT 200, VGT203, among others

Capstan & loading
unit: Call parts
supplier with make
and model for part
number & price.

Fast-forward &
rewind idler:
VXP0329 (Pan)
= 152674 (RCA)

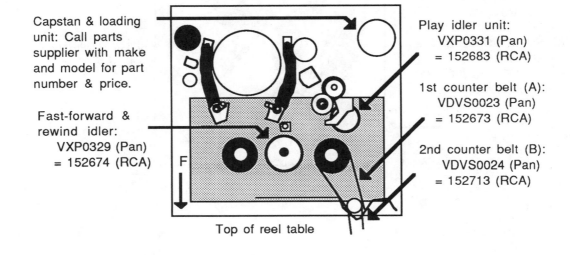

Top of reel table

Play idler unit:
VXP0331 (Pan)
= 152683 (RCA)

1st counter belt (A):
VDVS0023 (Pan)
= 152673 (RCA)

2nd counter belt (B):
VDVS0024 (Pan)
= 152713 (RCA)

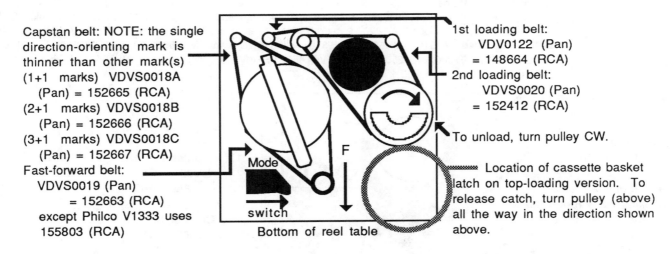

Capstan belt: NOTE: the single direction-orienting mark is thinner than other mark(s)
(1+1 marks) VDVS0018A (Pan) = 152665 (RCA)
(2+1 marks) VDVS0018B (Pan) = 152666 (RCA)
(3+1 marks) VDVS0018C (Pan) = 152667 (RCA)
Fast-forward belt:
VDVS0019 (Pan)
= 152663 (RCA)
except Philco V1333 uses 155803 (RCA)

Mode switch

F

Bottom of reel table

1st loading belt:
VDV0122 (Pan)
= 148664 (RCA)
2nd loading belt:
VDVS0020 (Pan)
= 152412 (RCA)

To unload, turn pulley CW.

Location of cassette basket latch on top-loading version. To release catch, turn pulley (above) all the way in the direction shown above.

Panasonic G
Includes Curtis Mathes JV729, JV753, JV792; GE 1VCR4002, 1VCR4012; JC Penney 686-5030, 686-5032, 686-5037; Magnavox VR8400; Panasonic PV 1220, PV 1320, PV 1520, NV8400; Philco V1002; Quasar VH5031, VH5032, VH5235; Sylvania VC2225LSL01, and others

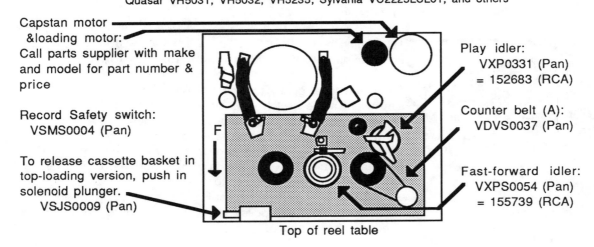

Capstan motor &loading motor:
Call parts supplier with make and model for part number & price

Record Safety switch:
VSMS0004 (Pan)

To release cassette basket in top-loading version, push in solenoid plunger.
VSJS0009 (Pan)

F

Top of reel table

Play idler:
VXP0331 (Pan)
= 152683 (RCA)

Counter belt (A):
VDVS0037 (Pan)

Fast-forward idler:
VXPS0054 (Pan)
= 155739 (RCA)

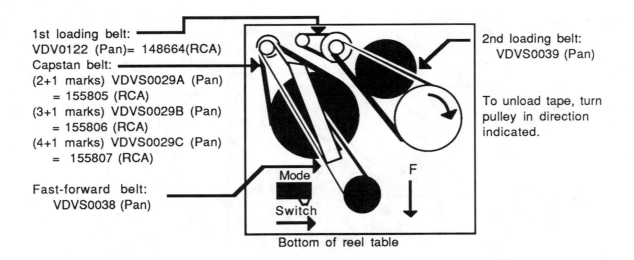

1st loading belt:
VDV0122 (Pan)= 148664(RCA)
Capstan belt:
(2+1 marks) VDVS0029A (Pan)
= 155805 (RCA)
(3+1 marks) VDVS0029B (Pan)
= 155806 (RCA)
(4+1 marks) VDVS0029C (Pan)
= 155807 (RCA)

Fast-forward belt:
VDVS0038 (Pan)

Mode Switch

F

Bottom of reel table

2nd loading belt:
VDVS0039 (Pan)

To unload tape, turn pulley in direction indicated.

Panasonic H

Includes Curtis Mathes HV754; JC Penney 686-5018; Magnavox VR8346; Panasonic PV 1480, PV 1780; Quasar VH5320, VH5623; RCA VGT300, VGT450, VGT650; Sylvania VC3620, among others.

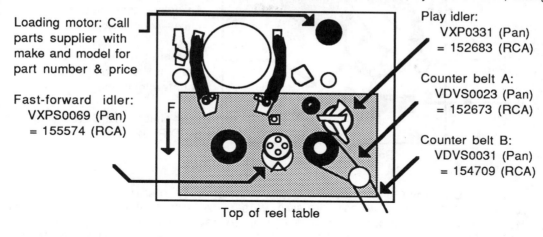

Loading motor: Call parts supplier with make and model for part number & price

Fast-forward idler: VXPS0069 (Pan) = 155574 (RCA)

Play idler: VXP0331 (Pan) = 152683 (RCA)

Counter belt A: VDVS0023 (Pan) = 152673 (RCA)

Counter belt B: VDVS0031 (Pan) = 154709 (RCA)

Top of reel table

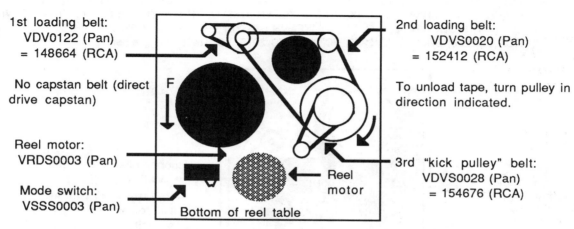

1st loading belt: VDV0122 (Pan) = 148664 (RCA)

No capstan belt (direct drive capstan)

Reel motor: VRDS0003 (Pan)

Mode switch: VSSS0003 (Pan)

2nd loading belt: VDVS0020 (Pan) = 152412 (RCA)

To unload tape, turn pulley in direction indicated.

3rd "kick pulley" belt: VDVS0028 (Pan) = 154676 (RCA)

Reel motor

Bottom of reel table

Panasonic I

Includes Curtis Mathes KV729, KV740, KV753, KV756; GE 1VCR5002, 1VCR5003; JC Penney 686-5053, 686-5060; Magnavox VR 8405, VR 8425SL01, VR 8435; Panasonic PV 1222, PV 1225, PV 1230, PV 1331R, PV 1430, PV 1525, PV 1530, PV 1630, PV 1730; Philco V1003, V1560; Quasar VH 5041, VH 5042, VH 5045, VH 5145, VH 5246, VH 5346, VH 5645, VH 5845, VH 5846XE; Sylvania VC 2230,VC 3140, among others

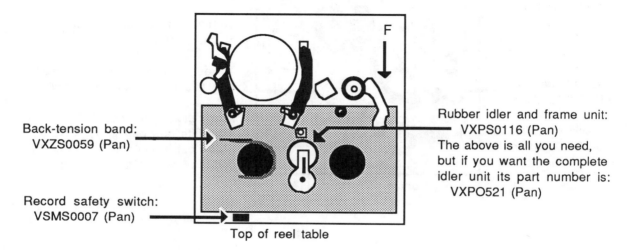

Back-tension band: VXZS0059 (Pan)

Record safety switch: VSMS0007 (Pan)

Rubber idler and frame unit: VXPS0116 (Pan) The above is all you need, but if you want the complete idler unit its part number is: VXPO521 (Pan)

Top of reel table

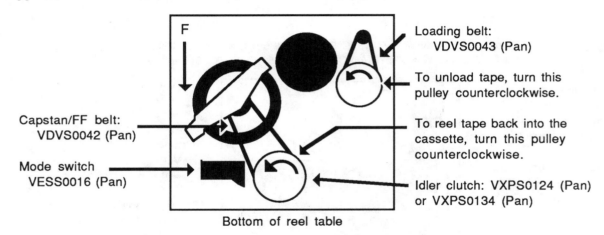

Loading belt:
VDVS0043 (Pan)

To unload tape, turn this
pulley counterclockwise.

Capstan/FF belt:
VDVS0042 (Pan)

To reel tape back into the
cassette, turn this pulley
counterclockwise.

Mode switch
VESS0016 (Pan)

Idler clutch: VXPS0124 (Pan)
or VXPS0134 (Pan)

Bottom of reel table

Panasonic J Includes J C Penney 686-5073; General Electric 1VCR6010X, 1VCR 6011X; Magnavox VR 8510, VR 8515, VR 8525, VR 8530, VR 8535, VR 8540, VR 8545, VR 8555; Panasonic PV 1330, PV 1330K, PV 1332, PV 1334R, PV 1334RK, PV 1340, PV 1340K, PV 1342, PV 1442, PV 1442K, PV 1535, PV 1535K, PV 1540, PV 1545, PV 1545K; Philco, V1561, V1670; Quasar VH 5151YQ, VH 5151K, VH 5152YQ, VH 5153YW, VH 5154K, VH 5154YQ, VH 5156YQ, VH 5245YQ, VH 5157YE, VH 5158, VH 5251, VH 5251K, VH 5254YQ, VH 5256YE, VH 5355YE, VH 5355K, VH 5655YE, VH 5655K, VH 5356; RCA VGT205; Sylvania VC 2215, VC 2975, VC 2976, VC 3146; among others

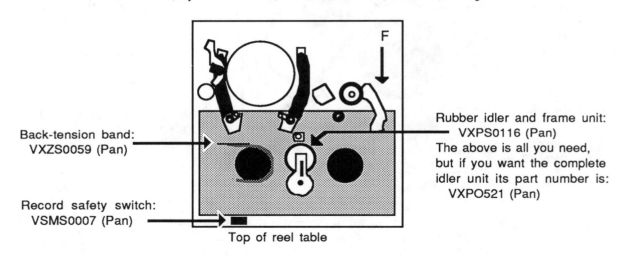

Back-tension band:
VXZS0059 (Pan)

Rubber idler and frame unit:
VXPS0116 (Pan)
The above is all you need,
but if you want the complete
idler unit its part number is:
VXPO521 (Pan)

Record safety switch:
VSMS0007 (Pan)

Top of reel table

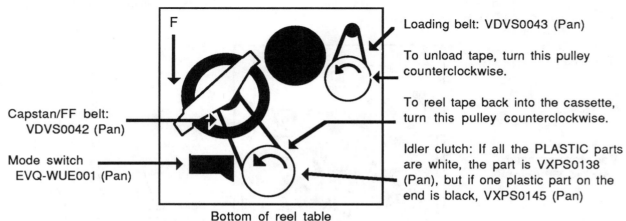

Loading belt: VDVS0043 (Pan)

To unload tape, turn this pulley
counterclockwise.

To reel tape back into the cassette,
turn this pulley counterclockwise.

Capstan/FF belt:
VDVS0042 (Pan)

Mode switch
EVQ-WUE001 (Pan)

Idler clutch: If all the PLASTIC parts
are white, the part is VXPS0138
(Pan), but if one plastic part on the
end is black, VXPS0145 (Pan)

Bottom of reel table

Panasonic K

Includes Panasonic PV 1640, PV 1740, PV 1740K; Quasar VH 5856YE, VH 5857YE, VH 5857K, among others.

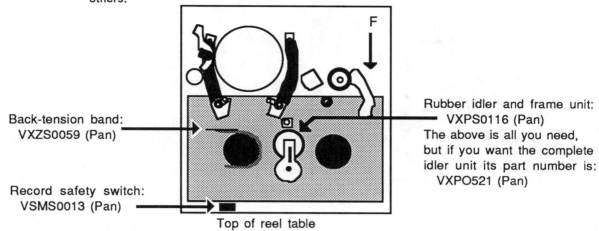

Back-tension band:
VXZS0059 (Pan)

Record safety switch:
VSMS0013 (Pan)

Rubber idler and frame unit:
VXPS0116 (Pan)
The above is all you need,
but if you want the complete
idler unit its part number is:
VXPO521 (Pan)

Top of reel table

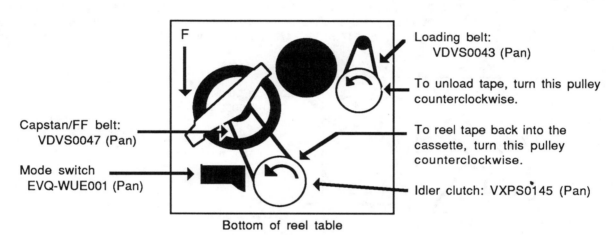

Loading belt:
VDVS0043 (Pan)

To unload tape, turn this pulley
counterclockwise.

Capstan/FF belt:
VDVS0047 (Pan)

To reel tape back into the
cassette, turn this pulley
counterclockwise.

Mode switch
EVQ-WUE001 (Pan)

Idler clutch: VXPS0145 (Pan)

Bottom of reel table

Panasonic L

Includes J C Penney 686-6570; Magnavox VR9522, VR9525, VR9530,VR9535, VR9540, VR9545, VR9555, VR9558; Panasonic PV 1061K, PV 1360, PV 1361, PV 1362, PV 1362K, PV 1364, PV 1364K, PV 1366, PV 1366K, PV 1460, PV 1461, PV 1461K, PV 1452, PV 1560, PV 1560K, PV 1561, PV 1561K, PV 1562, PV 1563, PV 1563K, PV 1564, PV 1566, PV 1567T, PV 1642, PV 1642K, PV 1742; Philco VT8762, VT8765; Quasar VH 5061, VH 5162, VH 5162, VH 5163, VH 5164, VH 5165, VH 5165K, VH 5168, VH 5168K, VH 5261, VH 5261K, VH 5262, VH 5268, VH 5268K, VH 5665, VH 5265K, VH 5865; Sylvania kVC 2975, VC8932, VC8940, VC8945, VC8960, among others

Toothed gear type idler:
VXLS0383 (Mat)

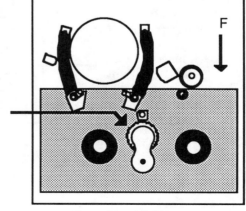

Top of reel table

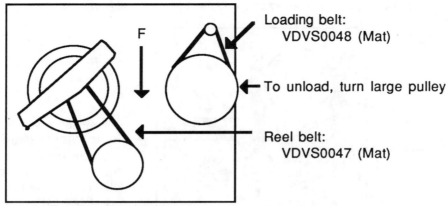

F

Loading belt:
VDVS0048 (Mat)

To unload, turn large pulley

Reel belt:
VDVS0047 (Mat)

Bottom of reel table

Panasonic M   Includes Panasonic PV 1700, PV 2700, PV 3700, PV 3720, PV 3770, PV 3770K, PV 4700, PV 4720, PV 4720K, PV 4750, PV 4750K, PV 4760, PV 4760K, PV 4767T, PV 4768, PV 4768K, PV 4770, PV 4770K, PV 4780, PV 4780K; Quasar VH5270, VH5370, VH5371, VH 5378, VH 5378K, VH 5379, VH 5470, VH 5470K, VH 5471, VH 5471K, VH 5479, VH 5479K, VH 5675, VH 5675K, VH 5975 among others

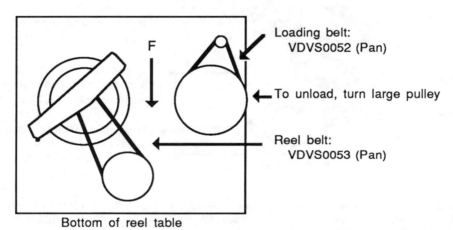

F

Loading belt:
VDVS0052 (Pan)

To unload, turn large pulley

Reel belt:
VDVS0053 (Pan)

Bottom of reel table

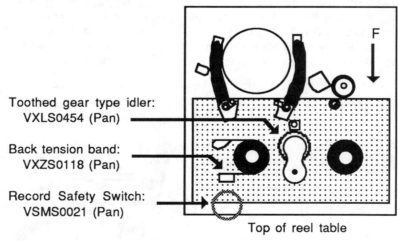

F

Toothed gear type idler:
VXLS0454 (Pan)

Back tension band:
VXZS0118 (Pan)

Record Safety Switch:
VSMS0021 (Pan)

Top of reel table

Panasonic N

Includes Panasonic PV 2800B, PV 2800BK, PV 2801G, PV 2801GK, PV 2802R, PV 2802RK, PV 2803W, PV 2803WK, PV 2812, PV 2812K, PV 2820, PV 2822, PV 2850, PV 2850K, PV 4820, PV 4820K, PV 4822, PV 4822K, PV 4850, PV 4850, PV 4852, PV 4860, PV 4860K, PV 4862, PV 4870, PV S4880, PV S4880K, PV 4970, PV S4986; Quasar VH 5280, VH 5280K, VH 5380, VH 5381, VH 5480, VH 5480K, VH 5485, and VH 5485K among others

To unload tape and eject cassette, push Change Lever forward once each time the gear mechanism stops as you continue to turn the pulley CCW.

Then again turn pully CCW lightly until gears stop. Again release lever and repeat process. Continue until the cassette basket comes to the fully ejected position

F

Main belt:
  VDVS0057 (Pan)
Found on top of chassis:
Back-tension band:
  VXZS0129 (Pan)

Record Safety Switch
VXMS0037 (Pan)

This is the pulley to turn CCW very gently to unload tape & eject cassette

Bottom view

Panasonic O

Includes Panasonic PV 4722, PV 4722K, PV S4764, PV S4764K, PV 4761, PV 4761K, PV 5480, PV 5481; Quasar VH 5180, VH 5381, VH 5477, VH 5477K, VH 5677, and VH 5677K among others

To unload tape and eject cassette, push Change Lever forward once each time the gear mechanism stops moving as you continue to turn the pulley CCW.

Then again turn pully CCW lightly until it stops. Again release lever and repeat process. Continue until the cassette basket comes to the fully ejected position

F

Main belt: VDVS0051 (Pan)
On top of VCR: Back-tension band:
  VXZS0129 (Pan)
Mode Switch: VSSS0081 (Pan)
Record Safety Switch
  VXMS0026 (Pan)
or VSMS0023 (Pan)

This is the pulley to turn CCW very gently to unload tape & eject cassette

Bottom view

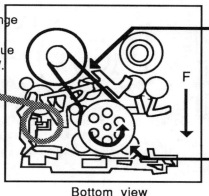

Penneys — Look under Hitachi, Panasonic, and Goldstar

Pentax — Look under Hitachi

Philco — Look under Panasonic

Quasar — Look under Panasonic

R C A — Look under Panasonic, Hitachi, and Samsung

Samsung A
Includes Samsung VT215T, VT221T, VT225T, VR2310, among others

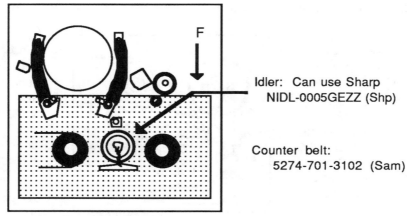

Idler: Can use Sharp
NIDL-0005GEZZ (Shp)

Counter belt:
5274-701-3102 (Sam)

Top of reel table

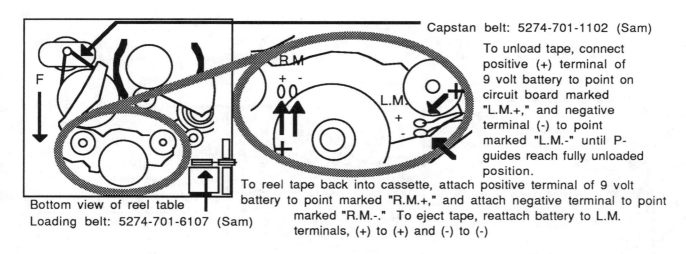

Capstan belt: 5274-701-1102 (Sam)

To unload tape, connect positive (+) terminal of 9 volt battery to point on circuit board marked "L.M.+," and negative terminal (-) to point marked "L.M.-" until P-guides reach fully unloaded position.

Bottom view of reel table
Loading belt: 5274-701-6107 (Sam)

To reel tape back into cassette, attach positive terminal of 9 volt battery to point marked "R.M.+," and attach negative terminal to point marked "R.M.-." To eject tape, reattach battery to L.M. terminals, (+) to (+) and (-) to (-)

Samsung B
Includes Samsung VT 311TA, among others

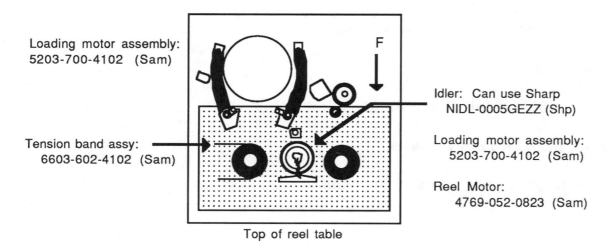

Loading motor assembly:
5203-700-4102 (Sam)

Tension band assy:
6603-602-4102 (Sam)

Idler: Can use Sharp
NIDL-0005GEZZ (Shp)

Loading motor assembly:
5203-700-4102 (Sam)

Reel Motor:
4769-052-0823 (Sam)

Top of reel table

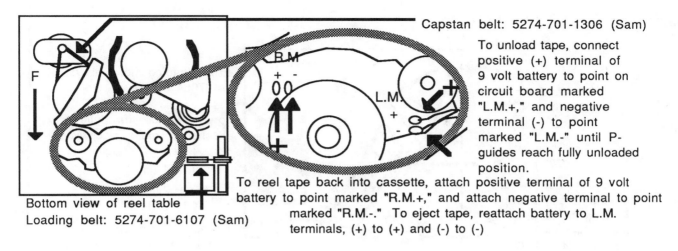

Bottom view of reel table
Loading belt: 5274-701-6107 (Sam)

Capstan belt: 5274-701-1306 (Sam)

To unload tape, connect positive (+) terminal of 9 volt battery to point on circuit board marked "L.M.+," and negative terminal (-) to point marked "L.M.-" until P-guides reach fully unloaded position.

To reel tape back into cassette, attach positive terminal of 9 volt battery to point marked "R.M.+," and attach negative terminal to point marked "R.M.-." To eject tape, reattach battery to L.M. terminals, (+) to (+) and (-) to (-)

Samsung C   Includes RCA VR250, VG7500, among others

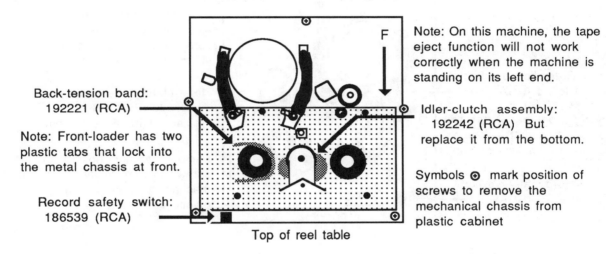

Back-tension band: 192221 (RCA)

Note: Front-loader has two plastic tabs that lock into the metal chassis at front.

Record safety switch: 186539 (RCA)

Note: On this machine, the tape eject function will not work correctly when the machine is standing on its left end.

Idler-clutch assembly: 192242 (RCA) But replace it from the bottom.

Symbols ⊙ mark position of screws to remove the mechanical chassis from plastic cabinet

Top of reel table

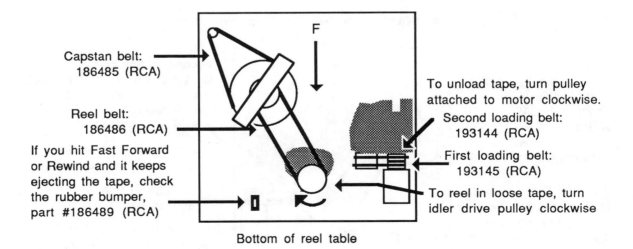

Capstan belt: 186485 (RCA)

Reel belt: 186486 (RCA)

If you hit Fast Forward or Rewind and it keeps ejecting the tape, check the rubber bumper, part #186489 (RCA)

To unload tape, turn pulley attached to motor clockwise.

Second loading belt: 193144 (RCA)

First loading belt: 193145 (RCA)

To reel in loose tape, turn idler drive pulley clockwise

Bottom of reel table

Samsung D   Includes RCA VPT 200

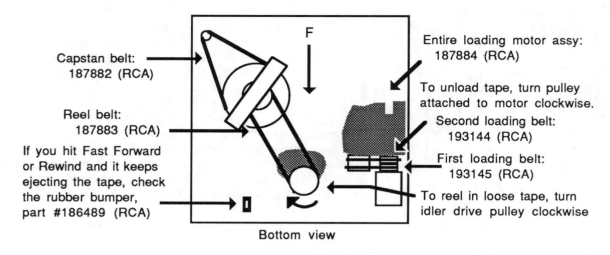

Capstan belt:
187882 (RCA)

Reel belt:
187883 (RCA)

If you hit Fast Forward
or Rewind and it keeps
ejecting the tape, check
the rubber bumper,
part #186489 (RCA)

Entire loading motor assy:
187884 (RCA)

To unload tape, turn pulley
attached to motor clockwise.

Second loading belt:
193144 (RCA)

First loading belt:
193145 (RCA)

To reel in loose tape, turn
idler drive pulley clockwise

Bottom view

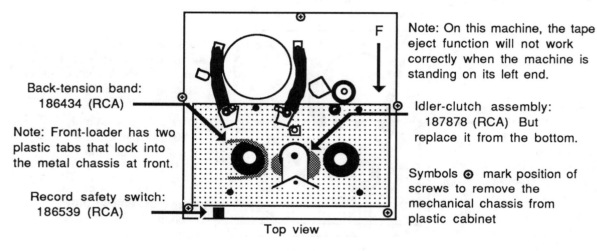

Back-tension band:
186434 (RCA)

Note: Front-loader has two
plastic tabs that lock into
the metal chassis at front.

Record safety switch:
186539 (RCA)

Note: On this machine, the tape
eject function will not work
correctly when the machine is
standing on its left end.

Idler-clutch assembly:
187878 (RCA) But
replace it from the bottom.

Symbols ⊕ mark position of
screws to remove the
mechanical chassis from
plastic cabinet

Top view

Samsung E   Includes RCA VG 7515, VG 7575, VR 280, VR 285,

Top view is identical to the Samsung C top view shown earlier.

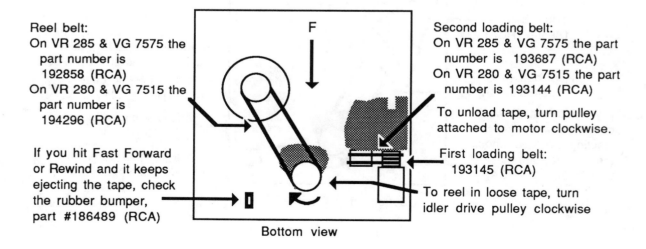

Reel belt:
On VR 285 & VG 7575 the
part number is
192858 (RCA)
On VR 280 & VG 7515 the
part number is
194296 (RCA)

If you hit Fast Forward
or Rewind and it keeps
ejecting the tape, check
the rubber bumper,
part #186489 (RCA)

Second loading belt:
On VR 285 & VG 7575 the part
number is 193687 (RCA)
On VR 280 & VG 7515 the part
number is 193144 (RCA)

To unload tape, turn pulley
attached to motor clockwise.

First loading belt:
193145 (RCA)

To reel in loose tape, turn
idler drive pulley clockwise

Bottom view

Samsung F

Includes RCA VPT 630HF, VPT 640HF, VPT 695HF, VR 270, VR 273, VR 273A, VR 275, VR 450, 451, VR 452, VR 470, VR 475, VR 595, VR 630HF, VR 640HF, VR 695HF

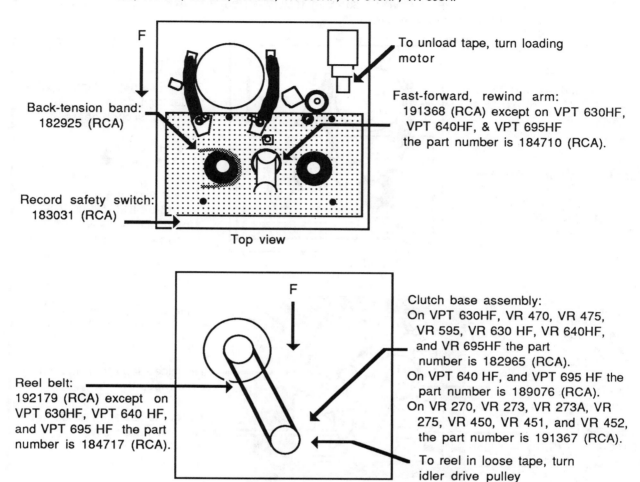

F

To unload tape, turn loading motor

Back-tension band: 182925 (RCA)

Fast-forward, rewind arm: 191368 (RCA) except on VPT 630HF, VPT 640HF, & VPT 695HF the part number is 184710 (RCA).

Record safety switch: 183031 (RCA)

Top view

Reel belt: 192179 (RCA) except on VPT 630HF, VPT 640 HF, and VPT 695 HF the part number is 184717 (RCA).

F

Clutch base assembly: On VPT 630HF, VR 470, VR 475, VR 595, VR 630 HF, VR 640HF, and VR 695HF the part number is 182965 (RCA). On VPT 640 HF, and VPT 695 HF the part number is 189076 (RCA). On VR 270, VR 273, VR 273A, VR 275, VR 450, VR 451, and VR 452, the part number is 191367 (RCA).

To reel in loose tape, turn idler drive pulley

Bottom view

Sanyo A

Includes Sanyo VCR 6400 among others

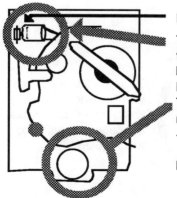

Loading motor belt

To unload tape, connect negative (-) terminal of 9 volt battery to the terminal marked "+" on the loading motor, and (+) terminal of battery to terminal marked "-" on loading motor -- or, turn loading motor CCW by hand to unload.

To reel tape back into cassette, attach 9 volt battery to reel motor in either polarity ("+" to "+" and "-" to "-," or vice versa -- either will work.

PART NOS. OF BELTS IN THIS VCR: 143-2-564T-03200 (SFC)
143-2-564T-03300 (SFC)
and   143-2-564T-03400 (SFC)

Bottom view of reel table   Reel motor: 4-529V-10800 (SFC)

Sanyo B

Includes Sanyo VHR 1250, VHR 1350, VHR 1550, VHR1600, VHR 1900, VHR 2250, VHR 2350, VHR 2550, VHR 2900, among others

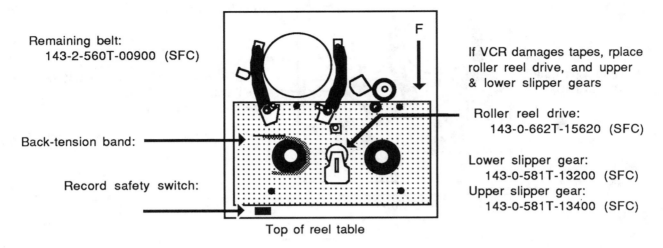

Remaining belt:
143-2-560T-00900 (SFC)

If VCR damages tapes, rplace roller reel drive, and upper & lower slipper gears

Back-tension band:

Roller reel drive:
143-0-662T-15620 (SFC)

Record safety switch:

Lower slipper gear:
143-0-581T-13200 (SFC)
Upper slipper gear:
143-0-581T-13400 (SFC)

Top of reel table

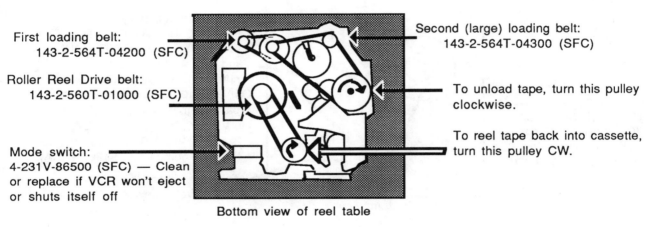

First loading belt:
143-2-564T-04200 (SFC)

Second (large) loading belt:
143-2-564T-04300 (SFC)

Roller Reel Drive belt:
143-2-560T-01000 (SFC)

To unload tape, turn this pulley clockwise.

Mode switch:
4-231V-86500 (SFC) — Clean or replace if VCR won't eject or shuts itself off

To reel tape back into cassette, turn this pulley CW.

Bottom view of reel table

Sanyo C

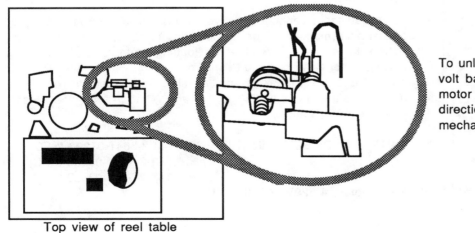

To unload tape, attach 9 volt battery to loading motor to make it rotate in direction that unloads mechanism

Top view of reel table

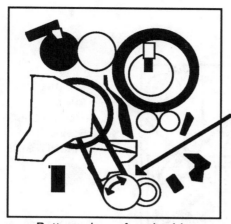

Bottom view of reel table

To reel in loose tape, turn reel pulley in either direction

Scott — Look under Emerson

Sears — Look under Hitachi and Fisher

Sharp A   Includes Sharp VC 8400 among others

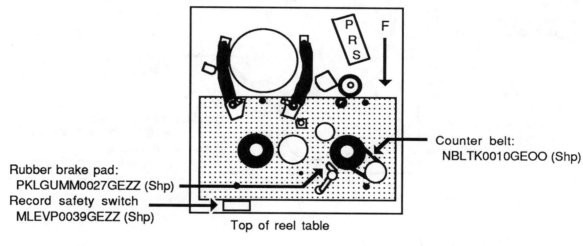

Counter belt:
NBLTK0010GEOO (Shp)

Rubber brake pad:
  PKLGUMM0027GEZZ (Shp)

Record safety switch
  MLEVP0039GEZZ (Shp)

Top of reel table

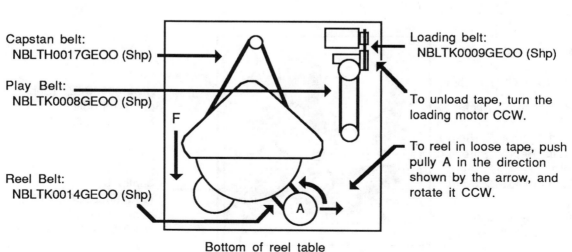

Capstan belt:
  NBLTH0017GEOO (Shp)

Play Belt:
  NBLTK0008GEOO (Shp)

Reel Belt:
  NBLTK0014GEOO (Shp)

Loading belt:
  NBLTK0009GEOO (Shp)

To unload tape, turn the loading motor CCW.

To reel in loose tape, push pully A in the direction shown by the arrow, and rotate it CCW.

Bottom of reel table

Sharp B

Includes Sharp VC9400 among others

Belt in cassette front-
loader assembly:
  NBLTK0009GEOO (Shp)

Record Safety switch
lever:
  MLEVP0034GEZZ (Shp)
Cassette Record Safety
switch:
  QSW-F0019GEZZ (Shp)

Back-tension band:
  LBNDK3010GEZZ (Shp)

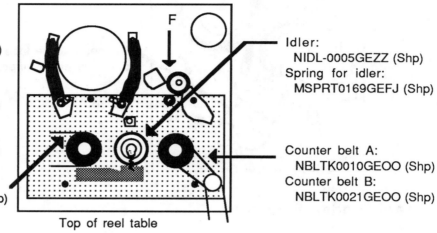

Top of reel table

Idler:
  NIDL-0005GEZZ (Shp)
Spring for idler:
  MSPRT0169GEFJ (Shp)

Counter belt A:
  NBLTK0010GEOO (Shp)
Counter belt B:
  NBLTK0021GEOO (Shp)

Complete cassette front-
loading assembly as a unit:
  CHLDX3010GE01 (Shp)

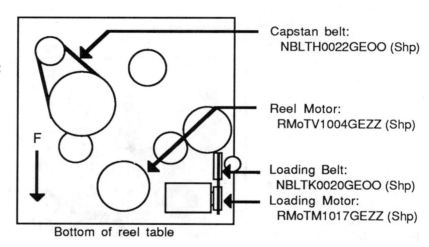

Bottom of reel table

Capstan belt:
  NBLTH0022GEOO (Shp)

Reel Motor:
  RMoTV1004GEZZ (Shp)

Loading Belt:
  NBLTK0020GEOO (Shp)
Loading Motor:
  RMoTM1017GEZZ (Shp)

Sharp C

Includes Sharp VC381 among others

Belt in cassette front-
loader assembly:
  NBLTK0009GEOO (Shp)

Record Safety switch
lever:
  MLEVP0034GEZZ (Shp)
Cassette Record Safety
switch:
  QSW-F0013GEZZ (Shp)

Back-tension band:
  LBNDK3011GEZZ (Shp)

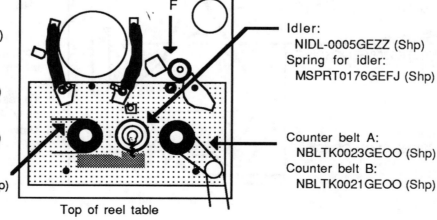

Top of reel table

Idler:
  NIDL-0005GEZZ (Shp)
Spring for idler:
  MSPRT0176GEFJ (Shp)

Counter belt A:
  NBLTK0023GEOO (Shp)
Counter belt B:
  NBLTK0021GEOO (Shp)

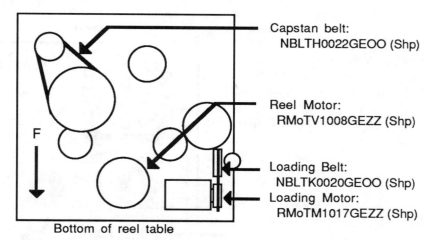

Complete cassette front-loading assembly as a unit:
CHLDX3014GE39 (Shp)

Capstan belt:
NBLTH0022GEOO (Shp)

Reel Motor:
RMoTV1008GEZZ (Shp)

Loading Belt:
NBLTK0020GEOO (Shp)

Loading Motor:
RMoTM1017GEZZ (Shp)

Bottom of reel table

Sharp D

Includes Montgomery Ward 10526, among others

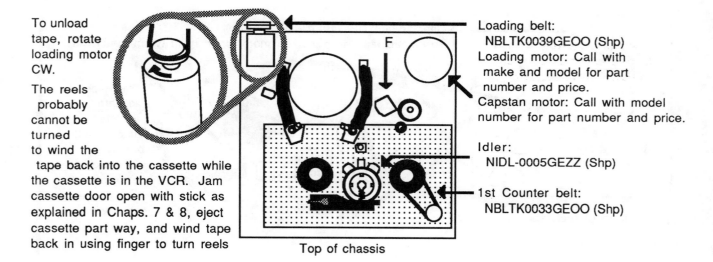

To unload tape, rotate loading motor CW.

The reels probably cannot be turned to wind the tape back into the cassette while the cassette is in the VCR. Jam cassette door open with stick as explained in Chaps. 7 & 8, eject cassette part way, and wind tape back in using finger to turn reels

Loading belt:
NBLTK0039GEOO (Shp)
Loading motor: Call with make and model for part number and price.
Capstan motor: Call with model number for part number and price.

Idler:
NIDL-0005GEZZ (Shp)

1st Counter belt:
NBLTK0033GEOO (Shp)

Top of chassis

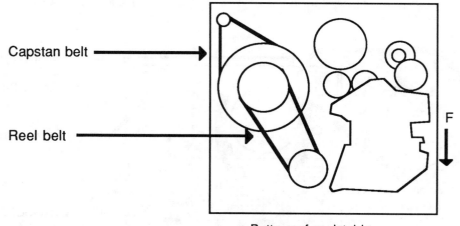

Capstan belt

Reel belt

Bottom of reel table

Sharp E

Includes Sharp VC481U, VC487U, VC489U, VC581U, VC584, VC586, VC5F7U; Montgomery Ward JSJ10550, JSJ10551 among others

To remove the front-loader, remove two screws ⊙ and slide front-loader assembly forward, then lift up and out.

Record safety switch is located in the front-loader.

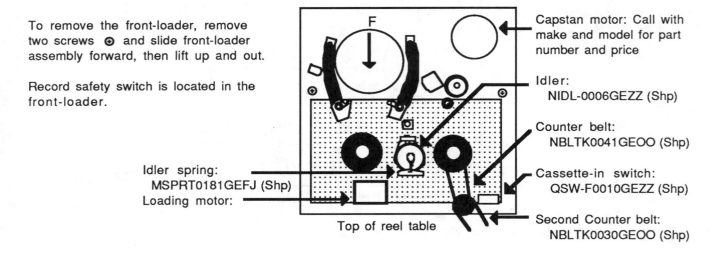

Capstan motor: Call with make and model for part number and price

Idler:
 NIDL-0006GEZZ (Shp)

Counter belt:
 NBLTK0041GEOO (Shp)

Cassette-in switch:
 QSW-F0010GEZZ (Shp)

Second Counter belt:
 NBLTK0030GEOO (Shp)

Idler spring:
 MSPRT0181GEFJ (Shp)
Loading motor:

Top of reel table

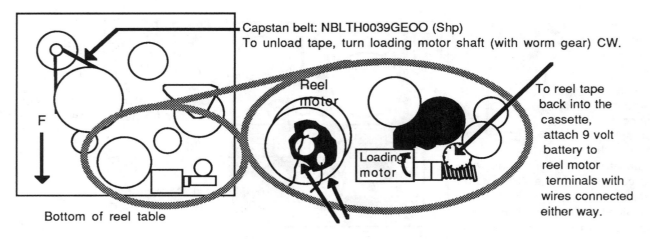

Capstan belt: NBLTH0039GEOO (Shp)
To unload tape, turn loading motor shaft (with worm gear) CW.

Reel motor

Loading motor

To reel tape back into the cassette, attach 9 volt battery to reel motor terminals with wires connected either way.

Bottom of reel table

Sharp F

Includes Sharp VC582U, VC583U, VC587U; Montgomery Ward JSJ10658, JSJ10659, JSJ10660, among others

To unload tape, rotate loading motor CW for a very long time.

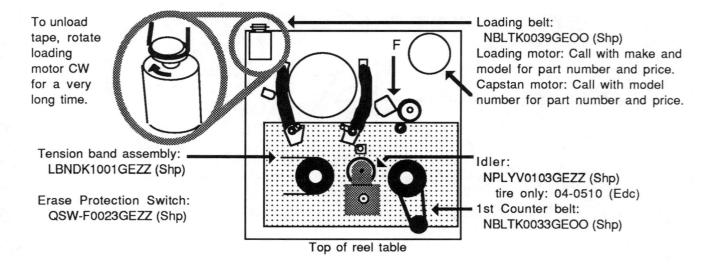

Loading belt:
 NBLTK0039GEOO (Shp)
Loading motor: Call with make and model for part number and price.
Capstan motor: Call with model number for part number and price.

Tension band assembly:
 LBNDK1001GEZZ (Shp)

Erase Protection Switch:
 QSW-F0023GEZZ (Shp)

Idler:
 NPLYV0103GEZZ (Shp)
 tire only: 04-0510 (Edc)
1st Counter belt:
 NBLTK0033GEOO (Shp)

Top of reel table

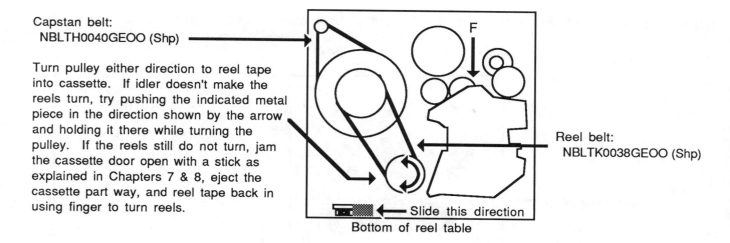

Capstan belt:
 NBLTH0040GEOO (Shp)

Turn pulley either direction to reel tape into cassette. If idler doesn't make the reels turn, try pushing the indicated metal piece in the direction shown by the arrow and holding it there while turning the pulley. If the reels still do not turn, jam the cassette door open with a stick as explained in Chapters 7 & 8, eject the cassette part way, and reel tape back in using finger to turn reels.

Reel belt:
 NBLTK0038GEOO (Shp)

Slide this direction
Bottom of reel table

Sharp G

Includes Sharp VC6730U, VC6833U, VC 683U, VC 6846UB, and VC 7843U among others

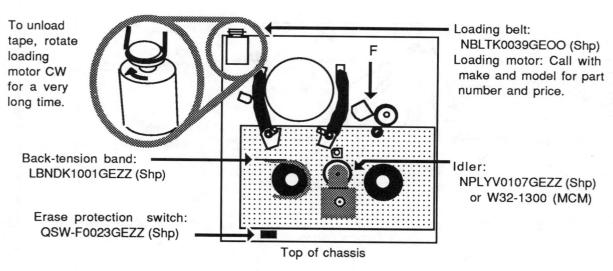

To unload tape, rotate loading motor CW for a very long time.

Loading belt:
 NBLTK0039GEOO (Shp)
Loading motor: Call with make and model for part number and price.

Back-tension band:
 LBNDK1001GEZZ (Shp)

Idler:
 NPLYV0107GEZZ (Shp)
 or W32-1300 (MCM)

Erase protection switch:
 QSW-F0023GEZZ (Shp)

Top of chassis

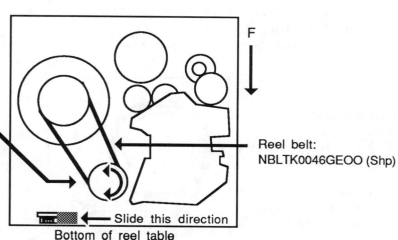

NOTE: This model has a direct-drive capstan. There is no capstan belt.

Turn pulley either direction to reel tape into cassette. If idler doesn't make the reels turn, try pushing the indicated metal piece in the direction shown by the arrow and holding it there while turning the pulley. If the reels still do not turn, jam the cassette door open with a stick as explained in Chapters 7 & 8, eject the cassette part way, and reel tape back in using finger to turn reels.

Reel belt:
NBLTK0046GEOO (Shp)

Slide this direction
Bottom of reel table

Shintom A

Includes Broksonic; Colt; Circuit City; KLH; Logik; Shintom LVA-2001, SVG-110, SVG-220, TN5900N, VCP 1000C, VCR-500,VCR-4530N, VCR-8900, VP-3540N, VP-3550N, VP-3560N, VP-5000N, VP-5540N, VP-5550; Singer; Toshiba M-P200/C

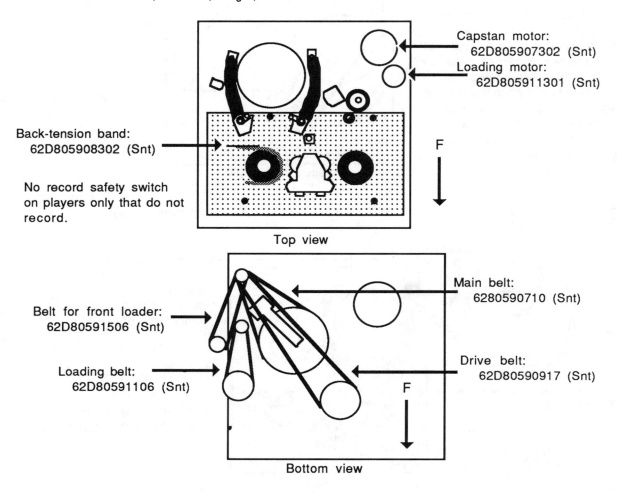

Capstan motor:
62D805907302 (Snt)

Loading motor:
62D805911301 (Snt)

Back-tension band:
62D805908302 (Snt)

No record safety switch on players only that do not record.

F

Top view

Main belt:
6280590710 (Snt)

Belt for front loader:
62D80591506 (Snt)

Loading belt:
62D80591106 (Snt)

Drive belt:
62D80590917 (Snt)

F

Bottom view

NOTE: Prime Electronics (See Appendix II for address) has generic complete belt replacement kits for older Sony Beta's.

Sony A

Includes Zenith KR9000 among others

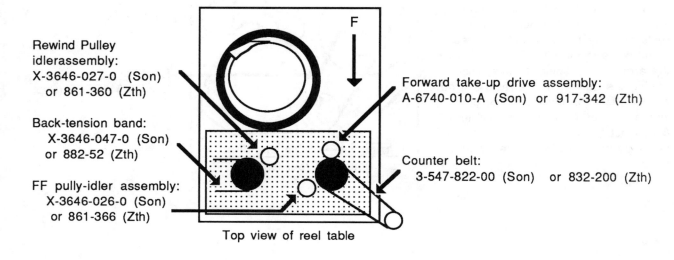

F

Rewind Pulley idlerassembly:
X-3646-027-0 (Son)
or 861-360 (Zth)

Back-tension band:
X-3646-047-0 (Son)
or 882-52 (Zth)

FF pully-idler assembly:
X-3646-026-0 (Son)
or 861-366 (Zth)

Forward take-up drive assembly:
A-6740-010-A (Son) or 917-342 (Zth)

Counter belt:
3-547-822-00 (Son)   or 832-200 (Zth)

Top view of reel table

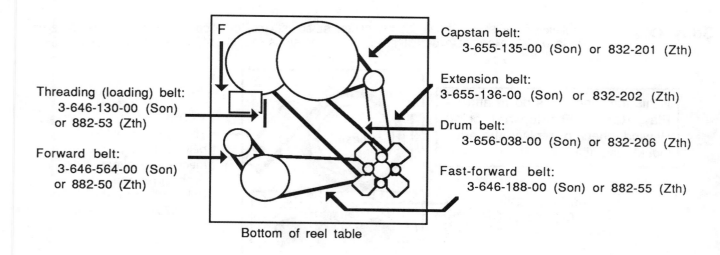

Threading (loading) belt:
   3-646-130-00 (Son)
   or 882-53 (Zth)

Forward belt:
   3-646-564-00 (Son)
   or 882-50 (Zth)

Capstan belt:
   3-655-135-00 (Son) or 832-201 (Zth)

Extension belt:
   3-655-136-00 (Son) or 832-202 (Zth)

Drum belt:
   3-656-038-00 (Son) or 832-206 (Zth)

Fast-forward belt:
   3-646-188-00 (Son) or 882-55 (Zth)

Bottom of reel table

Sony B

Includes Sony SL 5000, Zenith VR8900, VR 8910 among others

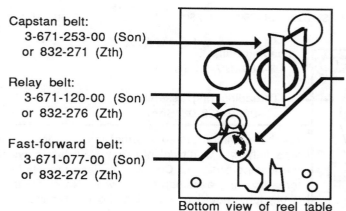

Capstan belt:
   3-671-253-00 (Son)
   or 832-271 (Zth)

Relay belt:
   3-671-120-00 (Son)
   or 832-276 (Zth)

Fast-forward belt:
   3-671-077-00 (Son)
   or 832-272 (Zth)

Bottom view of reel table

To unload tape, first release catch on top side of reel table holding loading ring. Then turn pulley on bottom side of reel table CCW until loading ring (on top) goes CW as far as it will go.
To wind tape back into cassette, jam cassette door open, eject cassette part way, and reach under cassette with finger to turn reels in cassette to reel tape in.

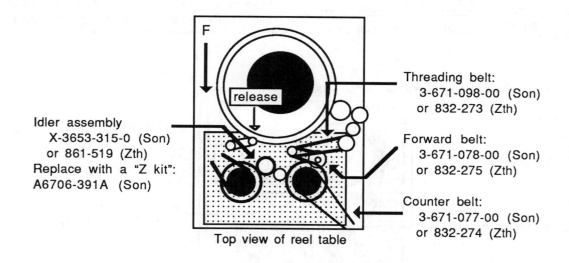

Idler assembly
   X-3653-315-0 (Son)
   or 861-519 (Zth)
Replace with a "Z kit":
   A6706-391A (Son)

Threading belt:
   3-671-098-00 (Son)
   or 832-273 (Zth)

Forward belt:
   3-671-078-00 (Son)
   or 832-275 (Zth)

Counter belt:
   3-671-077-00 (Son)
   or 832-274 (Zth)

Top view of reel table

Sony C

Includes Sony SL-10, SL-20, SL-25, SL-30, SL-60, SL-100, SL-2300, SL-2301, SL-2305, SL-2400, SL-2401, SL-2405, SL-2406, SL-3030, S-120, S-250, S-370, S-570, S-700, S-770 among others.

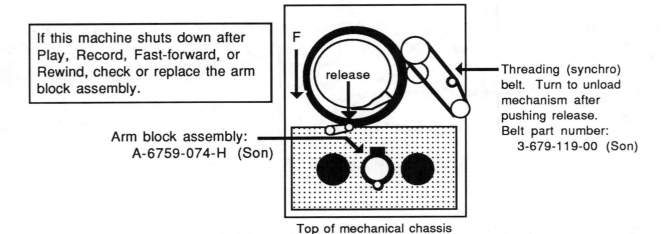

If this machine shuts down after Play, Record, Fast-forward, or Rewind, check or replace the arm block assembly.

F

release

Threading (synchro) belt. Turn to unload mechanism after pushing release. Belt part number: 3-679-119-00 (Son)

Arm block assembly: A-6759-074-H (Son)

Top of mechanical chassis

Sony D

Includes Sony SL-2000, SL-2001, SL-2005, SL-2500, SL-2501, SL-27000, SL-2700B, SL-2710 among others

Loading half-way micro switch:
1-552-664-00 (Son)

Loading belt: 3-669-327-00 (Son)

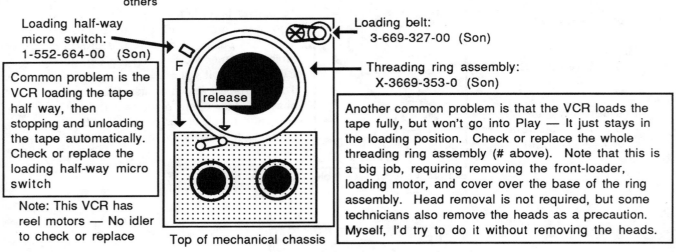

F

release

Threading ring assembly: X-3669-353-0 (Son)

Common problem is the VCR loading the tape half way, then stopping and unloading the tape automatically. Check or replace the loading half-way micro switch

Note: This VCR has reel motors — No idler to check or replace

Top of mechanical chassis

Another common problem is that the VCR loads the tape fully, but won't go into Play — It just stays in the loading position. Check or replace the whole threading ring assembly (# above). Note that this is a big job, requiring removing the front-loader, loading motor, and cover over the base of the ring assembly. Head removal is not required, but some technicians also remove the heads as a precaution. Myself, I'd try to do it without removing the heads.

Sony E

Includes Sony SLV-P30HF among others

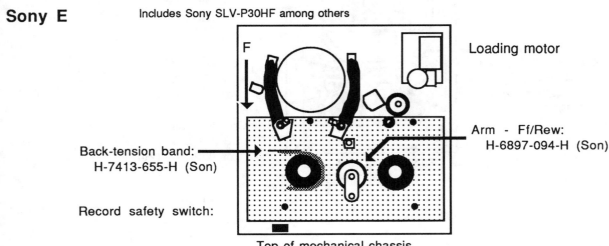

F

Loading motor

Back-tension band: H-7413-655-H (Son)

Arm - Ff/Rew: H-6897-094-H (Son)

Record safety switch:

Top of mechanical chassis

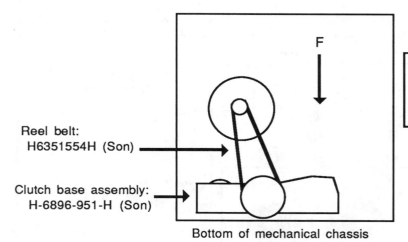

Reel belt:
H6351554H (Son)

Clutch base assembly:
H-6896-951-H (Son)

F

Common problem is shutting down, or refusing to fast-forward or rewind. Check or change the clutch base assembly and idler both.

Bottom of mechanical chassis

Sound Design — Look under Funai

Supra — Look under Akai, Samsung and Funai

Sylvania — Look under Panasonic

Symphonic — Look under Funai and N E C

Teac — Look under J V C and Funai

Technics — Look under Panasonic

Toshiba A   Includes Toshiba M5100, among others

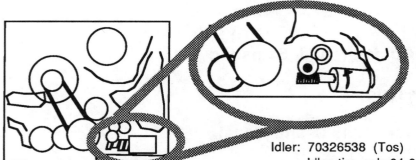

To unload tape, connect postive terminal of 9 volt battery to right terminal on loading motor, and (-) terminal on battery to left terminal on loading motor. Motor turns CCW to unload.

Bottom view of reel table

Idler: 70326538 (Tos)   or 08-5440 (Edc)
Idler tire only:04-0480 (Edc)
Antenna switcher: 70909201 (Tos)

Toshiba B

Includes Toshiba DX-3, DX-3C, DX-7, DX-7C, DX-400, DX-400C, DX-800, DX-800C, DX-900, DX-900C, M-1283, M-1287/C, M-2283, M-2287, M5300, M-5320, M-5330, among others

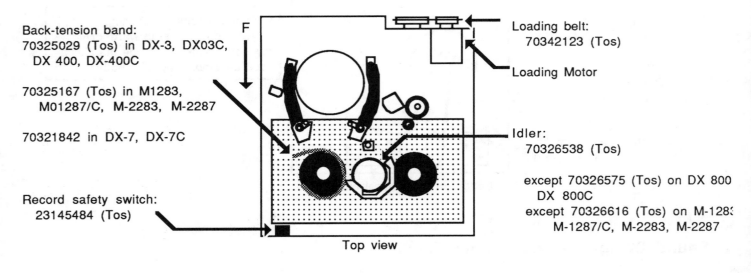

Back-tension band:
70325029 (Tos) in DX-3, DX03C, DX 400, DX-400C

70325167 (Tos) in M1283, M01287/C, M-2283, M-2287

70321842 in DX-7, DX-7C

F

Loading belt:
70342123 (Tos)

Loading Motor

Idler:
70326538 (Tos)

except 70326575 (Tos) on DX 800 DX 800C
except 70326616 (Tos) on M-128? M-1287/C, M-2283, M-2287

Record safety switch:
23145484 (Tos)

Top view

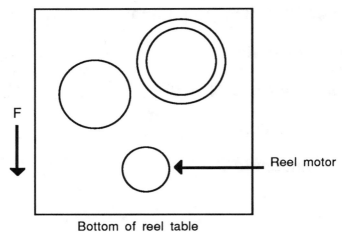

F

Reel motor

Bottom of reel table

Totevision — Look under Goldstar and Samsung

Unitech — Look under Samsung

Vector Research — Look under N E C and Funai

Video Concepts — Look Under Mitsubishi

Zenith — Look under JVC (for VHS) and Sony (for Beta)

Appendix VII

How to Remove and Realign Cassette Loading Assemblies on Common Models

Read this First

The procedures shown in the following pages apply to the front-loading mechanisms that most frequently need realignment in the indicated categories. But bear in mind that each of the manufacturers represented here has other front-loaders with different designs requiring different alignment procedures. Check to make absolutely certain that your front-loader is identical to the version pictured here before trying to use the procedure to realign it.

Fisher-category

Picture showing only the wiper gear on the right side (that is, the motor-side) of common Fisher front-loading assemblies.

Make certain that the pin from the moving basket assembly has not come out of the hole for it in the wiper arm.

When the cassette basket is in the fully-ejected position, this switch should be "closed" — that is, making contact.

Part of what is shown here is hidden behind outside gears here drawn as transparent, but you can look around the edges of them to see the pin and switches.

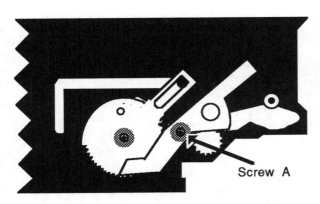

When the cassette basket pin is all the way forward on the right side, make certain that it is all the way forward on the left side too. Insert a cassette far enough to release latches holding the cassette basket, and with your finger turn the cassette loading motor on the right side to put the basket through the motions of loading. Watch carefully to make certain that both the right and left sides of the cassette basket go around the L-corner at exactly the same time. If they do not, remove screw marked "A" in the diagram, and lift off the plastic front-door opener lever to get at the gear underneath.

Screw A

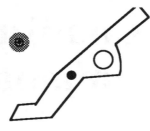

With screw A and plastic arm removed, note the appearance of the outward surface of the little intermediate gear, then carefully lift it off its axle.

Left side big wiper gear

With intermediate gear removed, move big wiper gear on this side to the same relative position as the wiper gear on the right side, then replace intermediate gear. Insert cassette and retest. If it is still not right, remove and retry, until both sides of the mechanism reach the ends of the L-tracks, and go around the corner, at the same time. Test by hand & with battery, as explained in Chapter 10

← Intermediate gear removed

Funai-category

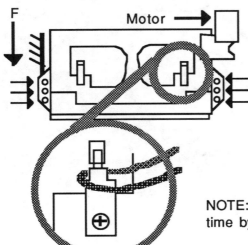

F

Motor →

1. To remove complete front-loading assembly from VCR, remove six screws indicated by six small arrows.
2.   Unplug 9-wire plug.
3. Gently lift basket from unit.
 NOTE: the front cabinet piece may need to be removed in order to remove front-loading assembly.

Return to Section 8 of Chapter 10.

NOTE: Both leaf switches must be closed at the same time by inserted cassette for proper operation.

Leaf switch part #:6401-01-108 (Multitech part number)

To remove a cassette stuck in the fully loaded position:
1. Remove the front-loading assembly from the VCR (see above)
2. Locate loading motor and worm gear on right side, as shown.
3. With thumb or scribe, rotate worm gear CCW until cassette can be removed.

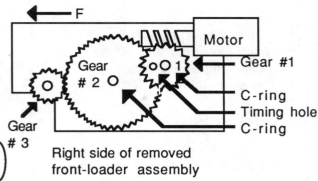

Right side of removed front-loader assembly

To retime front-loading assembly:
 Tools: #1 Philips, needle-nosed pliers, small flat-bladed screwdriver
1. After removing front-loading assembly, remove motor worm gear assembly by taking out two screws in positions indicated by arrows.

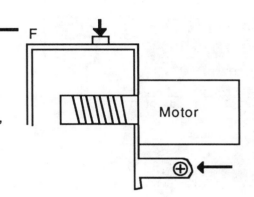

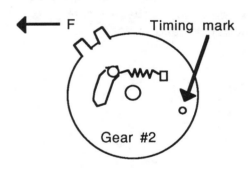

2. Notice the position of the timing mark on gear #1 (see previous diagram), then remove gear #1 by removing the C-ring, being careful of the spring-loaded gear behing gear #1.
3. Notice the position of the prongs on the top of gear #2, and the timing mark in the gear (see picture), then remove gear #2 by removing the C-ring.

4. Turn gear #3 fully CCW until you feel a spring release gear slowly.
5. Check to make certain that the three cassette basket slider rollers are fully forward in their slides.

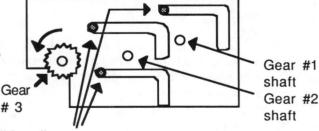

Cassette basket slide rollers

6. Replace gear #2 so that its timing mark is closest to the gear #1 shaft, and so that the prongs on gear #2 straddle and grip cassette basket roller.

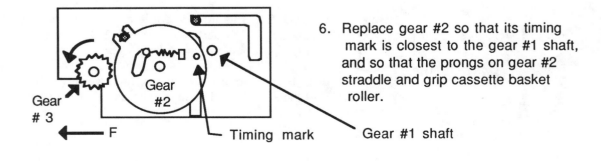

7. Replace gear #1 with its timing mark over gear #2's timing mark. Replace C-ring.
8. Reinstall motor and two screws.

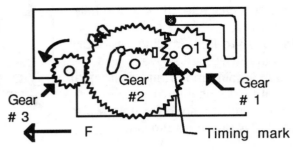

9. Remove gear #4 by removing C-ring. Be careful not to push the transverse axle to which it is attached back out of the hole; hold the axle transverse axle (see Section 12 of Chapter 10 for explanation of terminology) where it is when removing gear #4.
10. Turn gear #5 clockwise until pin just closes leaf switch.
11. Reinstall gear #4, mesh it with gear #5, and reinstall C-ring to hold it on.
12. Reinstall front-loading assembly into VCR with six screws, and reattach 9-wire plug.
13. Plug in, power up, and retest.

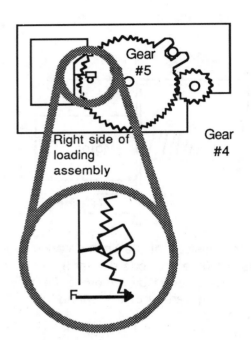

Hitachi-category

NOTE: Two metal catches latch at each side of the top near the front to hold the basket in the fully unloaded position until a cassette is inserted. They are supposed to be unlatched by the action of inserting a cassette. Check to make certain they are not damaged or bent, and that both unlatch when a tape is inserted, before proceeding further.

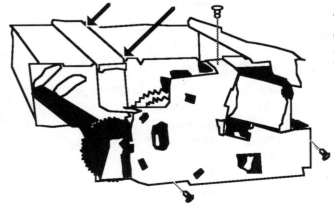

1. Remove complete front-loader assembly from VCR. Read through entire following procedure to find out how to check the three pairs of triangular marks and position of switches to determine whether the front-loader is presently out of alignment.

   NOTE: This front-loader is available as a complete assembly. Call parts supplier with VCR model number to get part number and price.

2. Remove one screw at top, and two at the bottom, of the sheet metal plate to gain access to enclosed gears.

3. Fold down the sub-assembly consisting of motor and one set of gears, leaving wires attached.

4. Find triangle shaped timing marks on big cam gear and smaller gear attached to transverse axle.

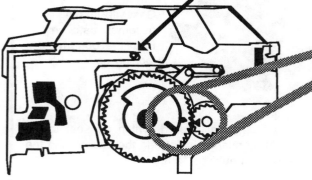

5. Align triangle-shaped timing marks on inside cam gear and smaller gear attached to transverse axle.

NOTE: Turning the cassette-loading motor clockwise unloads the cassette. Turning the motor CCW loads the cassette.

Be careful not to lose the spacer parts from here.

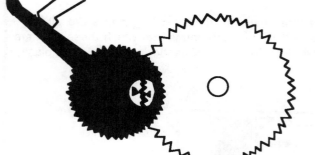

6. Make certain that the two gears on the right side of the front-loader are meshed so that the two triangle-shaped timing marks will point to each other when the gears move around. When the two marks described in step 5 are pointing to each other, these two marks also need to be pointing to each other. If they are not, remove the big white plastic gear and remesh so that they are correctly aligned. (To find out how to remove and reinstall the C-ring holding this gear on, see Sections 9 and 10 of Appendix I.)

7. Before reattaching motor sub-assembly, turn the worm gear attached to the motor until the triangle molded in the plastic that is visible through the opening in the metal aligns with the point formed in the metal at the bottom left side of the opening.

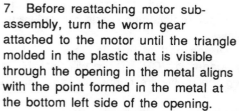

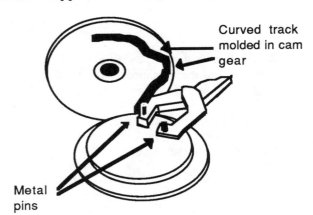

Curved track molded in cam gear

Metal pins

8. Reattach subassembly with motor, being careful to get the metal pins on both little plastic arms both into the curved track on the cam gear. When the subassembly is up against the main assembly as you put it back on, these pins can be guided into the track on the cam gear by pushing from below on arms.

9. Before putting screws back in, double check that the three pairs of triangle timing marks are all still aligned, that the pins are in the track on the cam gear, and that the plastic arms into which these pins are molded are not pinched or caught, but free to push the two buttons on the cassette in switch in the plastic enclosure. Then put back the four screws.

Right side of loading assembly

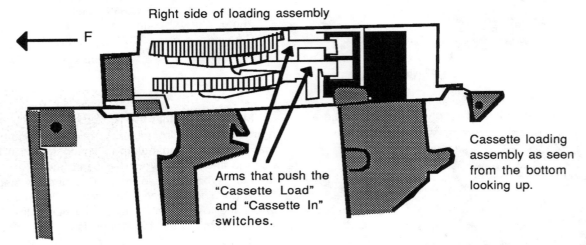

F

Arms that push the "Cassette Load" and "Cassette In" switches.

Cassette loading assembly as seen from the bottom looking up.

When the cassette basket ("tray") is in the fully-unloaded position, the inboard switch (closest to the basket) should be pushed in, and the outboard button should not be pushed in. When the basket is in the fully-loaded position, neither button should be pushed in — both should be "out."

Panasonic-category

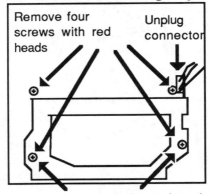

Remove four screws with red heads

Unplug connector

These two screws have threads like a wood screw. (Other two are machine threads.)

A front loader like this is used in the Panasonic category J, K, L, and M machines, although some numbers for replacement parts are different in the last two listed. Proceed as follows:
1. Remove any circuit board covering part of front-loader.
2. Remove screws holding front-loader in VCR. There will probably be four of them, with redish heads. Unplug the multi-wire connector, and remove assembly from VCR.
The following procedure covers both the case in which you only need to retime the front-loader mechanism, and also the case in which you need to replace the right side plate because it has been warped from the motor getting hot. We will disassemble all the way down to where we can replace the side-piece and motor, and then reassemble and realign the mechanism as we reassemble it. If you do not need to replace these parts, follow the same disassembly procedure down to the last gear removal, skip the further disassembly steps in which the motor and side are removed, and pick up at the point in the reassembly procedure where the gears are put back on.

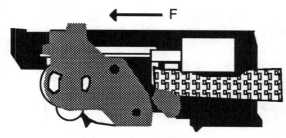

← F

These procedures apply to a Panasonic-category front-loader whose right side looks roughly like this. A common problem is that bad connections (sometimes caused by cigarette smoke in the air) in the mode switch cause the VCR to continue to send power to the cassette loading motor after the basket has stopped up against the furthest end of its excursion, instead of cutting off the power to the motor like it should, thereby overheating the motor, which in turn buckles the plastic side-piece enough to make the front-loader not work right. Then, when it doesn't work, someone may try to force it to take a tape, breaking off plastic parts inside. Also, sometimes this mechanism just gets knocked out of alignment.

To determine whether you will need to replace the right side plate and cassette loading motor, look carefully at the inside surface of the side plate on the right side (as you face the VCR), toward the rear — the place where the motor is just on the other side. If you see the slightest bulge in the plastic, you need to replace both parts. This will require unsoldering & resoldering the motor.

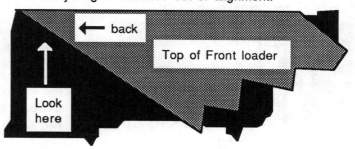

← back

Top of Front loader

Look here

While you are looking at the front-loader from this rear view, study the position of the black plastic blade pointing forward from where it is attached to the right rear edge of the cassette basket, because you will need to know the position it should be in when reassembling.

If you see a bulge on the inner side of the sideplate, the motor and the sideplate itself are among the parts needing replacement. The motor may still run, but when it got hot enough to melt the plastic, it got damaged and will stall again sooner or later. Parts with the same part numbers will work in all the various brands of machines in the Panasonic J and K category that use this front-loader. (Call for the different part numbers for L and M category units.) Continue disassembly up to step 25 to see what other parts are needed.

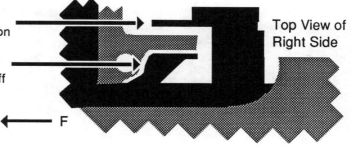

Motor

VEMS0099 (Pan) used in J,K,&L categories

Sideplate: VXAS0742 (Pan) used in J & K category VCRs

Notice how this plastic "bayonet" points forward when the cassette tray is in the fully ejected position

Make certain that this part of the right-side holder guide lever (VMLS0375) is not broken off

Picture is top view of the middle of the right side of the front loading assembly.

Top View of Right Side

← F

Make certain that wiper gear and sliding basket are all the way forward

↓

Left side of front-loader

3. Make certain that the cassette basket inside the front-loading assembly, and the large white plastic wiper gear that controls its movement, are both all the way forward, in the fully ejected position. The picture shows how the left side of the front-loader looks when it is in the fully-ejected position.
If it is not in the fully-ejected position, rotate the motor shaft on the right side by hand clockwise until they move all the way forward. If the mechanism is jammed up so tightly that this cannot be done, skip this step for the moment, and return to it after the motor has been removed.

NOTE: Remember that to move the cassette tray in the direction to load, a cassette must actually be sitting on the tray to unlatch both catches that otherwise lock it at the fully ejected position, and again lock up when the cassette is half-loaded.

Left side of front-loader

4. Examine closely the way the cassette compartment door opener lever catches on the front door of the front loader. Memorize it so that you will be able to put it back in the same way. Put a mark with a felt-tip pen on the side of it facing you. Then remove by pulling back the tiny black molded plastic clip that holds it on the pin molded into the black plastic that it turns on.

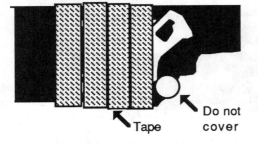

Tape   Do not cover

5. With the cassette basket is in the all-the-way forward, fully ejected position, a trick that saves time later is to wrap the whole left side plate and basket with masking tape, or something similar, to prevent the basket from falling away from the side plate when you remove the right side. Do not put tape over the little round plastic gear attached to the transverse axle, because you will need to remove it later.

Right side of front-loader

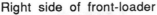
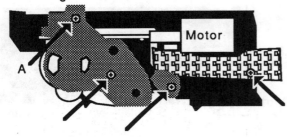

A   Motor

To Disassemble the Front Loader:

6. Study exactly how the wires are routed through the plastic piece in the region below the motor and circuit board so that you will be able to put them back in the same way later on reassembly.
7. Remove the four screws located in the positions indicated. Note the location of any screw(s) that are different in thread size or length from the others. (Often screw A is smaller.)

Right side of front-loader

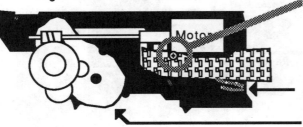

Motor

8. Remove the screw beside the motor and the ground wire that it attaches, if your version has this screw.
9. Study carefully how the wires are routed through the place for them in the plastic side-plate, especially where the arrow is. Maybe make a little sketch of it, so that you will know how to put them back the same way. Then remove wires from plastic locking tab.
10. Remove sheet metal plate to expose plastic gears underneath.

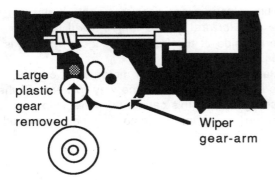

Large plastic gear removed

Wiper gear-arm

11. Slide off the large plastic gear that meshes with the worm gear on the shaft from the motor. Slide it off the axle as shown.
If you are only retiming the front-loader and not replacing the motor and/or side-plate, skip on down to step #33 below in the reassembly procedure where the timing of the smaller plastic gear attached to the transverse axle is adjusted relative to the large wiper gear-arm.

12. Remove the smaller round gear below it by using the tip of a small screwdriver to pull back the locking clip that is molded into the plastic gear and catches in a groove in the axle that the gear is on, as explained in Section 12 of Chapter 10.

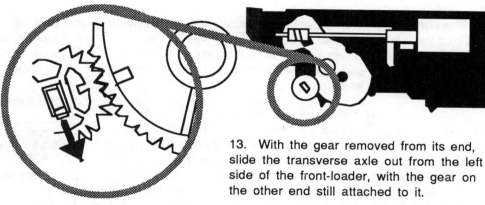

13. With the gear removed from its end, slide the transverse axle out from the left side of the front-loader, with the gear on the other end still attached to it.

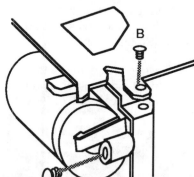

14. To remove the motor and worm gear shaft, begin by removing screw A, and put it in a separate container. (Some versions do not have this screw).

15. Remove screw B and put it in a separate container.

16. Make a mark with a felt-tip pen on the top of the motor.

17. You will need to free from their clips the wires running across the top at the right rear, and pry the sheet metal top piece that was held down by screw B up enough to be able to clear the motor. The plastic arm pressing against the backside of the motor needs to be pried out a tiny bit (not too much!) and the motor pulled to the rear a tiny bit in order to free the front side of the motor from a plastic mounting pin that is going a short distance into a hole on the front of the motor.

To remove the motor and worm gear shaft, you must release the front of the motor from the mounting pin on which it is impaled by inserting a small flat screwdriver at the front of the motor and pushing it slightly to the rear, bending the holder away from the motor, and pulling out on the worm gear shaft at its universal joint (located where shown in the picture).

Pull out here to remove worm gear shaft

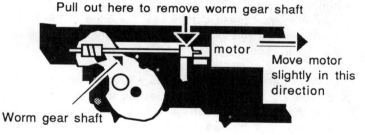

motor

Move motor slightly in this direction

Worm gear shaft

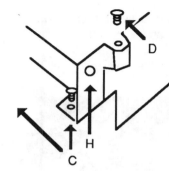

18. Remove the right-side wiper gear by sliding it off its mounting pin. Notice how it goes over the pin from the cassette basket on the other side coming out through the L-shaped slot. If this plastic pin is broken off from the cassette tray side-piece, you will need to order a new "cassette holder guide" part number VMDS0259 (Pan).

19. To remove the side-plate, first draw an upward-pointing arrow on the thin piece of sheet metal wrapped around the right rear edge of the side-plate, then remove this metal piece and set it aside.

20. Study the front flap hinge at its right end, and look carefully at how the little spring is positioned, so that you can put it back in this position if necessary later. Then put a piece of masking tape around the end of the door flap in such a way as to hold the spring to the door in its present position. Note how the other end of the spring is hooked through hole H.

21. Remove the two little screws C (one at each end) holding the front cross-piece to the side pieces. Put in separate container.

22. Note how the plastic "bayonet" piece is self-attached to the slider. Then remove the screw D holding the top piece of sheet metal to the top of the right side-piece. Put in separate container.

23. The right side plate and the front door flap should now come loose. Look at the grooves in the side plate that the cassette basket travels in as the side comes off. Try to keep the basket together with the left side piece that you taped together with it earlier, as the right side piece comes off.

24. Remove the thin piece of sheet metal from the front edge of the old right side piece, and put it onto the new side piece in the same position.

If the plastic pin is NOT broken off the "cassette holder guide" and you do not need to replace the right side holder guide lever (see below) and the gray plastic slider is not cracked, then GO DIRECTLY TO THE NEXT STEP #25. But if any of these plastic parts is broken, unscrew the screw attaching it to the cassette tray at the bottom, release the plastic tab holding it at the top, and separate it from the metal parts of the cassette tray to which it is attached. Carefully check the top and bottom of the right-side holder guide lever for (see illustration below) nicks or broken edges - replace if necessary: Part #VMLS0375 (Pan).

Gray plastic slider part # VXAS0736 (Pan). Correct position of bayonet.

Location of plastic pin that sometimes gets broken off from black plastic holder guide part #VMDS0259 (Pan).

Turn over the square gray plastic piece with the black sliding rectangular piece inside it. Slide the black part down until it catches and stops. DO NOT FORCE! Push in the gray plastic button protruding through the hole in the black part, and slide the black part down a little farther until the tab on the left edge aligns with the square opening on the edge of the gray piece, and lift out the black piece. If you've installed a new "cassette holder guide," remove and reattach the two little internal plastic parts and the the two springs that hold them, and then reattach to the gray plastic piece back using the same square opening to get it in. DO NOT TRY TO BEND THE PLASTIC PARTS - THEY'LL BREAK! When reattaching this to the sheet metal tray, hook the top piece of the tray into the slot on the "cassette holder guide" first, then make certain that the edge of the tray bottom is hooked under the edge of the little plastic tab on the rear end of the bottom of the black plastic piece before screwing down the bottom screw.

Check for damage to right side holder guide lever: VMLS0375 (Pan)

Push this

Release hole for tab

25. Work the new side piece into alignment with the edge of the cassette basket and the horizontal sheet metal piece running across the top. Get the front door into its hinge with the spring correctly positioned in hole H. This is the hardest part of the job. It will take several trials and three hands in order to get all the pieces in correct position for reassembly.

26.   Check to make certain that the cassette basket is correctly positioned in the groove that it travels in, and that the black plastic cassette door opener with the blade-like part again is positioned with the blade pointing forward, as it was before you disassembled it, its little pin positioned in the track for it on the interior side of the big right side piece, and that it is self-attached by its own little plastic piece. Replace the screw (D) you removed in step 22. Reattach the horizontal metal part running under the door flap, using screws (C).

27. Look at the back of the old motor and you will see a '+' and a '-' molded into the plastic. Write down which color wire goes to the terminal closest to the '+' , and which goes to the terminal closest to the '-'.

28. Use solder wick with a soldering iron to remove as much solder as possible from the two terminals on the old motor, without melting the insulation on the wire. Then reheat the terminals and work the old wire loose by wiggling it while the old solder is soft. (Holding the motor in a vise helps, if you've got one.)

29. Solder the colored wires to the corresponding terminals with the same marking on the new motor. (On all of them that I've ever seen, the red wire goes to the '+' terminal.)

Pin

30. Make certain that the pin attached to the cassette basket is all the way to the front of the L-shaped slot.

31. Reinstall the wiper gear, making certain that the pin from the cassette basket goes into the slot in the end of the wiper gear.

32. The worm gear shaft and motor (with wires and the circuit board just hanging from it) must be popped into place simultaneously. To install them, first make a mark with felt tip on the new motor in the same place you made it on the old motor to figure out what side of the new motor needs to go on top so that it will be positioned like the old motor, and so that the plastic pins will go into the holes for them on the front plate of the motor. Next, place the forward (left) end of the worm gear shaft into the hole for it in the new plastic side piece, leaving the other (universal joint) end tipped out away from the plastic side piece, while putting the rear end of the motor against the projection from the plastic side-piece that holds it. Join the two halves of the universal joint and then press the motor and worm drive shaft so that they go straight up against the plastic side piece. You will need to spread the plastic holding arm away from the motor in front, lift the top metal piece slightly, and spread the plastic arm at the rear very slightly to get the motor in. When the motor pops into place, you may still need to rotate it slightly, and fuss with it a little bit, to make certain that the mounting pins go into the holes for them on the front of the motor.

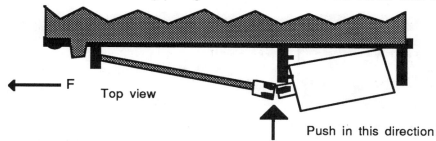

← F   Top view

Push in this direction

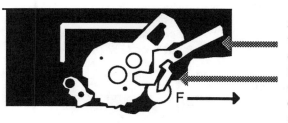

Timing mark

Timing mark

33. Remove masking tape. Insert the transverse axle (that should have one small plastic gear still atttached to it) into the hole for it on the left side plate, so that the other end of the axle emerges through the hole in the right side plate. Mesh the gears on the left side so that the tooth on the small gear closest to little projection molded into this gear meshes with the grove immediately above the little square mark molded into the left-side wiper gear.

(If you ever need to move the basket inward, REMEMBER that two little plastic protrusions on top of the movable tray latch and hold the basket in its fully-ejected position whenever it is put there. You can unlatch them both with your fingernail to move the basket inward until they latch again midway, or better: just insert a cassette to keep them unlatched.)

34. Reinstall the front-loader's door opener, making certain that the part on the top of it slides into the opening in front of the edge of the door flap where it was prior to disassembly. Press in on the opener so that the hold-down tab snaps over it to hold it on. Push down on the bottom edge of it to insure that it opens the door, and that the spring on right-end of the door closes the door again.

35. Reinstall the small round gear on the transverse axle, making sure that the projections on the two gears that serve as timing marks are aligned with each other. Push down until it snaps in place.

36. With the timing marks still aligned, install the large plastic gear so that the timing mark aligns with the place where a molded side-edge comes out from the black plastic side-piece. (It's like a little wall of plastic that projects from the side-piece where the end of the worm gear shaft goes in.) It will be held on by the next part(s).

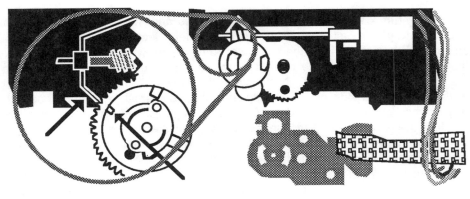

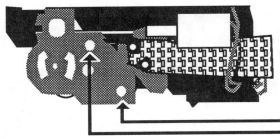

37. Reinstall the sheet metal side piece with its attached circuit board with the wires on the rear end running over the side of the circuit board closest to you, and going through the place for them under the black plastic clip at in the new side at the rear below the motor, in the same path you observed before disassembly. The leaf switches attached to the circuit board should be positioned so that the little pointed tips on the plastic pieces on the end of them are pushed up the plastic projections on the white wiper gear as viewed through the holes (A) and (B) in the next picture. Attach with four screws that were removed in Step 7.

38. Reinstall the very thin piece of shaped sheet metal removed from the rear edge of the black plastic side-plate, and attach it with the screw at the top right rear above the motor removed as screw B in step 15. Attach screw A at rear of motor removed in step 14.
39. Look through holes in metal plate to confirm that the triangular point of the switch that detects when the cassette is fully-loaded and when fully-unloaded, rests on the midpoint of projection A on the wiper gear.
40. Confirm that the triangular point of the switch that detects that a cassette has been inserted is resting on the midpoint of projection B on the wiper gear.
41. While looking at the right side of the front loader, push a cassette into the front opening, and confirm that first the triangle on the cassette loaded/unloaded switch leaves projection A, and then the triangle on the switch that detects cassette insertion leaves projection B. Insert a cassette and load & unload with a battery as explained in Sections 13-14 of Chapter 8.

(B) Switch that detects that a cassette has been inserted

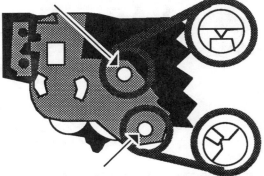

(A) Switch that detects when cassette is fully loaded or fully unloaded

43. After checking both ends of each wire on the front loader to verify that none has gotten broken off at the end, reattach to VCR, plug in the multi-wire connector, power up, and test. If it will accept and eject cassettes, then to insure that the same failure will not recur in the future, unplug and remove the mode switch on the BOTTOM of the mechanical chassis (after marking its position with a scribe as explained in Chapter 6), and clean it. It can be cleaned without disconnecting the wires by working the white plastic sliding parts all the way to the right, sliding the white plastic spacer underneath farther to the right opening up a space and spraying head cleaner into the opening(s) indicated by the arrows, and moving the sliding part of the switch repeatedly, and then moving it all the way to the left and repeating the process. If you can work some white litium grease into the interior where the moving part slides across the contacts, this would help prevent a recurrence of the problem. As explained in Chapter 6, after cleaning, the switch must be returned to correct position by trial & error positioning starting from the scribe mark.

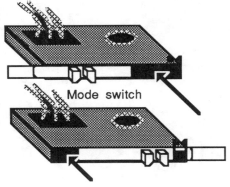

Mode switch

Sharp-category

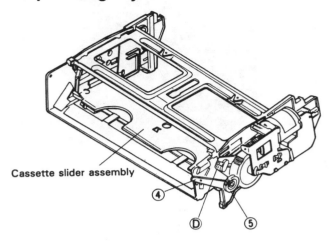

Cassette slider assembly

④

Ⓓ   ⑤

This cassette front-loader is used in Sharp VC 582, VC 583U/UB, and Montgomery Ward 10658, 10659, 10660, among others.
It is available as a complete assembly from Sharp as part #CHLDX3029GE03 (Shp).

To Remove the Front-loader assembly from VCR:
1. Remove the top and bottom of the cabinet, and the front cabinet piece. See Appendix III for help if needed.
2. Put the unit in the fully-ejected position, if it is not already in this position. To do this, turn the worm gear attached to the little motor on the right side CW until the basket comes all the way up and out.
3. Unplug the multi-wire connector at the right side of the front-loader, taking care not to break the little pins that go into the plug.
4. Remove the two screws attaching the front-loader.
5. Slide the whole assembly to the rear slightly, in the direction of arrow B, then when it comes free, lift it up and out of the VCR.

Hold-down screws

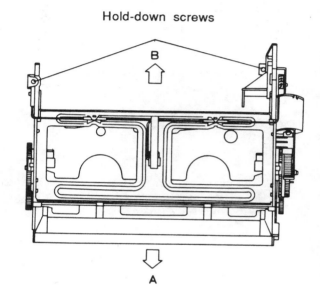

B

A

NOTE: If you are only replacing the push-button on the spring that unlatches the cassette door, go directly to the following steps, "B," under the heading "Replacing the Cassette Door Latch Release Push-button." If you are realigning the assembly or replacing the motor, skip directly to the steps beginning with number 6 below. Information on removing and replacing the troublesome sliding plastic door-opener appears as steps marked "A" under the heading: "R & R Sliding Plastic Cassette Door Opener."

Removing and Replacing the Sliding Plastic Cassette Door Opener
A-1. In case you need to replace the parts that open the cassette door, their part numbers are:
(412) Cassette Door Opening
        Lever MLEVP0081GEZZ (Shp)
(414) Spring for Cassette Door Opening
        Lever MSPRD0066GEFJ (Shp)
(415) Cassette Housing Frame (Right,
        Plastic) LHLDX1005GEOO (Shp)

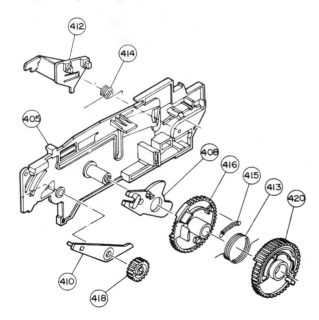

A-2. To install spring and lever, orient spring as shown, catch the little end with the L-bend in the "bin" for it at the rear end of the slot for it on the lower part of the side housing, push the round part over the pin for it in the plastic side housing, and bend the long, straight end of the spring counter-clockwise around toward the end of the sidepiece, holding it in this position temporarily with your finger.

A-3. Bring door opener up close to side plate, and catch the straight end of the spring behind the little tab for it on the back side of the sliding door opener.

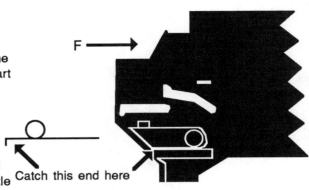

F →

Catch this end here

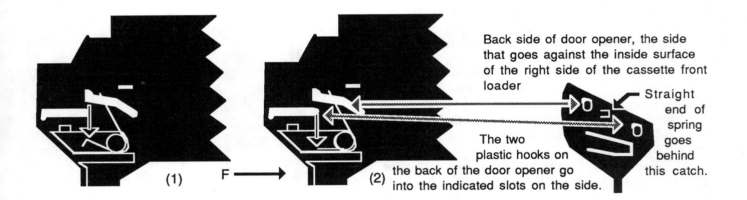

(1)   F →   (2)

Back side of door opener, the side that goes against the inside surface of the right side of the cassette front loader

Straight end of spring goes behind this catch.

The two plastic hooks on the back of the door opener go into the indicated slots on the side.

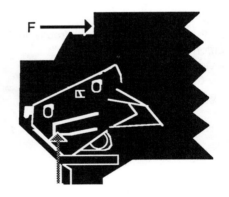

F →

A-4. Bring opener up to side plate with two tabs going into enlargements of slots at rear. After tabs have gone into slots for them, slide the opener toward the front of the side plate until the plastic latch catches on the raised edge for it on the side plate. (If you need to remove it again for any reason, you will need to pry up the plastic catch piece slightly at the point marked with an arrow, so that you can push the opener all the way back, so that the hooks on its other side can come through the enlargements at the ends of the two slots in which they travel.)

A-5.   Verify that the opener slides to the rear when pressed on, and then springs back to its forward position when you stop pushing on it. The spring pushes it to its forward position, in which it is shown in this illustration.

F →

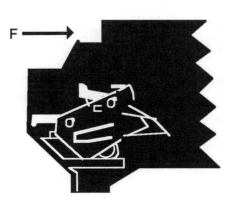

Replacing the Cassette Door-Latch Release Push-button:

B-1. To disassemble, turn the worm gear (1) attached to the motor CCW by hand until the cassette basket (2) reaches the bottom, fully-loaded position.

B-2. Spread the right and left sides of the assembly with your hands slightly, just enough to permit the tabs (A) of the cassette basket to come out of the holes for them on the right and left sides.

B-3. Pushing the tabs (B) on the right side of the cassette basket holder (4), pull it out of the cassette basket (5).

B-4. Remove the cassette door-latch release lever and button (6) from the basket holder.

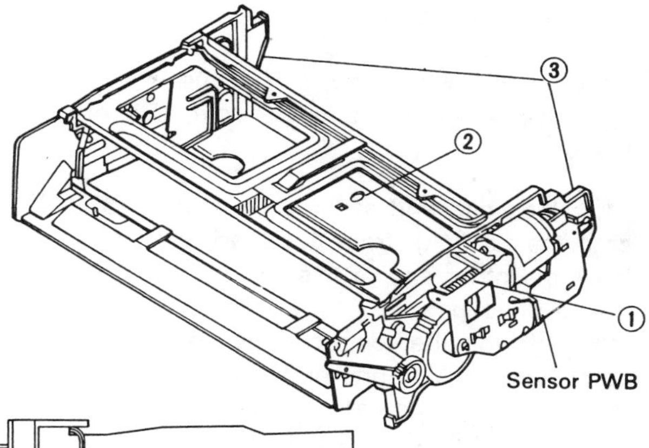
Sensor PWB

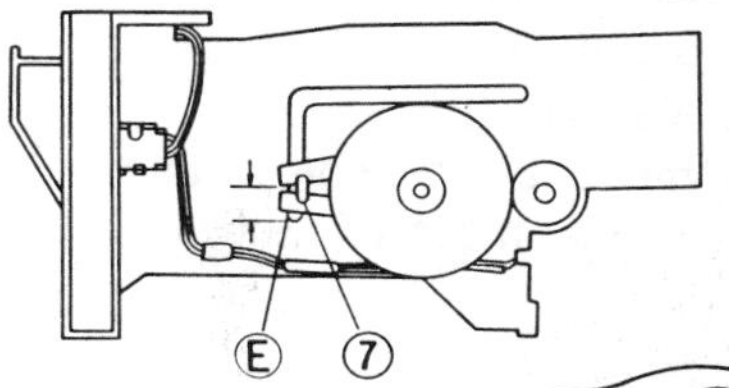

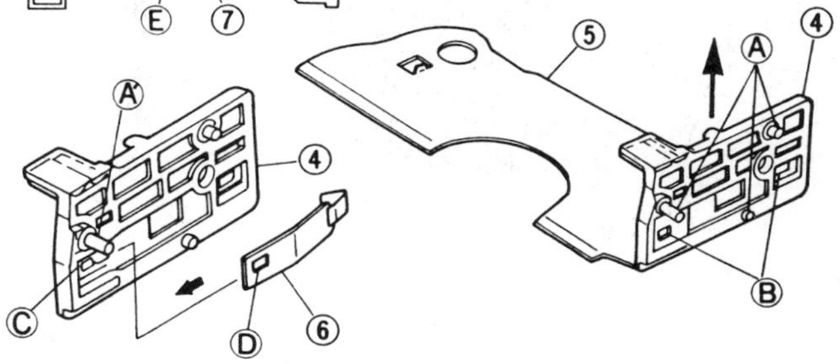

To reassemble and reinstall:

B-5. Install the new or rebent cassette door-latch release lever (6) in the basket holder (4). Make sure that the tab (C) of the basket holder is firmly engaged with the hold, (D), of the latch release lever.

B-6. Position the latch release lever so that it is positioned inside the tab of the cassette basket, (5).

B-7. Spread the right and left sides of the assembly, and engage the right and left tabs, (A), of the cassette basket engage in the grooves in the right and left sides. To facilitate installation, raise the wiper arms about 3/16" (5 mm) from the bottommost end point by turning the worm gear attached to the motor CW.

B-8. Make sure that the front pin, (A') of the cassette basket assembly is properly engaged in the place for it, (E), in the middle of the wiper arm, (7).

To Disassemble the Cassette Front-loader:

6. Push the two plastic locking tabs, A, holding on the sensor printed wiring circuit board (PWB).

7. Remove the screw, (2), from the cassette motor bracket, and remove the cassette motor assembly together with the sensor circuit board from the cassette assembly.

8. Pull off the worm wheel assembly (3).

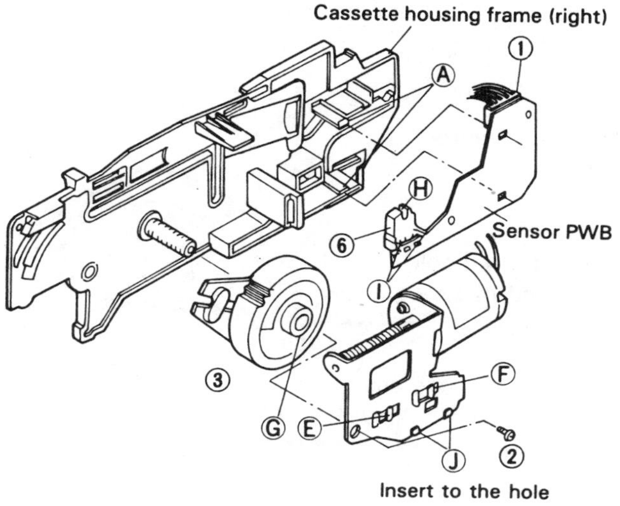
Cassette housing frame (right)
Sensor PWB
Insert to the hole

© 1990 Stephen N. Thomas

Reassembly and Alignment

9. Move the cassette basket to its fully unloaded position, make certain that the wiper arm on the left side is engaged and all the way forward, and set these parts aside ready for when you need them after the next steps.

10. Put the plastic tab, (A), of the right-side wiper arm into the hole, (B), of the drive gear, (2).

11. Hook one end of the drive spring, (3), onto the tab, (A), from the wiper arm, and hook the other end of this spring to tab, (C), of the drive gear.

12. Position the reciprocating spring, (4), as shown, hooking one end onto tab, (D), of the drive gear and catching the other end, (F), temporarily on tab (G) of the drive gear. You will need to hold the spring in this position with your thumb until you complete the next step, which is the most difficult.

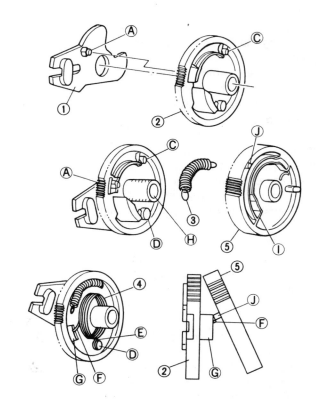

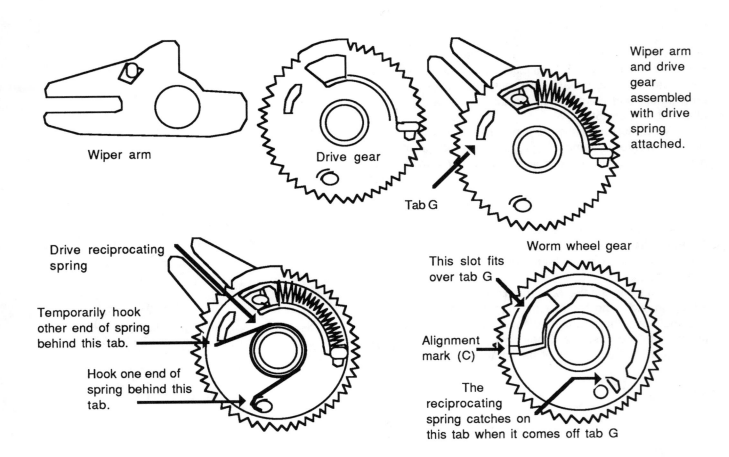

Wiper arm

Drive gear

Tab G

Wiper arm and drive gear assembled with drive spring attached.

Drive reciprocating spring

Temporarily hook other end of spring behind this tab.

Hook one end of spring behind this tab.

Worm wheel gear

This slot fits over tab G

Alignment mark (C)

The reciprocating spring catches on this tab when it comes off tab G

© 1990 Stephen N. Thomas

13. Place the worm wheel gear, (5), up against the drive gear, (2), in such a way that the male tab, (J) of the worm wheel gear comes up to tab, (G), of the drive gear, (2). Use a small screwdriver, or any other technique that works, to unhook the end, (F), of the spring, (3), from (G) and move it to male tab, (J), of the worm wheel gear, (5). With both ends of the spring, (3), hooked onto the correct tabs, push the two gears together, and while holding them together, rotate the worm wheel gear, (5), clockwise until the tab, (G), of the drive gear, (2), goes into the slot, (I), for it in the worm wheel gear, (5). Hold it together with your fingers in this position. It may take a few tries to get the idea of how this thing is spring-loaded. The engaged tab (G) will prevent it from unwinding CCW, as long as you hold the two gears pressed together.

14. After preparing the worm wheel assembly as described in 13, slide it onto the pin for it in the plastic side-piece in such a way that the alignment mark at the end of the slot, (C), aligns with the timing mark on the gear attached to the transverse axle (the "phase gear"), and in such a way that the pin from the cassette basket goes into the hole for it in the end of the wiper gear. Hold the worm wheel assembly pushed against the side-plate so that it does not come apart from the force of the spring.

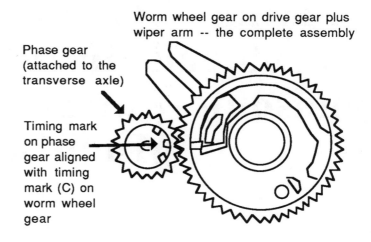

Worm wheel gear on drive gear plus wiper arm -- the complete assembly

Phase gear (attached to the transverse axle)

Timing mark on phase gear aligned with timing mark (C) on worm wheel gear

15. Align the tab, (E), of the sliding timing lever in the motor bracket with the groove or track, (G), in the outward side of the worm wheel gear assembly, and fasten the whole thing together with screw, (2).

16. Align the tab, (F), of the same timing lever with the notch, (H), in the switch, (6), on the sensor circuit board, and fasten the circuit board to the front-loader with the two tabs, (A). Make certain that the two tabs from the cassette switch are squarely engaged with the notches, (J), in the cassette motor bracket

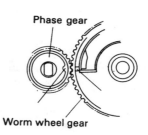

Phase gear

Worm wheel gear

Reinstalling the Front-loader in the VCR
D-1. Connect the multi-wire connector at the right side of the assembly.
D-2. Insert the feet of the front-loader into the holes for them in the reel table metal, and slide the assembly forward to latch it in place.
D-3. When it is in the correct position, fasten it with the two screws, part # XHPS330P06WSO (Shp).

Supplement

Fixing Common Ejection Problems in Late Model Panasonic-Category N and O Machines

Warning

The machines in this category are infamously difficult to work on. The problem is created by the fact that one motor (the capstan motor) is used to perform the functions that formerly were performed by three motors (cassette loading motor, tape loading motor, and capstan motor), by means of an extremely complicated set of gears and levers. This are difficult to realign if removed, so whenever possible, do only the minimum required to fix the problem.

In particular, try to avoid having to remove the front-loading mechanism, as once it is removed, it is a challenge to get it reinstalled with the gears that drive it correctly meshed. The repair of the most common problem described below can be completed without removing the front-loader.

Failure to Eject

This problem commonly shows up as a refusal of the machine to eject a cassette. The most common cause of this problem is that the soft metal "Release Lever Unit" (also called the "Change Lever" in the pictures of the undersides of these machine in Appendix VI) has gotten slightly bent. To check this part, remove the bottom cover of the VCR. It will look as shown in the first picture.

If a bent Change Lever is the problem, you can see that the part visible from underneath is not running straight up-and-down, but is tipping slightly, usually tipping toward the front of the VCR. It is difficult to remove and replace this lever, so the simplest solution is usually just to bend it back slightly, just enough to get it straight again. Usually,

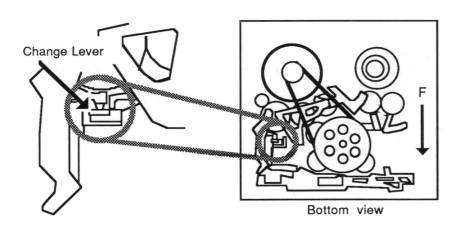

Change Lever

F

Bottom view

if it needs to be bent, the direction to bend it is back toward the rear of the machine, just enough to make it be straight. Then plug in the machine and test whether it will accept and eject a cassette.

The change lever can be replaced with a new one (part number VXAS1051 (Pan)), but this requires putting the machine into the alignment position as described in the following sections, and removing the snap rings and gears between you and it, as you view the machine from underneath. This can be done without having to realign the whole machine, if you are very careful not to move anything else as you disassemble and reassemble, but it is a lot easier just to bend the change lever already in the machine.

If the problem is not due to a bent change lever, the next most common cause of the problem is failure of a small plastic part located on the top side of the chassis, up by the pinch roller. It is called the "P5 Sector Gear" and its part number is VDGS0124 (Pan). You may want to order it up before you begin disassembly, as you will probably need one sooner or later anyway.

Here is what the part that causes most of these troubles looks like, by itself, out of the machine:

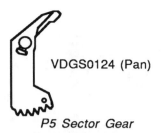

VDGS0124 (Pan)

P5 Sector Gear

It cannot be repaired, but must be replaced.

The following procedure is required to replace it.

Manually Put the VCR Front Loader into the Fully-Ejected Position first, and then into the Alignment Position

It is crucial first to put the VCR into the fully-ejected procedure, and then move the mechanism back into the position for checking alignment, in order to run the following procedure. This must be done even if there is no cassette inside the VCR. Follow the steps given for Panasonic N and O category machines

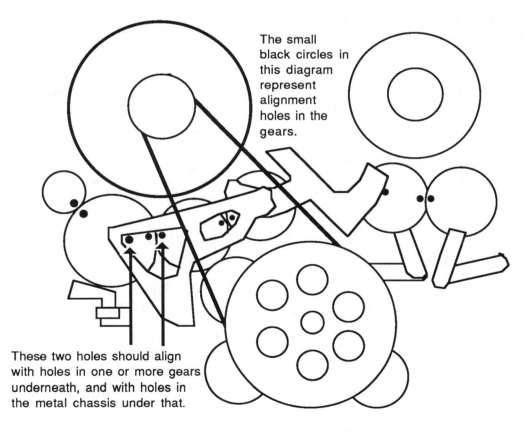

The small black circles in this diagram represent alignment holes in the gears.

These two holes should align with holes in one or more gears underneath, and with holes in the metal chassis under that.

Correct Alignment position of bottom gears of Panasonic N and O Category Machines

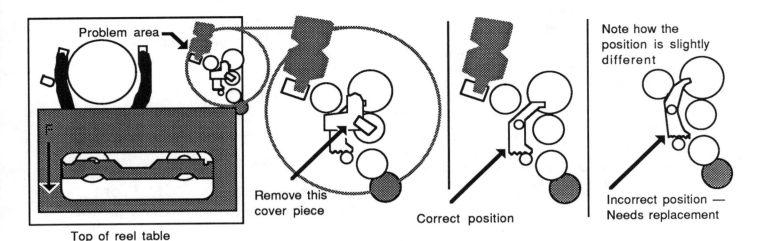

Location and Diagnosis of P5 Sector Gear problem

in Appendix VI to eject a tape manually. Push the change lever, turn the pulley counter-clockwise until the gear mechanism stops moving (while the pulley continues to turn), push the change lever again, and repeat the process until the front-loader goes all the way into the fully-ejected position.

Next, we need to run the mechanism back in part way until we get it to the alignment position. Turn the same pulley this time clockwise until the gear mechanism stops turning. Push the Change Lever once, and then again rotate the same pulley clockwise until the mechanism again stops turning. This should put it into the position it must be in to check alignment.

Making a Preliminary Check of Proper Alignment of Gears on Bottom of Chassis

Verify that the gears on the bottom are lined up with their timing holes matching the corresponding holes in the gears with which they mesh, as shown in the alignment diagrams. (For simplicity, the gears are shown as round circles with no teeth in the picture.) If the gears have jumped a tooth, or someone has messed with them, and they are out of alignment, you have got a big job ahead of you, and you will need to send for a service manual from the manufacturer to get the complete, extensive, multi-page procedure to realign the mechanism.

Turn over the VCR to work on the top side. Open up the top of the machine as described in the next section.

Getting Access to the Top of the Chassis

1. After removing the screws at the back of the cabinet, and the two screws on the bottom that hold on the top, lift the cabinet top up and off.

2. Free three locking tabs at top of front cabinet piece, and rotate it down and off.

3. Remove two screws fastening the top circuit board at the rear of the front-loader assembly, and unscrew six other screws holding down the other edges of the top circuit board. Free any additional locking tabs holding it down. Be careful with wires from the audio/control head as you lift the top circuit board up and slide it back far enough to gain access to the area around the rubber pinch roller shown in the next picture. You probably will not be able to see as much of the upper cylinder assembly as shown in the picture, because the top circuit board, held by its attached cables, will be in the way, but this does not present a problem.

Checking Alignment of Gears on Top of Chassis

When you have exposed the area around the pinch roller to get at the P5 Sector Gear as described in the next section, check to make certain that the alignment holes are lined up as shown in the next illustration:

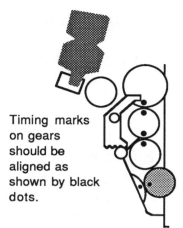

Timing marks on gears should be aligned as shown by black dots.

Correct Alignment Position of Top Gears

Removing the P5 Sector Gear

The secret to doing this repair easily is to remove this part without disturbing or moving the rest of the mechanism in the slightest. If you accidentally cause some ot the gears to move, you may need to retime the entire mechanism, which is a major job, so pretend that you are a bomb squad expert carefully dismantling a live bomb to prevent it from exploding.

Carefully pry off the plastic part that rests above this part, as shown in the illustration captioned "Location and Diagnosis of P5 Sector Gear Problem."

When the P5 Sector Gear is causing the problem, usually you will observe it to be resting in a position rotated a few degrees counter-clockwise from where

it should be, but this is difficult to measu accurately, or to recognize by eye if you've neve done this job before, so since the part is cheap and you've already got it (I hope!), I'd recommend going ahead and replacing it in any case, and seeing whether that fixes the problem.

Very carefully pry the P5 Sector Gear up from the shaft on which it is self-attached by its small plastic locking tab. Without moving anything, immediately slide the new P5 Sector Gear down the shaft aligned in such a way that the most clockwise tooth on the gear to the front of it goes into the notch closest to the little hole molded in the P5 Sector Gear.

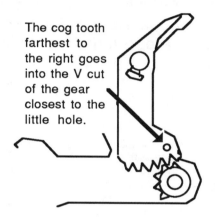

The cog tooth farthest to the right goes into the V cut of the gear closest to the little hole.

Slide the P5 Sector Gear down until it latches itself in position. Now replace the plastic part over it that you removed earlier. When reattaching the circuit board, be careful to push the jwires from the audio/control head up close to the top circuit board out of the way of the tape path when the machine is playing.

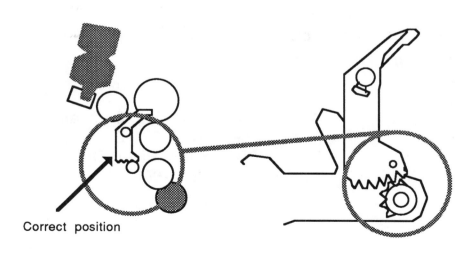

Correct position

Form for Ordering this Book by Mail

To order your own personal book, **xerox or copy** this page, fill in blanks, and mail to the address below.

Yes, please send me _____ copies of the latest edition of *How to Keep Your VCR Alive.* Price: $24.95 each.* Books are shipped postpaid, with all shipping and handling charges paid by the publisher. Allow four weeks for delivery. For volume discounts, write for Discount Schedule.

A professional head-cleaning stick is included with each book.

Book Return Policy: Satisfaction Guaranteed. Books in resalable condition are returnable for full refund.

Name _____

Address _____

_____

City _____ State _____ Zip _____

Tel. ( ) _____

My Method of Payment is: ❏ Check or money order is enclosed. (Do not send cash.through the mails!)

Please ship by: ❏ US Mail ❏ UPS

Mail this completed form to:

Retail Book Sales
Worthington Publishing Company
Post Office Box 16691-B
6907-202B Halifax River Drive
Tampa, FL 33687-6691

tel. (813) 988-5751

---

* $28.95 in Canadian dollars.